D0498914

PHILOSOPHY OF SCIENCE

Philosophy of Science: Contemporary Readings is a comprehensive anthology that draws together leading philosophers writing on the major themes in the philosophy of science. Yuri Balashov and Alex Rosenberg have carefully chosen articles under the headings:

- Science and Philosophy
- Explanation, Causation, and Laws
- Scientific Theories and Conceptual Change
- Scientific Realism
- Testing and Confirmation of Theories
- Science in Context

Each section is prefaced by an introductory essay by the editors which guides students gently into each topic.

Articles by the following leading philosophers are included:

Achinstein	Van Fraassen	Kuhn	Nagel	Salmon
Anderson	Gutting	Laudan	Popper	Schlick
Bloor	Hanson	Leplin	Quine	Shapere
Earman	Hempel	Mackie	Rosenberg	
Feyerabend	Kitcher	McMullin	Russell	

The book is highly accessible and user-friendly and provides a broad-ranging exploration of the subject, drawing from classic and contemporary articles. Ideal for any philosophy student, this book will prove essential reading for any philosophy of science course. The readings are designed to complement Alex Rosenberg's textbook, *Philosophy of Science: A Contemporary Introduction* (Routledge, 2000), though the anthology can also be used as a stand-alone volume.

Yuri Balashov teaches Philosophy at the University of Georgia, USA. He has published extensively in the leading Philosophy and Philosophy of Science journals. **Alex Rosenberg** teaches Philosophy at Duke University. His books include *Philosophy of Science: A Contemporary Introduction* (Routledge, 2000) and *The Philosophy of Social Science* (1995). In 1993 he won the Lakatos Prize in the Philosophy of Science.

Routledge Contemporary Readings in Philosophy

Series Editor: Paul K. Moser
Loyola University of Chicago

Routledge Contemporary Readings in Philosophy is a major new series of philosophy anthologies aimed at undergraduate students taking core philosophy disciplines. It is also a companion series to the highly successful Routledge Contemporary Introductions to Philosophy. Each book of readings will provide an overview of a core general subject in philosophy, offering students an accessible transition from introductory to higher-level undergraduate work in that subject. Each chapter of readings will be carefully selected, edited, and introduced. They will provide a broad overview of each topic and will include both classic and contemporary readings.

Philosophy of Science
Yuri Balashov and Alex Rosenberg

Metaphysics
Michael J. Loux

Epistemology
Michael Huemer with introduction by Robert Audi

PHILOSOPHY OF SCIENCE

Contemporary Readings

Edited by
Yuri Balashov and Alex Rosenberg

London and New York

First published 2002
by Routledge
11 New Fetter Lane, London EC4P 4EE

Simultaneously published in the USA and Canada
by Routledge
29 West 35th Street, New York, NY 10001

Routledge is an imprint of the Taylor & Francis Group

Typeset in Sabon by RefineCatch Limited, Bungay, Suffolk
Printed and bound in Great Britain by
St Edmundsbury Press, Bury St Edmunds, Suffolk

British Library Cataloguing in Publication Data
A catalogue record for this book is available from the British Library

Library of Congress Cataloging in Publication Data
Philosophy of science: contemporary readings / [edited by] Yuri Balashov and Alex Rosenberg.
p. cm. – (Routledge contemporary readings in philosophy)
Includes bibliographical references and index.
1. Science – Philosophy. I. Balashov, Yuri, 1960– . II. Rosenberg, Alexander, 1946– . III. Series.
Q175.3.P49 2001
501 – dc21 2001031916

ISBN 0–415–25781–6 (hbk)
ISBN 0–415–25782–4 (pbk)

CONTENTS

Preface ix
Acknowledgements x

PART I: SCIENCE AND PHILOSOPHY

Introduction 3
1 Moritz Schlick, "The Future of Philosophy" 8
2 Alex Rosenberg, "Biology and Its Philosophy" 22
Questions 34
Further Reading 35

PART II: EXPLANATION, CAUSATION, AND LAWS

Introduction 39
3 Carl Hempel, "Two Models of Scientific Explanation" 45
4 Bas van Fraassen, "The Pragmatics of Explanation" 56
5 Philip Kitcher, "Explanatory Unification and the Causal Structure of the World" 71
6 Wesley C. Salmon, "Scientific Explanation: Causation *and* Unification" 92
7 J.L. Mackie, "The Logic of Conditionals" 106
8 John Earman, "Laws of Nature" 115
Questions 125
Further Reading 126

PART III: SCIENTIFIC THEORIES AND CONCEPTUAL CHANGE

Introduction 129
9 Ernest Nagel, "Experimental Laws and Theories" 132

10 Paul Feyerabend, "Explanation, Reduction, and Empiricism" 141
11 Philip Kitcher, "Theories, Theorists and Theoretical Change" 163
Questions 188
Further Reading 189

PART IV: SCIENTIFIC REALISM

Introduction 193
12 Ernest Nagel, "The Cognitive Status of Theories" 197
13 Larry Laudan, "A Confutation of Convergent Realism" 211
14 Gary Gutting, "Scientific Realism versus Constructive Empiricism: A Dialogue" 234
15 Ernan McMullin, "A Case for Scientific Realism" 248
Questions 281
Further Reading 281

PART V: TESTING AND CONFIRMATION OF THEORIES

Introduction 285
16 Bertrand Russell, "On Induction" 289
17 Karl Popper, "Science: Conjectures and Refutations" 294
18 Karl Popper, "Darwinism as a Metaphysical Research Programme" 302
19 Charles Darwin, "Difficulties of the Theory" 305
20 Peter Achinstein, "The Grue Paradox" 307
21 N. Russell Hanson, "Seeing and Seeing As" 321
22 W.V. Quine, "Two Dogmas of Empiricism" 340
23 Larry Laudan and Jarrett Leplin, "Empirical Equivalence and Underdetermination" 362
24 Wesley Salmon, "Bayes's Theorem and The History of Science" 385
Questions 402
Further Reading 403

PART VI: SCIENCE IN CONTEXT: THE CHALLENGE OF HISTORY AND SOCIOLOGY

Introduction 407
25 Dudley Shapere, "The Structure of Scientific Revolutions" 410
26 Thomas Kuhn, "Objectivity, Value Judgment, and Theory Choice" 421

CONTENTS

27 David Bloor, "The Strong Programme in the Sociology of Knowledge" 438
28 Elizabeth Anderson, "Feminist Epistemology: An Interpretation and a Defense" 459
29 Ernan McMullin, "The Social Dimensions of Science" 489
Questions 500
Further Reading 501

Bibliography 502
Index 507

PREFACE

This anthology can be used in two ways: (1) as a companion to Alex Rosenberg's textbook *Philosophy of Science: A Contemporary Introduction* (Routledge, 2000) and (2) as a stand-alone volume for those who prefer to teach directly from an anthology, with or without another accompanying textbook.

Accordingly, the composition of the anthology follows the order in which the material is introduced and discussed in Rosenberg's book and includes figures whose work the textbook reports. But the anthology covers a somewhat wider range of issues.

Each part begins with a brief Introduction setting the readings in context and ends with a list of study questions and suggestions for further reading. All references are to the general Bibliography at the end of the volume. The detailed Index provides a useful guide to the book.

Our collection is intended for a wide audience of readers interested in the philosophy of science. It may be used in a typical introductory graduate course in the philosophy of science, but it is accessible to undergraduates. In fact, one of our goals in undertaking this project was to address the demands of the undergraduate student by providing a suitable alternative to more technical and specialized sources in the philosophy of science.

The issues covered in this volume are traditional to the post-positivist philosophy of science. Carving this general subject "at the joints," however, has never been an easy task. For one thing, some issues significantly overlap. For another, the very agenda of the philosophy of science is in flux. This is part of what makes the subject so exciting. We do believe that, to a reasonable approximation, we have carved the beast at the joints—at least for now!

Yuri Balashov and
Alex Rosenberg

ACKNOWLEDGEMENTS

We are grateful to Siobhan Pattinson, Ruth Jeavons, and Tony Bruce, of Routledge, for their help, advice, and encouragement. Our thanks also go to the referees who made very useful suggestions at the early stages of work on the anthology.

Special acknowledgements are due to the publishers and the authors for their kind permission to reprint or publish their materials below. Every effort has been made to trace all the copyright holders. If any have been inadvertently overlooked, the publishers will be pleased to make the necessary arrangements at the first opportunity.

PART I: SCIENCE AND PHILOSOPHY

M. Schlick, "The Future of Philosophy," in M. Schlick, *Philosophical Papers*, vol. 2, ed. Henk L. Mulder and Barbara F.B. van de Velde-Schlick, 1979, pp. 210–24. Dordrecht: Reidel. Reprinted with permission of G.M.H. van de Velde and Vienna Circle Foundation, Amsterdam.

A. Rosenberg, "Biology and Its Philosophy," in A. Rosenberg, *The Structure of Biological Science*, Chapter 1, 1985, pp. 1–11. Reprinted with permission of Cambridge University Press, Cambridge.

PART II: EXPLANATION, CAUSATION, AND LAWS

C. Hempel, "Two Models of Scientific Explanation," in *Frontiers of Science and Philosophy*, ed. R.G. Colodny, 1962, pp. 9–19. Reprinted with permission of the University of Pittsburgh Press, Pittsburgh.

B. van Fraassen, "The Pragmatics of Explanation," *American Philosophical Quarterly*, 1977, 14: 143–50. Reprinted with the author's permission.

P. Kitcher, "Explanatory Unification and the Causal Structure of the World," in *Scientific Explanation*, ed. Philip Kitcher and Wesley C. Salmon, 1989, pp. 410–505 (excerpts). Reprinted with permission of the University of Minnesota Press, Minneapolis.

W.C. Salmon, "Scientific Explanation: Causation *and* Unification," *Critica. Revista Hispanoamericana de Filosofía*, 1990, 22(66): 3–21.

J.L. Mackie, "The Logic of Conditionals," in J.L. Mackie, *Truth, Probability and Paradox*, Chapter 3, §10, 1973, pp. 109–19, Oxford: Clarendon Press.

J. Earman, "Laws of Nature," in J. Earman, *A Primer on Determinism*, 1986, pp. 81–90, Dordrecht: Reidel. Reprinted with permission of Kluwer Academic Publishers.

PART III: SCIENTIFIC THEORIES AND CONCEPTUAL CHANGE

E. Nagel, "Experimental Laws and Theories," in E. Nagel, *The Structure of Science: Problems in the Logic of Scientific Explanation*, 1961, pp. 79–105 (excerpts). New York and Burlingame: Harcourt, Brace & World, Inc. Reprinted with permission of Hackett Publishing Company, Inc. All rights reserved.

P. Feyerabend, "Explanation, Reduction, and Empiricism," in *Minnesota Studies in the Philosophy of Science*, vol. 3, ed. H. Feigl and G. Maxwell, 1962, pp. 29–97 (excerpts). Reprinted with permission of the University of Minnesota Press, Minneapolis.

P. Kitcher, "Theories, Theorists and Theoretical Change," *Philosophical Review*, 1978, 87: 519–47. © Cornell University, 1978. Reprinted with permission of the publisher and the author.

PART IV: SCIENTIFIC REALISM

E. Nagel, "The Cognitive Status of Theories," in E. Nagel, *The Structure of Science: Problems in the Logic of Scientific Explanation*, 1961, pp. 106–52 (excerpts). New York and Burlingame: Harcourt, Brace & World, Inc.

L. Laudan, "A Confutation of Convergent Realism," *Philosophy of Science*, 1981, 48: 19–38, 45–9. Reprinted with permission of the author and the University of Chicago Press.

G. Gutting, "Scientific Realism versus Constructive Empiricism: A Dialogue," *The Monist*, 1982, 65: 336–49. Copyright © 1982, THE MONIST, Peru, Illinois 61354. Reprinted with permission.

E. McMullin, "A Case for Scientific Realism," in *Scientific Realism*, ed. Jarrett Leplin, 1984, pp. 8–40. Copyright © 1984, The Regents of the University of California (University of California Press), Berkeley.

PART V: TESTING AND CONFIRMATION OF THEORIES

B. Russell, "On Induction," in B. Russell, *The Problems of Philosophy*, 1959 (first published 1912), pp. 60–9 (with some cuts). Reprinted with permission of Oxford University Press, Oxford.

K. Popper, "Science: Conjectures and Refutations," in K. Popper, *Conjectures and Refutations*, 1963, pp. 33–9. London: Routledge and Kegan Paul. Reprinted with permission of the Karl Popper estate.

K. Popper, "Darwinism as a Metaphysical Research Programme," in *The Philosophy of Karl Popper*, vol. 1, ed. P.A. Schilpp, 1974, pp. 134, 136–8. Reprinted with permission of Open Court Publishing Company, a division of Carus Publishing Company, Peru, IL.

C. Darwin, "Difficulties of the Theory," extract from Chapter 6 of *The Origin of Species*, first published in 1859.

P. Achinstein, "The Grue Paradox," in P. Achinstein, *The Book of Evidence*, Chapter 9 (with modifications), 2001, New York: Oxford University Press.

N. Russell Hanson, "Seeing and Seeing As," in N. Russell Hanson, *Perception and Discovery*, Chapter 6, 1969, pp. 91–110. San Francisco: Freeman. Figures included: From *Philosophical Investigations* by L. Wittgenstein, c. 1953. Reprinted with permission of Pearson Education, Inc., Upper Saddle River, NJ 07458.

W.V. Quine, "Two Dogmas of Empiricism," in W.V. Quine, *From a Logical Point of View*, 1953, pp. 20–46. Reprinted by permission of the publisher from *From a Logical Point of View* by Willard V. Quine, pp. 20–46, Cambridge, Mass.: Harvard University Press, © 1953, 1961, 1980 by the President and Fellows of Harvard College, renewed 1989 by W.V. Quine.

L. Laudan and J. Leplin, "Empirical Equivalence and Underdetermination," *Journal of Philosophy*, 1991, 88: 449–72.

W. Salmon, "Bayes's Theorem and The History of Science," in *Historical and Philosophical Perspectives of Science*, ed. R. Stuewer, 1970, pp. 68–86. Reprinted with permission of the University of Minnesota Press, Minneapolis.

PART VI: SCIENCE IN CONTEXT: THE CHALLENGE OF HISTORY AND SOCIOLOGY

D. Shapere, "The Structure of Scientific Revolutions," *Philosophical Review*, 1964, 73: 383–94. Reprinted with permission of Cornell University Press.

T. Kuhn, "Objectivity, Value Judgment, and Theory Choice," in Thomas S. Kuhn, *The Essential Tension: Selected Studies in Scientific Tradition and*

Change, 1977, pp. 320–39. Reprinted with permission of the University of Chicago Press, Chicago.

D. Bloor, "The Strong Programme in the Sociology of Knowledge," in D. Bloor, *Knowledge and Social Imagery*, 2nd edn, 1991, pp. 3–23. Reprinted with permission of the University of Chicago Press, Chicago.

E. Anderson, "Feminist Epistemology: An Interpretation and a Defense," *Hypatia*, 1995, 10: 50–84 (with some cuts). Indiana University Press.

E. McMullin, "The Social Dimensions of Science," in *The Social Dimensions of Science*, ed. E. McMullin, 1992, pp. 12–26. Reprinted with permission of the University of Notre Dame Press, Notre Dame.

PART I

SCIENCE AND PHILOSOPHY

INTRODUCTION

Philosophy of science is a difficult subject to define, in large part because philosophy is difficult to define. But on at least one controversial definition of philosophy, the relation between the sciences—physical, biological, social, and behavioral—and philosophy is so close that philosophy of science must be a central concern of both philosophers and scientists. According to this definition, philosophy deals with two sets of questions:

- First, the questions that the sciences—physical, biological, social, behavioral—cannot answer now and perhaps may never be able to answer.
- Second, the questions about why the sciences cannot answer the first lot of questions.

The history of science from the Greeks through Newton and Darwin to the present century reveals these (as yet) scientifically unanswered questions. Indeed, the history of science from the Greeks to the present is the history of one compartment of philosophy after another breaking away from philosophy and emerging as a separate discipline. Thus, by the third century BC, Euclid's work had made geometry a "science of space" separate from but still taught by philosophers in Plato's Academy. Galileo, Kepler, and finally Newton's revolution in the seventeenth century made physics a subject separate from metaphysics. To this day, the name of some departments in which physics is studied is "natural philosophy." In 1859 *The Origin of Species* set biology apart from philosophy (and theology) and at the turn of the twentieth century, psychology broke free from philosophy as a separate discipline. In the past fifty years, philosophy's millennium-long concern with logic has given rise to computer science.

All of these disciplines, which have spun off from philosophy, have left to philosophy a set of distinctive problems: issues they cannot resolve, but must leave either permanently or at least temporarily for philosophy to deal with. Consider an example. Newton's second law tells us that $F = ma$, force equals the product of mass and acceleration. Acceleration, in turn, is *dv/dt*, the first derivative of velocity with respect to time. But what is time? Here is a concept we all understand, and one which physics requires. Yet both ordinary people, who surely know what time is, and physicists, for whom the concept is indispensable, would be hard-pressed to tell us what exactly time is, or give a definition of it. Note that to define time in terms of hours, minutes, and seconds is to mistake the units of time for what they measure. It would be like defining space in terms of meters or inches. On the other hand, we cannot say that time is duration, because duration is just the passage of, well, . . . time. Our definition would presuppose the very notion we set out to define. Explaining exactly what time is or defining it is a problem which science left to philosophy at least 300 years ago. With the advent of the general theory of relativity, physicists may well be taking back this question and finally addressing it themselves.

Similarly, many biologists and not a few philosophers have held that after Darwin, evolutionary biology took back from philosophy the problem of identifying the nature of man or the purpose or meaning of life. And some hold that what it shows is that man's nature is only different by degrees from that of other animals, and that there is no purpose and meaning to life. It is for this reason that evolutionary theory is so widely resisted; it purports to answer questions that should be left to philosophy.

What the history of science and the legacy of problems it leaves to philosophy show is that the two intellectual inquiries have always been inextricably linked. And the legacy may help us to define philosophy. One of the oddities about philosophy is that it seems to be a heterogeneous subject without the unity that characterizes, say, economics or chemistry. Among its subdisciplines, there is logic—the study of valid forms of reasoning; esthetics—the study of the nature of beauty; ethics and political philosophy, which concern themselves with the basis of moral value and justice; epistemology—the study of the nature, extent and justification of knowledge; and metaphysics, which seeks to identify the fundamental kinds of things that really exist. What brings all these diverse questions together in one discipline? If philosophy consists in attempts to answer the two questions noted in the first paragraph, then all these subdisciplines will be part of philosophy.

Suppose one holds that in fact there are no questions that the sciences cannot now or cannot ever answer. One might claim that any question which is forever unanswerable is really a pseudo-question, a bit of meaningless noise masquerading as a legitimate question, like the question "Do green ideas sleep furiously?" or "When it's noon at Greenwich, what time is it on the sun?" Scientists and others impatient with the apparently endless pursuit of philosophical questions that seems to eventuate in no settled answers may hold this view. They may grant that there are questions the sciences cannot yet answer, such as "What was happening before the big bang which began the universe?" or

"How did inorganic molecules give rise to life?" or "Is consciousness merely a brain-process?" But they hold that, given enough time and money, enough theoretical genius and experimentation, all these questions can be answered, and the only ones left unanswered at the end of scientific inquiry will be pseudo-questions which intellectually responsible persons need not concern themselves with. Of course, sapient creatures like us may not be around long enough in the history of the universe to complete science, but that is no reason to conclude that science and its methods cannot in principle answer all meaningful questions.

The claim that it can do so, however, needs an argument, or evidence. The fact that there are questions like "What is time?" which have been with us, unanswered, for centuries is surely some evidence that serious questions may remain permanently unanswered by science. Could these really be pseudo-questions? We should only accept such a conclusion on the basis of an argument or a good reason. Suppose one wanted to argue that any questions still left over at the "end of inquiry," when all the facts that science should attend to are in, must be pseudo-questions. There may well be good arguments in favor of this conclusion. But these arguments will all have two related features: first, they draw substantially on an understanding of the nature of science itself which science does not provide; second, these arguments are not ones science can construct by itself: they are philosophical arguments. And this is because they invoke normative premises, and not just the factual ones that science could provide. For example, the argument trades on the assumption that there are some considerations science *should, ought to, is obliged to* attend to, as opposed to some things it can safely ignore. What are the factors that science *should* take into account when deciding which questions are answerable, and what are the answers to these questions, and which questions are not answerable? This is a matter for epistemology—the study of the nature, extent and justification of knowledge. And this means that philosophy is unavoidable, even in an argument that there are no questions science cannot answer, either now or eventually or perhaps just "in principle."

In the twentieth century, an important philosophical movement, Logical Positivism (or Logical Empiricism, as it was otherwise called), advanced the thesis that the only questions science (including mathematics and logic) could not answer were pseudo-questions. The agenda of Logical Positivism involved a radical reconsideration of the scope and content of the philosophical activity as a whole and of the relationship between philosophy and science. In the lecture "The Future of Philosophy" delivered in 1931 in Stockton, California, one of the founding fathers of Logical Positivism, Moritz Schlick, argues that while science is the pursuit of truth, the proper task of philosophy must be the pursuit of *meaning*: first and foremost a systematic attempt to clarify the meaning of scientific concepts, problems and their proposed solutions by using the resources of logic and conceptual analysis (hence the name of the doctrine). On this view, much of traditional philosophy has been on the wrong track: for the most part, it has aimed at constructing a "world view," a set of true propositions about the ultimate structure of reality, and has, in this regard, perceived itself as a continuation of science, or even a "superscience," the "Queen of Sciences." This attitude generated a historical sequence of grandiose but useless metaphysical systems, from the Greeks to Hegel and beyond, none of which is better than any other, because all of them are, in the opinion of the positivists, not just false, but meaningless. They attempted to do something impossible: to usurp the function of science and form propositions about reality, while lacking the empirical method required to produce knowledge of the world. According to Schlick, the *true* philosophy must be part and parcel of science all right, but it has first to recognize itself, not as a form of knowledge about the world but as an *activity* of uncovering the meaningful or genuine scientific problems, so that scientists can settle them. Central to this activity is a criterion purporting to distinguish scientifically meaningful from meaningless propositions: roughly and broadly, a proposition is meaningful when one is able to specify the observable circumstances under which it would be true or false. The attempt to formulate precisely such a criterion, or *principle of verification*, formed the core of the positivist program.

This ambitious attempt failed. Alex Rosenberg's critical discussion of Logical Positivism, its decline and aftermath, particularly with reference to biology, identifies the difficulties positivist philosophers have faced in their attempts to deprive philosophy of its inherent desire to be a continuation of science and a special form of knowledge about the world.

Nowadays, philosophers of science recognize the legitimacy of this desire. This of course does not mean that philosophers have some sort of special standing or perspective from which to ask and answer a range of questions that scientists cannot consider. These questions about science, its scope and limits, are as much questions that scientists can contribute to answering as they are questions for philosophers. Indeed, in many cases scientists are better placed to answer these questions, or the theories and findings they have uncovered have an essential role in answering the questions about science and its limits. But the conclusion here is that philosophy is inescapable, even by those who hold that in the end all real questions, all questions worth answering, can only be answered by science. Only a philosophical argument can underwrite this claim.

Perhaps the best indication of the relevance of philosophy to science is found in the history of science and in its contemporary state. Many great physicists and biologists were led to pose and reflect on genuinely philosophical questions by the internal logic of their theories, and the same is true of other scientific disciplines. Many of such questions would undoubtedly be relegated to the category of "pseudo-scientific" by positivists. What this shows is that the hard and fast lines and the peculiar division of labor between science and philosophy imposed by them are simply inappropriate. Contemporary philosophy of science is a flourishing discipline that benefits from interaction with other areas of philosophy, including but not restricted to metaphysics, epistemology, and the philosophy of language. Even more importantly, it is a joint effort involving philosophers as well as scientists.

It is worth adding that the positivist project was by no means in vain. First, it has set the standards of rigor and clarity in philosophical argumentation, which neither its opponents nor its successors could fail to appreciate and try to emulate. Second, it would not be exaggerating to say that the present multifarious agenda of the philosophy of science was largely shaped by the historical development of Logical Positivism in the first half of the twentieth century and by the reasons for which positivists subsequently surrendered their program in the 1950s and 1960s.

1

Moritz Schlick, "The Future of Philosophy"

Moritz Schlick (1882–1936) was a leading Logical Positivist philosopher and one of the founders of the Vienna Circle, a group of philosophers and scientists who put forward a project of reshaping philosophy according to the positivist ideal. The group, which included the philosophers Rudolf Carnap, Otto Neurath and Herbert Feigl, the mathematicians Kurt Gödel and Hans Hanh, and the physicist Philipp Frank, met in Vienna from 1922 to 1938. Schlick was a prolific writer and enthusiastic lecturer. His lecture reprinted here was given in Stockton, California, in 1931 and first published in *College of the Pacific Publications in Philosophy* I, Stockton, CA, 1932, pp. 45–62.

The study of the history of philosophy is perhaps the most fascinating pursuit for anyone who is eager to understand the civilization and culture of the human race, for all of the different elements of human nature that help to build up the culture of a certain epoch or a nation mirror themselves in one way or another in the philosophy of that epoch or of that nation.

The history of philosophy can be studied from two distinct points of view. The first point of view is that of the historian; the second one is that of the philosopher. They will each approach the study of the history of philosophy with different feelings. The historian will be excited to the greatest enthusiasm by the great works of the thinkers of all times, by the spectacle of the immense mental energy and imagination, zeal and unselfishness which they have devoted to their creations, and the historian will derive the highest enjoyment from all of these achievements. The philosopher, of course, when he studies the history of philosophy will also be delighted, and he cannot help being inspired by the wonderful display of genius throughout all the ages. But he will not be able to rejoice at the sight that philosophy presents

M. Schlick, *Philosophical Papers*, vol. 2, ed. H.L. Mulder and B.F.B. van de Velde-Schlick, 1979, pp. 210–24. Dordrecht: Reidel.

to him with exactly the same feelings as the historian. He will not be able to enjoy the thoughts of ancient and modern times without being disturbed by feelings of an entirely different nature.

The philosopher cannot be satisfied to ask, as the historian would ask of all the systems of thought—are they beautiful, are they brilliant, are they historically important? and so on. The only question which will interest him is the question, "What truth is there in these systems?" And the moment he asks it he will be discouraged when he looks at the history of philosophy because, as you all know, there is so much contradiction between the various systems—so much quarreling and strife between the different opinions that have been advanced in different periods by different philosophers belonging to different nations—that it seems at first quite impossible to believe that there is anything like a steady advance in the history of philosophy as there seems to be in other pursuits of the human mind, for example, science or technique.

The question which we are going to ask tonight is "Will this chaos that has existed so far continue to exist in the future?" Will philosophers go on contradicting each other, ridiculing each other's opinions, or will there finally be some kind of universal agreement, a unity of philosophical belief in the world?

All of the great philosophers believed that with their own systems a new epoch of thinking had begun, that they, at last, had discovered the final truth. If they had not believed this they could hardly have accomplished anything. This was true of Descartes, for instance, when he introduced the method which made him "the father of modern philosophy" as he is usually called; of Spinoza when he tried to introduce the mathematical method into philosophy; or even of Kant when he said in the preface to his greatest work that from now on philosophy might begin to work as securely as only science had worked thus far. They all believed that they had been able to bring the chaos to an end and start something entirely new which would at last bring about a rise in the worth of philosophical opinions. But the historian cannot usually share such a belief; it may even seem ridiculous to him.

We want to ask the question, "What will be the future of philosophy?" entirely from the point of view of the philosopher. However, to answer the question we shall have to use the method of the historian because we shall not be able to say what the future of philosophy will be except in so far as our conclusions are derived from our knowledge of its past and its present.

The first effect of a historical consideration of philosophical opinions is that we feel sure we cannot have any confidence in any one system. If this is so—if we cannot be Cartesians, Spinozists, Kantians, and so forth—it seems that the only alternative is that we become skeptics and we become inclined to believe that there can be no true system of philosophy because if there were any such system it seems that at least it must have been suspected and would have shown itself in some way. However, when we examine the history of philosophy honestly, it seems as if there were no traces of any discovery that might lead to unanimous philosophical opinion.

This skeptical inference, in fact, has been drawn by a good many historians, and even some philosophers have come to the conclusion that there is no such thing as philosophical advancement, and that philosophy itself is nothing but the history of philosophy. This view was advocated by more than one philosopher in the beginning of the century and it has been called "historicism." That philosophy consists only of its own history is a strange view to take, but it has been advocated and defended with apparently striking arguments. However, we shall not find ourselves compelled to take such a skeptical view.

We have thus far considered two possible alternatives that one may believe in. First, that the ultimate truth is really presented in some one system of philosophy and secondly, that there is no philosophy at all, but only a history of thought. I do not tonight propose to choose either of these two alternatives; but I should like to propose a third view which is neither skeptical nor based on the belief that there can be any system of philosophy as a system of ultimate truths. I intend to take an entirely different view of philosophy and it is, of course, my opinion that this view of philosophy will some time in the future be adopted by everybody. In fact, it would seem strange to me if philosophy, that noblest of intellectual pursuits, the tremendous human achievement that has so often been called the "queen of all sciences" were nothing at all but one great deception. Therefore it seems likely that a third view can be found by careful analysis and I believe that the view which I am going to advance here will do full justice to all the skeptical arguments against the possibility of a philosophical system and yet will not deprive philosophy of any of its nobility and grandeur.

Of course, the mere fact that thus far the great systems of philosophy have not been successful and have not been able to gain general acknowledgment is no sufficient reason why there should not be some philosophical system discovered in the future that would universally be regarded as the ultimate solution of the great problems. This might indeed be expected to happen if philosophy were a "science." For in science we continually find that unexpected satisfactory solutions for great problems are found, and when it is not possible to see clearly in any particular point on a scientific question we do not despair. We believe that future scientists will be more fortunate and discover what we have failed to discover. In this respect, however, the great difference between science and philosophy reveals itself. Science shows a gradual development. There is not the slightest doubt that science has advanced and continues to advance, although some people speak skeptically about science. It cannot be seriously doubted for an instant that we know very much more about nature, for example, than people living in former centuries knew. There is unquestionably some kind of advance shown in science, but if we are perfectly honest, a similar kind of advance cannot be discovered in philosophy.

The same great issues are discussed nowadays that were discussed in the time of Plato. When for a time it seemed as though a certain question were definitely settled, soon the same question comes up again and has to be discussed and reconsidered. It was characteristic of the work of the philosopher that he always had to begin at the beginning again. He never takes anything for granted. He feels that every solution to any philosophical problem is not certain or sure enough, and he feels that he must begin all over again in settling the problem. There is, then, this difference between science and philosophy which makes us very skeptical about any future advance of philosophy. Still we might believe that times may change, and that we might possibly find the true philosophical system. But this hope is in vain, for we can find reasons why philosophy has failed, and must fail, to produce lasting scientific results as science has done. If these reasons are good then we shall be justified in not trusting in any system of philosophy, and in believing that no such system will come forward in the future.

Let me say at once that these reasons do not lie in the difficulty of the problems with which philosophy deals; neither are they to be found in the weakness and incapacity of human understanding. If they lay there, it could easily be conceived that human understanding and reason might develop, that if we are not intelligent enough now our successors might be intelligent enough to develop a system. No, the real reason is to be found in a curious misunderstanding and misinterpretation of the nature of philosophy; it lies in the failure to distinguish between the scientific attitude and the philosophical attitude. It lies in the idea that the nature of philosophy and science are more or less the same, that they both consist of systems of true propositions about the world. In reality philosophy is never a system of propositions and therefore quite different from science. The proper understanding of the relationship between philosophy on one side and of the sciences on the other side is, I think, the best way of gaining insight into the nature of philosophy. We will therefore start with an investigation of this relationship and its historical development. This will furnish us the necessary facts in order to predict the future of philosophy. The future, of course, is always a matter of historical conjecture, because it can be calculated only from past and present experiences. So we ask now: what has the nature of philosophy been conceived to be in comparison with that of the sciences, and how has it developed in the course of history?

In its beginnings, as you perhaps know, philosophy was considered to be simply another name for the "search for truth"—it was identical with science. Men who pursued the truth for its own sake were called philosophers, and there was no distinction made between men of science and philosophers.

A little change was brought about in this situation by Socrates. Socrates, one might say, despised science. He did not believe in all the speculations about astronomy and about the structure of the universe in which the early philosophers indulged. He believed one could never gain any certain knowledge about these matters and he restricted his investigations to the nature of human character. He was not a man of science, he had no faith in it, and yet we all acknowledge him to be one of the greatest philosophers who ever lived. It is not Socrates, however, who created the antagonism that we find to exist later on between science and philosophy. In fact, his successors combined very well the study of human nature with the science of the stars and of the universe.

Philosophy remained united with the various sciences until gradually the latter branched off from philosophy. In this way, perhaps, mathematics, astronomy, mechanics and medicine became independent one after the other and a difference between philosophy and science was created. Nevertheless some kind of unity or identity of the two persisted, we might say, almost to modern times, i.e., until the nineteenth century. I believe we can say truthfully that there are certain sciences—I am thinking particularly of physics—which were not completely separated from philosophy until the nineteenth century. Even now some university chairs for theoretical physics are officially labelled chairs of "natural philosophy."

It was in the nineteenth century also that the real antagonism began, with a certain feeling of unfriendliness developing on the part of the philosopher toward the scientist and the scientist toward the philosopher. This feeling arose when philosophy claimed to possess a nobler and better method of discovering truth than the scientific method of observation and experiment. In Germany at the beginning of the nineteenth century Schelling, Fichte, and Hegel believed that there was some kind of royal path leading to truth which was reserved for the philosopher, whereas the scientist walked the pathway of the vulgar and very tedious experimental method, which required so much merely mechanical technique. They thought that they could attain the same truth that the scientist was trying to find but could discover it in a much easier way by taking a short cut that was reserved for the very highest minds, only for the philosophical genius. About this, however, I will not speak because it may be regarded, I think, as having been superseded.

There is another view, however, which tried to distinguish between science and philosophy by saying that philosophy dealt with the most general truths that could be known about the world and that science dealt with the more particular truths. It is this last view of the nature of philosophy that I must discuss shortly tonight as it will help us to understand what will follow.

This opinion that philosophy is the science that deals with those most general truths which do not belong to the field of any special science is the most common view that you find in nearly all of the text books; it has been adopted by the majority of philosophical writers in our present day. It is generally believed that as, for example, chemistry concerns itself with the true propositions about the different chemical compounds and physics with the truth about physical behavior, so philosophy deals with the most general questions concerning the nature of matter. Similarly, as history investigates the various chains of single happenings which determine the fate of the human race, so philosophy (as "philosophy of history") is supposed to discover the general principles which govern all those happenings.

In this way, philosophy, conceived as the science dealing with the most general truths, is believed to give us what might be called a universal picture of the world, a general world-view in which all the different truths of the special sciences find their places and are unified into one great picture—a goal which the special sciences themselves are thought incapable of reaching as they are not general enough and are concerned only with particular features and parts of the great whole.

This so-called "synoptic view" of philosophy, holding as it does that philosophy is also a science, only one of a more general character than the special sciences, has, it seems to me, led to terrible confusion. On the one hand it has given to the philosopher the character of the scientist. He sits in his library, he consults innumerable books, he works at his desk and studies various opinions of many philosophers as a historian would compare his different sources, or as a scientist would do while engaged in some particular pursuit in any special domain of knowledge; he has all the bearing of a scientist and really believes that he is using in some way the scientific method, only doing so on a more general scale. He regards philosophy as a more distinguished and much nobler science than the others, but not as essentially different from them.

On the other hand, with this picture of the philosopher in mind we find a very great contrast when we look at the results that have been really achieved by philosophical work carried on in this manner. There is all the outward appearance of the scientist in the philosopher's mode of work but there is no similarity of results. Scientific results go on developing, combining themselves with other achievements, and receiving general acknowledgment, but there is no such thing to be discovered in the work of the philosopher.

What are we to think of the situation? It has led to very curious and rather ridiculous results. When we open a text book on philosophy or when we view one of the large works of a present-day philosopher we often find an immense amount of energy devoted to the task of finding out what philosophy is. We do not find this in any of the other sciences. Physicists or historians do not have to spend pages to find out what physics or history are. Even those who agree that philosophy in some way is the system of the most general truths explain this generality in rather different ways. I will not go into detail with respect to these varying definitions. Let me just mention that some say that philosophy is the "science of values" because they believe that the most general issues to which all questions finally lead have to do with value in some way or another. Others say that it is epistemology, i.e. the theory of knowledge, because the theory of knowledge is supposed to deal with the most general principles on which all particular truths rest. One of the consequences usually drawn by the adherents of the view we are discussing is that philosophy is either partly or entirely metaphysics. And metaphysics is supposed to be some kind of a structure built over and partly resting on the structure of science but towering into lofty heights which are far beyond the reach of all the sciences and of experience.

We see from all this that even those who adopt the definition of philosophy as the most general science cannot agree about its essential nature. This is certainly a little ridiculous and some future historian a few hundred or a thousand years from now will think it very curious that discussion about the nature of philosophy was taken so seriously in our days. There must be something wrong when a discussion leads to such confusion. There are also very definite positive reasons why "generality" cannot be used as the characteristic that distinguishes philosophy from the "special" sciences, but I will not dwell upon them, but try to reach a positive conclusion in some shorter way.

When I spoke of Socrates a little while ago I pointed out that his thoughts were, in a certain sense, opposed to the natural sciences; his philosophy, therefore, was certainly not identical with the sciences, and it was not the "most general" one of them. It was rather a sort of Wisdom of Life. But the important feature which we should observe in Socrates, in order to understand his particular attitude as well as the nature of philosophy, is that this wisdom that dealt with human nature and human behavior consists essentially of a special *method*, different from the method of science and, therefore, not leading to any "scientific" results.

All of you have probably read some of Plato's Dialogues, wherein he pictures Socrates as giving and receiving questions and answers. If you observe what was really done—or what Socrates tried to do—you discover that he usually did not arrive at certain definite truths which would appear at the end of the dialogue but the whole investigation was carried on for the primary purpose of making clear what was meant when certain questions were asked or when certain words were used. In one of the Platonic Dialogues, for instance, Socrates asks "What is Justice?"; he receives various answers to his question, and in turn he asks what was meant by these answers, why a particular word was used in this way or that way, and it usually turns out that his disciple or opponent is not at all clear about his own opinion. In short, Socrates' philosophy consists of what we may call "The Pursuit of Meaning." He tried to clarify thought by analyzing the meaning of our expressions and the real sense of our propositions.

Here then we find a definitive contrast between this philosophic method, which has for its object the discovery of *meaning*, and the method of the sciences, which have for their object the discovery of *truth*. In fact, before I go any farther, let me state shortly and clearly that I believe Science should be defined as the "*pursuit of truth*" and Philosophy as the "*pursuit of meaning*." Socrates has set the example of the true philosophic method for all times. But I shall have to explain this method from the modern point of view.

When we make a statement about anything we do this by pronouncing a sentence and the sentence stands for the proposition. This proposition is either true or false, but before we can know or decide whether it is true or false we must know what this proposition says. We must know the meaning of the proposition first. After we know its sense we may be able to find out whether it is true or not. These two things, of course, are inseparably connected. I cannot find out the truth without knowing the meaning, and if I know the meaning of the proposition I shall at least know the beginning of some path that will lead to the discovery of the truth or falsity of the proposition even if I am unable to find it at present. It is my opinion that the future of philosophy hinges on this distinction between the discovery of sense and the discovery of truth.

How do we decide what the sense of a proposition is, or what we mean by a sentence which is spoken, written, or printed? We try to present to ourselves the significance of the different words that we have learned to use, and then endeavor to find sense in the proposition. Sometimes we can do so and sometimes we cannot; the latter case happens, unfortunately, most frequently with propositions which are supposed to be "philosophical."—But how can we be quite sure that we really know and understand what we mean when we make an assertion? What is the ultimate criterion of its sense? The answer is this: We know the meaning of a proposition when we are able to indicate exactly the circumstances under which it would be true (or, what amounts to the same, the circumstances which would make it false). The description of these circumstances is absolutely the only way in which the meaning of a sentence can be made clear. After it has been made clear we can proceed to look for the actual circumstances in the world and decide whether they make our proposition true or false. There is no vital difference between the ways we decide about truth and falsity in science and in everyday life. Science develops in the same ways in which does knowledge in daily life. The method of verification is essentially the same; only the facts by which scientific statements are verified are usually more difficult to observe.

It seems evident that a scientist or a philosopher when he propounds a proposition must of necessity know what he is talking about before he proceeds to find out its truth. But it is very remarkable that oftentimes it has happened in the history of human thought that thinkers have tried to find out whether a certain proposition was true or false before being clear about the meaning of it, before really knowing what it was they were desirous of finding out. This has been the case sometimes even in scientific investigations, instances of which I will quote shortly. And it has, I am almost tempted to say, nearly always been the case in traditional philosophy. As I have stated, the scientist has two tasks. He must find out the truth of a proposition and he must also find out the meaning of it, or it must be found out for him, but usually he is able to find it for himself. In so far as the scientist does find out the hidden meaning of the propositions which he uses in his science he is a philosopher. All of the great scientists have given wonderful examples of this philosophical method. They have discovered the real significance of words which were used quite commonly in the beginning of science but of which nobody had ever given a perfectly clear and definite account. When Newton discovered the concept of *mass* he was at that time really a philosopher. The greatest example of this type of discovery in modern times is Einstein's analysis of the meaning of the word "simultaneity" as it is used in physics. Continually, something is happening "at the same time" in New York and San Francisco, and although people always thought they knew perfectly well what was meant by such a statement, Einstein was the first one who made it really clear and did away with certain unjustified assumptions concerning time that had been made without anyone being aware of it. This was a real philosophical achievement—the discovery of meaning by a logical clarification of a proposition. I could give more instances, but perhaps these two will be sufficient. We see that meaning and truth are linked together by the process of verification; but the first is found by mere reflection about possible circumstances in the world, while the second is decided by really discovering the existence or nonexistence of those circumstances. The reflection in the first case is the philosophic method of which Socrates' dialectical proceeding has afforded us the simplest example.

From what I have said so far it might seem that philosophy would simply have to be defined as the science of meaning, as, for example, astronomy is the science of the heavenly bodies, or zoology the science of animals, and that philosophy would be a science just as other sciences, only its subject would be different, namely, "Meaning." This is the point of view taken in a very excellent book, *The Practice of Philosophy*, by Susan K. Langer. The author has seen quite clearly that philosophy has to do with the pursuit of meaning, but she believes the pursuit of meaning can lead to a science, to "a set of true propositions"—for that is the correct interpretation of the term "science." Physics is nothing but a system of truths about physical bodies. Astronomy is a set of true propositions about the heavenly bodies, etc.

But philosophy is not a science in this sense. There can be no science of meaning, because there cannot be any set of true propositions about meaning. The reason for this is that in order to arrive at the meaning of a sentence or of a proposition we must go beyond propositions. For we cannot hope to explain the meaning of a proposition merely by presenting another proposition. When I ask somebody, "What is the meaning of this or that?," he must answer by a sentence that would try to describe the meaning. But he cannot ultimately succeed in this, for his answering sentence would be but another proposition and I would be perfectly justified in asking "What do you mean by *this*?" We would perhaps go on defining what he meant by using different words, and repeat his thought over and over again by using new sentences. I could always go on asking "But what does this new proposition mean?" You see, there would never be any end to this kind of inquiry, the meaning could never be clarified, if there were no other way of arriving at it than by a series of propositions.

An example will make the above clear, and I believe you will all understand it immediately. Whenever you come across a difficult word for which you desire to find the meaning you look it up in the *Encyclopaedia Britannica*. The definition of the word is given in various terms. If you don't happen to know them you look up these terms. However, this procedure can't go on indefinitely. Finally you will arrive at very simple terms for which you will not find any explanation in the encyclopedia. What are these terms? They are the terms which cannot be defined any more. You will admit that there are such terms. If I say, e.g., that the lamp shade is yellow, you might ask me to describe what I mean by yellow—and I could not do it. I should have to show you some color and say that this is yellow, but I should be perfectly unable to explain it to you by means of any sentences or words. If you had never seen yellow and I were not in a position to show you any yellow color it would be absolutely impossible for me to make clear what I meant when I uttered the word. And the blind man, of course, will never be able to understand what the word stands for.

All of our definitions must end by some demonstration, by some activity.

There may be certain words at the meaning of which one may arrive by certain mental activities just as I can arrive at the signification of a word which denotes color by showing the color itself. It is impossible to define a color—it has to be shown. Reflection of some kind is necessary so that we may understand the use of certain words. We have to reflect, perhaps, about the way in which we learn these words, and there are also many ways of reflection which make it clear to us what we mean by various propositions. Think, for example, of the term "simultaneity" of events occurring in different places. To find what is really meant by the term we have to go into an analysis of the proposition and discover how the simultaneity of events occurring in different places is really determined, as was done by Einstein; we have to point to certain actual experiments and observations. This should lead to the realization that philosophical activities can never be replaced and expressed by a set of propositions. The discovery of the meaning of any proposition must ultimately be achieved by some act, some immediate procedure, for instance, as the showing of yellow; it cannot be given in a proposition. Philosophy, the "pursuit of meaning," therefore cannot possibly consist of propositions; it cannot be a science. The pursuit of meaning consequently is nothing but a sort of mental activity.

Our conclusion is that philosophy was misunderstood when it was thought that philosophical results could be expressed in propositions, and that there could be a system of philosophy consisting of a system of propositions which would represent the answers to "philosophical" questions. There are no specific "philosophical" truths which would contain the solution of specific "philosophical" problems, but philosophy has the task of finding the meaning of *all* problems and their solutions. It must be defined as *the activity of finding meaning*.

Philosophy is an activity, not a science, but this activity, of course, is at work in every single science continually, because before the sciences can discover the truth or falsity of a proposition they have to get at the meaning first. And sometimes in the course of their work they are surprised to find, by the contradictory results at which they arrive, that they have been using words without a perfectly clear meaning, and then they will have to turn to the philosophical activity of clarification, and they cannot go on with the pursuit of truth before the pursuit of meaning has been successful. In this way philosophy is an extremely important factor within science and it very well deserves to bear the name of "The Queen of Sciences."

The Queen of Sciences is not itself a science. It is an activity which is needed by all scientists and pervades all their other activities. But all real problems are scientific questions, there are no others.

And what was the matter with those great questions that have been looked upon—or rather looked up to—as specific "philosophical problems" for so many centuries? Here we must distinguish two cases. In the first place, there are a great many questions which look like questions because they are formed according to a certain grammatical order but which nevertheless are not real questions, since it can easily be shown that the words, as they are put together, do not make logical sense.

If I should ask, for instance: "Is blue more identical than music?," you would see immediately that there is no meaning in this sentence, although it does not violate the rules of English grammar. The sentence is not a question at all, but just a series of words. Now, a careful analysis shows that this is the case with most so-called philosophical problems. They look like questions and it is very difficult to recognize them as nonsensical, but logical analysis proves them none the less to be merely some kind of confusion of words. After this has been found out the question itself disappears and we are perfectly peaceful in our philosophical minds; we know that there can be no answers because there were no questions, the problems do not exist any longer.

In the second place, there are some "philosophical" problems which prove to be real questions. But of these it can always be shown by proper analysis that they are capable of being solved by the methods of science, although we may not be able to apply these methods at present for merely technical reasons. We can at least say what would have to be done in order to answer the question even if we cannot actually do it with the means at our disposal. In other words: problems of this kind have no special "philosophical" character, but are simply scientific questions. They are always answerable in principle, if not in practice, and the answer can be given only by scientific investigation.

Thus the fate of all "philosophical problems" is this: Some of them will disappear by being shown to be mistakes and misunderstandings of our language and the others will be found to be ordinary scientific questions in disguise. These remarks, I think, determine the whole future of philosophy.

Several great philosophers have recognized the essence of philosophical thinking with comparative clarity, although they have given no elaborate expression to it. Kant, e.g., used to say in his lectures that philosophy cannot be taught. However, if it were a science such as geology or astronomy, why then should it not be taught? It would then, in fact, be quite possible to teach it. Kant therefore had some kind of a suspicion that it was not a science when he stated "The only thing I can teach is philosophizing." By using the verb and rejecting the noun in this connection Kant indicated clearly, though almost involuntarily, the peculiar character of philosophy as an activity, thereby to a certain extent contradicting his books, in which he tries to build up philosophy after the manner of a scientific system.

A similar instance of the same insight is afforded by Leibniz. When he founded the Prussian Academy of Science in Berlin and sketched out the plans for its constitution, he assigned a place in it to all the sciences, but Philosophy was not one of them. Leibniz found no place for philosophy in the system of the sciences because he was evidently aware that it is not a pursuit of a particular kind of truth, but an activity that must pervade *every* search for truth.

The view which I am advocating has at the present time been most clearly expressed by Ludwig Wittgenstein; he states his point in these sentences:[1] "The object of philosophy is the logical clarification of thoughts. Philosophy is not a theory but an activity. The result of philosophy is not a number of 'philosophical propositions,' but to make propositions clear." This is exactly the view which I have been trying to explain here.

We can now understand historically why philosophy could be regarded as a very general science: it was misunderstood in this way because the "meaning" of propositions might seem to be something very "general," since in some way it forms the foundation of all discourse. We can also understand historically why in ancient times philosophy was identical with science: this was because at that time all the concepts which are used in the description of the world were extremely vague. The task of science was determined by the fact that there were no clear concepts. They had to be clarified by slow development, the chief endeavor of scientific investigation had to be directed towards this clarification, i.e., it had to be philosophical, no distinction could be made between science and philosophy.

At the present time we also find facts which prove the truth of our statements. In our days certain specific fields of study such as ethics and aesthetics are called "philosophical" and are supposed to form part of philosophy. However, philosophy, being an activity, is a unit which cannot be divided into parts or independent disciplines. Why, then, are these pursuits called philosophy? Because they are only at the beginnings of the scientific stage; and I think this is true to a certain extent also of psychology. Ethics and esthetics certainly do not yet possess sufficiently clear concepts, most of their work is still devoted to clarifying them, and therefore it may justly be called philosophical. But in the future they will, of course, become part of the great system of the sciences.

It is my hope that the philosophers of the future will see that it is impossible for them to adopt, even in outward appearance, the methods of the scientists. Most books on philosophy seem to be, I must confess, ridiculous when judged from the most elevated point of view. They have all the appearance of being extremely scientific books because they seem to use the scientific language. However, the finding of meaning cannot be done in the same way as the finding of truth. This difference will come out much more clearly in the future. There is a good deal of truth in the way in which Schopenhauer (although his own thinking seems to me to be very imperfect indeed) describes the contrast between the real philosopher and the academic scholar who regards philosophy as a subject of scientific pursuit. Schopenhauer had a very clear instinct when he spoke disparagingly of the "professorial philosophy of the professors of philosophy." His opinion was that one should not try to teach philosophy at all but only the history of philosophy and logic; and a good deal may be said in favor of this view.

I hope I have not been misunderstood as though I were advocating an actual separation of scientific and philosophical work. On the contrary, in most cases future philosophers will have to be scientists because it will be necessary for them to have a certain subject matter on which to work—and they will find cases of confused or vague meaning particularly in the foundations of the sciences. But, of course, clarification of meaning will be needed very badly also in a great many questions with which we are concerned in our ordinary human life. Some thinkers, and perhaps some of the strongest minds among them, may be especially gifted in this practical field. In such instances, the philosopher may not have to be a scientist—but in all cases he will have to be a man of deep understanding. In short, he will have to be a wise man.

I am convinced that our view of the nature of philosophy will be generally adopted in the future; and the consequence will be that it will no longer be attempted to teach philosophy as a system. We shall teach the special sciences and their history in the true philosophical spirit of searching for clarity and, by doing this, we shall develop the philosophical mind of future generations. This is all we can do, but it will be a great step in the mental progress of our race.

Note

1 *Tractatus Logico-Philosophicus*, London 1922, 4. 112.

2

Alex Rosenberg, "Biology and Its Philosophy"

Chapter 1 of Alex Rosenberg's *The Structure of Biological Science* (Cambridge University Press, 1985).

In August of 1838, after hitting upon a mechanism for evolution, Charles Darwin confided to his notebook: "Origin of man now proved.—Metaphysics must flourish.—He who understands baboon would do more towards metaphysics than Locke" (Barrett 1974: 281). Any philosopher—and many a biologist—coming upon this prediction over the next century and more would certainly have thought it quite false. Metaphysics, the philosophical examination of the ultimate nature of reality, did not flourish during the hundred years after Darwin published *On the Origin of Species*. Indeed, it came close to vanishing. And the causes for the disappearance of philosophical and theological speculation throughout this period were to be found in the influence of Darwin's own theory.

If ever there was a theory that put an end to traditional philosophizing, it was the one Darwin expounded. By providing a single, unified scientific theory of "the origin of man" and of biological diversity generally, Darwin made scientifically irrelevant a host of questions that philosophers and scientists had taken seriously since long before the time of John Locke. The theory of natural selection has put an end to much speculation about the purpose of the universe, the meaning of life, the nature of man, and the objective grounds of morality. It has grievously undermined the theologian's most compelling grounds for the existence of God, the argument from the earth's design to the existence of a designer. Philosophers and biologists certainly recognized this effect of Darwinism, and over the course of the decades after 1859 some of them made great efforts to refute the theory as much on philosophical grounds as on biological ones. Among biologists,

A. Rosenberg, *The Structure of Biological Science*, Chapter 1, 1985, pp. 1–11. Cambridge: Cambridge University Press.

this work has had ever-diminishing influence, and antievolutionary philosophy has almost completely disappeared within biology. As indeed has almost all philosophy as traditionally conceived.

By making the traditional questions of philosophy biologically irrelevant, Darwin also helped make them philosophically disreputable. But when the grand questions of metaphysics were expunged from philosophy, there seemed to be nothing left to the subject but "logic chopping" and "mere semantics." Thus philosophy as a whole lost its interest for most scientists. The conclusion seems inescapable that Darwin put an end to philosophizing, at least about biological matters. By and large, Darwinians and anti-Darwinians have agreed on one thing: If Darwin was right about the origin of man, metaphysics should vanish, not flourish. For a long time, therefore, Darwin's prediction about his revolution's effects on philosophy seemed quite wrong.

But the more recent history of philosophy, and especially the philosophy of science, has vindicated Darwin after all. This chapter traces the course of the reflections that did so. This brief history of how traditional philosophical issues became respectable again in philosophy is at the same time the best argument for biologists taking the philosophical examination of their subject matter with the utmost seriousness. The history to be briefly surveyed is that of Logical Positivism—or Logical Empiricism, as some of its proponents called it. The rise and fall of this movement in the philosophy of science has revealed that the philosophy of a science is part and parcel of that science itself. The questions philosophers deal with do not differ in kind from those scientists face. Some differ in generality and in urgency, but none is a question that scientists can ignore as irrelevant to their discipline and its agenda. This means that the justification for pursuing the philosophy of science is nothing more or less than the justification for science itself.

. . .

1. The Rise of Logical Positivism

Logical Positivism has certainly been the most important movement in the twentieth-century philosophy of science. Let us trace its motives, chief doctrine, and gravest difficulties. The motives were laudable, the doctrines striking, and the difficulties insurmountable. In surrendering the doctrines of Logical Positivism while honoring its motives, the philosophy of science transformed itself into something indistinguishable from science itself.

It is convenient to begin our exposition of Positivism with an important achievement of nonevolutionary biology. Throughout the latter half of the nineteenth century, embryology was at the forefront of experimental research. Among the most important of embryological experimentalists was

Hans Driesch. Two striking laboratory discoveries are associated with his name. Working with sea-urchin eggs and embryos, he was able to demonstrate that the physical deformation of the egg and the subsequent rearrangement of the blastomeres—the cells produced in the first few stages of fission—had no effect on the normal development of the embryo. This experiment suggests that spatial relations among early blastomeres are irrelevant to normal development. Even more strikingly, Driesch went on to show that a single blastomere isolated from the rest at the two- or four-cell stage can give rise to a complete sea-urchin embryo normal in every respect except size.

Driesch is honored in every account of embryology for these crucial experimental discoveries. But he is ridiculed for the explanatory theory that he offered to account for them. The fact that an embryo, or indeed a single cell, can regulate its development to compensate for missing cells suggested to Driesch the operation of an organizing principle, which he dubbed an "entelechy" (after a similar notion in Aristotle's philosophy), and which he held to determine the harmonious development of living things and to distinguish them from inanimate ones. Because spatiotemporal location and physical mass seemed irrelevant to development, physics could not account for embryological phenomena. Their causes must, he thought, be sought in nonmaterial forces. Therefore he adopted the view that entelechies have a nonspatiotemporal mode of existence, although they act "into" space and time. Entelechies are elementary "whole-making" factors that have no quantitative characteristics, are unanalyzable, and, according to Driesch, are knowable to the scientist only by reflection on the orderliness of direct human experience. It was perhaps inevitable that the temptations that led this important experimentalist to adopt such speculative explanations for the startling observations he made eventually overcame his biological interests altogether. Driesch ended his days as a professor of philosophy. Contemporary works still reprint his most important experimental papers but add cautions like the following: "Most embryologists, however, have had no difficulty in explaining regulation in terms of known physiological processes, making superfluous Driesch's mystical interpretations" (Gabriel and Fogel 1955:210).

Driesch's entelechy is just the sort of occult entity that has long bedeviled all the natural sciences. The Logical Positivist philosophers of the first half of the century expounded a philosophy of science that would eliminate such speculative metaphysics from legitimate science, that would enable us to objectively distinguish empirical claims from disguised pseudoscience like astrology and antiscience like special-creationism, and that would also determine the scope and form of intellectually respectable philosophical examinations of science. Because, according to these Logical Empiricists, knowledge is either based on observation and experiment, as in the sciences, or on formal deduction from definitions, as in mathematics, whatever transcended these limits could be safely disregarded as scientifically, or cognitively, meaningless—indeed, in the view of some, as quite literally nonsense. In the view of some of these philosophers, a claim like Driesch's that nonphysical entelechies control the development of embryos was on a par with Lewis Carroll's nonsense verse from Alice in Wonderland: "Twas Brillig and the Slithy toves did gyre and gimble in the wabe . . ."

What Logical Positivists required to eliminate metaphysical nonsense from empirical science was an objective principle or test that could be applied to statements and terms from any discipline and that would decide about the cognitive significance of the claim or concept. These philosophers searched for a principle of meaningfulness that made no demands on the specific *content* of scientifically legitimate statements but required them to have a specified relation to actual and possible empirical evidence that could test them. The history of the school of Logical Positivism is the history of attempts to find the correct formulation of such a principle. Positivists knew roughly what it had to look like, and they knew broadly what systems of statements clearly passed its standard as meaningful and what sets of statements plainly failed as meaningless. Paradigm cases, of meaninglessness like Driesch's entelechies on the one hand, and meaningfulness like Rutherford's electrons on the other, were employed to calibrate varying candidates for a satisfactory principle of "cognitive significance." Such a principle had to rule the former as meaningless and the latter as meaningful. Because the mark of science is that its claims are controlled and justified by experiment, observation, and other forms of data collection, Positivists held that, to be meaningful, expressions have to be empirically testable by observation and experiment. Those that are not have no more role to play in science than the statement that "green ideas sleep furiously." They may look respectable, and satisfy the rules of grammar of the languages they are couched in, but these pseudosentences on whose truth or falsity the empirically ascertainable facts cannot bear are literally *non*sense, or at any rate without scientific significance.

Problems arose for Positivists in formulating a manageable principle that operated along these lines and gave the right answers for the calibrating samples. Consider what is required for empirical testability. If complete verification by observations is required for testability, almost no sentences except those reporting immediate sensations are testable. Statements of physics about unobservable entities like electrons and quarks will turn out to be meaningless. Even general laws about regularities among observable phenomena will fail the test because they cannot be strictly verified, expressing as they do a claim about an indefinitely large number of events.

Accordingly, the notion of empirical testability was revised and weakened to allow for the theoretical entities of science and for the generality of its most characteristic claims, its laws and theories. Instead of strict and direct verifiability, Positivists opted for indirect confirmability: A statement is scientifically meaningful if and only if there is actual or possible empirical evidence that tends to confirm, though perhaps not completely verify, the statement. But the notion of confirmation is an unsuitably vague one, so vague that Driesch's entelechy theory might even pass its muster. Therefore many philosophers, as well as sympathetic scientists, were attracted to another formulation of cognitive meaningfulness, one due originally to Karl Popper. Its particular attraction is its ability to pass the general laws and theories characteristic of science as meaningful while excluding Driesch's entelechy theory. Verifying a law requires an indefinitely large number of positive instances, but only one negative instance seems required to falsify a law. By contrast, on Driesch's own exposition of his theory, claims about entelechies are unfalsifiable by experiments because entelechies have no quantitative properties, nor even a spatiotemporal location for that matter. Thus, it has long and widely been held, especially by scientists themselves, that the mark of a scientifically respectable proposition is that there be actual or possible empirically detectable states of affairs that could *falsify* it.

2. The Consequences for Philosophy

Following through on Positivist strictures on the meaningfulness of statements had the profoundest consequences for philosophy and especially for the philosophy of science: These disciplines were restricted largely to the treatment of purely "semantic" questions, in the most pejorative sense of that term. Philosophy is not an experimental science; it can claim neither a special range of facts as its subject matter nor any nonempirical mode of knowledge of the facts the "real" sciences study. It must, in the Positivist view, limit itself to the provision and examination of definitions, stipulations, and conventions about language, and to the study of their formal relations. Any other philosophical enterprise was condemned to intellectual disreputability, to the cognitive meaninglessness that characterized so much pre-twentieth-century metaphysics. It was for this reason that twentieth-century philosophy became largely the philosophy of language and that the philosophy of science became the study of the implicit and explicit definitions of the terms ubiquitous in science—like "law," "theory," and "explanation"—and of the terms of the special sciences—like "mass," "element," and "phenotype." The outcome of such investigations could at most be increased clarity about usage or proposed improvements in terminology, justified by considerations of convenience and simplicity.

So circumscribed, the philosophy of science has little to offer the sciences. It may show that the way in which physicists employ the term "law" differs from the way biologists do, or that what the latter call "explanations" differ from what the former do. But it can hardly assess or adjudicate substantive matters within or between the sciences. According to Positivist teachings, even the linguistic differences philosophy might uncover, and the distinct patterns of reasoning it can reveal, have no factual import, for they reflect conventions utterly independent of any fact of the matter. Such linguistic differences between sciences cannot constitute or reflect anything about the nature of the sciences' subject matter.

Philosophy, along with mathematics and logic, had long been a priori disciplines, domains in which truths have always been deemed *necessary* ones. It is just because of the necessity of mathematical truths, and for that matter philosophical ones, that they had to be known a priori: Experience never reveals the necessity of any truth it communicates. This is because claims of experience are falsifiable: Things can always be conceived to be different from the way they are experienced. But now the Logical Positivists thought they knew why mathematics and philosophy were a priori and necessary. It was not because the philosopher and mathematician had a special faculty of insight into necessary truths more firmly fixed, more secure, and more important than the merely contingent findings of empirical science. The truths of mathematics, and those philosophical claims left after the banishment of metaphysics, are necessary and a priori because they are disguised or undisguised *definitions* and the logical consequences of definitions. These truths are necessary because they have no content, restrict no factual possibilities, and merely express our conventions to use words in certain ways. They are vacuous trivialities. Philosophy provides a priori knowledge because it provides linguistic knowledge, not factual knowledge. As such, it does not compete with or cooperate with the sciences in providing factual knowledge. Because its only legitimate claims are not falsifiable, philosophy was condemned to a derivative role of clarifying and reconstructing the expression of factual knowledge, but not adding anything to it.

Positivists were willing to bear the high cost of casting down philosophy from its throne as queen of the sciences mainly because in doing so they were also ending the baleful effect of metaphysical speculation and pseudoscience on the real advance of knowledge.

For all its neatness and rigor, the Positivists' program fell apart in the immediate postwar period. It did not come unstuck through the attacks of its opponents and detractors, disgruntled metaphysicians who thought that philosophy did provide an alternative route to real knowledge that science could not reveal. The Positivists' program came apart at the hands of the Positivists themselves and of their students. They found that its fundamental distinctions could not be justified by Positivism's own standards of adequacy. The collapse of Logical Positivism is best illustrated for our purposes by examining more closely the claim that scientific knowledge must be falsifiable. More than any other slogan, this one has become the outstanding shibboleth of contemporary biological methodology.

3. Problems of Falsifiability

A proposition is scientific if and only if it is falsifiable. This is the criterion or principle of falsifiability. Falsifiability must be distinguished from falsity, of course. To *falsify* a proposition, that is, to show it is false, it is sufficient to infer from it some implication that is in fact not borne out by observation or experiment. For a proposition to be *falsifiable* it must only be logically possible to do this, not actually, physically possible; otherwise we should have to say that a true empirical law is unfalsifiable because it cannot in fact be shown to be false.

Consider such an expression as, say, Ohm's law, which states the relation between resistance, voltage, and amperage: $R = E/I$. To test the simple claim that, for a potential-difference of E volts, and a current of I amperes, the resistance, R, in ohms, is equal to E/I, we require an ammeter, a voltmeter, an ohmmeter, a conductor, a resistor, and a source of electrical potential. Testing Ohm's law by setting up the appropriate circuit and observing the deflection of the point on the ohmmeter while varying the voltage and amperage requires a host of subsidiary, auxiliary hypotheses be true: not just assumptions about the presence of an electrical potential, or that the meters are functioning properly. What is assumed when Ohm's law is put to the test is the whole body of physical and electrical theory that, first, underwrites the construction and reliability of the meters; second, enables us to alter the amperage and voltage; third, assures us we can ignore certain forces acting on the circuit; and, fourth, adjusts for other forces. In particular, trusting the voltmeter involves embracing Maxwell's equations, which describe how the electric field generates a magnetic field, which twists the needle on the meter's dial. Additionally, we must implicitly appeal to Newtonian mechanics, which governs the needle's resistance to a spring and its deflection of a pointer. Accordingly, all these assumptions, hypotheses, and background theories meet the test in a body, together with the law we set out to test. Science meets experience not sentence by sentence, but in large blocks of theories and laws, blocks that are themselves divided from others only by constraints of practical manageability. Adopting these constraints constitutes substantial contingent theoretical commitments.

Suppose, now, that in our test of Ohm's law the meters do not read as the law predicts. Where does the fault lie: What proposition is falsified? Ohm's law? The assumptions about the construction and reliability of the meters? The assumption that there are no relevant intervening forces, or that they can be neglected? Are Newton's laws or Maxwell's equations at fault, or is the special theory of relativity that lies behind them? Of course it will be replied that none of these wider theories is thrown into doubt by such a test. Good sense directs that we check the wiring, the conductance of the metal it is made of, the springs in the meters, etc. So far as practical matters are concerned, once defects at this level are excluded, it is Ohm's law that would be suspect. But so far as matters of *strict* falsifiability are concerned, we see there is no such thing. For a disconfirmation does not point the finger at one particular statement under test; there is no one statement under test, for the entire conjunction of propositions is required for the prediction that fails. We are free to give up any one of the conjuncts and preserve all the rest. And this is not a mere matter of logic; the actual practice of scientists interpreting their data often reflects this freedom. Indeed, the most radical of scientific revolutions results from a scientist finding the fault to which an experimental anomaly points deeply in the center of a research program, instead of at its peripheral assumptions about the accuracy of measuring instruments.

How deeply can the falsification of a test, or of several of them, point? In the history of science it has certainly pointed at least as far as the falsity of Newtonian physics and its "philosophical" assumption of causal determinism. The discovery of the irreducibly random phenomena of radioactivity in effect falsified the belief behind Newtonian mechanics that every event has a cause that produces it in accordance with strict and exceptionless laws. Quantum mechanics rests on the rejection of a Newtonian principle that physicists and philosophers spent two hundred years attempting to prove as a necessary truth of metaphysics. In fact, difficulties in reconciling quantum mechanics with the most fundamental aspects of physical theory and its mathematical structure have led to the questioning of even more central and more "metaphysical" assumptions. In particular, some philosophers and physicists view the Heisenberg uncertainty principle of quantum mechanics as good reason to surrender the logical principle of bivalence, that every meaningful proposition is either true or false. Even more radically, responsible physicists have held that recent experiments require either the surrender of quantum mechanics or the "metaphysical" thesis that there is a world of enduring physical objects that exist independently of our knowledge of them. If these two proposals are coherent, then the experimental evidence that tests quantum mechanics can lead us to surrender, for factual reasons, principles of logic and mathematics we supposed to be necessarily true, and metaphysical theses Positivists supposed to be without empirical significance.

If any proposition can be surrendered as a result of a falsifying experiment, and if in the actual history of science the most central and firmly held of our beliefs have sometimes been surrendered, then we cannot identify propositions as necessarily true—as propositions we embrace *come what may*—that are known a priori. We cannot draw a contrast between such statements and contingent factual propositions—statements that may or may not survive attempts at falsifications—and so have scientifically significant empirical content. Similarly, any proposition, no matter how apparently factual, no matter how apparently vulnerable to falsification, can be preserved in the face of any possible falsifying experiment. We may in all consistency maintain that the earth is flat, attributing all apparent evidence against this belief to the falsity of one or another of the auxiliary assumptions that, together with it, are jointly falsified in photographs of the earth taken by an astronaut. Similarly, claims that Positivists stigmatized as pure metaphysics may also be surrendered in the aftermath of a falsifying experiment. Is the thesis of thoroughgoing universal determinism one of metaphysics? Is it scientifically empty speculation to assert that every event has a cause? It has certainly been a traditional thesis of philosophy, and yet it is one that has certainly come to be doubted as a result of the discovery of quantum-mechanical phenomena.

If we are to conclude that quantum phenomena have falsified metaphysical determinism, then we must conclude that metaphysical principles are testable after all and therefore cognitively significant. The only way to deny this power to experiment and observation is to deny that they ever falsify any single proposition at all. Either way, falsifiability no longer distinguishes between meaningless metaphysics and factual science.

Testing Ohm's law involves adopting Maxwell's equations for electromagnetism, and adopting these involves buying into the relativistic electrodynamics that accounts for them. And behind this theory stands the post-Newtonian "world picture," the research program that has animated modern science since the seventeenth century. It would of course be fatuous to hold that all this is at risk when an experiment does not corroborate Ohm's law. Any concern that would give an experimentalist real pause must be livelier than this abstract possibility. Even a theorist need not lose any sleep over the furthest mathematical, conceptual, and logical foundations of modern science. But the theorist cannot hold them logically irrelevant to his or the experimentalists' day-to-day concerns, and he has assuredly taken sides on their truth. What is more, at least sometimes in the history of science, and the lives of scientists, these broadest theoretical concerns do take a serious turn—either because they are called into question or because they suggest a direction for research.

These conclusions provide cognitive legitimacy to the speculative philosophy from which the Logical Positivists thought themselves to have freed "real" science. The justification for eliminating or embracing such notions as Driesch's entelechy is no different in kind from that employed to assess claims about the existence of electrons, magnets, or virons. It differs from them by degree, and very great degree at that. But ridding biology of such notions is not after all a matter of applying some rule against useless metaphysics. For deciding on the existence or nonexistence of entelechies is nothing less than questioning the adequacy of competing embryological theories altogether. But because this question is surely not an excursion into cognitively meaningless speculation, it follows that disputes about entelechies are not scientifically idle after all. Driesch's vitalism or the mechanism it opposed are indeed metaphysical theories, but they do not stand apart from "real" science. For better or worse, they stand on a continuum from sheer speculation through research programs and grand unifying theory to general theory and special models, all the way across to particular empirical findings. Unpalatable as this conclusion may be for empiricist philosophers and empirical scientists, to deny it without providing a workable distinction somewhere along the continuum would be unprincipled dogmatism—a dogmatism that the Positivists and their students would not accept.

4. Philosophy of Science Without Positivism

The end of Positivism means an end to philosophy's proscriptions against either treading on the subject matter of the empirical sciences or engaging in empty metaphysics. For metaphysics can no longer be distinguished from theoretical science. And neither can be distinguished from logic, linguistic conventions, or their analysis. For the necessity and unrevisability that was supposed to mark these subjects also fails to distinguish them from science or metaphysics. Although the fall of Positivism frees philosophers (and scientists for that matter) to turn their attention to more exciting activities than the study of language, it also transforms the significance of the very study. It turns the linguistic and logical analysis Positivists produced into the kind of metaphysical and epistemological exploration of the foundations of science to which philosophy has traditionally attended. It reveals that the analysis of concepts is just metaphysics carried out under a different name.

This change is well illustrated in the philosophical problems generated by the apparent goal-directedness, or purposiveness, of living things. The *teleology* (from the Greek words for "ends" or "goals" and their study) of the animate world has always been a focus of philosophical debate. Vitalists held that the purposiveness of things could only be the result of special forces, like Driesch's entelechies; mechanists insisted that teleology was only a special and complex form of mechanical causality, ultimately to be understood through the application of physics and chemistry alone. Materialism is of course just as metaphysical a thesis as vitalism. So Positivism invoked a plague on both these houses and enjoined philosophers to turn their attention to the purely linguistic question of giving the *meaning* of characteristic teleological expressions of biology. A cottage industry sprang up, in which philosophers provided definitions of terms like "goal," "purpose," and especially "function"; these definitions were in turn rejected by other philosophers on the strength of counterexamples—clear cases of teleology that did not satisfy the definition or, still worse, nonteleological phenomena that did; the result was a cycle of revisions, qualifications, and reformulations that elicited another round of counterexamples, and so on.

With hindsight, however, philosophers came to see that the question of whether teleological expressions are definable in nonteleological physical terms is really just the ancient debate between vitalists and materialists carried out under the guise of linguistic analysis. If teleological statements can be translated into nonteleological, causal ones, then teleological processes are causal ones. If there is no difference between the formal claim about translation and materialists' allegedly factual one that living systems are just physical systems, then the linguistic question is identical to the metaphysical question of whether vitalism or materialism is correct.

In fact, the distinction between linguistic, metaphysical, and methodological problems and empirical issues is groundless. Biologists' attempts to uncover the purely causal mechanism of an apparently goal-directed activity like photosynthesis may or may not succeed. If it does, then this may strengthen a materialist metaphysical view. It will certainly encourage the continued exploitation of a methodology of searching for causal mechanisms to explain teleological behavior. But, of course, success in any one area of investigation cannot establish the general claim that all purposive phenomena are really causal. Nor does it establish the universal propriety of the methodology of searching for such mechanisms. What would? Well, nothing can ever be *established* in science. Nevertheless, a cogent explanation of why this method works will certainly strengthen the confidence of one biologist's particular account of photosynthesis.

On the other hand, suppose no causal mechanism for some goal-directed phenomena is detected, despite great effort. Under such conditions, biologists would be within their rights to insist that, nevertheless, further industry—better experimental materials and techniques—will eventually reveal such nonteleological mechanisms. Their conviction in this case may even be stigmatized as unfalsifiable. Nothing will convince them that the phenomena are irreducibly teleological. To this extent, the biologists' convictions about the facts and the appropriateness of their methods are in effect metaphysical. What if they succeed? Shall we withdraw the charge that the biologists' convictions were empty metaphysics, or shall we recognize that in the last analysis metaphysics is either unavoidable or indistinguishable from empirical science?

By providing and criticizing various definitions of teleological phenomena in terms of causal mechanisms, philosophers are in effect taking part in biologists' searches for such mechanisms. If biologists succeed, then the explanation for their success will appeal to the adequacy of philosophers' definitions of the phenomena. If biologists fail, then the explanation of the failure may appeal to philosophical criticisms of the definition under which the phenomena are described. Failure must in the long run support vitalists' alternative theories as much as success sustains materialists' views.

5. Speculation and Science

The decline of Positivism has thus led to a recognition of alternative philosophies of biology as explanatory theories of extremely high levels of generality. Philosophies of science both explain the most general characteristics of the objects of scientific investigation and also explain the successes and failures, the limits and prospects of alternative theories and methods in the sciences. Grand speculations in the philosophy of biology motivate and justify research programs. Their influence extends right through the hierarchy of theories, models, and experimental designs all the way down to findings of particular fact. Such commitments and convictions can retard as well as foster scientific progress by wrongly excluding lines of research, potential explanatory hypotheses, or improvements in intertheoretical depth and unity. Whether they foster or retard advance, they cannot be written off.

Nevertheless, because they are so far removed from the day-to-day work of biologists, it would be a mistake to think that all the matters broached in such speculations need be settled before biological progress can be made, recognized, or certified. It would be equally silly to suppose that the day-to-day progress of the field can be expected to settle these questions once and for all. To a large extent, therefore, the biologist is right not to keep one eye firmly focused on controversies in philosophy, and the philosopher should not expect that metaphysics and epistemology require the minutest attention to the latest biological results. If the importance of the philosophy of science for biology rested only on the distant relations between speculation and observation, biologists would probably do better to err on the side of neglecting philosophy rather than the side of absorption in it to the neglect of narrower matters.

But, in fact, specific biological and philosophical results are in much closer contact than it might be supposed, as the rest of this work will show.

The area of contact is in fact so wide that biologists and philosophers find themselves stumbling into one another's field, sometimes inadvertently and sometimes intentionally. At these points of contact, what counts as philosophy and what counts as biology become matters of arbitrary labeling at best. Because it is impossible to say where one field stops and the other starts, decisions about the significance of an inquiry in the no-man's-land between these two subjects must be made on their merits and not by appeal to jurisdictional determinations or dicta about the irrelevance of these subjects to one another. In fact, at least some of what contemporary philosophers have accomplished in their study of biology is not recognizably philosophy and so may slip by unnoticed in any discussion of the bearing of their subject and its methods on biology.

. . .

In the final analysis, however, the justification for pursuing the philosophy of biology rests on the fact that biologists cannot avoid the great questions that transcend their day-to-day concerns. For if there are correct answers to the questions faced every day in the lab and the field, and if the theories biologists propound are definitely true or false as a matter of the objective facts about the way the world works, then there must also be correct answers to the great questions of metaphysics and epistemology as well. If there is objective knowledge in biology, there is objective knowledge in its philosophy as well, for the two subjects are indistinguishable and inseparable. This is why Darwin was correct after all. For no theory has had greater impact on providing biological knowledge than his, and this expansion of knowledge must result in the flourishing of metaphysics.

References

Barrett, P.H. 1974. "Darwin's Early and Unpublished Notebooks." In H.E. Gruber (ed.), *Darwin on Man*. New York: Dutton.

Gabriel, M., and Fogel, S. 1955. *Great Experiments in Biology*. Englewood Cliffs, N.J.: Prentice-Hall.

QUESTIONS

1 Schlick thinks that

> philosophy was misunderstood when it was thought that philosophical results could be expressed in propositions, and that there could be a system of philosophy consisting of a system of propositions which would represent the answers to "philosophical" questions. There are

> no specific "philosophical" truths which would contain the solution of specific "philosophical" problems, but philosophy has the task of finding the meaning of *all* problems and their solutions. It must be defined as *the activity of finding meaning*.

Do you agree with this claim? Doesn't "the activity of finding meaning" itself result in propositions representing answers to philosophical questions?

2 In the first part of his lecture Schlick suggests that philosophers should drastically reconsider the goals of their discipline and its relationship with science. Toward the end, he claims that a major portion of the past history of philosophy (the only "reasonable" portion) could actually be read along the lines he suggests, as the activity of finding meanings of scientific problems. Do you agree with his reading of the history of philosophy?

3 Rosenberg argues that, contrary to the positivist dicta, there are no hard and fast lines between empirical questions and questions about the ultimate structure of reality ("metaphysical" questions, those that positivists deemed meaningless) and that the philosophical issues arising in science are not different in kind from the scientific issues. Thus the development of quantum mechanics undermined not only the empirical propositions of classical physics but also its underlying *metaphysics*, Driesch's entelechy theory can be meaningfully discussed in biology rather than being dismissed out of hand as just so much gibberish, etc. How would a staunch Logical Positivist, such as Schlick, reply to these claims?

FURTHER READING

Ayer (1946) is a concise exposition of Logical Positivism written by one of its leading proponents and widely read by generations of students since. Ayer (1959) contains a useful collection of articles by major twentieth-century positivists. Nagel (1961) and Hempel (1966) are two other classics written at the time of crisis for Logical Positivism: both are a must for those wishing to familiarize themselves with the "received" (i.e., positivist or inspired by Positivism) view of scientific theories, explanation, confirmation, and other related issues. Another work based on a 1969 symposium (Suppe 1977) greatly contributed to the advent of the post-positivist era in the philosophy of science; it contains both a very useful account of the "received view" and thorough criticism of it. A recent revival of historical interest in Logical Positivism is reflected in Giere and Richardson (1997) and Friedman (1999).

Those wishing to explore ways in which great twentieth-century physicists were led to engage in profound epistemological and methodological debates could begin with a seminal collection of articles by physicists and philosophers focused on Einstein's work (Schilpp 1949) and with Heisenberg's philosophical essays discussing the foundational problems involved in the emergence and development of quantum mechanics (Heisenberg 1958). Contemporary philosophy of physics is a flourishing industry. The reader interested in the philosophical issues raised by two cornerstones of twentieth-century physics, relativity theory and quantum mechanics, could start with Kosso (1998).

The philosophy of biology is another growth industry involving both philosophers (Hull 1974; Rosenberg 1985; Ruse 1988a) and biologists (Dawkins 1986; Levins and Lewontin 1985). Hull and Ruse (1998) is a recent collection of articles discussing some of the most controversial issues in the discipline.

McMullin (1970) provides a detailed analysis of the disciplinary structure of the post-positivist philosophy of science and its relation to the history of science.

PART II

EXPLANATION, CAUSATION, AND LAWS

INTRODUCTION

Philosophy, said Aristotle, begins with wonder. And by philosophy Aristotle meant science. Aristotle was right. Science seeks explanations to satisfy the wonder. But so do other human enterprises. The difference between science and other enterprises that seek explanations of why things are the way they are can be found in the sorts of standards that science sets itself for what will count as an explanation, a good explanation, and a better explanation. The philosophy of science seeks to uncover those standards, and the other rules that govern "scientific methods." It does so in part by examining the sorts of explanations scientists advance, accept, criticize, improve, and reject. But what scientists accept or not as explanations cannot be the sole source of standards for what scientific explanation should be. After all, scientists are not infallible in their explanatory judgments; what is more, scientists themselves disagree about the adequacy of particular explanations, and about what explanation in science is like overall. If the philosophy of science were just a matter of collating the decisions of scientists about what explanations are, it could not be a source for advice about how scientific explanation *should* proceed. Yet in fact, in many disciplines, especially the social and behavioral sciences, scientists turn to philosophy of science for "prescriptions"—rules about how explanations ought to proceed if they are going to be truly scientific.

If the philosophy of science is to do more than merely describe what some or even many scientists take to be scientific explanations—if it is to endorse one or another recipe for scientific explanation as correct—it will have to do more than merely report what scientists themselves think about the matter. In addition to learning what explanations scientists actually accept and reject, the philosophy of science will have to assess these choices against philosophical theories, especially theories in epistemology—the study of the nature, extent and justification of knowledge. But this means that the philosophy of science cannot escape the most central, distinctive and hardest questions that have vexed philosophers since the time of Socrates and Plato.

Traditionally the philosophy of science has sought a definition for "scientific explanation," but not a dictionary definition. A dictionary definition merely reports how scientists and others actually use the words "scientific explanation." Traditional philosophy of science seeks a checklist of conditions that any scientific explanation should satisfy. When all are satisfied, the checklist guarantees the scientific adequacy of an explanation. In other words, the traditional approach seeks a set of conditions individually necessary and jointly sufficient for something being a scientific explanation. This "explicit" definition, or as it was sometimes called, this "explication" or "rational reconstruction" of the dictionary definition, would render the concept of scientific explanation precise and philosophically well founded. Carl Hempel offers such definitions or rational reconstructions of what he considers to be two basic types of scientific explanation: Deductive-Nomological and Probabilistic or Probabilistic-Statistical Explanation. His definitions make explicit what has long been assumed: that explanation in science is driven by laws of nature or our best hypotheses about what these laws are.

An explicit definition, such as the one Hempel seeks, gives the necessary and sufficient conditions for a thing, event, state, process, property to be an instance of the term defined. For example, *triangle* is explicitly defined as "plane figure having three sides." Since the conditions are together sufficient we know that everything which fulfills them is a Euclidean triangle and since the conditions are individually necessary, we know that if just one is not satisfied by an item, it is not a Euclidean triangle. The beauty of such definitions is that they remove vagueness, and provide for maximally precise definitions.

An explicit definition or "explication" of the notion of a scientific explanation could serve the prescriptive task of a litmus test or yardstick for grading and improving explanations in the direction of increasing scientific adequacy. The demand that philosophical analysis result in such a precise and complete definition is in part a reflection of the influence of mathematical logic on the Logical Positivists and their immediate successors in the philosophy of science, like Hempel. For in mathematics concepts are introduced in just this way—by providing explicit definitions in terms of already understood previously introduced concepts. The advantage of such definitions is clarity: there will be no borderline cases and no unresolvable arguments about whether some proposed explanation is "scientific" or not. The disadvantage is that it is often impossible to give such a complete definition or "explication" for most concepts of interest. One way to show that the concept of explanation is not amenable to such rational reconstruction is to devise counter-examples—bona fide scientific explanations that do not satisfy the definition, and non-explanations that do satisfy it, thus showing that the concept in question fails to capture the essence of scientific explanation. This is van Fraassen's strategy in "The Pragmatics of Explanation." He goes on to argue that explanation is not as Hempel assumes, a logical relation between propositions, but the provision of information appropriate to an inquirer's interests and background knowledge.

But this approach raises the question of how scientific explanations differ from other sorts of explanation. On this point, there seem nowadays to be two prominent candidates. One view is that explanation is scientific and scientifically warranted if it *unifies* disparate phenomena under a small number of principles. This view is defended in Philip Kitcher's chapter. Its leading competitor relates scientific explanation to identifying the *causes* which bring about the phenomenon to be explained. This position has long been defended by Wesley Salmon. Salmon's and Kitcher's approaches to explanation are not necessarily mutually exclusive. Indeed, there are strong reasons to believe that successful unification of phenomena is not just an artifact of human ingenuity but provides access to the causal structure of the world.

If scientific explanation is causal explanation then understanding its nature is hardly possible without getting clear on the nature of causal relations—the relations between causes and effects. And these hinge on the operation of laws of nature. Or at least it is widely held that causation is law-governed sequence. If this is so, then scientific explanations require laws—the point well appreciated by Hempel and his followers. But what a law of nature is raises some of the most profound questions of metaphysics. To see what they are consider the features of a law on which there has continued to be wide agreement: laws are universal statements of the form "all *A*s are *B*s" or "if event *E* happens, then invariably, event *F* occurs." For example, "all pure samples of iron conduct electric currents at standard temperature and pressure" or "if an electric current is applied to a sample of iron under standard temperature and pressure, then the sample conducts the current." These are terminological variants of the same law. Philosophers tend to prefer the "if . . . then . . ." conditional version to express their form. Laws are general in yet another sense: they don't refer to particular objects, places or times, implicitly or explicitly. But these two conditions are not sufficient to distinguish laws from other statements grammatically similar to laws but without explanatory force. Compare the two following statements of the same universal form:

- All solid spherical masses of pure plutonium weigh less than 100,000 kilograms.
- All solid spherical masses of pure gold weigh less than 100,000 kilograms.

We have good reason to believe that the first statement is true: quantities of plutonium spontaneously explode long before they reach this mass. Nuclear warheads rely on this fact. There is also good reason to think that the second statement is true. But it is true just as a matter of cosmic coincidence. There could have been such a quantity of gold so configured somewhere in the universe. Presumably the former statement reports a natural law, while the latter describes a mere fact about the universe that might have been otherwise. What this shows is that universality of form is not enough to make a statement a law of nature.

One symptom of the difference between real laws and accidental generalizations philosophers have hit upon involves grammatical constructions known as "counterfactual conditionals," or "counterfactuals" for short. A counterfactual is a particular sort of if/then statement, one expressed in the subjunctive mood, instead of the indicative mood in which laws are expressed: We employ such statements often in everyday life: "If I had known you were coming, I would have baked a cake." Two examples of such counterfactual statements relevant for distinguishing laws from non-laws of the same grammatical "if . . . then . . ." form are the following:

- If it were the case that the moon is made of pure plutonium, it would be the case that it weighs less than 100,000 kilograms.
- If it were the case that the moon is made of pure gold, it would be the case that it weighs less than 100,000 kilograms.

Notice that the antecedents (the sentences following the "ifs") and the consequents (the sentences following the "thens") of both counterfactuals are false, yet the first counterfactual is true and the second is false.

These two statements are claims not about actualities, but about possibilities—the possible states of affairs that the moon is composed of plutonium and gold respectively. Each says that if the antecedent obtained (which it doesn't), the consequent would have obtained (even though as a matter of fact, neither does actually obtain). Now, we hold that the counterfactual about gold is false. But we believe that the counterfactual about plutonium expresses a truth. And the reason for this difference between these two grammatically identical statements about non-actual states of affairs is that there is a law about plutonium that supports the plutonium counterfactual, while the universal truth about gold masses is not a law, but merely an accidental generalization. So, it does not support the gold counterfactual.

Thus, we may add to our conditions on laws that besides being universal in form, they support counterfactuals. But unless we understand what makes true counterfactuals true independently of the laws which support them, the fact that laws support counterfactuals won't help explain the difference between them and accidental generalizations.

We know that laws support their counterfactuals, while accidental generalizations do not. But we don't know what it is about laws that makes for this difference. Presumably, they support their counterfactuals because laws express some real connection between their antecedents and their consequents that is missing between the antecedent and the consequent of an accidental generalization. Thus, there is something about being a sphere of pure plutonium that brings it about, or necessitates the fact that it cannot be 100,000 kilograms in mass, whereas there is nothing about being a sphere of gold that makes it impossible to be that massive. But what could this real connection between the antecedent and the consequent of a law be, which reflects the necessitation of the latter by the former? Certainly, laws do not express logical necessities. It is impossible to conceive of the violation of a logically necessary truth. It is easy to conceive of the violation of a natural law: there would be nothing contradictory about gravity varying as the cube of the distance between objects instead of as the square of the distance between them.

It is no explanation of the necessity of laws to say they reflect "nomological" or "physical" or "natural" instead of logical necessity. For what is it for a statement to be physical or natural necessity except that it is required to be the case by the laws of physics or nature? If this is what natural or physical necessity consists in, then grounding the necessity of laws on physical or natural necessity is grounding the necessity of laws on itself! This is reasoning in a circle, and it can lead nowhere.

These and related issues are explored in Mackie's chapter "The Logic of Conditionals." This paper motivates and makes accessible a large body of work by philosophers on the nature of laws. There are two major rival views of what a law of nature is in the current literature. One view identifies the ground of nomological or physical necessity with some facts about the world that are different from mere regularities: relations between universals or causal powers. On the first version of this view, widely known as the Dretske–Tooley–Armstrong theory, a lawful regularity, such as the fact that all metals are electric conductors, obtains—and the corresponding statement "all metals are electric conductors" is true, because *being a metal* "nomologically necessitates" *being an electric conductor*. Although such a relation between the two universals, "metallicity" and "conductivity," is itself contingent (could have failed to take place), its actual presence confers on particular facts falling under it the right sort of necessity (i.e., the nomological or physical necessity) and supports the relevant counterfactuals. On the contrary, no relation of necessitation obtains between *being made of gold* and *weighing less than 100,000 kilograms*. The corresponding cosmic regularity is still there, but not as a matter of nomological necessity; it is due to a historical accident. The suspect counterfactual mentioned above lacks ontological ground and thus cannot be truly asserted.

To uphold such a theory, however, one has to believe in the existence of universals. One has to believe, at the minimum, that, in addition to metals, there is such a thing as "metallicity," that besides the class of all red objects, there is "redness," and so on. In fact, one also has to believe that some universals stand in the relation of nomic necessitation to one another. Many philosophers find this mysterious and prefer to do without such heavy metaphysical commitments. It is, then, incumbent on them to provide an account of nomological necessity without appeal to the realm of abstract Platonic universals or some special sort of causal powers or necessitation operating in the world. They have to find a way of privileging certain generalizations, to show what makes them lawful and thus distinguishes from other, accidental generalizations without resorting to the Platonic riches of universals and second-order relations between them. One such proposal, first sketched by John Stuart Mill and elaborated by Frank Ramsey and David Lewis, is discussed and defended in John Earman's chapter.

The question of what kind of necessity laws have, and accidental generalizations lack, is exactly the sort of "metaphysical" question that the Logical Positivists hoped to avoid by not invoking the notion of causality in their analysis of explanation. But nomological necessity just turns out to be the same thing as the necessity that connects causes and their effects. One of the attractions of Kitcher's chapter, and its advocacy of explanation as unification under scientific principles which need not be laws as traditionally understood, is that it will enable us to avoid many of these hard philosophical questions.

3

Carl Hempel, "Two Models of Scientific Explanation"

Carl Gustav Hempel's (1905–1997) classic account of explanation is widely known as the "covering-law model." It includes two species: the Deductive-Nomological Explanation and the Probabilistic-Statistical Explanation. Since its introduction in the seminal paper co-authored by Hempel and Paul Oppenheim (1948), the model has been intensely debated and criticized in the philosophy of science literature.

For although certain experiments are always necessary to serve as a basis for reasoning, nevertheless, once these experiments are given, we should derive from them everything which anyone at all could possibly derive; and we should even discover what experiments remain to be done for the clarification of all further doubts. That would be an admirable help, even in political science and medicine, to steady and perfect reasoning concerning given symptoms and circumstances. For even while there will not be enough given circumstances to form an infallible judgment, we shall always be able to determine what is most probable on the data given. And that is all that reason can do.

(*The General Science*, Leibniz, 1677)

1. Introduction

Among the divers factors that have encouraged and sustained scientific inquiry through its long history are two pervasive human concerns which provide, I think, the basic motivation for all scientific research. One of these is man's persistent desire to improve his strategic position in the world by means of dependable methods for predicting and, whenever possible, controlling the events that occur in it. The extent to which science has been able to satisfy this urge is reflected impressively in the vast and steadily widening

C. Hempel: "Explanation in Science and History," in *Frontiers of Science and Philosophy*, ed. R.G. Colodny, 1962, pp. 9–19. Pittsburgh: The University of Pittsburgh Press.

range of its technological applications. But besides this practical concern, there is a second basic motivation for the scientific quest, namely, man's insatiable intellectual curiosity, his deep concern to *know* the world he lives in, and to *explain*, and thus to *understand*, the unending flow of phenomena it presents to him.

In times past questions as to the *what* and the *why* of the empirical world were often answered by myths; and to some extent, this is so even in our time. But gradually, the myths are displaced by the concepts, hypotheses, and theories developed in the various branches of empirical science, including the natural sciences, psychology, and sociological as well as historical inquiry. What is the general character of the understanding attainable by these means, and what is its potential scope? In this paper I will try to shed some light on these questions by examining what seem to me the two basic types of explanation offered by the natural sciences, and then comparing them with some modes of explanation and understanding that are found in historical studies.

First, then, a look at explanation in the natural sciences.

2. Two Basic Types of Scientific Explanation

2.1. *Deductive-Nomological Explanation*

In his book, *How We Think*,[1] John Dewey describes an observation he made one day when, washing dishes, he took some glass tumblers out of the hot soap suds and put them upside down on a plate: he noticed that soap bubbles emerged from under the tumblers' rims, grew for a while, came to a standstill, and finally receded inside the tumblers. Why did this happen? The explanation Dewey outlines comes to this: In transferring a tumbler to the plate, cool air is caught in it; this air is gradually warmed by the glass, which initially has the temperature of the hot suds. The warming of the air is accompanied by an increase in its pressure, which in turn produces an expansion of the soap film between the plate and the rim. Gradually, the glass cools off, and so does the air inside, with the result that the soap bubbles recede.

This explanatory account may be regarded as an argument to the effect that the event to be explained (let me call it the explanandum-event) was to be expected by reason of certain explanatory facts. These may be divided into two groups: (i) particular facts and (ii) uniformities expressed by general laws. The first group includes facts such as these: the tumblers had been immersed, for some time, in soap suds of a temperature considerably higher than that of the surrounding air; they were put, upside down, on a plate on which a puddle of soapy water had formed, providing a connecting soap film, etc. The second group of items presupposed in the argument includes the gas laws and various other laws that have not been explicitly suggested concerning the exchange of heat between bodies of different temperature, the elastic behavior of soap bubbles, etc. If we imagine these various presuppositions explicitly spelled out, the idea suggests itself of construing the explanation as a deductive argument of this form:

$$\text{(D)} \qquad \frac{\begin{array}{c} C_1, C_2, \ldots, C_k \\ L_1, L_2, \ldots, L_r \end{array}}{E}$$

Here, $C_1, C_2, \ldots, C_k$ are statements describing the particular facts invoked; $L_1, L_2, \ldots, L_r$ are general laws: jointly, these statements will be said to form the explanans. The conclusion E is a statement describing the explanandum-event; let me call it the explanandum-statement, and let me use the word "explanandum" to refer to either E or to the event described by it.

The kind of explanation thus characterized I will call *deductive-nomological explanation*; for it amounts to a deductive subsumption of the explanandum under principles which have the character of general laws: it answers the question "*Why* did the explanandum event occur?" by showing that the event resulted from the particular circumstances specified in $C_1, C_2, \ldots, C_k$ in accordance with the laws $L_1, L_2, \ldots, L_r$. This conception of explanation, as exhibited in schema (D), has therefore been referred to as the covering law model, or as the deductive model, of explanation.[2]

A good many scientific explanations can be regarded as deductive-nomological in character. Consider, for example, the explanation of mirror-images, of rainbows, or of the appearance that a spoon handle is bent at the point where it emerges from a glass of water: in all these cases, the explanandum is deductively subsumed under the laws of reflection and refraction. Similarly, certain aspects of free fall and of planetary motion can be accounted for by deductive subsumption under Galileo's or Kepler's laws.

In the illustrations given so far the explanatory laws had, by and large, the character of empirical generalizations connecting different observable aspects of the phenomena under scrutiny: angle of incidence with angle of reflection or refraction, distance covered with falling time, etc. But science raises the question "why?" also with respect to the uniformities expressed by such laws, and often answers it in basically the same manner, namely, by subsuming the uniformities under more inclusive laws, and eventually under comprehensive theories. For example, the question, "Why do Galileo's and Kepler's laws hold?" is answered by showing that these laws are but special consequences of the Newtonian laws of motion and of gravitation; and these, in turn, may be explained by subsumption under the more comprehensive general theory of relativity. Such subsumption under broader laws or theories usually increases both the breadth and the depth of our scientific understanding. There is an increase in breadth, or scope, because the new explanatory principles cover a broader range of phenomena; for example, Newton's principles govern free fall on the earth and on other celestial bodies, as well as the motions of planets, comets, and artificial satellites, the movements of pendulums, tidal changes, and various other phenomena. And the increase thus effected in the depth of our understanding is strikingly reflected in the fact that, in the light of more advanced explanatory principles, the original empirical laws are usually seen to hold only approximately, or within certain limits. For example, Newton's theory implies that the factor g in Galileo's law, $s = \frac{1}{2} gt^2$, is not strictly a constant for free fall near the surface of the earth; and that, since every planet undergoes gravitational attraction not only from the sun, but also from the other planets, the planetary orbits are not strictly ellipses, as stated in Kepler's laws.

One further point deserves brief mention here. An explanation of a particular event is often conceived as specifying its *cause*, or causes. Thus, the account outlined in our first illustration might be held to explain the growth and the recession of the soap bubbles by showing that the phenomenon was *caused* by a rise and a subsequent drop of the temperature of the air trapped in the tumblers. Clearly, however, these temperature changes provide the requisite explanation only in conjunction with certain other conditions, such as the presence of a soap film, practically constant pressure of the air surrounding the glasses, etc. Accordingly, in the context of explanation, a cause must be allowed to consist in a more or less complex set of particular circumstances; these might be described by a set of sentences: $C_1, C_2, \ldots, C_k$. And, as suggested by the principle "Same cause, same effect," the assertion that those circumstances jointly caused a given event—described, let us say, by a sentence E—implies that whenever and wherever circumstances of the kind in question occur, an event of the kind to be explained comes about. Hence, the given causal explanation implicitly claims that there are general laws—such as $L_1, L_2, \ldots, L_r$, in schema (D)—by virtue of which the occurrence of the causal antecedents mentioned in $C_1, C_2, \ldots, C_k$ is a sufficient condition for the occurrence of the event to be explained. Thus, the relation between causal factors and effect is reflected in schema (D): causal explanation is deductive-nomological in character. (However, the customary formulations of causal and other explanations often do not explicitly specify all the relevant laws and particular facts: to this point, we will return later.)

The converse does not hold: there are deductive-nomological explanations which would not normally be counted as causal. For one thing, the subsumption of laws, such as Galileo's or Kepler's laws, under more comprehensive principles is clearly not causal in character: we speak of causes only in reference to *particular* facts or events, and not in reference to *universal facts* as expressed by general laws. But not even all deductive-nomological explanations of particular facts or events will qualify as causal; for in a causal explanation some of the explanatory circumstances will temporally precede the effect to be explained: and there are explanations of type (D) which lack this characteristic. For example, the pressure which a gas of specified mass possesses at a given time might be explained by reference to its temperature and its volume at the same time, in conjunction with the gas law which connects simultaneous values of the three parameters.[3]

In conclusion, let me stress once more the important role of laws in deductive-nomological explanation: the laws connect the explanandum event with the particular conditions cited in the explanans, and this is what confers upon the latter the status of explanatory (and, in some cases, causal) factors in regard to the phenomenon to be explained.

2.2. *Probabilistic Explanation*

In deductive-nomological explanation as schematized in (D), the laws and theoretical principles involved are of *strictly universal form*: they assert that in *all* cases in which certain specified conditions are realized an occurrence of such and such a kind will result; the law that any metal, when heated under constant pressure, will increase in volume, is a typical example; Galileo's, Kepler's, Newton's, Boyle's, and Snell's laws, and many others, are of the same character.

Now let me turn next to a second basic type of scientific explanation. This kind of explanation, too, is nomological, i.e., it accounts for a given phenomenon by reference to general laws or theoretical principles; but some or all of these are of *probabilistic-statistical form*, i.e., they are, generally speaking, assertions to the effect that if certain specified conditions are realized, then an occurrence of such and such a kind will come about with such and such a statistical probability.

For example, the subsiding of a violent attack of hay fever in a given case might well be attributed to, and thus explained by reference to, the administration of 8 milligrams of chlor-trimeton. But if we wish to connect this antecedent event with the explanandum, and thus to establish its explanatory significance for the latter, we cannot invoke a universal law to the effect that the administration of 8 milligrams of that antihistamine will invariably terminate a hay fever attack: this simply is not so. What can be asserted is only a generalization to the effect that administration of the drug will be followed by relief with high statistical probability, i.e., roughly speaking, with a high relative frequency in the long run. The resulting explanans will thus be of the following type:

> John Doe had a hay fever attack and took 8 milligrams of chlor-trimeton.
>
> The probability for subsidence of a hay fever attack upon administration of 8 milligrams of chlor-trimeton is high.

Clearly, this explanans does not deductively imply the explanandum, "John Doe's hay fever attack subsided"; the truth of the explanans makes the truth of the explanandum not certain (as it does in a deductive-nomological explanation) but only more or less likely or, perhaps "practically" certain.

Reduced to its simplest essentials, a probabilistic explanation thus takes the following form:

$$\text{(P)} \qquad \left.\begin{array}{c} Fi \\ p(O, F) \text{ is very high} \\ \hline\hline Oi \end{array}\right\} \text{makes very likely}$$

The explanandum, expressed by the statement "Oi," consists in the fact that in the particular instance under consideration, here called i (e.g., John Doe's allergic attack), an outcome of kind O (subsidence) occurred. This is explained by means of two explanans-statements. The first of these, "Fi," corresponds to $C_1, C_2, \ldots, C_k$ in (D); it states that in case i, the factors F (which may be more or less complex) were realized. The second expresses a law of probabilistic form, to the effect that the statistical probability for outcome O to occur in cases where F is realized is very high (close to 1). The double line separating explanandum from explanans is to indicate that, in contrast to the case of deductive-nomological explanation, the explanans does not logically imply the explanandum, but only confers a high likelihood upon it. The concept of likelihood here referred to must be clearly distinguished from that of statistical probability, symbolized by "p" in our schema. A statistical probability is, roughly speaking, the long-run relative frequency with which an occurrence of a given kind (say, F) is accompanied by an "outcome" of a specified kind (say, O). Our likelihood, on the other hand, is a relation (capable of gradations) not between kinds of occurrences, but between statements. The likelihood referred to in (P) may be characterized as the strength of the inductive support, or the degree of rational credibility, which the explanans confers upon the explanandum; or, in Carnap's terminology, as the *logical*, or *inductive*, (in contrast to statistical) *probability* which the explanandum possesses relative to the explanans.

Thus, probabilistic explanation, just like explanation in the manner of schema (D), is nomological in that it presupposes general laws; but because these laws are of statistical rather than of strictly universal form, the resulting explanatory arguments are inductive rather than deductive in character. An inductive argument of this kind *explains* a given phenomenon by showing that, in view of certain particular events and certain statistical laws, its occurrence was to be expected with high logical, or inductive, probability.

By reason of its inductive character, probabilistic explanation differs from its deductive-nomological counterpart in several other important respects; for example, its explanans may confer upon the explanandum a more or less high degree of inductive support; in this sense, probabilistic explanation admits of degrees, whereas deductive-nomological explanation appears as an either-or affair: a given set of universal laws and particular statements either does or does not imply a given explanandum statement. A fuller examination of these differences, however, would lead us far afield and is not required for the purposes of this paper.[4]

One final point: the distinction here suggested between deductive-nomological and probabilistic explanation might be questioned on the ground that, after all, the universal laws invoked in a deductive explanation can have been established only on the basis of a finite body of evidence, which surely affords no exhaustive verification, but only more or less strong probability for it; and that, therefore, all scientific laws have to be regarded as probabilistic. This argument, however, confounds a logical issue with an epistemological one: it fails to distinguish properly between the *claim* made by a given law-statement and the *degree of confirmation*, or *probability*, which it possesses on the available evidence. It is quite true that statements expressing laws of either kind can be only incompletely confirmed by any given finite set—however large—of data about particular facts; but law-statements of the two different types make claims of different kind, which are reflected in their logical forms: roughly, a universal law-statement of the simplest kind asserts that *all* elements of an indefinitely large reference class (e.g., copper objects) have a certain characteristic (e.g., that of being good conductors of electricity); while statistical law-statements assert that in the long run, a specified proportion of the members of the reference class have some specified property. And our distinction of two types of law and, concomitantly, of two types of scientific explanation, is based on this difference in claim as reflected in the difference of form.

The great scientific importance of probabilistic explanation is eloquently attested to by the extensive and highly successful explanatory use that has been made of fundamental laws of statistical form in genetics, statistical mechanics, and quantum theory.

3. Elliptic and Partial Explanations: Explanation Sketches

As I mentioned earlier, the conception of deductive-nomological explanation reflected in our schema (D) is often referred to as the covering law model, or the deductive model, of explanation: similarly, the conception underlying schema (P) might be called the probabilistic or the inductive-statistical, model of explanation. The term "model" can serve as a useful reminder that the two types of explanation as characterized above constitute ideal types or theoretical idealizations and are not intended to reflect the manner in which working scientists actually formulate their explanatory accounts. Rather, they are meant to provide explications, or rational reconstructions, or theoretical models, of certain modes of scientific explanation.

In this respect our models might be compared to the concept of mathematical proof (within a given theory) as construed in meta-mathematics. This concept, too, may be regarded as a theoretical model: it is not intended to provide a descriptive account of how proofs are formulated in the writings of mathematicians: most of these actual formulations fall short of rigorous and, as it were, ideal, meta-mathematical standards. But the theoretical model has certain other functions: it exhibits the rationale of mathematical proofs by revealing the logical connections underlying the successive steps; it provides standards for a critical appraisal of any proposed proof constructed within the mathematical system to which the model refers; and it affords a basis for a precise and far-reaching theory of proof, provability, decidability, and related concepts. I think the two models of explanation can fulfill the same functions, if only on a much more modest scale. For example, the arguments presented in constructing the models give an indication of the sense in which the models exhibit the rationale and the logical structure of the explanations they are intended to represent.

I now want to add a few words concerning the second of the functions just mentioned; but I will have to forgo a discussion of the third.

When a mathematician proves a theorem, he will often omit mention of certain propositions which he presupposes in his argument and which he is in fact entitled to presuppose because, for example, they follow readily from the postulates of his system or from previously established theorems or perhaps from the hypothesis of his theorem, if the latter is in hypothetical form; he then simply assumes that his readers or listeners will be able to supply the missing items if they so desire. If judged by ideal standards, the given formulation of the proof is elliptic or incomplete; but the departure from the ideal is harmless: the gaps can readily be filled in. Similarly, explanations put forward in everyday discourse and also in scientific contexts are often *elliptically formulated*. When we explain, for example, that a lump of butter melted because it was put into a hot frying pan, or that a small rainbow appeared in the spray of the lawn sprinkler because the sunlight was reflected and refracted by the water droplets, we may be said to offer elliptic formulations of deductive-nomological explanations; an account of this kind omits mention of certain laws or particular facts which it tacitly takes for granted, and whose explicit citation would yield a complete deductive-nomological argument.

In addition to elliptic formulation, there is another, quite important, respect in which many explanatory arguments deviate from the theoretical model. It often happens that the statement actually included in the explanans, together with those which may reasonably be assumed to have been taken for granted in the context at hand, explain the given explanandum only *partially*, in a sense which I will try to indicate by an example. In his *Psychopathology of Everyday Life*, Freud offers the following explanation of a slip of the pen that occurred to him: "On a sheet of paper containing principally short daily notes of business interest I found, to my surprise, the incorrect date, 'Thursday, October 20th,' bracketed under the correct date of the month of September. It was not difficult to explain this anticipation as the expression of a wish. A few days before I had returned fresh from my vacation and felt ready for any amount of professional work, but as yet there were few patients. On my arrival I had found a letter from a patient announcing her arrival on the 20th of October. As I wrote the same date in September I may certainly have thought 'X. ought to be here already; what a pity about that whole month!,' and with this thought I pushed the current date a month ahead."[5]

Clearly, the formulation of the intended explanation is *at least incomplete* in the sense considered a moment ago. In particular, it fails to mention any laws or theoretical principles in virtue of which the subconscious wish, and the other antecedent circumstances referred to, could be held to explain Freud's slip of the pen. However, the general theoretical considerations Freud presents here and elsewhere in his writings suggests strongly that his explanatory account relies on a hypothesis to the effect that when a person has a strong, though perhaps unconscious, desire, then if he commits a slip of pen, tongue, memory, or the like, the slip will take a form in which it expresses, and perhaps symbolically fulfills, the given desire.

Even this rather vague hypothesis is probably more definite than what Freud would have been willing to assert. But for the sake of the argument let us accept it and include it in the explanans, together with the particular statements that Freud did have the subconscious wish he mentions, and that he was going to commit a slip of the pen. Even then, the resulting explanans permits us to deduce only that the slip made by Freud would, *in some way or other*, express and perhaps symbolically fulfill Freud's subconscious wish.

But clearly, such expression and fulfillment might have been achieved by many other kinds of slip of the pen than the one actually committed.

In other words, the explanans does not imply, and thus fully explain, that the particular slip, say s, which Freud committed on this occasion, would fall within the narrow class, say W, of acts which consist in writing the words "Thursday, October 20th"; rather, the explanans implies only that s would fall into a wider class, say F, which includes W as a proper subclass, and which consists of all acts which would express and symbolically fulfill Freud's subconscious wish *in some way or other*.

The argument under consideration might be called a *partial explanation*: it provides complete, or conclusive, grounds for expecting s to be a member of F, and since W is a subclass of F, it thus shows that the explanandum, i.e., s falling within W, accords with, or bears out, what is to be expected in consideration of the explanans. By contrast, a deductive-nomological explanation of the form (D) might then be called *complete* since the explanans here does imply the explanandum.

Clearly, the question whether a given explanatory argument is complete or partial can be significantly raised only if the explanandum sentence is fully specified; only then can we ask whether the explanandum does or does not follow from the explanans. Completeness of explanation, in this sense, is relative to our explanandum sentence. Now, it might seem much more important and interesting to consider instead the notion of a complete explanation of some *concrete event*, such as the destruction of Pompeii, or the death of Adolf Hitler, or the launching of the first artificial satellite: we might want to regard a particular event as completely explained only if an explanatory account of deductive or of inductive form had been provided for all of its aspects. This notion, however, is self-defeating; for any particular event may be regarded as having infinitely many different aspects or characteristics, which cannot all be accounted for by a finite set, however large, of explanatory statements.

In some cases, what is intended as an explanatory account will depart even further from the standards reflected in the model schemata (D) and (P) above. An explanatory account, for example, which is not explicit and specific enough to be reasonably qualified as an elliptically formulated explanation or as a partial one, can often be viewed as an *explanation sketch*: it may suggest, perhaps quite vividly and persuasively, the general outlines of what, it is hoped, can eventually be supplemented so as to yield a more closely reasoned argument based on explanatory hypotheses which are indicated more fully, and which more readily permit of critical appraisal by reference to empirical evidence.

The decision whether a proposed explanatory account is to be qualified as an elliptically formulated deductive or probabilistic explanation, as a partial explanation, as an explanation sketch, or perhaps as none of these is a matter of judicious interpretation; it calls for an appraisal of the intent of the given argument and of the background assumptions that may be assumed to have been tacitly taken for granted, or at least to be available, in the given context. Unequivocal decision rules cannot be set down for this purpose any more than for determining whether a given informally stated inference which is not deductively valid by reasonably strict standards is to count nevertheless as valid but enthymematically formulated, or as fallacious, or as an instance of sound inductive reasoning, or perhaps, for lack of clarity, as none of these.

. . .

Notes

1 See Dewey, John. *How We Think*. Boston, New York, Chicago, 1910; Chapter VI.

2 For a fuller presentation of the model and for further references, see, for example, Hempel, C. G. and P. Oppenheim, "Studies in the Logic of Explanation," *Philosophy of Science* 15: 135–175 (1948). (Secs. 1–7 of this article, which contain all the fundamentals of the presentation, are reprinted in Feigl, H. and M. Brodbeck (eds.), *Readings in the Philosophy of Science*. New York, 1953.)—The suggestive term "covering law model" is W. Dray's; cf. his *Laws and Explanation in History*. Oxford, 1957; Chapter I. Dray characterizes this type of explanation as "subsuming what is to be explained under a general law" (*loc. cit.*, p. 1), and then rightly urges, in the name of methodological realism, that "the requirement of a *single* law be dropped" (*loc. cit.*, p. 24; italics, the author's): it should be noted, however, that, like the schema (D) above, several earlier publications on the subject (among them the article mentioned at the beginning of this note) make explicit provision for the inclusion of more laws than one in the explanans.

3 The relevance of the covering-law model to causal explanation is examined more fully in sec. 4 of Hempel, C. G., "Deductive-Nomological *vs*. Statistical Explanation." In Feigl, H., et al. (eds.), *Minnesota Studies in the Philosophy of Science*, vol. III. Minneapolis, 1962.

4 The concept of probabilistic explanation, and some of the peculiar logical and methodological problems engendered by it, are examined in some detail in Part II of the essay cited in note 3.

5 Freud, S. *Psychopathology of Everyday Life*. Translated by A. A. Brill. New York (Mentor Books) 1951; p. 64.

4

Bas van Fraassen, "The Pragmatics of Explanation"

A sustained critique, not only of the Hempelian model of explanation, but of what the author takes to be three common misconceptions about explanation, the chief among them being that explanation is a relation between a proposition or a set of propositions constituting a theory and phenomena or facts. According to Bas van Fraassen, explanation essentially involves *pragmatic* aspects, those having to do with the *use* of theoretical statements by epistemic agents in a particular *context*.

There are two problems about scientific explanation. The first is to describe it: when is something explained? The second is to show why (or in what sense) explanation is a virtue. Presumably we have no explanation unless we have a good theory; one which is independently worthy of acceptance. But what virtue is there in explanation over and above this? I believe that philosophical concern with the first problem has been led thoroughly astray by mistaken views on the second.

I. False Ideals

To begin I wish to dispute three ideas about explanation that seem to have a subliminal influence on the discussion. The first is that explanation is a relation simply between a theory or hypothesis and the phenomena or facts, just like truth for example. The second is that explanatory power cannot be logically separated from certain other virtues of a theory, notably truth or acceptability. And the third is that explanation is the overriding virtue, the end of scientific inquiry.

When is something explained? As a foil to the above three ideas, let me propose the simple answer: *when we have a theory which explains*. Note

B. van Fraassen, "The Pragmatics of Explanation," *American Philosophical Quarterly*, 1977, 14: 143–50.

first that "have" is not "have on the books"; I cannot claim to have such a theory without implying that this theory is acceptable all told. Note also that both "have" and "explains" are tensed; and that I have allowed that we can have a theory which does not explain, or "have on the books" an unacceptable one that does. Newton's theory explained the tides but not the advance in the perihelion of Mercury; we used to have an acceptable theory, provided by Newton, which bore (or bears timelessly?) the explanation relationship to some facts but not to all. My answer also implies that we can intelligibly say that the theory explains, and not merely that people can explain by means of the theory. But this consequence is not very restrictive, because the former could be an ellipsis for the latter.

There are questions of usage here. I am happy to report that the history of science allows systematic use of both idioms. In Huygens and Young the typical phrasing seemed to be that phenomenon may be explained *by means of* principles, laws and hypotheses, or *according to* a view.[1] On the other hand, Fresnel writes to Arago in 1815 "Tous ces phénomènes . . . sont réunis et expliqués par la même théorie des vibrations," and Lavoisier says that the oxygen hypothesis he proposes *explains* the phenomena of combustion.[2] Darwin also speaks in the latter idiom: "In scientific investigations it is permitted to invent any hypothesis, and if it explains various large and independent classes of facts it rises to the rank of a well-grounded theory"; though elsewhere he says that the facts of geographical distribution are *explicable on* the theory of migration.[3]

My answer did separate acceptance of the theory from its explanatory power. Of course, the second can be a reason for the first; but *that* requires their separation. Various philosophers have held that explanation logically requires true (or acceptable) theories as premises. Otherwise, they hold, we can at most mistakenly believe that we have an explanation.

This is also a question of usage, and again usage is quite clear. Lavoisier said of the phlogiston hypothesis that it is too vague and consequently "s'adapte a toutes les explications dans lesquelles on veut le faire entrer."[4] Darwin explicitly allows explanations by false theories when he says "It can hardly be supposed that a false theory would explain, in so satisfactory a manner as does the theory of natural selection, the several large classes of facts above specified."[5] More recently, Gilbert Harman has argued similarly: that a theory explains certain phenomena is part of the evidence that leads us to accept it. But that means that the explanation-relation is visible beforehand. Finally, we criticize theories selectively: a discussion of celestial mechanics around the turn of the century would surely contain the assertion that Newton's theory does explain many planetary phenomena, though not the advance in the perihelion of Mercury.

There is a third false ideal, which I consider worst: that explanation is the *summum bonum* and exact aim of science. A virtue could be overriding in one of two ways. The first is that it is a minimal criterion of acceptability. Such is consistency with the facts in the domain of application (though not necessarily with all data, if these are dubitable!). Explanation is not like that, or else a theory would not be acceptable at all unless it explained all facts in its domain. The second way in which a virtue may be overriding is that of being required when it can be had. This would mean that if two theories pass other tests (empirical adequacy, simplicity) equally well, then the one which explains more must be accepted. As I have argued elsewhere,[6] and as we shall see in connection with Salmon's views below, a precise formulation of this demand requires hidden variables for indeterministic theories. But of course, hidden variables are rejected in scientific practice as so much "metaphysical baggage" when they make no difference in empirical predictions.

II. A Biased History

I will outline the attempts to characterize explanation of the past three decades, with no pretense of objectivity. On the contrary, the selection is meant to illustrate the diagnosis, and point to the solution, of the next section.

1. *Hempel*

In 1966, Hempel summarized his views by listing two main criteria for explanation. The first is the criterion of *explanatory relevance*: "the explanatory information adduced affords good grounds for believing that the phenomenon to be explained did, or does, indeed occur."[7] That information has two components, one supplied by the scientific theory, the other consisting of auxiliary factual information. The relationship of providing good grounds is explicated as (a) implying (D–N case), or (b) conferring a high probability (I–S case), which is not lowered by the addition of other (available) evidence.

As Hempel points out, this criterion is not a sufficient condition for explanation: the red shift gives us good grounds for believing that distant galaxies are receding from us, but does not explain why they do. The classic case is the *barometer example*: the storm will come exactly if the barometers fall, which they do exactly if the atmospheric conditions are of the correct sort; yet only the last factor explains. Nor is the criterion a necessary condition; for this the classic case is the *paresis example*. We explain why the mayor, alone among the townsfolk, contracted paresis by his history of latent, contracted syphilis; yet such histories are followed by paresis in only a small percentage of cases.

The second criterion is the requirement of *testability*; but since all serious candidates for the role of scientific theory meet this, it cannot help to remove the noted defects.

2. *Beckner, Putnam, and Salmon*

The criterion of explanatory relevance was revised in one direction, informally by Beckner and Putnam and precisely by Salmon. Morton Beckner, in his discussion of evolution theory, pointed out that this often explains a phenomenon only by showing how it could have happened, given certain possible conditions.[8] Evolutionists do this by constructing models of processes which utilize only genetic and natural selection mechanisms, in which the outcome agrees with the actual phenomenon. Parallel conclusions were drawn by Hilary Putnam about the way in which celestial phenomena are explained by Newton's theory of gravity: celestial motions could indeed be as they are, given a certain possible (though not known) distribution of masses in the universe.[9]

We may take the paresis example to be explained similarly. Mere consistency with the theory is of course much too weak, since that is implied by logical irrelevance. Hence Wesley Salmon made this precise as follows: to explain is to exhibit (the) statistically relevant factors.[10] (I shall leave till later the qualifications about "screening off.") Since this sort of explication discards the talk about modelling and mechanisms of Beckner and Putnam, it may not capture enough. And indeed, I am not satisfied with Salmon's arguments that his criterion provides a sufficient condition. He gives the example of an equal mixture of Uranium 238 atoms and Polonium 214 atoms, which makes the Geiger counter click in interval $(t, t+m)$. This means that one of the atoms disintegrated. Why did it? The correct answer will be: because it was a Uranium 238 atom, if that is so—although the probability of its disintegration is much higher relative to the previous knowledge that the atom belonged to the described mixture.[11] The problem with this argument is that, on Salmon's criterion, we can explain not only why there was a disintegration, but also why *that* atom disintegrated *just then*. And surely that is exactly one of those facts which atomic physics leaves unexplained?

But there is a more serious general criticism. Whatever the phenomenon is, we can amass the statistically relevant factors, as long as the theory does not rule out the phenomenon altogether. "What more could one ask of an explanation?" Salmon inquires.[12] But in that case, as soon as we have an empirically adequate theory, we have an explanation of every fact in its domain. We may claim an explanation as soon as we have shown that the phenomenon can be embedded in some model allowed by the theory—that is, does not throw doubt on the theory's empirical adequacy.[13] But surely that is too sanguine?

3. *Global Properties*

Explanatory power cannot be identified with empirical adequacy; but it may still reside in the performance of the theory as a whole. This view is accompanied by the conviction that science does not explain individual facts but general regularities and was developed in different ways by Michael Friedman and James Greeno. Friedman says explicitly that in his view, "the kind of understanding provided by science is global rather than local" and consists in the simplification and unification imposed on our world picture.[14] That S_1 explains S_2 is a conjunction of two facts: S_1 implies S_2 relative to our background knowledge (and/or belief) K, and S_1 unifies and simplifies the set of its consequences relative to K. Friedman will no doubt wish to weaken the first condition in view of Salmon's work.

The precise explication Friedman gives of the second condition does not work, and is not likely to have a near variant that does.[15] But here we may look at Greeno's proposal.[16] His abstract and closing statement subscribe to the same general view as Friedman. But he takes as his model of a theory one which specifies a single probability space Q as the correct one, plus two partitions (or random variables) of which one is designated *explanandum* and the other *explanans*. An example: sociology cannot explain why Albert, who lives in San Francisco and whose father has a high income, steals a car. Nor is it meant to. But it does explain delinquency in terms of such other factors as residence and parental income. The degree of explanatory power is measured by an ingeniously devised quantity which measures the information I the theory provides of the explanandum variable M on the basis of explanans S. This measure takes its maximum value if all conditional probabilities $P(M_i/S_j)$ are zero or one (D–N case), and its minimum value zero if S and M are statistically independent.

Unfortunately, this way of measuring the unification imposed on our data abandons Friedman's insight that scientific understanding cannot be identified as a function of grounds for rational expectation. For if we let S and M describe the behavior of the barometer and coming storms, with P (barometer falls) = P (storm comes) = 0.2, P (storm comes/barometer falls) = 1, and P (storm comes/barometer does not fall) = 0, then the quantity I takes its maximum value. Indeed, it does so whether we designate M or S as explanans.

It would seem that such asymmetries as exhibited by the red shift and barometer examples must necessarily remain recalcitrant for any attempt to strengthen Hempel's or Salmon's criteria by global restraints on theories alone.

4. *The Major Difficulties*

There are two main difficulties, illustrated by the old paresis and barometer examples, which none of the examined positions can handle. The first is that there are cases, clearly in a theory's domain, where the request for explanation is nevertheless rejected. We can explain why John, rather than his brothers contracted paresis, for he had syphilis; but not why he, among all those syphilitics, got paresis. Medical science is incomplete, and hopes to find the answer some day. But the example of the uranium atom disintegrating just then rather than later, is formally similar and we believe the theory to be complete. We also reject such questions as the Aristotelians asked the Galileans: why does a body free of impressed forces retain its velocity? The importance of this sort of case, and its pervasive character, has been repeatedly discussed by Adolf Grünbaum.

The second difficulty is the asymmetry revealed by the barometer: even if the theory implies that one condition obtains when and only when another does, it may be that it explains the one in terms of the other and not vice versa. An example which combines both the first and second difficulty is this: according to atomic physics, each chemical element has a characteristic atomic structure and a characteristic spectrum (of light emitted upon excitation). Yet the spectrum is explained by the atomic structure, and the question why a substance has that structure does not arise at all (except in the trivial sense that the questioner may need to have the terms explained to him).

5. *Causality*

Why are there no longer any Tasmanian natives? Well, they were a nuisance, so the white settlers just kept shooting them till there were none left. The request was not for population statistics, but for the story; though in some truncated way, the statistics "tell" the story.

In a later paper Salmon gives a primary place to causal mechanisms in explanation.[17] Events are bound into causal chains by two relations: spatio-temporal continuity and statistical relevance. Explanation requires the exhibition of such chains. Salmon's point of departure is Reichenbach's *principle of the common cause*: every relation of statistical relevance ought to be explained by one of causal relevance. This means that a correlation of simultaneous values must be explained by a prior common cause. Salmon gives two statistical conditions that must be met by a common cause *C* of events *A* and *B*:

(a) $P(A \,\&\, B/C) = P(A/C)P(B/C)$
(b) $P(A/B \,\&\, C) = P(A/C)$ "*C screens off B from A.*"

If $P(B/C) \neq 0$ these are equivalent, and symmetric in A and B.

Suppose that explanation is typically the demand for a common cause. Then we still have the problem: when does this arise? Atmospheric conditions explain the correlation between barometer and storm, say; but are still prior causes required to explain the correlation between atmospheric conditions and falling barometers?

In the quantum domain, Salmon says, causality is violated because "causal influence is not transmitted with spatio-temporal continuity." But the situation is worse. To assume Reichenbach's principle to be satisfiable, continuity aside, is to rule out all genuinely indeterministic theories. As example, let a theory say that C is invariably followed by one of the incompatible events A, B, or D, each with probability 1/3. Let us suppose the theory complete, and its probabilities irreducible, with C the complete specification of state. Then we will find a correlation for which only C could be the common cause, but it is not. Assuming that A, B, D are always preceded by C and that they have low but equal prior probabilities, there is a statistical correlation between $\phi = (A \text{ or } D)$ and $\psi = (B \text{ or } D)$, for $P(\phi/\psi) = P(\psi/\phi) = 1/2 \neq P(\phi)$. But C, the only available candidate, does not screen off ϕ from ψ; $P(\phi/C \,\&\, \psi) = P(\phi/\psi) = 1/2 \neq P(\phi/C)$ which is 2/3. Although this may sound complicated, the construction is so general that almost any irreducibly probabilistic situation will give a similar example. Thus Reichenbach's *principle of the common cause* is in fact a demand for hidden variables.

Yet we retain the feeling that Salmon has given an essential clue to the asymmetries of explanation. For surely the crucial point about the barometer is that the atmospheric conditions screen off the barometer fall from the storm? The general point that the asymmetries are totally bound up with causality was argued in a provocative article by B. A. Brody.[18] Aristotle certainly discussed examples of asymmetries: the planets do not twinkle because they are near, yet they are near if and only if they do not twinkle (*Posterior Analytics*, I, 13). Not all explanations are causal, says Brody, but the others use a second Aristotelian notion, that of essence. The spectrum angle is a clear case: sodium has that spectrum because it has this atomic structure, which is its essence.

Brody's account has the further advantage that he can say when questions do not arise: other properties are explained in terms of essence, but the request for an explanation of the essence does not arise. However, I do not see how he would distinguish between the questions why the uranium atom disintegrated and why it disintegrated just then. In addition there is the problem that modern science is not formulated in terms of causes and essences, and it seems doubtful that these concepts can be redefined in terms which do occur there.

6. *Why-Questions*

A why-question is a request for explanation. Sylvain Bromberger called *P* the *presupposition* of the question *Why-P*? and restated the problem of explanation as that of giving the conditions under which proposition *Q* is a correct answer to a why-question with presupposition *P*.[19] However, Bengt Hannson has pointed out that "Why was it John who ate the apple?" and "Why was it the apple which John ate?" are different why-questions, although the comprised proposition is the same.[20] The difference can be indicated by such phrasing, or by emphasis ("Why did *John* . . . ?") or by an auxiliary clause ("Why did John rather than . . . ?"). Hannson says that an explanation is requested, not of a proposition or fact, but of an *aspect* of a proposition.

As is at least suggested by Hannson, we can cover all these cases by saying that we wish an explanation of why *P* is true in contrast to other members of a set *X* of propositions. This explains the tension in our reaction to the paresis-example. The question why the mayor, in contrast to other townfolk generally, contracted paresis *has* a true correct answer: because of his latent syphilis. But the question why he did in contrast to the other syphilitics in his country club, has no true correct answer. Intuitively we may say: *Q* is a correct answer to *Why P in contrast to X*? only if *Q* gives reasons to expect that *P*, in contrast to the other members of *X*. Hannson's proposal for a precise criterion is: the probability of *P* given *Q* is higher than the average of the probabilities of *R* given *Q*, for members *R* of *X*.

Hannson points out that the set *X* of alternatives is often left tacit; the two questions about paresis might well be expressed by the same sentence in different contexts. The important point is that explanations are not requested of propositions, and consequently a distinction can be drawn between answered and rejected requests in a clear way. However, Hannson makes *Q* a correct answer to *Why P in contrast to X*? when *Q* is statistically irrelevant, when *P* is already more likely than the rest; or when *Q* implies *P* but not the others. I do not see how he can handle the barometer (or red shift, or spectrum) asymmetries. On his precise criterion, that the barometer fell is a correct answer to why it will storm as opposed to be calm. The difficulty is very deep: if *P* and *R* are necessarily equivalent, according to our accepted theories, how can *Why P in contrast to X*? be distinguished from *Why R in contrast to X*?

III. The Solution

1. *Prejudices*

Two convictions have prejudiced the discussion of explanation, one methodological and one substantive.

The first is that a philosophical account must aim to produce necessary and sufficient conditions for theory T explaining phenomenon E. A similar prejudice plagued the discussion of counterfactuals for twenty years, requiring the exact conditions under which, if A were the case, B would be. Stalnaker's liberating insight was that these conditions are largely determined by context and speaker's interest. This brings the central question to light: what *form* can these conditions take?

The second conviction is that explanatory power is a virtue of theories by themselves, or of their relation to the world, like simplicity, predictive strength, truth, empirical adequacy. There is again an analogy with counterfactuals: it used to be thought that science contains, or directly implies, counterfactuals. In all but limiting cases, however, the proposition expressed is highly context-dependent, and the implication is there at most relative to the determining contextual factors, such as speakers' interest.

2. *Diagnosis*

The earlier accounts lead us to the format: C explains E relative to theory T exactly if (a) T has certain global virtues, and (b) T implies a certain proposition $\phi(C, E)$ expressible in the language of logic and probability theory. Different accounts directed themselves to the specification of what should go into (a) and (b). We may add, following Beckner and Putnam, that T explains E exactly if there is a proposition C consistent with T (and presumably, background beliefs) such that C explains E relative to T.

The significant modifications were proposed by Hannson and Brody. The former pointed out that the explanadum E cannot be reified as a proposition: we request the explanation of something F in contrast to its alternatives X (the latter generally tacitly specified by context). This modification is absolutely necessary to handle some of our puzzles. It requires that in (b) above we replace "$\phi(C, E)$" by the formula form "$\psi(C, F, X)$." But the problem of asymmetries remains recalcitrant, because if T implies the necessary equivalence of F and F' (say, atomic structure and characteristic spectrum), then T will also imply $\psi(C, F', X)$ if and only if it implies $\psi(C, F, X)$.

The only account we have seen which grapples at all successfully with this, is Brody's. For Brody points out that even properties which we believe to be constantly conjoined in all possible circumstances, can be divided into essences and accidents, or related as cause and effect. In this sense, the asymmetries were no problem for Aristotle.

3. The Logical Problem

We have now seen exactly what logical problem is posed by the asymmetries. To put it in current terms: how can we distinguish propositions which are true in exactly the same possible worlds?

There are several known approaches that use impossible worlds. David Lewis, in his discussion of causality, suggests that we should look not only to the worlds theory *T* allows as possible, but also to those it rules out as impossible, and speaks of counterfactuals which are counterlegal. Relevant logic and entailment draw distinctions between logically equivalent sentences and their semantics devised by Routley and Meyer use both inconsistent and incomplete worlds. I believe such approaches to be totally inappropriate for the problem of explanation, for when we look at actual explanations of phenomena by theories, we do not see any detours through circumstances or events ruled out as impossible by the theory.

A further approach, developed by Rolf Schock, Romane Clark, and myself distinguishes sentences by the facts that make them true. The idea is simple. That it rains, that it does not rain, that it snows, and that it does not snow, are four distinct facts. The disjunction that it rains or does not rain is made true equally by the first and second, and not by the third or fourth, which distinguishes it from the logically equivalent disjunction that it snows or does not snow.[21] The distinction remains even if there is also a fact of its raining or not raining, distinct or identical with that of its snowing or not snowing.

This approach can work for the asymmetries of explanation. Such asymmetries are possible because, for example, the distinct facts that light is emitted with wavelengths λ, μ, . . . conjointly make up the characteristic spectrum, while quite different facts conjoin to make up the atomic structure. So we have shown how such asymmetries *can* arise, in the way that Stalnaker showed how failures of transitivity in counterfactuals *can* arise. But while we have the distinct facts to classify asymmetrically, we still have the non-logical problem: whence comes the classification? The only suggestion so far is that it comes from Aristotle's concepts of cause and essence; but if so, modern science will not supply it.

4. *The Aristotelian Sieve*

I believe that we should return to Aristotle more thoroughly, and in two ways. To begin, I will state without argument how I understand Aristotle's theory of science. Scientific activity is divided into two parts, *demonstration* and *explanation*, the former treated mainly by the *Posterior Analytics* and the latter mainly by Book II of the *Physics*. Illustrations in the former are mainly examples of explanations in which the results of demonstration are *applied*; this is why the examples contain premises and conclusions which are not necessary and universal principles, although demonstration is only to and from such principles. Thus the division corresponds to our pure versus applied science. There is no reason to think that principles and demonstrations have such words as "cause" and "essence" in them, although looking at pure science from outside, Aristotle could say that its principles state causes and essences. In applications, the principles may be filtered through a conceptual sieve originating outside science.

The doctrine of the four "causes" (*aitiai*) allows for the systematic ambiguity or context-dependence of why-questions.[22] Aristotle's example (Physics II, 3; 195a) is of a lantern. In a modern example, the question why the porch light is on may be answered "because I flipped the switch" or "because we are expecting company," and the context determines which is appropriate. Probabilistic relations cannot distinguish these. Which factors are explanatory is decided not by features of the scientific theory but by concerns brought from outside. This is true even if we ask specifically for an "efficient cause," for how far back in the chain should we look, and which factors are merely auxiliary contributors?

Aristotle would not have agreed that essence is context-dependent. The essence is what the thing *is*, hence, its sum of classificatory properties. Realism has always asserted that ontological distinctions determine the "natural" classification. But which property is counted as explanatory and which as explained seems to me clearly context dependent. For consider Bromberger's flagpole example: the shadow is so long because the pole has this height, and not conversely. At first sight, no contextual factor could reverse this asymmetry, because the pole's height is a property it has in and by itself, and its shadow is a very accidental feature. The general principle linking the two is that its shadow is a function $f(x, t)$ of its height x and the time t (the latter determining the sun's elevation). But imagine the pole is the pointer on a giant sundial. Then the values of f have desired properties for each time t, and we appeal to these to explain why it is (had to be) such a tall pole.

We may again draw a parallel to counterfactuals. Professor Geach drew my attention to the following spurious argument: If John asked his father for money, then they would not have quarreled (because John is too proud to ask after a quarrel). Also if John asked and they hadn't quarreled, he would receive. By the usual logic of counterfactuals, it follows that if John asked his father for money, he would receive. But we know that he would not, because they have in fact quarreled. The fallacy is of equivocation, because "what was kept constant" changed in the middle of the monologue. (Or if you like, the aspects by which worlds are graded as more or less similar to this one.) Because science cannot dictate what speakers decide to "keep constant" it contains no counterfactuals. By exact parallel, *science contains no explanations.*

5. *The Logic of Why-Questions*

What remains of the problem of explanation is to study its logic, which is the logic of why-questions. This can be put to some extent, but not totally, in the general form developed by Harrah and Belnap and others.[23]

A question admits of three classes of response, *direct answers*, *corrections*, and *comments*. A *presupposition*, it has been held, is any proposition implied by all direct answers, or equivalently, denied by a correction. I believe we must add that the question "Why *P*, in contrast to *X*?" also presupposes that (a) *P* is a member of *X*, (b) *P* is true and the majority of *X* are not. This opens the door to the possibility that a question may not be uniquely determined by its set of direct answers. The question itself should decompose into factors which determine that set: the *topic P*, the *alternatives X*, and a *request specification* (of which the doctrine of the four "causes" is perhaps the first description).

We have seen that the propositions involved in question and answer must be individuated by something more than the set of possible worlds. I propose that we use the facts that make them true (see footnote 21). The context will determine an asymmetric relation among these facts, of *explanatory relevance*; it will also determine the theory or beliefs which determine which worlds are *possible*, and what is *probable* relative to what.

We must now determine what direct answers are and how they are evaluated. They must be made true by facts (and only by facts forcing such) which are explanatorily relevant to those which make the topic true. Moreover, these facts must be statistically relevant, telling for the topic in contrast to the alternatives generally; this part I believe to be explicable by probabilities, combining Salmon's and Hannson's account. How strongly the answers count for the topic should be part of their evaluation as better or worse answers.

The main difference from such simple questions as "Which cat is on the mat?" lies in the relation of a why-question to its presuppositions. A why-question may fail to arise because it is ill-posed (*P* is false, or most of *X* is true), or because only question-begging answers tell probabilistically for *P* in contrast to X generally, or because none of the factors that do tell for *P* are explanatorily relevant in the question-context. Scientific theory enters mainly in the evaluation of possibilities and probabilities, which is only part of the process, and which it has in common with other applications such as prediction and control.

IV. Simple Pleasures

There are no explanations in science. How did philosophers come to mislocate explanation among semantic rather than pragmatic relations? This was certainly in part because the positivists tended to identify the pragmatic with subjective psychological features. They looked for measures by which to evaluate theories. Truth and empirical adequacy are such, but they are weak, being preserved when a theory is watered down. Some measure of "goodness of fit" was also needed, which did not reduce to a purely internal criterion such as simplicity, but concerned the theory's relation to the world. The studies of explanation have gone some way toward giving us such a measure, but it was a mistake to call this explanatory power. The fact that seemed to confirm this error was that we do not say that we *have* an explanation unless we have a theory which is acceptable, and victorious in its competition with alternatives, whereby we can explain. Theories are applied in explanation, but the peculiar and puzzling features of explanation are supplied by other factors involved. I shall now redescribe several familiar subjects from this point of view.

When a scientist campaigns on behalf of an advocated theory, he will point out how our situation will change if we accept it. Hitherto unsuspected factors become relevant, known relations are revealed to be strands of an intricate web, some terribly puzzling questions are laid to rest as not arising at all. We shall be in a much better position to explain. But equally, we shall be in a much better position to predict and control. The features of the theory that will make this possible are its empirical adequacy and logical strength, not special "explanatory power" and "control power." On the other hand, it is also a mistake to say explanatory power is nothing but those other features, for then we are defeated by asymmetries having no "objective" basis in science.

Why are *new* predictions so much more to the credit of a theory than agreement with the old? Because they tend to bring to light new phenomena which the older theories cannot explain. But of course, in doing so, they throw doubt on the empirical adequacy of the older theory: they show that a pre-condition for explanation is not met. As Boltzmann said of the radiometer, "the theories based on older hydrodynamic experience can never describe" these phenomena.[24] The failure in explanation is a by-product.

Scientific inference is inference to the best explanation. That does not rule at all for the supremacy of explanation among the virtues of theories. For we evaluate how good an explanation is given by how good a theory is used to give it, how close it fits to the empirical facts, how internally simple and coherent the explanation. There is a further evaluation in terms of a prior judgment of which kinds of factors are explanatorily relevant. If this further evaluation took precedence, overriding other considerations, explanation would be the peculiar virtue sought above all. But this is not so: instead, science schools our imagination so as to revise just those prior judgments of what satisfies and eliminates wonder.

Explanatory power is something we value and desire. But we are as ready, for the sake of scientific progress, to dismiss questions as not really arising at all. Explanation is indeed a virtue; but still, less a virtue than an anthropocentric pleasure.[25]

Notes

1 I owe these and following references to my student Mr. Paul Thagard. For instance see C. Huygens, *Treatise on Light*, tr. by S. P. Thompson (New York, 1962), pp. 19, 20, 22, 63; Thomas Young, *Miscellaneous Works*, ed. by George Peacock (London, 1855), Vol. I, pp. 168, 170.

2 Augustin Fresnel, *Oeuvres Complètes* (Paris, 1866), Vol. I, p. 36 (see also pp. 254, 355); Antoine Lavoisier, *Oeuvres* (Paris, 1862), Vol. II, p. 233.

3 Charles Darwin, *The Variation of Animals and Plants* (London, 1863), Vol. I, p. 9; *On the Origin of the Species* (Facs. of first edition, Cambridge, Mass., 1964), p. 408.

4 Antoine Lavoisier, *op. cit.*, p. 640.

5 *Origin* (sixth ed., New York, 1962), p. 476.

6 "Wilfrid Sellars on Scientific Realism," *Dialogue*, vol. 14 (1975), pp. 606–616.

7 C. G. Hempel, *Philosophy of Natural Science* (Englewood Cliffs, New Jersey, 1966), p. 48.

8 *The Biological Way of Thought* (Berkeley, 1968), p. 176; this was first published in 1959.

9 In a paper of which a summary is found in Frederick Suppe (ed.), *The Structure of Scientific Theories* (Urbana, Ill., 1974).

10 "Statistical Explanation," pp. 173–231 in R. G. Colodny (ed.), *The Nature and Function of Scientific Theories* (Pittsburgh, 1970); reprinted also in Salmon's book cited below.

11 *Ibid.*, pp. 207–209. Nancy Cartwright has further, unpublished, counter-examples to the necessity and sufficiency of Salmon's criterion.

12 *Ibid.*, p. 222.

13 These concepts are discussed in my "To Save the Phenomena," *The Journal of Philosophy*, vol. 73 (1976), forthcoming.

14 "Explanation and Scientific Understanding," *The Journal of Philosophy*, vol. 71 (1974), pp. 5–19.

15 See Philip Kitcher, "Explanation, Conjunction, and Unification," *The Journal of Philosophy*, vol. 73 (1976), pp. 207–212.

16 "Explanation and Information," pp. 89–103 in Wesley Salmon (ed.), *Statistical Explanation and Statistical Relevance* (Pittsburgh, 1971). This paper was originally published with a different title in *Philosophy of Science*, vol. 37 (1970), pp. 279–293.

17 "Theoretical Explanation," pp. 118–145 in Stephan Körner (ed.), *Explanation* (Oxford, 1975).

18 "Towards an Aristotelian Theory of Scientific Explanation," *Philosophy of Science*, vol. 39 (1972), pp. 20–31.

19 "Why-Questions," pp. 86–108 in R. G. Colodny (ed.), *Mind and Cosmos* (Pittsburgh, 1966).

20 "Explanations—Of What?" (mimeographed: Stanford University, 1974).

21 Cf. my "Facts and Tautological Entailments," *The Journal of Philosophy*, vol. 66 (1969), pp. 477–487 and in A. R. Anderson, *et al.* (ed.), *Entailment* (Princeton, 1975); and "Extension, Intension, and Comprehension" in Milton Munitz (ed.), *Logic and Ontology* (New York, 1973).

22 Cf. Julius Moravcsik, "Aristotle on Adequate Explanations," *Synthese*, vol. 28 (1974), pp. 3–18.

23 Cf. N. D. Belnap, Jr., "Questions: Their Presuppositions, and How They Can Fail to Arise," *The Logical Way of Doing Things*, ed. by Karel Lambert (New Haven, 1969), pp. 23–39.

24 Ludwig Boltzmann, *Lectures on Gas Theory*, tr. by S. G. Brush (Berkeley, 1964), p. 25.

25 The author wishes to acknowledge helpful discussions and correspondence with Professors N. Cartwright, B. Hannson, K. Lambert, and W. Salmon, and the financial support of the Canada Council.

5

Philip Kitcher,* "Explanatory Unification and the Causal Structure of the World"

Drawing on some earlier work (Friedman 1974), Philip Kitcher (1981) developed his unification theory of explanation. Central to his approach is the idea that science enhances understanding by unifying disparate phenomena. This feature is overlooked by the covering-law model and this creates a number of problems for it—the problems that, Kitcher argues, are nicely handled by the unification theory. The selection below is excerpted from a later and more comprehensive paper in which the author explores, among other things, the connections between his theory and other approaches to explanation, in particular, Salmon's causal approach (see next selection).

1. Introduction

The modern study of scientific explanation dates from 1948, the year of the publication of the pioneering article by C. G. Hempel and Paul Oppenheim. Nearly forty years later, philosophers rightly continue to appreciate the accomplishments of the covering-law models of explanation and the classic sequence of papers in which Hempel articulated his view. Even though it has become clear that the Hempelian approach to explanation faces difficulties of a number of types, the main contemporary approaches to explanation attempt to incorporate what they see as Hempelian insights (with distinct facets of the covering-law models being preserved in different cases), and they usually portray themselves as designed to accommodate one or more of the main problems that doomed the older view. My aim in this essay is to compare what I see as the chief contemporary rivals in the theory of explanation, to understand their affiliations to the covering-law models and their

P. Kitcher, "Explanatory Unification and the Causal Structure of the World," in *Scientific Explanation*, ed. P. Kitcher and W.C. Salmon, 1989, pp. 410–505 (excerpts). Minneapolis: University of Minnesota Press.

efforts to address the troubles of those models, and to evaluate their success in doing so. Ecumenical as this may sound, the reader should be forewarned that I shall also be interested in developing further, and defending, an approach to explanation that I have championed in previous essays (1981, 1985).

1.1. Hempel's Accounts

Let us start with Hempel. The principal features of Hempel's account of explanation are (i) that explanations are arguments, (ii) that the conclusion of an explanation is a sentence describing the phenomenon to be explained, and (iii) that among the premises of an explanation there must be at least one law of nature. Although the original treatment (1948) focused on cases in which the argument is deductive and the conclusion a singular sentence (a sentence in which no quantifiers occur), it was clear from the beginning that the account could be developed along two different dimensions. Thus there can be covering-law explanations in which the argument is nondeductive or in which the conclusion is general. *D–N* explanations are those explanations in which the argument is deductive and the conclusion is either a singular sentence or a nonstatistical generalization. Hempel assigned deductive explanations whose conclusion is a statistical generalization a special category—*D–S* explanations—but their kinship with the official cases of D–N explanation suggests that we should broaden the D–N category to include them (see Salmon 1984 . . .). Finally, *I–S* explanations are those explanations in which the argument is inductive and the conclusion a singular sentence to which the premises assign high probability.

The motivation for approaching explanation in this way stems from the character of the explanations given in scientific works, particularly in those texts that are intended to introduce students to the main ideas of various fields. Expository work in physics, chemistry, and genetics (and, to a less obvious extent, in other branches of science) often proceeds by deriving descriptions of particular events—or, more usually, descriptions of empirical regularities—from sets of premises in which statements identified as laws figure prominently. Among the paradigms, we may include: the demonstration that projectiles obtain maximum range on a flat plain when the angle of projection is 45°, the Newtonian derivation of Galileo's law of free fall, Bohr's argument to show that the frequencies of the lines in the hydrogen spectrum satisfy the formulas previously obtained by Balmer and others, the kinetic-theoretic deduction of the Boyle–Charles law, computations that reveal the energy required for particular chemical reactions, and the derivation of expected distributions of traits among peas from specifications of the crosses and Mendel's laws. In all these cases, we can find scientific texts that contain arguments that come very close indeed to the ideal form of explanation that Hempel describes.

1.2. Hempel's Problems

There are four major types of objection to the Hempelian approach. The first is the obverse of the motivational point just canvassed. Although we can identify some instances in which full-dress covering-law explanations are developed, there seem to be many occasions on which we accept certain statements as explanatory without any ability to transform them into a cogent derivation of a sentence describing the phenomenon to be explained. This objection, made forcefully in a sequence of papers by Michael Scriven (1959, 1962, 1963), includes several different kinds of case, of which two are especially important for our purposes here. One source of trouble lies in our propensity to accept certain kinds of historical narrative—both in the major branches of human history and in evolutionary studies—as explaining why certain phenomena obtain, even though we are unable to construct any argument that subsumes the phenomena under general laws. Another results from the existence of examples in which we explain events that are very unlikely. Here the paradigm is Scriven's case (later elaborated by van Fraassen) of the mayor who contracts paresis. Allegedly, we hold that the question "Why did the mayor get paresis?" can be answered by pointing out that he had previously had untreated syphilis, despite the fact that the frequency of paresis among untreated syphilitics is low.

A second line of objection to the covering-law models is based on the difficulty in providing a satisfactory analysis of the notion of a scientific law. Hempel is especially forthright in acknowledging the problem (1965, 338). The challenge is to distinguish laws from mere accidental generalizations, not only by showing how to characterize the notion of a projectible predicate (and thus answer the questions raised by Goodman's seminal 1956) but also by diagnosing the feature that renders pathological some statements containing only predicates that are intuitively projectible (for example, "No emerald has a mass greater than 1000 kg.").

The first objection questions the necessity of Hempel's conditions on explanation. The third is concerned with their sufficiency. As Sylvain Bromberger made plain in the early 1960s (see especially his 1966), there are numerous cases in which arguments fitting one of Hempel's preferred forms fail to explain their conclusions. One example will suffice for the present. We can explain the length of the shadow cast by a high object (a flagpole or a building, say) by deriving a statement identifying the length of the shadow from premises that include the height of the object, the elevation of the sun, and the laws of the propagation of light. That derivation fits Hempel's D–N model and appears to explain its conclusion. But, equally, we can derive the height of the object from the length of the shadow, the elevation of the sun and the laws of the propagation of light, and the latter derivation intuitively *fails* to explain its conclusion. Bromberger's challenge is to account for the asymmetry.

A close cousin of the asymmetry problem is the difficulty of debarring Hempelian arguments that appeal to irrelevant factors. If a magician casts a spell over a sample of table salt, thereby "hexing" it, we can derive the statement that the salt dissolved on being placed in water from premises that include the (apparently lawlike) assertion that all hexed salt dissolves on being placed in water. (The example is from Wesley Salmon's seminal 1970; it originally comes from Henry Kyburg [1965]). But, it is suggested, the derivation does not explain why the salt dissolved.

Finally, Hempel's account of statistical explanation was also subject to special problems. One trouble, already glimpsed in the paresis example, concerns the requirement of high probability. Among the guiding ideas of Hempel's account of explanation is the proposal that explanation works by showing that the phenomenon to be explained was to be expected. In the context of the statistical explanation of individual events, it was natural to formulate the idea by demanding that explanatory arguments confer high probability on their conclusions. But, as was urged by both Richard Jeffrey (1969) and Wesley Salmon (1970), this entails a whole class of counterintuitive consequences, generated by apparently good explanations of improbable occurrences. Moreover, the high-probability requirement itself turns out to be extremely hard to formulate (see Hempel 1965 for the surmounting of preliminary difficulties, and Coffa 1974 for documentation of residual troubles). Indeed, critics of Hempel's I–S model have charged that the high-probability requirement can only be sustained by supposing that all explanation is fundamentally deductive (Coffa 1974; Salmon 1984, 52–53).

Even a whirlwind tour of that region of the philosophical landscape occupied by theories of explanation (a region thick with syphilitic mayors, flagpoles, barometers, and magicians) can help to fix our ideas about the problems that an adequate account of scientific explanation must overcome. Contemporary approaches to the subject rightly begin by emphasizing the virtues of Hempel's work, its clarity, its connection with parts of scientific practice, its attention to the subtleties of a broad range of cases. When we have assembled the familiar difficulties, it is appropriate to ask "What went wrong?" The main extant rivals can be viewed as searching for the missing ingredient in the Hempelian approach, that crucial factor whose absence allowed the well-known troubles that I have rehearsed. I shall try to use the four main problem-types to chart the relations among Hempel's successors, and to evaluate the relative merits of the main contemporary rivals.

2. The Pragmatics of Explanation

Not all of the problem-types need be viewed as equally fundamental. Perhaps there was a basic mistake in Hempel's account, a defect that gave rise directly to one kind of difficulty. Solve that difficulty, and we may discover that the remaining troubles vanish. The suggestion is tantalizing, and it has encouraged some important proposals.

One approach is to regard the first type of problem as fundamental. Hempel clearly needed an account of the pragmatics of explanation. As his own detailed responses to the difficulties raised by Scriven (Hempel 1965, 359–64, 427) make entirely clear, he hoped to accommodate the plausible suggestion that narratives can serve an explanatory function even when we have no idea as to how to develop the narrative into an argument that would accord with one of the models. The strategy is to distinguish between what is said on an occasion in which explanatory information is given and the ideal underlying explanation.[1] Although the underlying explanation is to be an argument including laws among its premises, what is said need not be. Indeed, we can provide some information about the underlying argument without knowing all the details, and this accounts for the intuitions of those (like Scriven) who insist that we can sometimes say explanatory things without producing a fully approved Hempelian argument (or without knowing much about what the fully approved argument for the case at hand would be).

Instead of backing into the question of how to relate explanations to what is uttered in acts of explaining, we can take the characterization of explanatory acts as our fundamental problem. This strategy has been pursued in different ways by Peter Achinstein and Bas van Fraassen, both of whom believe that the main difficulties of the theory of explanation will be resolved by gaining a clear view of the pragmatics of explanation. Because van Fraassen's account introduces concepts that I take to be valuable to any theory of explanation, I shall consider his version.[2]

2.1. Van Fraassen's Pragmatics

Van Fraassen starts with the claim that explanations are answers to why-questions. He proposes that why-questions are essentially contrastive: the question "Why *P*?" is elliptical for "Why *P* rather than *P**, *P***, . . .?" In this way he can account for the fact (first noted in Dretske 1973 and further elaborated in Garfinkel 1981) that the same form of words can pose different contrastive why-questions. When Willie Sutton told the priest that he robbed banks because that is where the money is, he was addressing one version of the question "Why do you rob banks?," although not the one that the priest intended.

With this in mind, van Fraassen identifies a why-question as an ordered triple $<P_k,X,R>$. P_k is the topic of the question, and an ordinary (elliptical) formulation of the question would be "Why P_k?" X is the contrast class, a set of propositions including the topic P_k. Finally R is the relevance relation. Why-questions arise in contexts, where a context is defined by a body of background knowledge K. The questions have presuppositions: each why-question presupposes that its topic is the only true member of the contrast class (intuitively, the question "Why P_k in contrast to the rest of X?" is inappropriate if P_k is false or if some other member of the contrast class is true), and also that there is at least one true proposition A that stands in the relation R to $<P_k,X>$. A why-question arises in a context K provided that K entails that the topic is the only true member of the contrast class and does not entail that there is no answer to the question (more exactly, that there is no true A bearing R to $<P_k,X>$).

Van Fraassen recognizes that the theory of explanation ought to tell us when we should reject questions rather than attempting to answer. His pragmatic machinery provides a convincing account. We reject the why-question Q in context K if the question does not arise in this context, and, instead of trying to answer the question, we offer corrections. If Q does arise in a context, then a direct answer to it takes the form "Because A," where A is a true proposition that bears R to $<P_k,X>$. The proposition A is the core of the direct answer.

2.2. *Why Pragmatics Is Not Enough*

Because he hopes to avoid the tangles surrounding traditional approaches to explanation, van Fraassen places no constraints on the relations that can serve as relevance relations in why-questions. In consequence, his account of explanation is vulnerable to trivialization. The trouble can easily be appreciated by noting that it is *prima facie* possible for any true proposition to explain any other true proposition. Let A, B both be true. Then, given van Fraassen's thesis that explanations are answers to why-questions, A will explain B in context K provided that there is a question "Why B?" that arises in K for which A is the core of a direct answer. We construct an appropriate question as follows: let $X = \{B, -B\}$, $R = \{<A, <B, X>>\}$. Provided that K entails the truth of B and does not contain any false proposition entailing the nonexistence of any truth bearing R to $<B, X>$, then the question $<B, X, R>$ arises in K, its topic is B, and its only direct answer is A.

Wesley Salmon and I have argued (Kitcher and Salmon 1987) that van Fraassen's account cannot avoid this type of trivialization. We diagnose the absence of constraints on the relevance relation as the source of the trouble. Intuitively, genuine why-questions are triples $<P_k, X, R>$ where R is a genuine

relevance relation, and a large part of the task of a theory of explanation is to characterize the notion of a genuine relevance relation.

. . .

Van Fraassen contends that his pragmatic approach to explanation solves the problem of asymmetry that arises for the Hempelian account. His solution consists in showing that there is a context in which the question "Why is the height of the tower *h*?" is answered by the proposition that the length of the shadow cast by the tower at a certain time of day is *s*. That proposition answers the question by providing information about the intentions of the builder of the tower. Thus it has seemed that van Fraassen does not touch the Hempelian problem of distinguishing the explanatory merits of two derivations (both of which satisfy the conditions of the D–N model), and that the claim to have solved the problem of asymmetry is incorrect (see Salmon 1984, 95, and Kitcher and Salmon 1987, for arguments to this effect).

. . .

I suggest that van Fraassen's illuminating discussion of why-questions is best seen not as a solution to all the problems of the theory of explanation, but as a means of tackling the problems of the first type (see section 1). Given solutions to the difficulties with law, asymmetry, irrelevance, and statistical explanation, we could embed these solutions in van Fraassen's framework, and thus handle the general topic of how to relate idealized accounts of explanation to the everyday practice of answering why-questions. This is no small contribution to a theory of explanation, but it is important to see that it cannot be the whole story.

2.3. *Possible Goals for a Theory of Explanation*

Van Fraassen's work also enables us to see how to concentrate the three residual problems that arise for Hempel's account into one fundamental issue. The central task of a theory of explanation must be to characterize the genuine relevance relations, and so delimit the class of genuine why-questions. To complete the task it will be necessary to tackle the problems of asymmetry and irrelevance, to understand the structure of statistical explanations, and, if we suppose that genuine relevance involves lawlike dependence, to clarify the concept of law.[3] However, the formulation of the task is ambiguous in significant respects. Should we suppose that there is a single set of genuine relevance relations that holds for all sciences and for all times? If not, if the set of genuine relevance relations is different from science to science and from epoch to epoch, should we try to find some underlying characterization that determines how the different sets are generated, or should we rest content with studying a particular science at a particular time and isolating the genuine relevance relations within this more restricted temporal and disciplinary area?[4]

It appears initially that Hempel sought a specification of the genuine relevance relations that was time-independent and independent also of the branch of science. However, in the light of our integration of Hempel's approach with van Fraassen's treatment of why-questions, I think we can achieve a more defensible view of the Hempelian task. Plainly, the set of ideal relevance relations (or of ideal why-questions) may be invariant across times and sciences, even though different actual questions become genuine in the light of changing beliefs. Thus one conception of the central problem of explanation—I shall call it the *Hempelian conception*—is the question of defining the class of genuine relevance relations that occur in the ideal why-questions of each and every science at each and every time. We can then suppose that variation in the why-questions arises partly from differing beliefs about which topics are appropriate, partly from differing views about the character of answers to underlying ideal why-questions, and partly from differing ideas about what would yield information about those answers.

. . .

Because philosophical attention to the history of science has exposed numerous important shifts in methodological ideals, the Hempelian conception of the theory of explanation may seem far too ambitious and optimistic. However, one way to respond to claims about shifting standards is to argue that there are overarching principles of *global* methodology that apply to all sciences at all times. As particular scientific fields evolve, the principles of global methodology are filled out in different ways, so that there are genuine modifications of *local* methodology.[5] The version of the Hempelian conception that I have just sketched assigns to global methodology a characterization of ideal why-questions. Shifts in admissible why-questions, corresponding to changes in local methodology, can occur against the background of constancy in the underlying ideals—witness my brief discussion of functional/teleological questions.

Perhaps this picture makes the Hempelian conception somewhat less at odds with current thinking about the modification of methodology in the history of science. But can anything positive be said in favor of that conception? I believe it can. The search for understanding is, on many accounts of science, a fundamental goal of the enterprise. That quest may take different forms in different historical and disciplinary contexts, but it is tempting to think that there is something that underlies the various local endeavors, something that makes each of them properly be seen as a striving after the same goal. The Hempelian conception proposes that there is an abstract conception of human understanding, that it is important to the development of science, and that it is common to the variety of ways in which understanding is sought and gained. Scientific explanations are intended to provide objective understanding of nature. The task of characterizing the ideal notions of explanation, why-question, and relevance is thus one of bringing into focus one of the basic aims of science.

I do not suppose that these remarks provide any strong reasons for thinking that the Hempelian conception is correct. It might turn out that there is nothing but ritual lip movements in the avowal of explanation as an aim of the sciences. Nonetheless, there is an obvious motivation for pursuing the Hempelian conception, for, if it is correct, then we can hope to obtain some insight into the rationality and progressiveness of science. Since I know of no conclusive reasons for abandoning my preferred version of the conception, I propose to consider theories of explanation that undertake the ambitious task of characterizing the ideal relevance relations. More modest projects can come once ambition has failed.

3. Explanation as Delineation of Causes

There are two main approaches to explanation that can be seen as undertaking the project just outlined. One of these can be motivated by considering the problems of asymmetry. Intuitively, the length of the shadow cast by a flagpole is causally dependent on the height of the flagpole, but the height is not causally dependent on the shadow-length. Thus we arrive at the straightforward proposal that Hempel's failure to solve problems of asymmetry (and irrelevance) stems from the fact that causal notions are avoided in his analyses. Diagnosis leads quickly to treatment: genuine relevance relations are causal relations, explanations identify causes.

Of course, the invocation of causal notions has its costs. Hempel's account of explanation was to be part of an empiricist philosophy of science, and it could therefore only draw on those concepts that are acceptable to empiricists. If causal concepts are not permissible as primitives in empiricist analyses, then either they must be given reductions to empiricist concepts or they must be avoided by empiricists. Hempel's work appears to stand in a distinguished tradition of thinking about explanation and causation, according to which causal notions are to be understood either in terms of the concept of explanation or in terms of concepts that are themselves sufficient for analyzing explanation. Empiricist concerns about the evidence that is available for certain kinds of propositions are frequently translated into claims about conceptual priority. Thus, the thesis that we can only gain evidence for causal judgments by identifying lawlike regularities generates the claim that the concept of law is prior to that of cause, with consequent dismissal of analyses that seek to ground the notion of law in that of cause.

One of Hume's legacies is that causal judgments are epistemologically problematic. For those who inherit Hume's theses about causation (either his positive or his negative views) there are obvious attractions in seeking an account of explanation that does not take any causal concept for granted. A successful analysis of explanation might be used directly to offer an analysis of causation—most simply, by proposing that one event is causally dependent on another just in case there is an explanation of the former that includes a description of the latter. Alternatively, it might be suggested that the primitive concepts employed in providing an analysis of explanation are just those that should figure in an adequate account of causation.

Because the invocation of causal dependency is so obvious a response to the problems of asymmetry and irrelevance, it is useful to make explicit the kinds of considerations that made that response appear unavailable. One central theme of the present essay is that there is a tension between two attractive options. Either we can have a straightforward resolution of asymmetry problems, at the cost of coming to terms with epistemological problems that are central to the empiricist tradition, or we can honor the constraints that arise from empiricist worries about causation and struggle to find some alternative solution to the asymmetries. The two major approaches to explanation respond to this tension in diametrically opposite ways. As we may anticipate, the central issues that arise concern the adequacy of proposed epistemological accounts of causation and of suggestions for overcoming problems of asymmetry and irrelevance without appealing to causal concepts.

. . .

4. Explanation as Unification

On both the Hempelian and the causal approaches to explanation, the explanatory worth of candidates—whether derivations, narratives, or whatever—can be assessed individually. By contrast, the heart of the view that I shall develop in this section (and which I shall ultimately try to defend) is that successful explanations earn that title because they belong to a set of explanations, the *explanatory store*, and that the fundamental task of a theory of explanation is to specify the conditions on the explanatory store. Intuitively, the explanatory store associated with science at a particular time contains those derivations which collectively provide the best systematization of our beliefs. Science supplies us with explanations whose worth cannot be appreciated by considering them one-by-one but only by seeing how they form part of a systematic picture of the order of nature.

. . .

All this is abstract and somewhat metaphorical. To make it more precise, let us begin with the proposal that *ideal* explanations are derivations. Here there is both agreement and disagreement with Hempel. An argument can be thought of as an ordered pair whose first member is a set of statements (the premises) and whose second member is a single statement (the conclusion). Hempel's proposal that explanations are arguments appears to embody this conception of arguments as *premise–conclusion* pairs. But, on the systematization account, an argument is considered as a derivation, as a sequence of statements whose status (as a premise or as following from previous members in accordance with some specified rule) is clearly specified. An ideal explanation does not simply list the premises but shows how the premises yield the conclusion.

. . .

For a derivation to count as an *acceptable* ideal explanation of its conclusion in a context where the set of statements endorsed by the scientific community is *K*, that derivation must belong to the explanatory store over *K*, *E(K)*. At present, I shall assume that *K* is both consistent and deductively closed, and that the explanatory store over a set of beliefs is unique. *E(K)* is to be the set of derivations that best systematizes *K*, and I shall suppose that the criterion for systematization is unification.[6] *E(K)*, then, is the set of derivations that best unifies *K*. The challenge is to say as precisely as possible what this means.

We should be clear about just what is to be defined. The set of derivations we are to characterize is the set of explanations that would be acceptable to those whose beliefs comprised the members of *K*.

. . .

The idea that explanation is connected with unification has had some important advocates in the history of the philosophy of science. It appears to underlie Kant's claims about scientific method[7] and it surfaces in classic works in the logical empiricist tradition (see Hempel [1965] 345, 444; Feigl [1970] 12). Michael Friedman (1974) has provided the most important defense of the connection between explanation and unification. Friedman argues that a theory of explanation should show how explanation yields understanding, and he suggests that we achieve understanding of the world by reducing the number of facts we have to take as brute.[8] Friedman's motivational argument suggests a way of working out the notion of unification: characterize *E(K)* as the set of arguments that achieves the best tradeoff between minimizing the number of premises used and maximizing the number of conclusions obtained.

Something like this is, I think, correct. Friedman's own approach did not set up the problem in quite this way, and it proved vulnerable to technical difficulties (see Kitcher 1976 . . .). I propose to amend the account of unification by starting from a slight modification of the motivational idea that Friedman shares with T. H. Huxley (see note 8). Understanding the phenomena is not simply a matter of reducing the "fundamental incomprehensibilities" but of seeing connections, common patterns, in what initially appeared to be different situations. Here the switch in conception from premise-conclusion pairs to derivations proves vital. *Science advances our understanding of nature by showing us how to derive descriptions of many phenomena, using the same patterns of derivation again and again, and, in demonstrating this, it teaches us how to reduce the number of types of facts we have to accept as ultimate* (*or brute*).[9] So the criterion of unification I shall try to articulate will be based on the idea that *E(K)* is a set of derivations that makes the best tradeoff between minimizing the number of patterns of derivation employed and maximizing the number of conclusions generated.

. . .

5. Asymmetry and Irrelevance

The time has now come to put all this abstract machinery to work. Problems of asymmetry and irrelevance take the following general form. There are derivations employing premises which are (at least plausible candidates for) laws of nature and that fail to explain their conclusions. The task is to show that the unwanted derivations do not belong to the explanatory store over our current beliefs. To complete the task we need to argue that any systematization of our beliefs containing these derivations would have a basis that fares worse . . . than the basis of the systematization that we actually accept. In practice, this task will be accomplished by considering a small subset of the explanatory store, the derivations that explain conclusions akin to that of the unwanted derivation, and considering how we might replace this subset and include the unwanted derivation. I want to note explicitly that there is a risk that we shall overlook more radical modifications of the explanatory store which would incorporate the unwanted derivation. If there are such radical modifications that do as well by the criteria of unifying power as the systematization we actually accept, then my account is committed to claiming that we were wrong to treat the unwanted derivation as nonexplanatory.

5.1. The "Hexed" Salt

Let us start with the classic example of explanatory irrelevance. A magician waves his hands over some table salt, thereby "hexing" it. The salt is then thrown into water, where it promptly dissolves. We believe that it is not an acceptable explanation of the dissolving of the salt to point out that the salt was hexed and that all hexed salt dissolves in water. What is the basis of this belief?

Suppose that *E(K)* is the explanatory store over our current beliefs, *K*, and that *S* is some set of derivations, acceptable with respect to *K*, that has the unwanted derivation of the last paragraph as a member. One of the patterns used to generate *E(K)* derives claims about the dissolving of salt in water from premises about the molecular composition of salt and water and about the forming and breaking of bonds. This pattern can be used to generate derivations whose conclusions describe the dissolving of hexed salt and the dissolving of unhexed salt. How does *S* provide similar derivations? Either the basis of *S* does not contain the standard pattern or it contains both the standard pattern and a nonstandard pattern that yields the unwanted derivation. In the former case, *S* fares less well than *E(K)* because it has a more restricted consequence set, and, in the latter case, it has inferior unifying power because its basis employs all the patterns of the basis of *E(K)* and one more besides.

It is obviously crucial to this argument that we exclude the gerrymandering of patterns. For otherwise the claim that the basis of *S* must contain either the nonstandard pattern alone or the nonstandard pattern plus the standard pattern would be suspect. The reason is that we could gerrymander a "pattern" by introducing some such Goodmanian predicate as "*x* is either hexed, or is unhexed and has molecular structure *NaCl*." Now we could recover derivations by starting from the claim that all table salt satisfies this predicate, by using the principle that all hexed table salt dissolves in water to generate the conclusion from one disjunct and by using the standard chemical derivation to generate the conclusion from the other disjunct. This maneuver is debarred by the requirement that the predicates used in patterns must be protectable from the perspective of *K*.

Consider next a refinement of the original example. Not all table salt is hexed, but presumably all of it is hexable. (For present purposes, we may assume that hexing requires only that an incantation be muttered with the magician's thoughts directed at the hexed object; this will obviate any concerns that some samples of table salt might be too large or too inaccessible to have the magician wave a hand over them.) Suppose now that it is proposed to explain why a given sample of table salt dissolves in water by offering the following derivation:

> *a* is a hexable sample of table salt.
> *a* was placed in water.
> Whenever a hexable sample of table salt is placed in water, it dissolves.
> *a* dissolved.

I take it that this derivation strikes us as nonexplanatory (although it is useful to point out that it is not as badly nonexplanatory as the derivation in the original example). Suppose that *S* is a systematization of *K* that contains the derivation. Can we show that *S* has less unifying power than *E(K)?*

Imagine that *S* had the same unifying power as *E(K)*. Now in *E(K)* the mini-derivation that is most akin to the one we want to exclude derives the conclusion that *a* dissolved from the premise that *a* is a sample of table salt, the premise that *a* was placed in water, and the generalization that samples of table salt that are placed in water dissolve. Of course, this mini-derivation is embedded within a much more exciting chemical derivation whose conclusion is the generalization that samples of table salt dissolve when placed in water. That derivation instantiates a general pattern that generates claims about the dissolving (or failure to dissolve) of a wide variety of substances from premises about molecular structure. In its turn, that general pattern is a specification of an even more general pattern that derives conclusions about chemical reactions and state changes for all kinds of substances from premises about molecular structures and energy distributions. If *S* is to rival *E(K)* then it must integrate the unwanted mini-derivation in analogous fashion.

That can be done. One way to proceed would be to use the standard chemical derivation to yield the conclusion that all samples of table salt dissolve when placed in water and then deduce that all hexable samples of table salt dissolve when placed in water. But now we can appeal to a principle of simplifying derivations to eliminate redundant premises or unnecessary steps. When embedded within the standard chemical derivation, the unwanted mini-derivation is inferior to its standard analog because the latter is obtainable more directly from the same premises. An alternative way of trying to save the unifying power of *S* would be to amend the standard chemical patterns to suppose that they apply only to hexable substances. But since it is supposed that *all* substances are hexable, and since this fact is used throughout *S* to generate derivations to rival those produced in *E(K)*, this option effectively generates a set of derivations that systematically contain idle clauses. Since it is believed that everything is hexable, the outcome is as if we added riders about objects being self-identical or being nameable to our explanations, and again a principle of simplification directs that the idle clauses be dropped.[10]

We can now achieve a diagnosis of the examples of explanatory irrelevance. Citation of irrelevant factors will either commit one to patterns of explanation that apply only to a restricted class of cases or the irrelevancies will be idle wheels that are found throughout the explanatory system. The initial hexing example illustrates the first possibility; the refinement shows the second.

5.2. *Towers and Shadows*

Let us now turn to the asymmetry problem, whose paradigm is the case of the tower and the shadow. Once again, let *K* be our current set of beliefs, and let us compare the unifying power of *E(K)* with that of some systematization *S* containing a derivation that runs from the premises about shadow length and sun elevation to a conclusion about the tower's height. As in the case of the irrelevance problem, there is a relatively simple argument for maintaining that *S* has less unifying power than *E(K)*. There are also some refinements of the original, troublesome story that attempt to evade this simple argument.

Within *E(K)* there are derivations that yield conclusions about the heights of towers, the widths of windows, the dimensions of artifacts and natural objects alike, which instantiate a general pattern of tracing the present dimensions to the conditions in which the object originated and the modifications that it has since undergone. Sometimes, as with flagpoles and towers, the derivations can be relatively simple: we start with premises about the intentions of a designer and reason to an intermediate conclusion about the dimensions of the object at the time of its origin; using further premises about the conditions that have prevailed between the origin and the present, we reason that the object has persisted virtually unaltered and thus reach a conclusion about its present dimensions. With respect to some natural objects, such as organisms, stars, and mountain ranges, the derivation is much more complex because the objects have careers in which their sizes are substantially affected. However, in all these cases, there is a very general pattern that can be instantiated to explain current size, and I shall call derivations generated by this pattern *origin-and-development* explanations.

Now if *S* includes *origin-and-development explanations*, then the basis of *S* will include the pattern that gives rise to these derivations. To generate the unwanted derivation in *S*, the basis of *S* must also contain another pattern that derives conclusions about dimensions from premises about the characteristics of shadows (the *shadow* pattern). In consequence, *S* would fare worse than *E(K)* according to our principles . . . because its basis would contain all the patterns in the basis of *E(K)* and one more. Notice that, once again, the "no gerrymandering" requirement comes into play to block the device of fusing some doctored version of the pattern that generates *origin-and-development* explanations with the shadow pattern. So *S* must foreswear *origin-and-development* explanations.

However, it now seems that *S* must have a consequence set that is more restricted than that of *E(K)*. The reason is that the *shadow* pattern cannot be instantiated in all the cases in which we provide *origin-and-development*

explanations. Take any unilluminated object. It casts no shadow. Hence we cannot instantiate the *shadow* pattern to explain its dimensions.

This is correct as far as it goes, but the asymmetry problem cuts deeper. Suppose that a tower is actually unilluminated. Nonetheless, it is possible that it should have been illuminated, and if a light source of a specified kind had been present and if there had been a certain type of surface, then the tower would have cast a shadow of certain definite dimensions. So the tower has a complex dispositional property, the disposition to cast a shadow of such-and-such a length on such-and-such a surface if illuminated by a light-source at such-and-such an elevation above the surface. From the attribution of this dispositional property and the laws of propagation of light we can derive a description of the dimensions of the tower. The derivation instantiates a pattern, call it the *dispositional-shadow* pattern, that is far more broadly applicable than the *shadow* pattern.

But can it be instantiated widely enough? To be sure it will provide surrogates for *origin-and-development* explanations in those cases in which we are concerned with ordinary middle-sized objects. But what about perfectly transparent objects (very thin pieces of glass, for example)? Well, they can be handled by amending the pattern slightly, supposing that such objects have a disposition to be coated with opaque material and then to cast a shadow. Objects that naturally emit light can be construed as having a disposition to have their own light blocked and then to cast a shadow. Objects that are so big that it is hard to find a surface on which their shadows could be cast (galaxies, for example) can be taken to have the dispositional property of casting a shadow on some hypothetical surface.

Yet more dispositional properties will be needed if we are to accommodate the full range of instances in which *origin-and-development* explanations are available. An embryologist might explain why the surface area of the primitive gut (archenteron) in an early embryo is of such-and-such a size by deriving a description of the gut from premises about how it is formed and how modified. To instantiate the *dispositional-shadow* pattern in such cases, we shall need to attribute to the gut-lining a dispositional property to be unrolled, illuminated, and thus to cast a shadow. A biochemist might explain the diameter in the double helix of a DNA molecule by identifying the constraints that the bonding pattern imposes on such molecules both as they are formed and as they persist.[11] Taking a clue from the principles of electronmicroscopy, the *dispositional-shadow* pattern can be instantiated by supposing that DNA molecules have a dispositional property to be coated and irradiated in specified ways and to produce absorption patterns on special surfaces. And so it goes.

Perhaps there are some objects that are too small, or too large, too light sensitive, or too energetic for us to attribute to them any disposition to cast anything like a shadow. If so, then even with the struggling and straining of the last paragraph, the *dispositional-shadow* pattern will still fail to generate derivations to rival those present in *E(K)*. But I shall assume that this is not so, and that for any object whose dimensions we can explain using our accepted patterns of derivation, it is possible to find a dispositional property that has something to do with casting a shadow.

However, if we now consider the critical predicate that appears in the *dispositional-shadow* pattern, we find that it is something like the following: "*x* has the disposition to cast a shadow if illuminated by a light source or *x* has the disposition to produce an absorption pattern if *x* is suitably coated and irradiated or *x* has the disposition to cast a shadow if *x* is covered with opaque material or *x* has the disposition to cast a shadow if *x* is sectioned and unrolled or *x* has the disposition to cast a shadow after *x* has been treated to block its own light sources or . . ." At this point it is surely plain that we are cutting across the distinctions drawn by the projectable predicates of our language. Any "pattern" that employs a predicate of the sort that I have (partially) specified is guilty of gerrymandering, for, from our view of the properties of things, the dispositions that are lumped together in the predicate are not homogeneous. I conclude that even if it is granted that we can find for each object some dispositional property that will enable us to derive a specification of dimensions from the ascription of the disposition, there is no *common* dispositional property that we can employ for *all* objects. To emulate the scope of *E(K)*, the basis of *S* would have to contain a multiplicity of patterns, and our requirement against gerrymandering prohibits the fusion of these into a single genuine pattern.

As in the case of the irrelevance problem, there is a natural diagnosis of the trouble that brings out the central features of the foregoing arguments. Explanation proceeds by tracing the less fundamental properties of things to more fundamental features, and the criterion for distinguishing the less from the more fundamental is that appeal to the latter can be made on a broader scale. Thus an attempt to subvert the order of explanation shows up in the provision of an impoverished set of derivations (as in our original example of the tower and the shadow) or in the attempt to disguise an artificial congeries of properties as a single characteristic (as in our more recent reflections).

. . .

6. Conclusions

As Railton clearly recognizes, . . . differences in views about scientific explanation connect to differences in metaphysics. The causal approach is wedded to a strong version of realism in which the world is seen as having a structure independent of our efforts to systematize it. It should be no surprise that the metaphysical extras bring epistemological problems in their train. . . . I have been trying to show that we can make sense of scientific explanation and our view of the causal structure of nature without indulging in the metaphysics. The aim has been to develop a simple, and, I think, very powerful idea. The growth of science is driven in part by the desire for explanation, and to explain is to fit the phenomena into a unified picture insofar as we can. What emerges in the limit of this process is nothing less than the causal structure of the world.

Notes

* I owe a long-standing debt to Peter Hempel, who first inspired my interest in the study of scientific explanation and whose writings on the topic seem to me paradigms of what is best in twentieth-century philosophy. My own thinking about explanation was redirected by Michael Friedman's seminal essay on explanation and scientific understanding, and I have also learned much from the comments, encouragement, and advice of Paul Churchland, Paul Humphreys, David Papineau, Kenneth Schaffner, and Stephen Stich. Above all I am deeply grateful to Wesley Salmon, for the depth and lucidity of his ideas and the kindness and patience of his conversation. The present essay continues a long dialogue, and, because that dialogue has been so pleasant and so instructive, I trust that it is not yet over.

1 This approach to pragmatic issues has been articulated with considerable sophistication by Peter Railton. See his (1981) and his unpublished doctoral dissertation.

2 Achinstein's theory of explanation, as presented in his (1983), is extremely complex. I believe that it ultimately suffers from the same general difficulty that I present below for van Fraassen. However, it is eminently possible that I have overlooked some subtle refinement that makes for a disanalogy between the two versions.

3 As should now be evident, the second of the four problem-types that beset the Hempelian account assumes a derivative status. We may be able to manage the theory of explanation without a characterization of laws if we can distinguish the genuine relevance relations without invoking the notion of lawlike dependence. I shall articulate an approach below on which this possible strategy is attempted.

4 I am extremely grateful to Isaac Levi for raising the issue of the goals of a theory of explanation, by inquiring whether we can expect there to be a single set of relevance relations that applies for all sciences at all times.

5 The distinction between global and local methodology is drawn in more detail in chapter 7 of (Kitcher 1983). It is only right to note that some

scholars have challenged the idea that there is any very substantive global methodology. See, for example, Laudan 1984.

6 We might think of the systematization approach as covering an entire family of proposals among which is that based on the view of systematization as unification. Since it appears to me that the latter view provides the best chances of success, I shall concentrate on it and ignore alternative possible lines of development.

7 See my (1986) for a reconstruction of Kant's views that tries to defend this attribution.

8 ". . . our total picture of nature is simplified via a reduction in the number of independent phenomena that we have to accept as ultimate" (Friedman 1974, 18). There is an interesting recapitulation here of T. H. Huxley's summary of Darwin's achievement. "In ultimate analysis everything is incomprehensible, and the whole object of science is simply to reduce the fundamental incomprehensibilities to the smallest possible number" (Huxley 1896, 165).

9 I think it entirely possible that a different system of representation might articulate the idea of explanatory unification by employing the "same way of thinking again and again" in quite a different—and possibly more revealing—way than the notions from logic that I draw on here. Kenneth Schaffner has suggested to me that there is work in AI that can be deployed to provide the type of account I wish to give, and Paul Churchland has urged on me the advantages of connectionist approaches. I want to acknowledge explicitly that the adaptation of ideas about logical derivation may prove to be a ham-fisted way of developing the idea of explanatory unification. But, with a relatively developed account of a number of facets of explanation available, others may see how to streamline the machinery.

10 Notice that derivations that systematically contain idle clauses are not so clearly nonexplanatory as the kind of irrelevant derivation with which we began. It seems to me that this is because the unwanted mini-derivations are viewed as giving us information about the structure of a full, ideal, derivation, and the natural implication is that the properties picked out in the premises will play a key role. Once we see that these properties are inessential, and that predicates referring to them figure throughout all our derivations, then we may feel that cluttering up the explanatory store does nothing more than add a harmless irrelevancy. The resultant derivations are untidy, but I think that there is reason to argue about whether they should be counted as nonexplanatory.

11 For this example, it is important to recognize that the *origin-and-development* pattern must allow for explanatory derivations in which we appeal to general constraints that keep a system close to an equilibrium state throughout its career. . . .

References

Achinstein, Peter. 1983. *The Nature of Explanation*. New York: Oxford University Press.

Bromberger, Sylvain. 1963. A Theory about the Theory of Theory and about the Theory of Theories. In *Philosophy of Science: The Delaware Seminar*, ed. W. L. Reese. New York: John Wiley.

——. 1966. Why-Questions. In *Mind and Cosmos*, ed. R. Colodny. Pittsburgh: University of Pittsburgh Press.

Coffa, J. Alberto. 1974. Hempel's Ambiguity, *Synthèse* 28: 141–64.

Dretske, Fred. 1973. Contrastive Statements, *Philosophical Review*, 82: 411–37.

Feigl, Herbert. 1970. The "Orthodox" View of Theories: Remakrs in Defense as well as Critique. In *Minnesota Studies in the Philosophy of Science*, Volume IV, eds. M. Radner and S. Winokur. Minneapolis: University of Minnesota Press.

Friedman, Michael. 1974. Explanation and Scientific Understanding, *Journal of Philosophy*, 71: 5–19.

Garfinkel, Alan. 1981. *Forms of Explanation*. New Haven: Yale University Press.

Goodman, Nelson. 1956. *Fact, Fiction, and Forecast*. Indianapolis: Bobbs-Merrill.

Hempel, C. G. 1965. *Aspects of Scientific Explanation*. New York: Free Press.

Hempel, C. G., and Oppenheim, P. 1948. Studies in the Logic of Explanation. Chapter 9 of Hempel (1965).

Huxley, Thomas Henry. 1896. *Darwiniana*. New York: Appleton.

Jeffrey, Richard. 1969. Statistical Explanation vs. Statistical Inference. In *Essays in Honor of Carl G. Hempel*, ed. Nicholas Rescher. Dordrecht, Holland: D. Reidel.

Kitcher, Philip. 1976. Explanation, Conjunction, and Unification, *Journal of Philosophy* 73: 207–12.

——. 1981. Explanatory Unification, *Philosophy of Science*, 48: 507–31.

——. 1983. *The Nature of Mathematical Knowledge*. New York: Oxford University Press.

——. 1985. Two Approaches to Explanation, *Journal of Philosophy*, 82: 632–39.

——. 1986. Projecting the Order of Nature. In *Kant's Philosophy of Physical Science*, ed. Robert Butts. Dordrecht: D. Reidel.

——, and Salmon, Wesley. 1987. Van Fraassen on Explanation, *Journal of Philosophy*, 84: 315–30.

Kyburg, Henry. 1965. Comment, *Philosophy of Science*, 35: 147–51.

Laudan, Larry. 1984. *Science and Values*. Berkeley: University of California Press.

Railton, Peter. 1981. Probability, Explanation, and Information, *Synthèse*, 48: 233–56.

Salmon, Wesley. 1970. Statistical Explanation. In *The Nature and Function of Scientific Theories*, ed. R. Colodny. Pittsburgh: University of Pittsburgh Press.

——. 1984. *Scientific Explanation and the Causal Structure of the World*. Princeton: Princeton University Press.

Scriven, Michael. 1959. Definitions, Explanations, and Theories. In *Minnesota Studies in the Philosophy of Science*, Volume II, eds. H. Feigl, M. Scriven, and G. Maxwell. Minneapolis: University of Minnesota Press.

——. 1962. Explanations, Predictions, and Laws. In *Minnesota Studies in the Philosophy of Science*, Volume III, eds. H. Feigl and G. Maxwell. Minneapolis: University of Minnesota Press.

——. 1963. The Temporal Asymmetry between Explanations and Predictions. In *Philosophy of Science. The Delaware Seminar*, Volume I, ed. B. Baumrin. New York: John Wiley.

6

Wesley C. Salmon, "Scientific Explanation: Causation *and* Unification"

To explain something is to tell a causal story about how it happened. Causal interactions and causal processes should figure centrally in such an explanatory account. Wesley Salmon (1984) developed his causal–mechanical theory of explanation; the paper reprinted below summarizes some of the main ideas of his theory and compares it with other approaches to explanation.

For the past few years I have been thinking about the philosophy of scientific explanation from the standpoint of its recent history. Many of these reflections have been published in *Four Decades of Scientific Explanation* (1990). They have, I believe, provided some new insight on some old problems, and they suggest that genuine progress has been made in this area of philosophy of science.

§1. Looking Back: Two Grand Traditions

The classic essay, "Studies in the Logic of Explanation," by Carl G. Hempel and Paul Oppenheim (1948) constitutes the fountainhead from which almost everything done subsequently on philosophical problems of scientific explanation flows. Strangely enough, it was almost totally ignored for a full decade. Although the crucial parts were reprinted in the famous anthology *Readings in the Philosophy of Science*, edited by Herbert Feigl and May Brodbeck (1953), it is not cited at all in R. B. Braithwaite's well-known book, *Scientific Explanation* (1953). During the first decade after publication of the Hempel–Oppenheim paper very little was published on scientific explanation in general—Braithwaite's book being the main exception. Most of the work on explanation during that

W.C. Salmon, "Scientific Explanation: Causation *and* Unification," *Critica. Revista Hispanoamericana de Filosofía*, 1990, 22(66): 3–21.

period focused either on explanation in history or on teleological/functional explanation.

In the years 1957 and 1958 the situation changed dramatically. At that time a deluge of work on scientific explanation began, much of it highly critical of the Hempel–Oppenheim view. Vigorous attacks came from Michael Scriven and N. R. Hanson among others. Sylvain Bromberger and Israel Scheffler offered important criticisms, but they were offered more in the spirit of friendly amendments than outright attacks on the Hempel–Oppenheim program (see Salmon 1990, pp. 33–46).

When we reflect on what happened we can see that two grand traditions emerged. Hempel advocated a view of scientific explanation according to which explanation consists in deductive or inductive subsumption of that which is to be explained (the explanandum) under one or more laws of nature. This tradition could find examples that had strong intuitive appeal—for instance, the explanation of the laws of optics by Maxwell's electrodynamics, or the explanation of the ideal gas law by the molecular-kinetic theory. These examples also illustrate what is often called "theoretical reduction" of one theory to another. Another example, if it could be worked out successfully, would be methodological individualism in the social sciences, for it would result in the reduction of the various social sciences to psychology.

Ironically, the very examples that furnish the strongest intuitive appeal for the subsumption approach are of a type that Hempel and Oppenheim found intractable. Although they offered an account of explanations of particular facts, they acknowledged in a notorious footnote (note 33), that they could *not* provide an account of explanations of general laws. To the best of my knowledge, Hempel never returned to this recalcitrant problem. It should also be noted that, while Hempel and Oppenheim casually identified their pattern of explanation (later known as the *deductive-nomological* or *D-N* model) with causal explanation, Hempel later argued emphatically that causality does not play any sort of crucial role in scientific explanation (1965, § 2.2).

The other major tradition was advanced primarily by Scriven, and it made a strong identification between causality and explanation. Roughly and briefly, to explain an event is to identify its cause. The examples that furnish the strongest intuitive basis for this conception are cases of explanations of particular occurrences—for instance, the sinking of the Titanic or the Chernobyl nuclear accident. The most serious problem with this approach has been the lack of any adequate analysis of causality on which to found it. Given Hume's searching critique of that concept, something more was needed.

As these two traditions developed over the years, there was often conflict, sometimes quite rancorous, between their advocates. At present, I believe, we have reached a stage in which a significant degree of rapprochement is entirely possible.

§2. Explanation as Unification

The idea that scientific explanation consists in showing that apparently disparate phenomena can be shown to be fundamentally similar has been around for a long time, long before 1948. However, Michael Friedman, in "Explanation and Scientific Understanding" (1974), seems to have been the first philosopher to articulate this conception clearly and to attempt to spell out the details. His basic thesis is that we increase our scientific understanding of the world to the extent that we can reduce the number of independently acceptable assumptions that are required to explain natural phenomena. By phenomena he means regularities in nature such as Kepler's first law (planets move in elliptical orbits) or Hooke's law (the amount of deformation of an elastic body is proportional to the force applied). It should be noted that Friedman is attempting to furnish an account of the explanation of laws, which is just the sort of explanation Hempel and Oppenheim found themselves unable to handle.

In order for Friedman's program to work, it is obviously necessary to be able to count the number of assumptions involved in any given explanation. In order to facilitate that procedure, Friedman offers a definition of a technical term, "K-atomic statement." This concept is relativized to a knowledge situation K. A statement is K-atomic provided it is not equivalent to two or more generalizations that are independently acceptable in knowledge situation K. A given statement is acceptable independently of another if it is possible to have evidence adequate for the acceptance of the given statement without *ipso facto* having evidence adequate to accept the other. The problem that arises for Friedman's program is that it seems impossible to have any K-atomic statements—at least, any that could plausibly be taken as fundamental laws of nature. For instance, Newton's law of universal gravitation, which prior to Einstein, was a good candidate for a fundamental law, can be partitioned into (1) "Between all pairs of masses in which both members are of astronomical dimensions there is a force of attraction proportional to the product of the masses and inversely proportional to the square of the distance between them," (2) "Between all pairs of masses in which one member is of astronomical dimensions and one is smaller there is a force of attraction . . .," and (3) "Between all pairs of masses in which both are of less than astronomic size there is a force of attraction. . . ." Statement (1) is supported by planetary motions and the motion of the moon. Statement (2) is supported by Newton's falling apple, and indeed, by all phenomena to which Galileo's law of falling bodies applies. Statement (3) is supported by the Cavendish torsion-balance experiment. It seems possible to partition virtually any universal statement into two or more independently acceptable generalizations.

If Friedman's program had worked it would have solved the Hempel–Oppenheim problem of footnote 33. It appears, however, not to be satisfactory in the form originally given. Although Philip Kitcher (1976) offered his own (different) critique of Friedman's paper, he accepted the basic idea of explanation as unification, and he has elaborated it in a different way in a series of papers, of which "Explanatory Unification and the Causal Structure of the World" (1989) is the most recent and most detailed.

§3. Causality and Mechanism

Around 1970, when I was trying to work out the details of the *statistical relevance* or *S-R* model of scientific explanation, I had hopes that the fundamental causal concepts could be explicated in terms of statistical concepts alone, and that, consequently, the *S-R* model could furnish what was chiefly lacking in the causal approach. By 1980, that no longer seemed possible, and I shifted my focus to an attempt to explicate certain causal mechanisms, in particular, causal interactions and causal processes (see Salmon 1984, chapters 5–6). I took as primitives the notion of a process and that of a spatio-temporal intersection of processes. The aim is to distinguish between processes that are causal and those that are not (causal processes vs. pseudo-processes) and to distinguish those intersections of processes (whether causal or pseudo) that are genuine causal interactions and those that are not.

The basic idea—stated roughly and briefly—is that an intersection of two processes is a *causal interaction* if both processes are modified in the intersection in ways that persist beyond the point of intersection, even in the absence of further intersections. When two billiard balls collide, for instance, the state of motion of each is modified, and those modifications persist beyond the point of collision. A *process is causal* if it is capable of transmitting a mark—that is, if it is capable of entering into a causal interaction. For example, a beam of white light becomes and remains red if it passes through a piece of red glass, and the glass absorbs some energy in the same interaction.

However, not all intersections of causal processes are causal interactions. If two light rays intersect they are superimposed on one another in the locus of intersection, but after they leave that place each of them continues on as if nothing had happened. A process—such as a light beam—is causal if it can be modified or marked in a way the persists beyond the point of intersection as a result of *some* intersection with another process. Causal processes are capable of transmitting energy, information, and causal influence from one part of spacetime to another. I have argued that causal processes are precisely the kinds of causal connections Hume sought, but was unable to find. I have also argued that such connections do not violate Hume's strictures against mysterious powers.

It is important to recognize that these causal mechanisms are not necessarily deterministic. In particular, causal processes can interact probabilistically. My favorite example is Compton scattering, in which an energetic photon collides with a virtually stationary electron. The angles at which the photon and electron emerge from the interaction are not strictly determined; there is, instead, a probability distribution over a whole range of pairs of angles. By conservation of momentum and energy, however, there is a strict correlation between the two scattering angles.

The causal mechanisms of interaction and transmission are strongly local; they leave no room for what Einstein called "spooky action-at-a-distance." Interactions occur in a restricted spacetime region, and processes transmit in a spatio-temporally continuous fashion. Regrettably (to me and many others), however, quantum mechanics appears to involve violations of local causality. There seems to be a quantum mechanism, often known as "the collapse of the wave function," which is radically nonlocal, and which is not really understood as yet.

I prefer to think of the conception of explanation that emerges from these considerations as causal/mechanical. The aim of explanations of this sort is to exhibit the ways in which nature operates; it is an effort to lay bare the mechanisms that underlie the phenomena we observe and wish to explain.

§4. Some New Perspectives

During the 1960s and 1970s the ideas developed by Hempel constituted a *received view* of scientific explanation. It was based on the Hempel–Oppenheim 1948 paper, and was articulated most fully in Hempel's "Aspects of Scientific Explanation" (1965). As a result of numerous criticisms, it is fair to say, the "received view" is no longer received. Its natural successor is the unification conception due chiefly to Friedman and Kitcher.

The causal conception as originally advocated by Scriven and others has also undergone transformation, primarily as a result of more careful and detailed analysis of causality, but also because of the admitted possibility that there are mechanisms of a noncausal type as well. It has involved an explicit recognition of the Humean critique of causality, and an attempt to overcome the Humean difficulties.

Given the history of opposition between the "received view" and the causal view of scientific explanation, it is not surprising that philosophers continue to find opposition between the successors. Friedman, for example, contrasted *local* and *global* accounts. According to the older views of both Hempel and Scriven, explanation is a local affair, in the sense that one could give a perfectly acceptable explanation of a small and isolated phenomenon without appeal to global theories. One could give a Hempelian explanation of the electrical conductivity of a particular penny by pointing out that it is made of copper, and copper is an electrical conductor. One could give a Scrivenesque explanation of a stain on a carpet by citing the fact that a clumsy professor bumped an open ink bottle off of the desk with his elbow. In contrast to both of the foregoing accounts, Friedman's unification view requires us to look at our entire body of scientific knowledge, to see whether a given attempt at explanation reduces the number of assumptions needed to systematize that body of knowledge. Friedman's conception is patently global.

Kitcher (1989) has made a related distinction between conceptions he characterizes as "bottom-up" and "top-down." The Hempelian approach illustrates the bottom-up way. We begin by explaining the conductivity of a penny by appeal to the generalization that copper is a conductor. We can explain why copper is a conductor in terms of the fact that it is a metal. We can explain why metals are conductors in terms of the behavior of their electrons. And so it goes from the particular fact to the more general laws until we finally reach the most comprehensive available theory. The causal/mechanical approach has the same sort of bottom-up quality. From relatively superficial causal explanations of particular facts we appeal to ever more general types of mechanisms until we reach the most ubiquitous mechanisms that operate in the universe. Kitcher's top-down approach, in contrast, looks to the most general explanatory schemes we can find, and works down from there to characterize such items as laws and causal relations.

In a spirit quite different from those of Friedman and Kitcher, Peter Railton has advocated an approach that makes the bottom-up and top-down, as well as the local and global, conceptions complementary rather than contrary. In "Probability, Explanation, and Information" (1981) he introduces the concept of an *ideal explanatory text* which is extremely global and detailed. He suggests, however, that we hardly ever seek to articulate fully such an ideal text. Rather, we focus on portions or aspects of the ideal text, and try to illuminate these. When we succeed we have furnished *explanatory information*. Different investigators, or groups of investigators, have different interests and work on different portions of the ideal text. Pragmatic considerations determine for a given individual or group what portion of the ideal text to look at, and in what depth of detail.

§5. Rapprochement?

My main purpose in this paper is to consider the possibility, suggested by Railton's work, that the successors of the "received view" and its causal opponent, are actually compatible and complementary. Let me begin by offering a couple of examples.

(1) A friend recounted the following incident. Awaiting take-off on a jet airplane, he found himself sitting across the aisle from a young boy who was holding a helium-filled balloon by a string. In order to pique the child's curiosity, he asked the boy what he thought the balloon would do when the airplane accelerated rapidly for takeoff. After considering for a few moments, the boy said he thought it would move toward the back of the cabin. My friend said *he* believed that it would move forward in the cabin. Several other passengers overheard this claim and expressed skepticism. A flight attendant even wagered a miniature bottle of Scotch that he was wrong—a wager he was happy to accept. In due course, the pilot received clearance for takeoff, the airplane accelerated, and the balloon moved toward the front of the cabin. And my friend enjoyed a free drink courtesy of the flight attendant.

Two explanations of the balloon's strange behavior can be given. First, it can be pointed out that, when the plane accelerates, the rear wall of the cabin exerts a force on the air molecules near the back, which produces a pressure gradient from rear to front. Given that the inertia of the balloon is smaller than that of the air it displaces, the balloon tends to move in the direction of less dense air. This is a straightforward causal explanation in terms of the forces exerted on the various parts of the physical system. Second, one can appeal to Einstein's principle of equivalence, which says that an acceleration is physically equivalent to a gravitational field. The effect of the acceleration of the airplane is the same as that of a gravitational field. Since the helium balloon tends to rise in air in the earth's gravitational field, it will tend to move forward in the air of the cabin in the presence of the aircraft's acceleration. This second explanation is clearly an example of a unification-type explanation, for the principle of equivalence is both fundamental and comprehensive.

(2) A mother leaves her active baby in a carriage in a hall that has a smooth level floor. She carefully locks the brakes on the wheels so that the carriage will not move in her absence. When she returns she finds, however, that by pushing, pulling, rocking, bouncing, etc., the baby has succeeded in moving the carriage some little distance. Another mother, whose education includes some physics, suggests that next time the carriage brakes be left unengaged. Though skeptical, the first mother tries the experiment and finds that the carriage has moved little, if at all, during her absence. She

asks the other mother to explain this lack of mobility when the brakes are off.

Two different explanations can be given; each assumes that the rolling friction of the carriage is negligible when the brakes are off. The first (at least in principle) possible explanation would involve an analysis of all of the forces exerted by the baby on the carriage and the carriage on the baby, showing how they cancel out. This would be a detailed causal explanation. The second explanation would appeal to the law of conservation of linear momentum, noting that the system consisting of the baby and the carriage is essentially isolated (with respect to horizontal motion) when the brake is off, but is linked with the floor, the building, and the earth when the brake is on. This is an explanation in the unification sense, for it appeals directly to a fundamental law of nature.

The first point I should like to emphasize in connection with both of these examples from physics is that both explanations are perfectly legitimate in both cases; neither is intrinsically superior to the other. Pragmatic considerations often determine which of the two types is preferable in any particular situation. Invocation of Einstein's principle of equivalence would be patently inappropriate for the boy with the balloon, and for the other adults in that situation, because it is far too sophisticated. All of them could, however, understand a clear explanation in terms of forces and pressures. The two examples are meant to show that explanations of the two different types are not antithetical, but rather, complementary.

I should like also to consider a famous example from biology, (3) the case of the peppered moth in the vicinity of Liverpool, England. This moth spends much of its life on the trunks of plane trees, which naturally have a light-colored bark. Prior to the industrial revolution the pale form of this moth was prevalent, for its light color matched the bark of the tree, and consequently provided protection against predators. During the industrial revolution in that area, air pollution darkened the color of the tree bark, and the dark (melanic) form of the peppered moth became prevalent, because the darker color then provided better protection. In the post-industrial-revolution period, since the pollution has been drastically reduced, the plane trees have again acquired their natural light-colored bark, and the light form of the peppered moth is again becoming dominant.

In this example, like the two preceding, two different explanations are available to account for the changes in color of the moth. The first has already been suggested in the presentation of the example; it involves such evolutionary considerations as natural selection, mutation, and the heritability of traits. This is the unification style of explanation in terms of basic and comprehensive principles of biology. The second kind of explanation is biochemical in nature; it deals with the nitty-gritty details of the causal processes and interactions involved in the behavior of DNA and RNA molecules and the synthesis of proteins leading up to the coloration of the moth. In order to explain the above-mentioned changes in color, it would have to take account also of the births, deaths, and reproductive histories of the individual moths. Although such a causal/mechanical explanation would be brutally complex, it is possible in principle. Again, there is nothing incompatible about the two kinds of explanation.

The use of this kind of biological example leads into a more general consideration regarding the status of functional explanations. In the case of the peppered moth, we were clearly concerned with a function of the coloration, namely, its function as camouflage for protection against predators. Although some philosophers have tried to cast doubt upon the legitimacy of functional explanations, I am strongly inclined to consider them scientifically admissible. In my opinion, Larry Wright, a student of Scriven, has given the most convincing theory (1976). Wright makes a distinction between *teleological explanations* and *functional ascriptions*, but his accounts of them are fundamentally similar; they involve what he calls a *consequence-etiology*. It is a *causal* account in which the cause of a feature's presence is the fact that in the past when it has been present it has had a certain result or consequence. It is not *just* that it has had such consequences in the past; in addition, the fact that it had such consequences is causally responsible for its coming into being in the present instance.

I shall use the term "functional explanation" to cover both teleological explanations and functional ascriptions in Wright's terminology. Although functional explanations in this sense are causal, they do not have a fine-grained causal character—that is to say, they do not go into the small details of the causal processes and interactions involved. They do, of course, appeal to the *mechanisms* of evolution—inheritance and natural selection—but these are coarse-grained mechanisms. Wright is, however, perfectly willing to admit that fine-grained causal explanations are also possible. Just as we can give a straightforwardly mechanistic account of the workings of a thermostat, whose function is to control temperature in a building, so also is it possible, at least in principle, to give a thoroughly physico-chemical account of some item that has a biological function, such as the color of the peppered moth. Although some philosophers have maintained that the mechanistic explanation, when it can be given, supersedes the functional explanation, Wright holds that they are completely compatible, and that the functional explanation need not give way to the mechanistic explanation. I think he is correct in this view.

The philosophical issue of the status of functional explanations is not confined to biology; the problem arises in psychology, anthropology, and the other social or behavioral sciences as well. Whether one regards

Freudian psychoanalysis as a science or not, the issue is well-illustrated in that discipline. According to Freud, the occurrence and the content of dreams can be explained functionally. The dream preserves sleep by resolving some psychological problem that might otherwise cause the subject to awaken. The content of the dream is determined by the nature of the problem. However, even if it is possible to provide a psychoanalytic explanation of a given dream, it may also be possible to give another explanation in completely neurophysiological terms. This would be a fine-grained causal explanation that incorporates the physical and chemical processes going on in the nervous system of the subject. I am suggesting that the two explanations need not conflict with one another, and I believe that, in this opinion, I am in agreement with Freud.

§6. Can Quantum Mechanics Explain?

Ever since the publication of the famous Einstein–Podolsky–Rosen paper (1935), there has been considerable controversy over the explanatory status of the quantum theory. Einstein seems to have taken a negative attitude, while Bohr appears to have adopted an affirmative one. As the discussion has developed, the question of local causality versus action-at-a-distance has become the crucial issue. The EPR paper showed that there could, in principle, be correlations between remote events that seem to defy explanation. Further work by David Bohm, John Bell, and A. Aspect have shown that such correlations actually exist in experimental situations, and that *local hidden-variable* causal explanations are precluded. A clear and engaging account of these issues can be found in N. David Mermin (1985). Because these fine-grained causal explanations are not possible, many philosophers, myself included, have concluded that quantum mechanics does not provide explanations of these correlations. As I suggested above, there seem to be mechanisms at the quantum level that are noncausal, and that are not well understood.

Other philosophers have taken a different attitude. On the basis of the undeniable claim that quantum mechanics is a highly successful theory in providing precise predictions and descriptions (they are statistical, but extremely successful), we need ask for no more. The quantum theory can be formulated on the basis of a small number of highly general principles, and it applies universally.

In terms of the distinct conceptions of scientific explanation we have been discussing, it seems that quantum theory provides explanations of the unification type, but it does not provide those of the causal/mechanical sort. This situation contrasts with that in other scientific disciplines where, as we have seen, explanations of both kinds are possible, at least in principle. The same circumstance may seem to occur in anthropological or sociological explanations of some human institutions, where we can give functional explanations of certain phenomena, but fine-grained causal explanations are far beyond our grasp. In contrast to quantum mechanics, however, there is no solid theoretical basis for claiming that fine-grained causal explanations are impossible in principle in these disciplines.

In answer to the question of this section, "Can quantum mechanics explain?" the answer must be, for the time being at least, "In a sense 'yes', but in another sense 'no'." In (W. Salmon 1984, pp. 242–59) I had admitted only the negative answer to this question.

§7. Two Concepts of Explanation

One of the chief aims and accomplishments of science is to enhance our understanding of the world we live in. In the past, it has often been said that this aim is beyond the scope of science—that science can describe, predict, and organize, but that it cannot provide genuine understanding. Among philosophers of science and philosophical scientists at present there seems to be a fair degree of consensus about the ability of science to furnish explanations, and therefore to contribute to our understanding of the world. As is obvious from the foregoing discussion, however, there is no great consensus on the nature of this understanding. I should like to suggest that it has at least two major aspects, corresponding to the two types of explanation that have been discussed above.

On the one hand, understanding of the world involves a general world-view—a *Weltanschauung*. To understand the phenomena in the world requires that they be fitted into the general world-picture. Although it is often psychologically satisfying to achieve this sort of agreement between particular happenings and the world-view, it must be emphasized that psychological satisfaction is not the criterion of success. To have *scientific* understanding we must adopt the world-view that is best supported by all of our scientific knowledge. The fundamental theories that make up this world-view must have stood up to scientific test; they must be supported by objective evidence. Perhaps we need not ask what makes a scientific world-picture superior to a mythic or religious or poetic world-view. Nevertheless, I would ask, and try to give an answer. The superiority of understanding based on a scientific world-view lies in the fact that we have much better reason to regard that world-view as true—even though some other world-view might have more psychological appeal.

The conception of understanding in terms of fitting phenomena into a comprehensive scientific world-picture is obviously connected closely with the unification conception of scientific explanation. It also corresponds closely to the goal of many contemporary scientists who are trying to find one unified theory of the physical world—for example, those who see in so-called "superstring theory" a TOE (theory of everything). Many scientists seem to believe that it is both feasible and desirable to try to discover some completely unified theory that will explain everything.

On the other hand, there is a different fundamental notion of scientific understanding that is essentially mechanical in nature. It involves achieving a knowledge of how things work. One can look at the world, and the things in it, as black boxes whose internal workings we cannot directly observe. What we want to do is open the black box and expose its inner mechanisms.

This conception of scientific explanation brings us face to face with the problem of realism versus anti-realism. Although one can open up a clock to find out how it works by direct observation of its parts, one cannot do so with a container full of a gas. Gases are composed of molecules or atoms (monatomic molecules), and these are too small to be observed by means of the naked eye, a magnifying glass, or a simple optical microscope. The search for mechanistic explanations often takes us into the realm of unobservables. Although some philosophers, past and present, have adopted a skeptical or agnostic attitude toward unobservables, I think it is possible to argue persuasively that we can have genuine knowledge of such micro-entities as bacteria and viruses, atoms and molecules, electrons and protons, and even quarks and neutrinos. I believe we can have compelling inductive evidence concerning the existence and nature of such entities (Salmon 1984, chapter 8). The ideal of this approach is to have the capacity to provide explanations of natural phenomena in terms of the most fundamental mechanisms and processes in the world.

Consideration of these two conceptions of scientific explanation suggests that there may be a kind of explanatory duality corresponding to the two approaches. To invoke Railton's terminology and Kitcher's metaphor, we can think in terms of reading the ideal explanatory text either from the bottom-up or from the top-down. There are, of course, intermediate stages between the two extremes—there are degrees of coarse- or fine-grainedness. The kinds of examples brought up by Wright in his comparison of the course-grained consequence-etiology explanations with the fine-grained mechanical explanations do not usually appeal to either the most general laws of nature or the most fundamental physical mechanisms. Moreover, we often give mechanical explanations of everyday contrivances, such as the hand-brake on a bicycle, without any appeal to unobservables.

It is extremely tempting to try to bring a linguistic distinction in English to bear on the explanatory duality I am discussing, but I fear it also holds certain risks. Sometimes we seek explanations by asking "How?" and sometimes by asking "Why?" Consider, for example, "How did the first large mammals get to New Zealand?" and "Why did the first large mammals go to New Zealand?" The answer to the first question is that they were humans, and they went in boats. I do not know the answer to the second question, but it undoubtedly involves human purposes and goals. The danger in making the distinction between how-questions and why-questions in terms of examples of this sort is that it easily leads to anthropomorphism—to the conclusion that "genuine" explanations always involve an appeal to goals or purposes. That would certainly be a step in the wrong direction. But not all examples have this feature. If one asks *why* a penny conducts electricity, one good answer is that it is made of copper, and copper is a good conductor. If one asks *how* this penny conducts electricity, it would seem that a mechanism is called for. A story about electrons that are free to move through the metal would be an appropriate answer. In this case, the why-question elicits an appeal to a general law; the how-question evokes a description of underlying mechanisms.

§8. Conclusion

The attempt to gain scientific understanding of the world is a complicated matter. We have succeeded to some extent in reaching this goal, but what we have achieved to date has taken several centuries of effort on the part of many people, some of whom were or are towering geniuses. Many of the explanations that have been found are extraordinarily difficult to understand. When we think seriously about the very concept of scientific understanding, it does not seem plausible to expect a successful characterization of scientific explanation in terms of any simple formal schema or simple linguistic formulation. It is not surprising that there might be the kind of duality I have been discussing.

The situation may be even more extreme. As one of my graduate students, Kenneth Gemes, has suggested, perhaps it is futile to try to explicate the concept of scientific explanation in a comprehensive manner. It might be better to list various explanatory virtues that scientific theories might possess, and to evaluate scientific theories in terms of them. Some theories might get high scores on some dimensions, but low scores on others—recall our brief consideration of quantum mechanics. I have been discussing two virtues, one in terms of unification, the other in terms of exposing underlying mechanisms. Perhaps there are others that I have not considered. The foregoing discussion might serve as motivation to search for additional scientific explanatory qualities.

References

Braithwaite, R. B., *Scientific Explanation*. Cambridge: Cambridge University Press, 1953.

Einstein, Albert, *et al.*, "Can Quantum-Mechanical Description of Reality Be Considered Complete?" *Physical Review*, Vol. 47 (1935), pp. 777–80.

Feigl, Herbert, and May Brodbeck (eds.), *Readings in the Philosophy of Science*. New York: Appleton-Century-Crofts, 1953.

Friedman, Michael, "Explanation and Scientific Understanding," *Journal of Philosophy*, Vol. 71 (1974), pp. 5–19.

Hempel, Carl G., "Aspects of Scientific Explanation," in *Aspects of Scientific Explanation and Other Essays in the Philosophy of Science* (New York: Free Press, 1965), pp. 331–496.

——, and Paul Oppenheim, "Studies in the Logic of Explanation," *Philosophy of Science*, Vol. 15 (1948), pp. 135–75. Reprinted in Carl G. Hempel, *Aspects of Scientific Explanation and Other Essays in the Philosophy of Science* (New York: Free Press, 1965).

Kitcher, Philip, "Explanation, Conjunction, and Unification," *Journal of Philosophy*, Vol. 73 (1976), pp. 207–12.

——, "Explanatory Unification and the Causal Structure of the World," in Kitcher and Salmon (eds.), 1989, pp. 410–505.

——, and Wesley C. Salmon (eds.), *Scientific Explanation*, Vol. XIII, *Minnesota Studies in the Philosophy of Science*. Minneapolis: University of Minnesota Press, 1989.

Mermin, N. David, "Is the Moon There When Nobody Looks? Reality and the Quantum Theory," *Physics Today*, Vol. 38, no. 4 (April, 1985), pp. 38–47.

Railton, Peter, "Probability, Explanation, and Information," *Synthèse*, Vol. 48 (1981), pp. 233–56.

Salmon, Wesley C., *Scientific Explanation and the Causal Structure of the World*. Princeton, N.J.: Princeton University Press, 1984.

——, *Four Decades of Scientific Explanation*. Minneapolis: University of Minnesota Press, 1990. Also published in Kitcher and Salmon (eds.), 1989.

Wright, Larry, *Teleological Explanations*. Berkeley & Los Angeles: University of California Press, 1976.

7

J.L. Mackie, "The Logic of Conditionals"

John Leslie Mackie (1917–1981) debates the nature of counterfactual conditional statements and their relation to the issue of laws of nature.

We can use the suppositional account to clear up some problems about the logic of conditionals and their contrasting relations with natural laws and accidental generalizations.

Can conditionals be contraposed? Is "If *P*, *Q*" equivalent to "If not-*Q*, not-*P*?" We should expect it to be. But we are also inclined to say that "If *P*, *Q*" is compatible with "If not-*P*, *Q*"—e.g. "If Boycott makes a century, England will win" is compatible with "If Boycott doesn't make a century, England will win"; there is even a standard form for the conjunction of two such conditionals: "England will win whether Boycott makes a century or not." On the other hand, "If *P*, *Q*" seems not to be compatible with what I shall call the contrary conditional, "If *P*, not-*Q*"—e.g. "If you go to Benidorm you'll have a lovely time" is not compatible with "If you go to Benidorm you won't have a lovely time." In general the point of saying "If *P*, *Q*" is to make a discrimination, to say that on the supposition that *P*, we shall have *Q* and we shall not have not-*Q*. But we cannot retain, without qualification, the compatibility of "If *P*, *Q*" and "If not-*P*, *Q*," the incompatibility of "If *P*, *Q*" and "If *P*, not-*Q*," and equivalent contraposition all together. For if "If *P*, *Q*" and "If not-*P*, *Q*" were compatible, so would their contrapositives "If not-*Q*, not-*P*" and "If not-*Q*, not not-*P*" be compatible; and these two are of the form "If *P*, *Q*" and "If *P*, not-*Q*" (with "not-*Q*" replacing "*P*" and "not-*P*" replacing "*Q*").

This problem does not, of course, arise with material conditionals, because "$P \supset Q$" and "$P \supset —Q$" are compatible; they merely together entail "—*P*."

J.L. Mackie, *Truth, Probability and Paradox*, Chapter 3, §10, 1973, pp. 109–19. Oxford: Clarendon Press.

What, then, are we to give up? This will depend on the purposes for which we are using the conditionals. Granted that every (primary) use of a conditional is tantamount to asserting something within the scope of some supposition, there may be different reasons for doing so, different illocutionary acts involved. If one is treating what is supposed as a genuine possibility—or in counterfactual cases as having been a genuine possibility—that is, as something not ruled out by the background assumptions in the light of which one is considering it, but as something which, in view of those background assumptions, would have—or would have had—some determinate outcome, then one will regard "If *P*, *Q*" and "If *P*, not-*Q*" as incompatible; they are not incompatible in themselves, as conditionals, but together they are incompatible with this which we may call the *straightforward* or *direct* use of conditionals. Alternatively, one may treat what is supposed as a possibility only in a weaker sense, as something which one's background assumptions rule out, but which can nevertheless be considered initially as possible with a view to showing, in the end, that it is impossible: in other words, there is an *indirect* or *reductio ad absurdum* way of using a supposition. In this sort of use, "If *P*, *Q*" and "If *P*, not-*Q*" are compatible: they do not conflict because, on the assumptions with respect to which one is considering possibilities, it is not really possible that *P*.

Someone who asserts a pair of conditionals of the form "If *P*, *Q*" and "If not-*P*, *Q*" (in their direct use) will be relying implicitly on some set of background assumptions "*S*" such that "*S*" and "*P*" together entail "*Q*" and equally "*S*" and "not-*P*" together entail "*Q*." But then "*S*" itself must entail "*Q*," and rule out "not-*Q*." Consequently no conditional of the form "If not-*Q* . . ." will be acceptable to such a speaker for direct use, and in particular he cannot accept for direct use the contrapositives of "If *P*, *Q*" and "If not-*P*, *Q*." But he can accept them for *reductio ad absurdum* use; he can assert both "If not-*Q*, not-*P*" and "If not-*Q*, not not-*P*" as a way of bringing out the impossibility of "not-*Q*" on the assumptions on which he is implicitly relying.

Thus the incompatibility of "If *P*, *Q*" and "If *P*, not-*Q*" holds for the direct use, and contraposition is accordingly restricted. It is not done away with even for the direct use. There will be many direct uses of the form "If *P*, *Q*" such that "not-*Q*" is not ruled out by the background assumptions, conditionals of the form "If not-*Q* . . ." will therefore be acceptable for direct use, and then using "If *P*, *Q*" will commit one to being prepared to use "If not-*Q*, not-*P*." As long as the antecedent "If not-*Q*" is admissible, the contraposition is valid. For the indirect use, contraposition holds without restriction, and "If *P*, *Q*" and "If *P*, not-*Q*" are compatible.

We can easily illustrate these general principles. Anyone who says "If Boycott makes a century, England will win," implicitly relying on assumptions which do not in themselves ensure that England will win without the premiss "Boycott makes a century"—so that he is *not* prepared to say also "If Boycott doesn't make a century, England will win"—will allow the real possibility of England's not winning and is logically committed to the contrapositive "If England does not win, Boycott will not have made a century." But someone who says both "If Boycott makes a century, England will win" and "If Boycott doesn't make a century, England will win" cannot coherently consider it as a possibility, in relation to whatever background assumptions he is for the moment relying on, that England should not win, and cannot therefore use directly either contrapositive, and he is not logically committed to either (since an indirect use would be unnatural in this context). If this speaker does begin a statement with the words "If England doesn't win . . ." he must be changing his ground, moving to some different background assumptions, and then there will be no simple logical connections with what he said in reliance on the previous assumptions. On the other hand, the contrary conditionals "If there were an infallible perception it would be instantaneous (in order to be infallible)" and "If there were an infallible perception it would not be instantaneous (being a perception)" are compatible if they are used together in a *reductio ad absurdum* argument to show that there could not be an infallible perception.

Such a pair of compatible contraries will naturally be expressed in a subjunctive or counterfactual form, since they are compatible only because the speaker is committed to rejecting their common antecedent. But we must not convert this rule, and say that pairs of contrary subjunctive and counterfactual conditionals are always compatible. They are not. Subjunctive and counterfactual conditionals are normally used directly, just as open conditionals are. "If you had struck that match it would have lit" and "If you had struck that match it would not have lit" are just as incompatible, in their ordinary direct use, as are "If you strike that match it will light" and "If you strike that match it will not light."

We should not confuse the two distinct points that a pair of contrary counterfactuals (i) will be compatible in an indirect use, and (ii) may be separately acceptable, with a shift of assumptions, in a direct use. "If you had struck that match, it would not have lit (because it was wet)" and "If you had struck that match, it would have lit (because you are a careful person and would have dried it first)" are separately acceptable for direct use. But since they rely on different beliefs retained for use in conjunction with the assumption, they cannot be conjoined for direct use to give "If you had struck that match it would both have lit and not lit", which would be nonsense in a direct use. This illustrates (ii). But, to go back to (i), there will be an indirect use of these two counterfactuals together: "That match was wet" and "You are (in the required sense) a careful person" are together incompatible with the supposition "You struck that match"; and this fact might be brought out by an indirect use of the conditional "If you had struck that match it would both have lit and not lit"—for the reasons given above—which relies on the retaining of both the assumptions which are together incompatible with the supposition. The conclusion drawn from this double counterfactual is, of course, "So you (as you are) could not have struck it (as it is)," not merely "So you did not strike it," which is conceded automatically by the counterfactual form.

This shows how we can deal with the problem of competing counterfactuals. "If Bizet and Verdi had been compatriots, Bizet would have been Italian" and "If Bizet and Verdi had been compatriots, Verdi would have been French" seem equally plausible. But they do compete; if we try to combine them we get the absurdity that if they had been compatriots, Bizet would have been Italian and Verdi French. In the first place, these counterfactuals are separately acceptable. If, where we introduce the belief-contravening supposition that the two composers were compatriots we retain the true belief that Verdi was Italian, we shall assert, within the scope of that supposition, that Bizet was Italian too. If, instead, we retain the equally true belief that Bizet was French, we shall assert within the scope of our supposition that Verdi was French. The counterfactuals can clearly be interpreted as of the condensed argument sub-species: in each case the consequent would follow from the supposition in conjunction with a true premiss and certain linguistic rules about the term "compatriots" and the nationality descriptions "French" and "Italian." But the three premisses "Verdi was Italian," "Bizet was French," and "Bizet and Verdi were compatriots" form, in the light of the linguistic rules, an inconsistent triad. There is therefore no *direct* use for the combined counterfactual, in which we should have to retain both the premisses "Verdi was Italian" and "Bizet was French" along with a supposition that conflicts with their conjunction. But there is still a possible *reductio ad absurdum* use. We could say "If Bizet and Verdi had been compatriots Bizet would have been Italian and Verdi would have been French (for the reasons indicated) and so they would not have been compatriots; that is, they couldn't have been compatriots." And of course they couldn't have been, so long as we are considering only possibilities which allow us to retain at once "Verdi was Italian" and "Bizet was French." And this is all there is to it. Since we have denied that non-material conditionals can be true, the question which of these competing counterfactuals is true does not arise. Neither is true; each is acceptable in certain circumstances; but they are co-acceptable only in an indirect use. Since in ordinary circumstances we have no reason for preferring one to the other of these, we are not likely to be very strongly tempted to use either.

These counterfactuals were competing on equal terms, but a similar logical pattern applies even where symmetry is absent, e.g. where one of the competing counterfactuals, but not the other, is sustained by a causal law. The statements "Cyanide is a deadly poison," "Jones is alive," and "Jones took cyanide" form an inconsistent triad, and the first and second of these sustain, respectively, the competing counterfactuals "If Jones had taken cyanide, he would not be alive" and "If Jones had taken cyanide, cyanide would not have been a deadly poison." But these are not on equal terms: we are much more prepared, when introducing the belief-contravening supposition that Jones took cyanide, to stick to the law that cyanide is a deadly poison than to the particular fact that Jones is alive. The point is not that the former generalization is "so secure that we are willing to retain it at all costs":[1] the fact that Jones is now alive may be equally "secure." The point is that the counterfactual form concedes that Jones did not in fact take cyanide, so that the supposition that he did take it introduces a *different* situation from the actual one, and there is no reason for taking the observation that Jones is alive in the actual situation as informing us about the different possible one. This is not because the law about cyanide is known or secure, but merely because we know that there are causal laws, that a difference in a temporal antecedent is often followed by a different outcome. That this is the point is confirmed by the fact that the *open* conditional "If Jones took cyanide, cyanide is not a deadly poison" is quite natural and plausible. This is so because we can quite reasonably retain the observed fact that Jones is now alive for use along with the supposition, considered as an *open* possibility, that he took cyanide; we are now considering a situation consisting of the whole of the actual one along with the fulfilment of the antecedent, which is being treated as neither known to be fulfilled nor known not to be fulfilled. In the possible situation thus constructed, the law that cyanide is a deadly poison cannot hold. The corresponding counterfactual is not plausible because the contrary-to-fact supposition, just by being contrary-to-fact, introduces a situation other than the actual one, and so does away with our reason for retaining, within the scope of this supposition, such particular features of the actual situation as that Jones is now alive.

This brings us to what is perhaps the greatest benefit resulting from our fuller understanding of conditionals: the light that this throws upon their relations with causal laws and accidental generalizations and consequently upon the nature of causal laws themselves. Why do causal laws entail or sustain counterfactuals whereas accidental generalizations do not? Does this fact show, as is widely believed among philosophers, that causal law statements include, in their meaning, something stronger than merely factual universality? Do these statements implicitly assert the existence of some sort of "natural necessity" in the events themselves? Is there some special virtue either in causal law statements, or in the objective laws which they report, which enables them to entail counterfactuals, mysterious truths that hold beyond the actual world and govern the realm of possibilities as well?

My contention is that this way of asking the questions is thoroughly misleading. Counterfactual conditionals are not to be taken literally as truths about possible worlds, but as a species of human procedure. They are just non-material conditionals plus a hint that their antecedents are unfulfilled, and non-material conditionals merely express the asserting of something within the scope of some supposition—which may be done for any one of a number of reasons, which may themselves be reasonable or unreasonable. All sorts of statements can sustain counterfactuals, including, as we have seen, such singular statements as "Bizet was French." The real problem is not to find any extra virtue in causal laws, but to find what special deficiency in accidental generalizations prevents them from sustaining counterfactuals. Or, more generally, to explain why some logically formulable counterfactuals are more acceptable than others.

Once we ask the right question it is comparatively easy to find the answer. Let us consider the accidental generalization "Everyone in this room understands Italian," established, presumably, by complete enumeration, by checking each individual in turn. To use this to sustain the counterfactual "If Mr. Chou En-Lai were in this room he would understand Italian" would be to introduce the supposition—admitted to be false—that Mr. Chou En-Lai is in this room, and then to assert, within its scope, that Mr. Chou En-Lai understands Italian, using the supposition and the enumeratively established universal together to yield this result. But it is not reasonable to use them together. Since our sole ground for believing this universal was the enumerative check, that ground collapses as soon as we add the supposition that someone *else* is in the room; someone who—as the counterfactual form concedes—is not in fact in the room and whose understanding of Italian has therefore not been checked by this enumeration. The adding of the contrary-to-fact supposition takes us from the actual situation to a different, merely possible, one, one in which we have not checked everyone's understanding of Italian. If the universal were true and Mr. Chou En-Lai were in the room then of course he would understand Italian; but since our reason for believing the universal evaporates as soon as we introduce the supposition, we cannot reasonably take this universal as we know it and this supposition as joint premisses in an argument, even a telescoped argument; and unless we do so we have no reason for asserting the counterfactual. Since the complete check was our only reason for believing the universal we are not justified in retaining it within the scope of our supposition, and in fact we are not prepared to do so.

This account is confirmed if we contrast the counterfactual with an open conditional. The accidental generalization "Everyone in this room understands Italian" does sustain the *open* conditional "If Mr. Chou En-Lai is in this room he understands Italian"—that is, one of the persons present may be Mr. Chou En-Lai disguised or unrecognized, and if so he has passed the test of his understanding of Italian. This is acceptable because the open supposition does not carry us to a different situation, but—like "If Jones took cyanide" in our previous example—only *adds* to the actual situation an item that is taken as neither known to be so nor known not to be so. Thus the open supposition that Mr. Chou En-Lai is here does not cancel our reason for retaining the enumeratively established universal, whereas the counterfactual supposition—that is, the supposition that he is here coupled with the admission that he is not here—does cancel it. It is the contrary-to-factness of the antecedent that makes us unable to use an enumeratively established universal within its scope.

This account can easily be extended to cover examples where the accidental generalization is known not by a complete enumeration but by some other but logically equivalent process. If we know that none of the stones in this box is radioactive because a Geiger counter near by shows no response, this universal does not sustain the counterfactual "If that other stone were in this box it would not be radioactive," again because the supposition that some *other* stone is in the box undermines the evidence of the Geiger counter as a reason for asserting the universal within the scope of the supposition.

On the other hand, a generalization sustains a counterfactual if our reason for adhering to it is such as to survive its being put within the scope of a belief-contravening supposition. Let us look at some contrasting examples before proceeding to a general explanation.

Suppose that the gathering in this room is a meeting of the Italian Poetry Circle, and this fact is clearly announced in a notice on the outside of the door in several languages, including Chinese. This would give pretty good grounds for saying that if Mr. Chou En-Lai had been here he would have understood Italian. Still stronger grounds would be provided by the presence of a doorkeeper who had been instructed to let in only those who proved their understanding of Italian. Similarly if this box were, say, the left-hand box of a pair attached to a collecting and sorting device which pushes all radioactive objects it encounters into the right-hand box, knowledge of this device would sustain the counterfactual "If that other stone were in this box it would not be radioactive."

Lying behind the grounds relied upon in these cases are, of course, causal laws: in the first example we have devices which (more or less efficiently) cause the exclusion of those who do not understand Italian, and in the second a device which causes the exclusion of radioactive objects. So the question is, why do causal laws work in the way they do? That is, why can we (i) combine a law with suppositions that go beyond cases for which the law has been checked, and so advance open or subjunctive conditionals, and (ii) combine it with suppositions which we take to be not fulfilled, which we regard as altering the extension of the law's subject term, and so advance counterfactual conditionals? The answer is simply, because we have what we take to be good inductive evidence for the law.

If we have good inductive evidence for the law "All *A*s are *B*," then this evidence supports the conclusion that an unobserved *A* is *B*; it therefore justifies an argument from the supposition that a certain object *X* is an *A* to the conclusion that *X* is *B*; it therefore justifies us in asserting that *X* is *B* within the scope of the supposition that *X* is an *A*, and hence for saying that if *X* is an *A*, it is *B*. Such evidence will, therefore, sustain the open conditional "If *X* is an *A*, it is *B*." But this evidence is logically related in exactly the same way to the argument from the supposition that *Y* is an *A* to the conclusion that *Y* is *B*, even if we happen to know or believe that *Y* is not an *A*; it therefore justifies us in asserting that *Y* is *B*, within the scope of the contrary-to-fact supposition that *Y* is an *A*, and hence for saying that if *Y* had been an *A*, it would have been *B*. Such evidence will therefore sustain also the counterfactual "If *Y* had been an *A*, it would have been *B*." Formally, all that is required to let a law sustain counterfactuals is that there should be the same logical relation (i) between the evidence and the proposed law (covering unobserved instances) as things are, and (ii) between the evidence and the proposed law with things otherwise the same but with additional instances of the law's subject term. And this holds for all ordinary inductive reasoning.

To enable a law to sustain counterfactuals, then, *all* that is needed is that it should be supported by what we take to be good inductive evidence. It is no part of my task, in offering this explanation, to say either what is good inductive evidence or why it is so. The hard fact is that we do reason inductively: given that we do, the sustaining of counterfactuals by laws which are (directly or indirectly) supported inductively is an automatic consequence, in view of the general account of conditionals and counterfactuals that I have offered. Inductive evidence is, by definition, projective; inductive evidence for a law provides a ground for believing that law which is not impaired either (i) by the supposition that there is an instance of the subject term which has not been included in the evidence, or (ii) by the supposition that there are additional (contrary-to-fact) instances of the subject term. The sustaining of counterfactuals by laws or "nomic universals" is nothing more than the projective force of inductive evidence in a new guise. This explanation has not made use of any notion of a special *content* in a causal law; unless it can be shown that the inductive reasoning itself requires a special content in its conclusions—and there is no plausibility in this suggestion—there is nothing in the sustaining of counterfactuals to show that causal laws have a content or meaning that in any way differs from that of a straightforward factual universal. Causal laws *may* also contain certain special sorts of information, but there is nothing in their distinctive relation to counterfactuals to require that they should. Their sustaining of counterfactuals is exhaustively accounted for, not by their content, but by the inductive character of the evidence which directly or indirectly confirms them. In so far as the term "nomic universal" suggests a special content, it should be discarded forthwith.

I conclude, then, that we need no longer lament that we do not know how to construe conditionals or that the exact analysis of counterfactuals in particular is an unsolved riddle. The suppositional account does satisfactorily elucidate all the standard uses of conditionals—peripheral as well as central—and also relates them intelligibly to other uses of the word "if." Puzzles about the logic of conditionals can then be resolved, and the power of causal law statements to sustain counterfactuals loses at once its mystery and its supposed profound significance.

Note

1 Cf. N. Rescher, "Belief-contravening Suppositions," *Philosophical Review* 70 (1961), 198.

8

John Earman, "Laws of Nature"

John Earman introduces and defends the Mill–Ramsey–Lewis theory of the laws of nature. *Modal* considerations—those having to do with the nature of possibility and necessity and the related notion of a *possible world*—occur in his discussion. For our purposes, a possible world is just a total way things might have been. And things might have been different from the way they actually are (i.e., from the way they are in the actual world) in an infinite variety of detail.

> The problem is sometimes put in the form that we all distinguish between uniformities due to natural law and those which are merely accidentally true, "historical accidents on the cosmic scale"; if natural laws are just uniformities, how can this distinction be made? It seems to me foolish to deny (as some Humeans do) that such a distinction is made in common speech; but it also seems perfectly sensible to try to give the rationale for this distinction within the ambit of a constant conjunction view.
>
> (R. B. Braithwaite, *Scientific Explanation*)

We have made a start on understanding what properties laws of nature must have if the world is to be deterministic, but nothing much has been said about what laws of nature are, about what distinguishes laws from non-laws. And, strictly speaking, we are in the embarrassing position of having no examples to work with, for none of the examples of so-called laws cited in previous chapters is truly a law since what is asserted has proven to be false (and, by meta-induction, a similar fate awaits every such example??). This realization need cause no undue alarm if we are willing to apply to the history of science a Principle of Respect, recommending that when we encounter a textbook example of a "law" we assume, unless there are specific contextual indications to the contrary, that (1) the scientists of the period had good reason to believe that what the "law" asserts is true (or

J. Earman, *A Primer on Determinism*, 1986, pp. 81–90. Dordrecht: Reidel.

approximately true), and (2) the scientists of the period were justified in believing that, if what the "law" asserts is true, then it does indeed express a law of nature. While I agree with the spirit of this principle, I think that some caution is required in applying it. In the young sciences it may be a struggle to find any informative generalization that works tolerably well, and so the standards of lawhood may be lax. We can avoid this problem by looking only to the mature sciences for our examples. But in the mature sciences the search for laws is constrained by the record of past successes and failures; research scientists assume, consciously or not, that the candidate laws must have a certain mathematical form, must incorporate certain variables, must conform to certain symmetry and invariance principles, must reduce in special cases to the old "laws," must mesh with "laws" in allied fields, etc. Here opposing snares await us. One is the vulgar relativism of seeing the notion of law so inextricably tied to a scientific community, a research tradition, or whatever that only historical *reportage* is possible. The other is the arrogant abstractionism of supposing that an analysis of laws amounts to no more and no less than finding a core concept that cuts across every branch of science and every period in the history of science. I will be careful to avoid the snare of relativism, but I will knowingly step into a mild form of the abstractionism snare as it applies to modern physics. For my focus in this chapter is on the attempts of philosophers of science to provide an abstractive analysis of laws of physics. My main concern will not be so much with the rather thin character of these attempts as with the discordance which has recently grown to the extent that it cannot be ignored. While unanimity is an unattainable and even undesirable goal in philosophy, something is amiss when we cannot agree even approximately on how to understand a notion that is fundamental to the study not only of determinism but to the methodology and content of the sciences in general.

When in doubt it is a good practice to return to the source. In this case the source is David Hume.

1. Hume's Definitions of "Cause"

Hume defined "cause" three times over. (Recall: The constant conjunction definition says that a cause is "an object precedent and contiguous to another, and where all the objects resembling the former are plac'd in a like relation of priority and contiguity to those objects, that resemble the latter." The felt determination definition takes a cause to be "an object precedent and contiguous to another, and so united with it in the imagination, that the idea of the one determines the mind to form the idea of the other, and the impression of the one to form a more lively idea of the other."[1] And finally, in the *Enquiry*, but not in the *Treatise*, Hume defines a cause as "an object followed by another . . . where, if the first object had not been, the second never had existed."[2])

The two principal definitions (constant conjunction, felt determination) provide the anchors for the two main strands of the modern empiricist accounts of laws of nature[3] while the third (the counterfactual definition) may be seen as the inspiration of the non-Humean necessitarian analyses. Corresponding to the felt determination definition is the account of laws that emphasizes human attitudes, beliefs, and actions. Latter day weavers of this strand include Nelson Goodman, A. J. Ayer, and Nicholas Rescher. In *Fact, Fiction and Forecast* Goodman writes: "I want only to emphasize the Humean idea that rather than a sentence being used for prediction because it is a law, it is called a law because it is used for prediction . . ." (1955, p. 26). In "What Is a Law of Nature?" Ayer explains that the difference between "generalizations of fact" and "generalizations of law" "lies not so much on the side of facts which make them true or false, as in the attitude of those who put them forward" (1956, p. 162). And in a similar vein, Rescher maintains that lawfulness is "mind-dependent"; it is not something which is discovered but which is supplied: "Lawfulness is not found in or extracted from the evidence, but it is superadded to it. *Lawfulness is a matter of imputation*" (1970, p. 107). By contrast, the constant conjunction definition promotes the view that laws are to be analyzed in terms of the *de re* characteristics of regularities, independently of the attitudes and actions of actual or potential knowers.

Hume himself gives passing acknowledgement to the fact that the two strands can diverge.[4] And where they diverge, I follow the constant conjunction strand and declare my starting assumptions that whatever our beliefs, we could be mistaken because there is something to be mistaken about—the distinction between uniformities due to natural laws and those which are merely cosmic accidents is to be drawn in terms of features of the uniformities and not in terms of our attitudes towards them.[5] At the same time I readily concede that this strand cannot be successfully woven into an account of laws by completely ignoring the other strand, for while ontology need not follow epistemology, our account of laws must explain how it is possible to form rational beliefs about what the laws of our world are. The hope is that this epistemological constraint can be met without becoming so entangled in the felt determination strand that we become captives of the Goodman–Ayer–Rescher web.

Against this hope I sense a rising sentiment among philosophers of science that the problem of giving a regularity analysis of laws bears an ominous resemblance to the problem of providing a criterion of "cognitive significance" to separate empirically meaningful assertions from metaphysical nonsense. It was initially an article of faith among the positivists and logical empiricists that such a criterion must exist and that providing it in a suitable form was only a matter of finding the appropriate technical formulation. But as attempt after attempt fell into the philosophical waste bin this faith has given way to an indifferent agnosticism or, worse, an insipid lip service. If a similar ignominious fate awaits the regularity account of laws, then it would seem best to redirect our efforts elsewhere.

A growing band of philosophers is already at work in the elsewhere, constructing a non-empiricist conception of laws. But before turning to their views, let us review the sources of dissatisfaction with the standard regularity account and explore the prospects of improving it within an empiricist framework.

2. The Naive Regularity Account

The crudest form of the regularity account puts laws of nature and Humean regularities into one–one correspondence. In the linguistic mode favored by the logical positivists, this account might be rendered thus:

(*H*) Laws are what are expressed by true lawlike sentences.

What makes the naive regularity account naive is the assumption that "lawlike" can be captured by syntactical and semantical conditions on individual sentences. E.g., *S* is lawlike just in case *S* is general in form (say, a universal condition $(x)\ (Fx \supset Gx)$ so dear to philosophers determined to make use of their required symbolic logic course) and the predicates are suitably kosher ("*F*" and "*G*" are non-positional, purely qualitative, non-Goodmanized, etc.). This is, to be sure, sloppy and vague, but the impression given by the older references was that all the mysteries of laws would disappear once the appropriate technical apparatus was applied to make notions like "generality" and "non-positional predicate" really precise.[6]

We do not need to await the outcome of the technical maneuvers. W. A. Suchting, David Armstrong, and other down-under philosophers have done such a thorough demolition job on the naive regularity account that we can be confident that no way of fiddling with the details of (*H*) will produce a defensible version. I will just remind you of some of the considerations and refer you to Armstrong (1983) for further details.

There is first the difficulty of uninstantiated lawlike generalizations. To exclude all such generalizations from law status is too severe; witness Newton's First Law ("If the net impressed force acting on a massive body is zero, then the body moves inertially") whose antecedent is very unlikely to be instanced in a universe well populated by massive particles obeying Newton's Law of Universal Gravitation.[7] Contrariwise, to welcome in all uninstanced lawlike generalizations has even more unwelcome consequences, for then the vacuity of the antecedent condition would mean that $(x)\ (Fx \supset Gx)$, $(x)\ (Fx \supset G'x)$, $(x)\ (Fx \supset G''x)$, etc., where Gx, $G'x$, $G''x$, etc., may be pairwise incompatible, are all laws. Such contrary "laws" are intuitively repugnant, and they pose difficulties for the widely accepted view that laws license subjunctive conditionals. If *o* (which as a matter of fact is non-*F*) were *F*, would it be *G*, or *G'*, or *G''*, etc.? A uniform treatment of uninstanced generalizations is unacceptable. But what basis does the naive regularity theorist have for treating such generalizations differentially?

The problem of uninstanced generalizations pales beside the problem of instanced lawlike generalizations which, by the judgments of philosophical intuition and the history of science, do not correspond to laws. Reichenbach's old example still suffices: "All bodies of pure gold have a mass of less than 10,000 kg." This statement is general in form; its predicates are surely kosher; and it is widely instanced. But even if we were assured that it is true, we would not regard it as expressing a law. Nor would it help to be given the further assurance that the known instances are not exhaustive or that there are an infinite number of instances (so that the generalization is not equivalent to a finite conjunction of singular statements). Such assurances would do nothing to convince us that Reichenbach's generalization is a generalization of law rather than of fact.

Can the separation of generalizations of law from generalizations of fact be effected by *de re* features of regularities, or as empiricists are we forced to grasp the safety cord of Hume's felt determination definition? My strategy for answering this question will be, first, to state general constraints on an empiricist account of laws and, second, to explore the prospects and problems of constructing a more appealing regularity account within the confines of these constraints.

3. The Empiricist Constraints

I will state the constraints in a form that may be distasteful to some empiricists. But to mix a metaphor, while I can genuflect before Hume's altar with the best of them, I am no knee-jerk empiricist. I see no reason to deny ourselves whatever analytical tools may help to shape the issues into a manageable form. Without further apology, I state the basic or 0-th empiricist constraint as

(E0) Laws are contingent, i.e., they are not true in all possible worlds.

Next, I propose two forms for further constraints:

(F1) For any possible worlds W_1, W_2, if W_1 and W_2 agree on ___, then W_1 and W_2 agree on laws.

(F2) For any possible worlds W_1, W_2, if W_1 and W_2 agree on laws, then W_1 and W_2 agree on ___.

The blanks are to be filled in by non-question-begging empirical features. "All Humean regularities" is such a feature, but if used as the filling in both blanks it seems that the conjunction of the resulting constraints forces us back to the naive regularity account.

The filling I prefer for the blank in (F1) produces the following constraint:

(E1) For any W_1, W_2, if W_1 and W_2 agree on all occurrent facts, then W_1 and W_2 agree on laws.

I will refer to (E1) as the empiricist loyalty test on laws, for I believe it captures the central empiricist intuition that laws are parasitic on occurrent facts. Ask me what an occurrent fact is and I will pass your query on to empiricists. But in lieu of a reply, I will volunteer that the paradigm form of a singular occurrent fact is: the fact expressed by the sentence $P(o, t)$, where "P" is again a suitably kosher predicate, "o" denotes a physical object or spatial location, and "t" denotes a time. . . . There may also be general occurrent facts (I think there are), but these presumably are also parasitic on the singular occurrent facts. Conservative empiricists may want to restrict the antecedent of (E1) so as to range only over observable facts while more liberal empiricists may be happy with unobservable facts such as the fact that quark q is charming and flavorful at t. In this way we arrive at many different versions of the loyalty test, one for each persuasion of empiricist.

The well-known motivations for (E1) fall into two related categories. There are ontological argument and sloganeering ("The world is a world of occurrent facts"), the two often being hard to distinguish. Then there are epistemological arguments and threatenings, the most widely used being the threat of unknowability, based on two premises: we can in principle know directly or non-inferentially only (some subset of) occurrent facts; what is underdetermined by everything we can in principle know non-inferentially is unknowable in principle. . . . The argument connects back to the ontological if we add the further premise that what isn't knowable in principle *isn't* in principle.[8]

Finding a filling for the blank in (F2) which produces a defensible but not toothless constraint is more difficult. Consider:

(E2) For any W_1, W_2, if W_1 and W_2 agree on laws, then W_1 and W_2 agree on regularities entailed by the laws.

This lacks bite in the case of non-probabilistic laws, but it is of some help in separating some of the views on the nature of physical probabilities. Hardcore frequency theorists would hold if W_1 and W_2 agree on lawful probabilities and if they both contain infinite repetitions of the relevant chance experiment, then they must agree on limiting relative frequencies; but the hardcore propensity theorist will counter that while agreement of relative frequencies is likely, it is not mandatory. However, a more important difference between frequency and propensity theorists concerns (E1) and the grounding of physical probabilities on occurrent facts. . . . Little use will be made of (E2) in what follows.

Two things remain uncaptured by (E0)—(E2). Neither can be stated in the form of tidy constraint, but nonetheless each is an important part of the empiricist conception of laws. The first is the intuition that appropriate qualitative and quantitative differences in particular occurrent fact and general regularity make for differences in laws (E3). The second intuition is that there is a democracy of facts and regularities in that each has a vote in electing the laws (E4). The worry about (E4), of course, is whether democracy can prevail without degenerating into the mob rule of the naive regularity view. And the problem with (E3) is that it seems impossible to specify ahead of time in a content and context free manner what counts as an appropriate difference. That (E3) and (E4) are painfully vague does not mean that they are useless; on the contrary, a good check on any proposed implementation of (E0) and (E1) is how well it makes sense of (E3) and (E4).

In the next section I will review what I take to be the most promising approach to laws which fulfills the above constraints and which maintains firm contact with Hume's constant conjunction idea. I will capitalize the *e* in "empiricism" to indicate my brand of empiricism. There are other and perhaps better brands, but this one recommends itself as a useful foil.

4. Mill, Ramsey, and Lewis

John Stuart Mill, as thoroughgoing an Empiricist as they come, was no naive regularity theorist. Humean uniformities are often called laws in common parlance; but scientific parlance is quite another thing:

> Scientifically speaking, that title [Laws of Nature] is employed in a more restricted sense to designate the uniformities when reduced to their most simple expression. (1904, p. 229)

This "restricted sense" is explained more fully a little further on:

> According to one mode of expression, the question, What are laws of

> nature? may be stated thus: What are the fewest and simplest assumptions, which being granted, the whole existing order of nature would result? Another mode of starting the question would be thus: What are the fewest general propositions from which all the uniformities which exist in the universe might be deductively inferred? (1904, p. 230)

When allowance is made for the fact that Mill assumed determinism, his conception of laws seems to correspond exactly to Frank Ramsey's, or rather to David Lewis' de-epistemologized version of Ramsey. Ramsey's dictum was that laws are "consequences of those propositions which we should take as axioms if we knew everything and organized it as simply as possible in a deductive system" (1978, p. 138). Lewis suggests we expunge the reference of knowledge in favor of conditions on deductive systems, known or unknown: ". . . a contingent generalization is a law of nature if and only if it appears as a theorem (or axiom) in each of the true deductive systems that achieves a best combination of simplicity and strength" (1973, p. 73). Deductive systems are

> deductively closed, axiomatizable sets of true sentences. Of these true deductive systems, some can be axiomatized more *simply* than others. Also some of them have more *strength*, or *information content*, than others. The virtues of simplicity and strength tend to conflict . . . What we value in a deductive system is a properly balanced combination of simplicity and strength—as much of both as truth and our way of balancing will permit. (1973, p. 73)

Many other forms of the idea that lawhood attaches to individual regularities only via their membership in a coherent system of regularities could be cited,[9] but for the moment let us stick with the Mill–Ramsey–Lewis version and enumerate its virtues.

I take it as evident that the M–R–L account does satisfy the basic Empiricist constraints (E0) and (E1), does provide for the democracy of facts and regularities (E4) without surrendering to the mob rule of the naive regularity account, and does provide a framework for understanding what sorts of differences in particular fact and general regularity make for differences in laws (E3). It also has the virtue of explaining why laws have or tend to have various "lawlike" characteristics, such as universality. . . . It allows in some vacuous generalizations without opening the floodgates to all. And it connects in a direct and natural way to the actual practice of scientific theorizing or at least to the most widely held reconstruction of the practice in the form of the hypothetico-deductive method. In fact, in much of the current literature on the structure and function of scientific theories, "theory" and "deductive system" can be freely interchanged.

5. Deductive Systematization: A Closer Look

It is no criticism of M–R–L to note that simplicity and allied notions such as coherence and systematization are vague and slippery, for so is the notion of laws of nature. The question is whether the vaguenesses and slippages match. That old nemesis, Reichenbach's gold lump generalization, gives pause. If this generalization is to be counted out as a law by the lights of M–R–L it is because it is not an axiom or theorem in the best (or each of the best) overall deductive systems for our world. Consider then what would happen if we were to add it as an additional axiom. There would, by hypothesis, be a gain in strength. And, presumably, there would also be a loss in simplicity. The loss must, *pace* M–R–L, outweigh the gain. I will not say otherwise. But I do say that it is not compellingly obvious that the scales tip in this way while it is compelling that Reichenbach's generalization is not to be counted as a law.

The trouble here may not lie with the squishy notion of simplicity but with the seemingly more solid notion of strength. Lewis suggests strength be measured by information content, and that is as good a measure as any if we are interested in strength *per se*. But actual scientific practice speaks in favor not of strength *per se* but strength in intended applications; for dynamical laws this means strength as measured by the amount of occurrent fact and regularity that is systematized or explained relative to appropriate initial and/or boundary conditions. The advantage offered by deterministic generalizations here is obvious: while they can be strengthened *per se*, they are, in their intended applications, as strong as strong can be; for given the state of the system at any instant, they entail everything true of the system, past, present, and future, and any other generalization is either incompatible or adds nothing to applied strength. This helps to explain why we feel confident that in having discovered a simple set of true deterministic relationships we have discovered laws. This is not to say that determinism is either necessary or sufficient for a good trade-off between simplicity and applied strength. If a deterministic set of generalizations can be constructed only at the price of very high complexity, then the scales may tip against determinism; but typically the complexity must be great indeed before the tip becomes pronounced. And when no set of true deterministic generalizations is available, many different compromises between simplicity and strength may recommend themselves. This helps to explain why, independently of ontological considerations, determinism has been prized as a methodological guide to scientific theorizing.

What started as an objection to the M–R–L account has turned into a plus. Another plus comes from reflection on the notion of chaos. It is tempting to define chaos as the absence of any pattern or regularity, but [later] discussion . . . will cast doubt on the coherence of this idea. However, chaos as the non-existence of laws is explicable on the M–R–L account. This form of chaos need not require that all regularity is absent but only that the existent regularities are sufficiently weak and messy that there are no good compromises between strength and simplicity.

In closing, I have to confess to a real worry about the M–R–L account, or rather to the linguistic version I have been reviewing. Given a choice of language—primitive predicates and logical apparatus—we may be able to identify a best overall deductive system. But different choices of language may promote different candidates for the role of best system. These candidates may be incommensurable, not admitting meaningful comparisons of simplicity and strength. Or else they may be commensurable and equally good in their different ways, forcing us to say either that there are no laws since there are no non-trivial axioms or theorems common to all the best systems, or that the laws are relative to a choice of language. These worries can be diminished by refusing to give in to the logical positivists' fear of the ontological and their flight to the linguistic. Recall that my canonical formulation of determinism assumes that the possible worlds can be characterized in terms of space-time magnitudes. Worlds are thus isomorphic to sets of basic propositions, each asserting that the value of such-and-such a magnitude takes a value of so-and-so at thus-and-such a spatio-temporal location. The laws of the actual world are then the propositions that appear in each of the deductively closed systems of general propositions that achieve a best systematization of the basic propositions true of the actual world. So while different systems may employ different concepts, there will of necessity be a strong common core.

. . .

Notes

1 These are the versions of the constant conjunction and felt determination definitions Hume gives in the *Treatise*. The definitions are repeated with some significant changes in the *Enquiry*.

2 This counterfactual definition does not appear in the first edition of the *Enquiry*.

3 Here I am following Suchting (1974).

4 See especially Secs. 13 and 15 of Bk. I of the *Treatise*.

5 For a more detailed discussion of this point, see Suchting (1974) and Armstrong (1983).

6 For an attempt to fill in some of the details, see Reichenbach (1954).
7 Unlikely but not impossible since the net impressed force acting on a particle can be zero even when other particles are present. But the point is that we do not want the lawfulness of Newton's First Law to turn on such a happenstance.
8 This last move would yield the stronger version of (E1); namely, if W_1 and W_2 agree on all occurrent facts, then they are the same world.
9 See, for example, Braithwaite (1960), Berofsky (1968) and Tondl (1973); see Suchting (1974) for a critical discussion.

References

Armstrong, D. M. (1983), *What is a Law of Nature?* (Cambridge: Cambridge University Press).

Ayer, A. J. (1956), "What is a Law of Nature?" *Revue Internationale de Philosophie 10*, 144–165.

Berofsky, B. (1968), "The Regularity Theory," *Noûs* 2, 315–340.

Braithwaite, R. B. (1960), *Scientific Explanation* (New York: Harper Torch Books).

Goodman, N. (1955), *Fact, Fiction and Forecast* (Cambridge, MA: Harvard University Press).

Hume, D. (1973), *Treatise of Human Nature* (Oxford: Oxford University Press).

Hume, D. (1975), *Enquiry Concerning Human Nature* (Oxford: Clarendon Press).

Lewis, D. (1973), *Counterfactuals* (Cambridge, MA: Harvard University Press).

Mill, J. S. (1904), *A System of Logic* (New York: Harper and Row).

Ramsey, F. P. (1978), *Foundations of Mathematics* (Atlantic Highlands, NJ: Humanities Press).

Reichenbach, H. (1954), *Nomological Statements and Admissible Operations* (Amsterdam: North-Holland).

Rescher, N. (1970), *Scientific Explanation* (New York: Free Press).

Suchting, W. A. (1974), "Regularity and Law," in R. S. Cohen and M. W. Wartofsky (eds.) *Boston Studies in the Philosophy of Science*, Vol. 14 (Dordrecht: D. Reidel).

Tondl, L. (1973), *Scientific Procedures* (Dordrecht: D. Reidel).

QUESTIONS

1 A famous French scientist, historian, and philosopher Pierre Duhem (1861–1916) maintained that explaining phenomena is metaphysical and hence not a proper part of science. Thus he wrote: "A physical theory is not an explanation. It is a system of mathematical propositions, deduced from a small number of principles, which aim to represent as simply, as completely, and as exactly as possible a set of experimental laws" (Duhem 1954, p. 19). Take issue with his claim.

2 Is *deduction* from laws and particular facts *necessary* for all scientific explanations? (Must all proper scientific explanations be such deductions?) Is deduction from laws and particular facts *sufficient* for scientific explanation? (Is anything that is a deduction from empirically true laws and statements of empirical fact a scientific explanation?)
3 Could you defend the spirit, if not the letter, of Hempel's model against some of the standard objections, such as the asymmetry problem, the low-probability problem (the "paresis problem"), and the relevance problem (the "hexed salt problem")?
4 Some philosophers and scientists think that explanatory power of theories is evidential: that the theory's ability to explain phenomena, not just to "get the facts right," gives one stronger reason to believe in the truth of the theory. Which of the authors represented in this section would agree with this claim? Which of them would disagree? Why?
5 What are counterfactuals and why are they such a headache? How are they relevant to the issue of explanation? How do they relate to the laws of nature?
6 Reflect on the question "What is a law of nature?" Are you satisfied with the answer that the Mill–Ramsey–Lewis theory provides?

FURTHER READING

Hempel's classic works on explanation are collected in his *Aspects of Scientific Explanation* (1965). Salmon and Kitcher (1989) discuss competing accounts of explanation. Useful anthologies on explanation include Pitt (1988) and Ruben (1993).

Lewis (1973) is a classic treatment of counterfactuals, which also contains a brief statement of the Mill–Ramsey–Lewis theory, which is elaborated in Lewis (1983). Armstrong (1983) is a devastating critique of the simple regularity theory of laws and an exposition of the theory of laws as relations among universals. Van Fraassen's provocative *Laws and Symmetries* (1989) defends the view that there aren't any laws of nature at all. Weinert (1995) is a recent collection of articles debating pros and cons of various accounts of scientific laws.

PART III

SCIENTIFIC THEORIES AND CONCEPTUAL CHANGE

INTRODUCTION

What is distinctive about a theory is that it goes beyond the explanations of particular phenomena to explain these explanations. When particular phenomena are explained by an empirical generalization, a theory will go on to explain why the generalization obtains, and to explain its exceptions—the conditions under which it fails to obtain. When a number of generalizations are uncovered about the phenomena in a domain of enquiry, a theory may emerge which enables us to understand the diversity of generalizations as all reflecting the operation of a single or small number of processes. Theories, in short, unify, and they do so almost always by going beyond, beneath and behind the phenomena empirical regularities report to find underlying processes that account for the phenomena we observe. This is probably the source of the notion that what makes an explanation scientific is the unification it effects. For theories are our most powerful explainers, and they operate by bringing diverse phenomena under a small number of fundamental assumptions.

How exactly do the parts of a theory work together to explain a diversity of different phenomena? One answer, reflected in the selection by Ernest Nagel, "Experimental Laws and Theories," has been traditional in science and philosophy since the time of Euclid. Indeed, it is modeled on Euclid's own presentation of geometry. Like almost all mathematicians and scientists before the twentieth century, Euclid held geometry to be the science of space and his *Elements* to constitute a theory about the relations among points, lines and surfaces in space.

Euclid's theory is an axiomatic system. That is, it consists of a small set of postulates or axioms—propositions not proved in the axiom system but assumed to be true within the system—and a large number of theorems derived from the axioms by deduction in accordance with rules of logic. Besides the axioms and theorems there are definitions of terms, such as straight line, nowadays usually defined as the shortest distance between two points, and circle, the locus of points equidistant from a given point. The definitions of course employ terms not defined in the axiomatic system, like point and distance. If every term in the theory were defined, the number of definitions would be endless, so some terms will have to be undefined or "primitive" terms.

It is critical to bear in mind that a statement which is an axiom in one axiomatic system may well be a theorem derived from other assumptions in another axiom system, or it may be justified independently of any other axiom system whatever. Indeed, one set of logically related statements can be organized in more than one axiom system, and the same statement might be an axiom in one system and a theorem in another. Which axiomatic system one chooses in a case like this cannot be decided by considerations of logic. In the case of Euclid's five axioms, the choice reflects the desire to adopt the simplest statements that would enable us conveniently to derive certain particularly important further statements as theorems. Euclid's axioms have always been accepted as so evidently true that it was safe to develop geometry from them. But, strictly speaking, to call a statement an axiom is not to commit oneself to its truth, but simply to identify its role in a deductive system.

According to Nagel, in the axiomatic system that constitutes the structure of a theory, the experimental laws are the empirical generalizations about observations, which can be derived as theorems, and thereby explained by the underived "axioms" of the theory, its theoretical laws, often expressed in terms that do not name or describe observational phenomena. It is these theoretical claims involving unobservable entities that describe underlying processes. But the distinction between theoretical laws and empirical ones, and between the theoretical vocabulary in which the former are expressed and the observational vocabulary in which the latter are stated, raises a profound philosophical problem, as this and the next set of readings reveal.

The problem is that of reconciling the indispensability of theoretical laws and concepts in scientific explanation with the fact that we cannot have direct knowledge of the objects and properties these concepts and the laws expressed in them refer to. On the one hand, we cannot explain the experimental laws we have uncovered, or explain the functioning of the technologically complex instruments we have created based on scientific knowledge without appealing to the molecules, atoms, and subatomic entities out of which they are composed. And yet, according to the standpoint of empiricism, "the official epistemology" of science, we can have no knowledge of these things and their properties. How can it be that for science we need to invoke concepts whose instances are in principle unknowable?

Nagel was sensitive to the problem and attempted to solve it by allowing that theories provide implicit definitions of their own technical terms in those statements of the theory that link the theoretical term to observational ones. These "bridge principles,"or "correspondence rules," are said to provide a "partial interpretation" of theoretical terms which enables us to subject theoretical laws to test because the observational terms to which theoretical ones are linked figure in experimental claims we can test empirically. (The issue of testing and confirmation of scientific theories is complex and is discussed in more detail in Part V.)

The axiomatic analysis of the structure of theories has the further advantage that it provides an account of how science advances and improves both its explanatory range and predictive power over time. Newer, broader, deeper, or more fundamental theories supersede older narrower, shallower theories by showing the phenomena they describe to be special cases, or obvious consequences of the processes which the broader and/or deeper theories describe. And the way they do this is by showing that the axioms or theoretical laws of older theories can be derived as theorems, and thereby explained by the newer theories along with the experimental laws which the narrower theory already explains. This relation between theories is referred to as the "reduction" of narrower theories to broader and deeper ones. It has long been held to characterize the sequence of theories in physics from the physics of Galileo and Kepler through Newtonian mechanics to Einstein's special and general theories of relativity. If the basic idea of reduction is correct, it gives a precise explication of the sense in which conceptual change in physics consists in the cumulation of knowledge producing a succession of theories, each a closer approximation to the truth.

Paul Feyerabend's "Explanation, Reduction, and Empiricism" explains how the axiomatic approach, and its distinction between theoretical laws and experimental ones, generates this reductionistic picture of scientific progress. But then he goes on to question its accuracy both as an account of the relations between theories and as a history of scientific progress. He poses a challenge to the very possibility of comparing competing scientific theories by bringing them to a "common denominator." The seriousness of this challenge will become clearer in Part VI. It also raises profound problems in the philosophy of language. Philip Kitcher's "Theories, Theorists and Theoretical Change" explores the central issues in involved in conceptual change and argues that some of the problems mentioned above may be exaggerated. This paper bears rereading, especially after a study of the selections in Part VI.

9

Ernest Nagel, "Experimental Laws and Theories"

This selection, excerpted from Ernest Nagel's (1901–1985) classic book (1961), introduces the reader to the "received view" (motivated by the development of Logical Positivism) of scientific theories as partially interpreted axiomatic systems.

I. Three Major Components in Theories

A reasonably good case can . . . be made for distinguishing experimental laws from theories, even if the distinction is not a precise one. We shall in any event adopt the distinction . . . in part also because it permits us to segregate under a convenient rubric important problems that pertain primarily to explanatory hypotheses having the generic characteristics of those we are calling "theories." We shall now look more closely at the articulation of theories, and examine in what manner they are related to matters that are usually regarded in scientific practice as objects of observation and experiment.

For the purpose of analysis, it will be useful to distinguish three components in a theory: (1) an abstract calculus that is the logical skeleton of the explanatory system, and that "implicitly defines" the basic notions of the system; (2) a set of rules that in effect assign an empirical content to the abstract calculus by relating it to the concrete materials of observation and experiment; and (3) an interpretation or model for the abstract calculus, which supplies some flesh for the skeletal structure in terms of more or less familiar conceptual or visualizable materials. We will develop these distinctions in the order just mentioned. However, they are rarely given explicit formulation in actual scientific practice, nor do they correspond to actual

E. Nagel, *The Structure of Science: Problems in the Logic of Scientific Explanation*, 1961, pp. 79–105 (excerpts). New York and Burlingame: Harcourt, Brace & World, Inc.

stages in the construction of theoretical explanations. The order of exposition here adopted must therefore not be assumed to reflect the temporal order in which theories are generated in the minds of individual scientists.

1. A scientific theory (such as the kinetic theory of gases) is often suggested by materials of familiar experience or by certain features noted in other theories. Theories are in fact usually so formulated that various more or less visualizable notions are associated with the nonlogical expressions occurring in them, that is, with "descriptive" or "subject matter" terms like "molecule" or "velocity," which, unlike logical particles such as "if-then" and "every," do not belong to the vocabulary of formal logic but are specific to discourse about some special subject matter. Nevertheless, the nonlogical terms of a theory can always be dissociated from the concepts and images that normally accompany them by ignoring these latter, so that attention is directed exclusively to the logical relations in which the terms stand to one another. When this is done, and when a theory is carefully codified so that it acquires the form of a deductive system (a task which, though often difficult in practice, is realizable in principle), the fundamental assumptions of the theory formulate nothing but an abstract relational structure. In this perspective, accordingly, the fundamental assumptions of a theory constitute a set of abstract or uninterpreted postulates, whose constituent nonlogical terms have no meanings other than those accruing to them by virtue of their place in the postulates, so that the basic terms of the theory are "implicitly defined" by the postulates of the theory. Moreover, insofar as the basic theoretical terms are only implicitly defined by the postulates of the theory, the postulates assert nothing, since they are statement-forms rather than statements (that is, they are expressions having the form of statements without being statements), and can be explored only with the view to deriving from them other statement-forms in conformity with the rules of logical deduction. In short, a fully articulated scientific theory has embedded in it an abstract calculus that constitutes the skeletal structure of the theory.

Some illustrations will help make clear what is meant by saying that the postulates of a theory implicitly define the terms occurring in them. . . .

[Thus] the assumptions that formulate a physical theory such as the kinetic theory of gases provide only an implicit definition for terms like "molecule" or "kinetic energy of molecules." For the assumptions state only the structure of relations into which these terms enter, and thereby stipulate the formal conditions to be satisfied by anything for which those terms can become labels. To be sure, these terms are commonly associated with a set of intuitively satisfying images and familiar notions. In consequence, the terms have a suggestive power that makes them appear meaningful independent of the postulates in which they occur. Nevertheless, what it is to be a molecule, for example, is prescribed by the assumptions of the theory. Indeed, there is no way of ascertaining what is the "nature" of molecules except by examining the postulates of the molecular theory. It is in any event the notion of "molecule" as implicitly defined by the postulates that does the work expected of the theory.

. . .

2. It is clear, however, that if a theory is to explain experimental laws, it is not sufficient that its terms be only implicitly defined. Unless something further is added to indicate how its implicitly defined terms are related to ideas occurring in experimental laws, a theory cannot be significantly affirmed or denied and in any case is scientifically useless. [For example, the] postulates of the kinetic theory of gases do not provide any hint as to what experimentally determinable matters its implicitly defined terms are supposed to signify—even when the term "molecule," for example, is taken to signify an imperceptible particle. If the theory is to be used as an instrument of explanation and prediction, it must somehow be linked with observable materials.

The indispensability of such linkages has been repeatedly stressed in recent literature, and a variety of labels have been coined for them: coordinating definitions, operational definitions, semantical rules, correspondence rules, epistemic correlations, and rules of interpretation. The ways in which theoretical notions are related to observational procedures are often quite complex, and there appears to be no single schema which adequately represents all of them. An example will nevertheless help bring out some important features of such correspondence rules.

The Bohr theory of the atom was devised in order to explain, among other things, experimental laws about the line spectra of various chemical elements. In brief outline the theory postulates the following. It assumes that there are atoms, each of which is composed of a relatively heavy nucleus carrying a positive electric charge and a number of negatively charged electrons with smaller mass moving in approximately elliptic orbits with the nucleus at one of the foci. The number of electrons circulating around the nucleus varies with the chemical elements. The theory further assumes that there are only a discrete set of permissible orbits for the electrons, and that the diameters of the orbits are proportional to h^2n^2, where h is Planck's constant (the value of the indivisible quantum of energy postulated in Max Planck's theory of radiation) and n is an integer. Moreover, the electromagnetic energy of an electron in an orbit depends on the diameter of the orbit. However, as long as an electron remains in any one orbit, its energy is constant and the atom emits no radiation. On the other hand, an electron may "jump" from an orbit with a higher energy level to an orbit with a lower energy level; and when it does so, the atom emits an electromagnetic radiation, whose wave length is a function of these energy differences. Bohr's theory is an eclectic fusion of Planck's quantum hypothesis and ideas borrowed from classical electrodynamic theory; and it has now been replaced by a more satisfactory theory. Nevertheless, the theory was successful in explaining a number of experimental laws of spectroscopy and for a time was a fertile guide to the discovery of new laws.

But how is the Bohr theory brought into relation with what can be observed in the laboratory? On the face of it, the electrons, their circulation in orbits, their jumps from orbits to orbits, and so on, are all conceptions that do not apply to anything manifestly observable. Connections must therefore be introduced between such theoretical notions and what can be identified by way of laboratory procedures. In point of fact, connections of this sort are instituted somewhat as follows. On the basis of the electromagnetic theory of light, a line in the spectrum of an element is associated with an electromagnetic wave whose length can be calculated, in accordance with the assumptions of the theory, from experimental data on the position of the spectral line. On the other hand, the Bohr theory associates the wave length of a light ray emitted by an atom with the jump of an electron from one of its permissible orbits to another such orbit. In consequence, the *theoretical* notion of an electron jump is linked to the *experimental* notion of a spectral line. Once this and other similar correspondences are introduced, the experimental laws concerning the series of lines occurring in the spectrum of an element can be deduced from the theoretical assumptions about the transitions of electrons from their permissible orbits.

3. This example of a rule of correspondence also illustrates what is meant by an interpretation or model for a theory. The Bohr theory is usually not presented as an abstract set of postulates, augmented by an appropriate number of rules of correspondence for the uninterpreted nonlogical terms implicitly defined by the postulates. It is customarily expounded, as in the above sketch, by way of relatively familiar notions, so that instead of being statement-forms the postulates of the theory appear to be statements, at least part of whose content can be visually imagined. Such a presentation is adopted, among other reasons, because it can be understood with greater ease than can an inevitably longer and more complicated purely formal exposition. But in any event, in such an exposition the postulates of the theory are embedded in [an] interpretation.

. . .

II. Rules of Correspondence

We must now call attention to certain features of rules of correspondence that have thus far not been explicitly mentioned.

1. The above example of a rule of correspondence for the Bohr theory of the atom provides a convenient point of departure for noting one such feature. It will be evident that the rule cited in the example does not provide an *explicit* definition of any theoretical notion in the Bohr theory, in terms of predicates used to characterize matters normally said to be observable. The example thus suggests that in general rules of correspondence do not supply such definitions.

Let us make clearer what is involved in this suggestion. When an expression is said to be "explicitly defined," the expression may always be eliminated from any context in which it occurs, since it can be replaced by the defining expression without altering the sense of the context. Thus, the expression "x is a triangle" is explicitly defined by the expression "x is a closed plane figure bounded by three straight line segments." The former (or defined) expression can therefore be eliminated from any context in favor of the latter (or defining) expression; for example, the statement "The area of a triangle is equal to one-half the product of its base and altitude," can be replaced by the logically equivalent statement "The area of a closed plane figure bounded by three straight line segments is equal to one-half of the product of its base and altitude." On the other hand, the theoretical expression in the Bohr theory "x is the wave length of the radiation emitted when an electron jumps from the next-to-the-smallest to the smallest permissible orbit of the hydrogen atom" is not being explicitly defined when it is coordinated with an expression having approximately the form of "y is the line occurring at a certain position in the spectrum of hydrogen." It is indeed patent that the two expressions have quite different connotations. Accordingly, although the rule of correspondence establishes a definite connection between the two expressions, the former cannot be replaced by the latter in such statements as "Transitions of electrons from their next-to-the-smallest to their smallest permissible orbits occur in about ten per cent of hydrogen atoms." Were the indicated replacement attempted, the result would in fact be nonsense.

. . .

Another reason of perhaps even greater weight is that theoretical notions are frequently coordinated by rules of correspondence with more than one experimental concept. As has already been argued, theoretical notions are only implicitly defined by the postulates of a theory (even when the theory is presented by way of a model). There are therefore an unlimited number of experimental concepts to which, as a matter of logical possibility, a theoretical notion may be made to correspond. For example, the theoretical notion of electron transition in the Bohr theory corresponds to the experimental notion of a spectral line; but that theoretical notion can also be coördinated (via Planck's radiation law, which is deducible from the Bohr theory) with experimentally determinable temperature changes in black-body radiation. Accordingly, in those cases in which a given theoretical notion is made to correspond to two or more experimental ideas (though presumably on different occasions and in the context of different problems), it would be absurd to maintain that the theoretical concept is explicitly defined by each of the two experimental ones in turn.

. . .

Let us look . . . more closely at the correspondence between the notion of wave length in the electromagnetic theory of light and the experimental notion of a spectral line. Even a cursory examination shows that the correspondence is not unique. For spectral lines are all of finite breadth, and the resolving power of optical instruments is limited. Accordingly, what is experimentally identified as a spectral line corresponds, not to a unique wave length, but to a vaguely bounded range of wave lengths. And conversely, a theoretically monochromatic beam of light (i.e., a beam of radiation composed of rays all with the same wave length) is coordinated in practice with experimentally determinable spectral lines that have a discernible width and that are therefore produced, from the standpoint of the theory, by polychromatic radiation.

The general point that emerges from these examples is that, though theoretical concepts may be articulated with a high degree of precision, rules of correspondence coordinate them with experimental ideas that are far less definite. The haziness that surrounds such correspondence rules is inevitable, since experimental ideas do not have the sharp contours that theoretical notions possess. This is the primary reason why it is not possible to formalize with much precision the rules (or habits) for establishing a correspondence between theoretical and experimental ideas.

. . .

2. A further point must now be made about the way rules of correspondence serve as links between theoretical and experimental ideas. The sketch given above of the Bohr theory of the atom will again serve to introduce the discussion. According to that account, although there are rules of correspondence for some of the notions employed in the theory, not all the theoretical notions are linked with experimental ideas. For example, there is a rule of correspondence for the theoretical notion of electrons in transition from one permissible orbit to another; but there is no such rule for the

notion of electrons moving with accelerated velocities on an orbit. Similarly, in the kinetic theory of gases, there is no correspondence rule for the theoretical notion of the instantaneous velocity of single molecules, although there is such a rule for the theoretically defined notion of the average kinetic energy of the molecules. Moreover, there is at present a correspondence rule for the notion of the number of molecules in a standard volume of gas under standard conditions of temperature and pressure (Avogadro's number); but Avogadro's number was not determined by experimental means until relatively late in the history of the kinetic theory, and until then there was no rule of correspondence for that theoretical notion.

. . .

Considerations of this . . . kind thus lead us to expect that not every constituent notion in a theory will be linked with some experimental idea by a correspondence rule. In any event, the primary role of many symbols occurring in theories is to facilitate the formulation of a theory with great generality, to make possible logical and mathematical transformations in a relatively simple manner, or to serve as heuristic aids for the extended application of the theory. Illustrations of such symbols are the continuous variables and differential quotients of mathematical physics; these are extensively used, despite the fact that theoretical notions such as mathematically continuous density functions or instantaneous velocities, when they are strictly construed, do not correspond to any experimental concepts. An indefinite number of further examples of such symbols can be found in the locutions used when a theory is embedded in some convenient model—for example, in the language of point-masses of analytical mechanics, of the ether of nineteenth-century electromagnetic theory, of valence bonds of analytic chemistry, or of "wavicles" of current quantum theory.

Since theories are constructed with a view toward explaining a wide variety of experimental laws, it is clear that such an end can in general be achieved only if a theory is so formulated that no reference is made in it to any set of specialized experimental concepts. For otherwise the theory would be limited in its application to situations to which just those concepts are relevant. Indeed, the more comprehensive the range of possible application of a theory, the more meager is its explicitly formulated content with respect to specialized details of some subject matter. Such details are left to be supplied by supplementary assumptions and correspondence rules, introduced as occasion requires when the theory is employed in different experimental contexts. This does not mean, however, that scientific theories tend in the limit to become empty of all content as their range of application becomes more inclusive. It does mean that a theory seeks to formulate a highly general structure of relations that is invariant in a wide variety of experimentally different situations but that can be specialized by augmenting the fundamental postulates of the theory with more restrictive assumptions, so as to yield systematically a series of diversified subordinate structures.

[An] example, though not fully typical of all scientific theories, will illustrate this point, and will thereby make clearer the architecture of at least some theories. The . . . example is taken from the Newtonian theory of mechanics. According to the theory, a change in the momentum of a body (when referred to a suitable spatial frame of reference) is equal to the force acting on the body. This can be written as $ma = F$, where "m" is the mass of the body, "a" its acceleration at an instant, and "F" is the force. A number of very general consequences about the motions of bodies can be formally derived from this fundamental postulate, even though the nature of the force that may act on a body is not stated. However, nothing can be inferred from the equation about the actual motion of a body unless, among other things, further assumptions are introduced about the force that is supposed to be acting—assumptions that in some cases at any rate include a rule of correspondence between the theoretical notion of force and certain experimental ideas. The fundamental postulates of Newtonian theory place very few formal restrictions on the kind of mathematical functions that may be used to express the character of forces. In practice, however, the functions are of a relatively simple kind. For example, in the study of vibratory motions, the general form of the force-function is: $F = Ar + Br^2 + Cr^3 + Dv + Ef(t)$, where "$r$" is the distance of the body from some designated point, "v" is the velocity of the body along this line, "$f(t)$" a function of the time t, and "A," "B," "C," "D," and "E" are arbitrary constants for which different numerical values are assigned according to the problem under consideration. Thus, if A is negative and the remaining constants are zero, the body undergoes simple harmonic motion without frictional resistance; if A and D are both negative and the remaining constants are zero, the body is undergoing damped harmonic motion; if A and D are both negative, E is not zero, B and C both zero, and $f(t)$ a periodic function of the time, the body is undergoing a forced vibration; and so on. In general, by specializing F in various ways, different experimental laws can be deduced from the fundamental equations of Newtonian mechanics.

Although these examples are not paradigmatic of all theories—since not all theories contain parameters that are specialized in the manner just indicated—the examples do illustrate one important way in which theories differ from experimental laws, as well as one technique by which some theories achieve comprehensive generality. For unlike terms occurring in experimental laws, theoretical notions employed in the basic assumptions of a theory may either not be associated with any experimental ideas whatever, or may be associated with experimental ideas that vary from context to context. The possibility of extending a theory to cover fresh subject matter depends in considerable measure upon this feature of theories. These examples also help to enforce the point that a theory remains otiose for scientific inquiry until it is linked by some correspondence rules to experimentally identifiable properties of a subject matter.

10

Paul Feyerabend, "Explanation, Reduction, and Empiricism"

Paul Feyerabend (1924–1994) was probably one of the most provocative philosophers of science in the twentieth century. In his book *Against Method: Outline of an Anarchistic Theory of Knowledge* (1975) he undertook a vigorous attack on the very idea that science has a distinctive methodology demarcating it from non-science and pseudo-science. Although his earlier (and nowadays largely neglected) works can be viewed as preparing the ground for this attack, they are more moderate and still valuable. This selection is excerpted from an 1962 article criticizing Nagel's theory of intertheoretic reduction.

1. Two Assumptions of Contemporary Empiricism

...

Nagel's theory of reduction is based upon two assumptions. The first assumption concerns the relation between the secondary science, i.e., the discipline to be reduced, on the one side, and the primary science, i.e., the discipline to which reduction is made, on the other. It is asserted that this relation is the relation of deducibility. Or, to quote Nagel,

(1) "The objective of the reduction is to show that the laws, or the general principles of the secondary science, are simply logical consequences of the assumptions of the primary science."[1]

The second assumption concerns the relation between the meanings of the primitive descriptive terms of the secondary science and the meanings of the primitive descriptive terms of the primary science. It is asserted that the

P. Feyerabend, "Explanation, Reduction, and Empiricism," in *Minnesota Studies in the Philosophy of Science*, vol. 3, ed. H. Feigl and G. Maxwell, 1962, pp. 29–97 (excerpts). Minneapolis: University of Minnesota Press.

former will not be affected by the process of reduction. Of course, this second assumption is an immediate consequence of (1), since a derivation is not supposed to influence the meanings of the statements derived. However, for reasons which will become clear later, it is advisable to formulate this invariance of meaning as a separate principle. This is also done by Nagel, who says: "It is of the utmost importance to observe that the expressions peculiar to a science will possess meanings that are fixed by its *own* procedures, and are therefore intelligible in terms of its own rules of usage, *whether or not the science has been, or will be, reduced to some other discipline*."[2] Or, to express it in a more concise manner:

(2) Meanings are invariant with respect to the process of reduction.

(1) and (2) admit of two different interpretations, just as does any theory of reduction and explanation: such a theory may be regarded either as a *description* of actual scientific practice, or as a *prescription* which must be followed if the scientific character of the whole enterprise is to be guaranteed. Similarly, (1) and (2) may be interpreted as *assertions* concerning actual scientific practice, or as *demands* to be satisfied by the theoretician who wants to follow the scientific method. Both of these interpretations will be scrutinized in the present paper.

Two very similar assumptions, or demands, play a decisive role in the orthodox theory of explanation, which may be regarded as an elaboration of suggestions that were first made, in a less definite form, by Popper.[3] The first assumption (demand) concerns again the relation between the explanandum, or the laws, or the facts to be explained, on the one side, and the explanans, or the discipline which functions as the basis of explanation, on the other. It is again asserted (required) that this relation is (be) the relation of deducibility. Or, to quote Hempel and Oppenheim

(3) "The explanandum must be a logical consequence of the explanans; in other words, the explanandum must be logically deducible from the information contained in the explanans, for otherwise the explanans would not constitute adequate grounds for the explanation."[4]

Considering what has been said in the case of reduction one would expect the assumption (demand) concerning meanings to read as follows:

(4) Meanings are invariant with respect to the process of explanation.

However, despite the fact that (4) is a trivial consequence of (3), this assumption has never been expressed in as clear and explicit a way as (2).[5] There was even a time when a consequence of (4), viz., the assertion that *observational* meanings are invariant with respect to the process of explanation, seemed to be in doubt. It is for this reason that I have separated (2) from (1), and (4) from (3).

. . .

To sum up: two ideas which are common to both the modern empiricist's theory of reduction and to his theory of explanation are:

(A) reduction or explanation is (or should be) by derivation;
(B) the meanings of (observational) terms are invariant with respect to both reduction and explanation.

. . .

2. Criticism of Reduction or Explanation by Derivation

The task of science, so it is assumed by those who hold the theory about to be criticized, is the explanation, and the prediction, of known singular facts and regularities with the help of more general theories. In what follows we shall assume T′ to be the totality of facts and regularities to be explained, D′ the domain in which T′ makes correct predictions, and T (domain $D' \subset D$) the theory which functions as the basis of explanation. Considering (3) we shall have to demand that T be either strong enough to contain T′ as a logical consequence, or at least compatible with T′ (inside D′, that is). Only theories which satisfy one or the other of the two demands just stated are admissible as explanatia. Or, taking the demand for explanation for granted,

(5) only such theories are admissible (for explanation and prediction) in a given domain which either *contain* the theories already used in this domain, or are at least *consistent* with them.

It is in this form that (A) will be discussed in the present section and in the sections to follow.

. . . [C]ondition (5) is an immediate consequence of the logical empiricist's theory of explanation and reduction, and it is therefore adopted—at least by implication—by all those who defend that theory. However, its correctness has been taken for granted by a much wider circle of thinkers, and it has also been adopted independently of the problem of explanation. Thus, in his essay "Studies in the Logic of Confirmation" C. G. Hempel demands that "every logically consistent observation report" be "logically compatible with the class of all the hypotheses which it confirms," and more especially, he has emphasized that observation reports do "not confirm any hypotheses which contradict each other."[6] If we adopt this principle, then a theory T (see the notation introduced at the beginning of the present section) will be confirmed by the observations confirming a more narrow theory T′ only if it is compatible with T′. Combining this with the principle that a theory is admissible only if it is confirmed to some degree by the evidence available, we at once arrive at (5).

. . .

This discussion will be conducted in three steps. It will first be argued that most of the cases which have been used as shining examples of scientific explanation *do not* satisfy (5) and that it is not possible to adapt them to the deductive scheme. It will then be shown that (5) *cannot* be defended on empirical grounds and that it leads to very unreasonable consequences. Finally, it will turn out that once we have left the domain of empirical generalizations, (5) *should not* be satisfied either. In connection with this last, methodological step, the elements of a positive methodology for theories will be developed, and the historical, psychological, and semantical aspects of such a methodology will be discussed. Altogether the three steps will show that (A) is in disagreement both with actual scientific practice and with reasonable methodological demands. I start now with the discussion of the *actual* inadequacy of (5).

3. The First Example

A favorite example of both reduction and explanation is the reduction of what Nagel calls the Galilean science to the physics of Newton,[7] or the explanation of the laws of the Galilean physics on the basis of the laws of the physics of Newton. By the Galilean science (or the Galilean physics) is meant, in this connection, the body of theory dealing with the motion of material objects (falling stone, penduli, balls on an inclined plane) near the surface of the earth. A basic assumption here is that the vertical accelerations involved are constant over any finite (vertical) interval. Using T′ to express the laws of this theory, and T to express the laws of Newton's celestial mechanics, we may formulate Nagel's assertion to the effect that the one is reducible to the other (or explainable on the basis of the other) by saying that

(6) $T \,\&\, d \vdash T'$

where d expresses, in terms of T, the conditions valid inside D′. In the case under discussion d will include description of the earth and its surroundings (supposed to be free from air; we shall also abstract from all those phenomena which are due to the rotation of the earth and whose inclusion would strengthen, rather than weaken our case), and reference will be made to the fact that the variation H of the height above ground level in the processes described is very small if compared with the radius R of the earth.

As is well known (6) cannot be correct: as long as H/R has some finite value, *however small*, T′ will not follow (logically) from T and d. What will follow will rather be a law, T″, which, while being experimentally indistinguishable from T′ (on the basis of the experiments which formed the inductive evidence for T′ in the first place), is yet inconsistent with T′. If, on the other hand, we want to derive T′ precisely, then we must replace d by a statement which is patently false, as it would have to describe the conditions in the close neighborhood of the earth as leading to a vertical acceleration that is constant over a finite interval of vertical distance. It is therefore impossible, *for quantitative reasons*, to establish a deductive relationship between T and T′, or even to make T and T′ compatible. This shows that the present example is not in agreement with (5) and is, therefore, also incompatible with (A), (1), and (3).

. . .

4. Reasons for the Failure of (5) and (3)

The basic argument is really very simple, and it is very surprising that it has not been used earlier. It is based upon the fact that *one and the same set of observational data is compatible with very different and mutually inconsistent theories*. This is possible for two reasons: first, because theories, which are universal, always go beyond any set of observations that might be available at any particular time; second, because the truth of an observation statement can always be asserted within a certain margin of error only. The first reason allows for theories to differ in domains where experimental results are not yet available. The second reason allows for such differences even in those domains where observations have been made, provided the differences are restricted to the margin of error connected with the observations. Both reasons taken together sometimes allow considerable freedom in the construction of our theories.

. . .

. . . [B]ecause of the latitude which experience allows the theoretician, and because of the different way in which this latitude will be exercised by thinkers of different tradition, temperament, and interests, it is to be expected that two different theories, and especially two theories of a different degree of generality, will be inconsistent with each other even in those cases where both are confirmed by the set [of experimental data]. In this argument it was assumed that the experimental evidence which inside D′

confirms T and T′ is the same in both cases. Although this may be so in the specific example discussed, it is certainly not true in general. Experimental evidence does not consist of facts pure and simple, but of facts analyzed, modeled, and manufactured according to some theory.

The first indication of this manufactured character of the evidence is seen in the corrections which we apply to the readings of our measuring instruments, and in the selection which is made among those readings. Both the corrections and the selection made depend upon the theories held, and they may be different for the theoretical complex containing T, and for the theoretical complex containing T′. Usually T will be more general, more sophisticated, than T′, and it will also be invented a considerable time after T′. New experimental techniques may have been introduced in the meantime. Hence, the "facts," within D′, which count as evidence for T will be different from the "facts," within D′, which counted as evidence for T′ when the latter theory was first introduced. An example is the very different manner in which the apparent brightness of stars was determined in the seventeenth century and is determined now. This is another important reason why T usually will not satisfy (5) with respect to T′: not only are T and T′ connected with different theoretical ideas leading to different predictions even in the domain where they overlap and are both confirmed, but the better experimental techniques and the improved theories of measurement will usually provide evidence for T which is different from the evidence for T′ even within the domain of common validity. In short: introducing T very often leads to recasting the evidence for T′. The demand that T should satisfy (5) with respect to T′ would in this case imply the demand that new and refined measurements not be used, which is clearly inconsistent with empiricism.

. . .

5. Second Example: The Problem of Motion[8]

. . .

. . . [T]he theory which was most influential in the Middle Ages . . . was Aristotle's theory of motion as the actualization of potentiality. According to Aristotle

(7) "motion is a process arising from the continuous action of a source of motion, or a 'motor,' and a 'thing moving.'"[9]

This principle, according to which any motion (and not only accelerated motion) is due to the action of some kind of force, can be easily supported by such common observations as a cart drawn by a horse and a chair pushed around by an angry husband. It gets into difficulties when one considers the motion of things thrown: stones continue to move despite the fact that contact with the motor apparently ceases when they leave the hand. Various theories have been suggested to eliminate this difficulty. From the point of view of later developments, the most important one of these theories is the impetus theory. The impetus theory retains (7) and the general background of the Aristotelian theory of motion. Its distinction lies in the specific assumptions it makes concerning the causes that are responsible for the motion of the projectile. According to the impetus theory, the motor (for example the hand) transfers upon the projectile an inner moving force which is responsible for its continuation of path, and which is continually decreased by the resisting air and by the gravity of the projectile. A stone in empty space would therefore either remain at rest or move (along a straight line)[10] with constant speed, depending on whether its impetus is zero or possesses a finite value.

At this point a few words must be said about the characterization of locomotion. The question as to its proper characterization was a matter of dispute. To us it seems quite natural to characterize motion by space transversed, and, as a matter of fact, one of the suggested characterizations did just this: it defined motion kinematically by reference to space transversed. This apparently very simple characterization needs further specification if an account is to be given of nonuniform movements where the distinction becomes relevant between average velocity and instantaneous velocity. Compared with the actual space transversed by a given body, the instantaneous velocity is a rather abstract notion since it refers to the space that would be transversed if the velocity were to retain constancy over a finite interval of time.

Another characterization of motion is the dynamical. It defines motion in terms of the forces which bring it about in accordance with (7). Adopting the impetus theory the motion of a stone thrown would have to be characterized by its inherent impetus, which pushes it along until it is exhausted by the opposing forces of friction and gravity.

Which characterization is the better one to take? From an operationalistic point of view (and we shall adopt this point of view, since we want to follow the empiricist as far as possible), the dynamical characterization is definitely to be preferred: while it is fairly easy to observe the impetus enclosed in a moving body by bringing it to a stop in an appropriate medium (such as soft wax) and then noting the effect of such a maneuver, it is much more difficult, if not nearly impossible, to arrange matters in such a way that from a given moment on, a non-uniformly moving object assumes a constant speed with a value identical with the value of the instantaneous velocity of the object at that moment and then to watch the effect of this procedure.

With the use of the dynamical characterization, the "inertial law" pronounced above reads as follows:

(8) The impetus of a body in empty space which is not under the influence of any outer force remains constant.

Now, in the case of inertial motions (8) gives correct predictions about the behavior of material objects. According to (3), explanation of this fact will involve derivation of (8) from a theory and suitable initial conditions. Disregarding the demand for explanation, we can also say, on the basis of (5), that any theory of motion that is more general than (8) will be adequate only if it contains (8) which, after all, is a very basic law. According to (2), the meanings of the key terms of (8) will be unaffected by such a derivation. Assuming Newton's mechanics to be the primary theory, we shall therefore have to demand that (8) be derivable from it *salva significatione*. Can this demand be satisfied?

At first sight it would seem that it is much easier to derive (8) from Newton's theory than it is to establish the correctness of (6): as opposed to Galileo's law (8) is not in quantitative disagreement with anything asserted by Newton's theory. Even better: (8) seems to be identical with Newton's first law so that the process of derivation seems to degenerate into a triviality.

In the remainder of the present section, it will be shown that this is not so and that it is impossible to establish a deductive relationship between (8) and Newton's theory. Later on this will be the starting point of our criticism of (B).

Let me repeat, before beginning the argument, that (8), *taken by itself*, cannot be attacked on empirical grounds. Indeed, we have indicated a primitive method of measurement of impetus, and the attempt to confirm (8) by using this method will certainly show that within the domain of error connected with such crude measurements, (8) is perfectly all right. It is, therefore, quite in order to ask for the explanation, or the reduction, of (8), and the failure to arrive at a satisfactory solution of this task cannot be blamed upon the empirical inadequacy of (8).

We now turn to an analysis of the main terms of (8). According to Nagel the meaning of these terms is to be regarded as "fixed" by the procedures and assumption of the impetus theory, and any one of them is "therefore intelligible in terms of its own rules of usage."[11] What are these meanings, and what are the rules which establish them?

Take the term "impetus." According to the theory of which (8) is a part, the impetus is the force responsible for the movement of the object that has ceased to be in direct contact, by push, or by pull, with the material mover. If this force did not act, i.e., if the impetus were destroyed, then the object would cease to move and fall to the ground (or simply remain where it is, in case the movement were on a frictionless horizontal plane). A moving object which is situated in empty space and which is influenced neither by gravity nor by friction is not outside the reach of any force. It is pushed along by the impetus, which may be pictured as a kind of inner principle of motion (similar, perhaps, to the vital force of an organism which is the inner principle of *its* motion).

We now turn to Newton's celestial mechanics and the description, in terms of this theory, of the movement of an object in empty space. (Newton's theory still retains the notion of absolute space and allows therefore for such a description to be formed.) Quantitatively, the same movement results. But can we discover in the description of this movement, or in the explanation given for it, anything resembling the impetus of (8)? It has been suggested that the momentum of the moving object is the perfect analogue of the impetus. It is correct that the measure of this magnitude (viz., mv) is identical with the measure that has been suggested for the impetus.[12] However, it would be very mistaken if we were, on that account, to identify impetus and momentum. For whereas the impetus is supposed to be something that pushes the body along,[13] the momentum is the result rather than the cause of its motion. Moreover, the inertial motion of classical mechanics is a motion which is supposed to occur by itself, and without the influence of any causes. After all, it is this feature which according to most historians, radical empiricists included, constitutes one of the main differences between the Aristotelian theory and the celestial mechanics of the seventeenth, eighteenth, and nineteenth centuries; in the Aristotelian theory, the natural state in which an object remains without the assistance of any causes is the state of rest. A body at rest (in its natural place, we should add) is not under the influence of any forces. In the Newtonian physics it is the state of being at rest or in uniform motion which is regarded as the natural state. This means, of course, the explicit denial of a force such as the impetus is supposed to represent.

Now this denial need not mean that the concept of such a force cannot be formed within Newton's mechanics. After all, we deny the existence of unicorns and use in this denial the very concept of a unicorn. Is it then perhaps possible to define a concept such as impetus in terms of the theoretical primitives of Newton's theory? The surprising fact is that any attempt to arrive at such a definition leads to disappointment (which shows, by the way, that theories such as Newton's are expressed in a language that is much more tightly knit than is the language of everyday life). I have already pointed out that the momentum, which would give us the correct mathematical value, is not what we want. What we want is a *force* that acts upon the isolated object and is responsible for its motion. The concept of such a force can of course be formed within Newton's theory. But considering (a) that the movement under review (the inertial movement) occurs with constant velocity, and (b) Newton's second law, we obtain in all relevant cases zero for the *value* of this force which is not the measure we want. A positive measure is obtained only if it is assumed that the movement occurs in a resisting medium (which is, of course, the original Aristotelian assumption), an assumption which is inconsistent with another feature of the case considered, i.e., with the fact that the inertial movement is supposed by Newton's theory to occur in empty space. I conclude from this *that the concept of impetus, as fixed by the usage established in the impetus theory, cannot be defined in a reasonable way within Newton's theory*. And this is not further surprising. For this usage involves laws, such as (7), which are inconsistent with the Newtonian physics.

In the last argument, the assumption that the concept *force* is the same in both theories played an essential role. This assumption was used in the transition from the assertion, made by the impetus theory, that inertial motions occur under the influence of forces to the calculation of the magnitude of these forces on the basis of Newton's second law. Its legitimacy may be derived from the fact that both the impetus theory and Newton's theory apply the concept *force* under similar circumstances (paradigm-case argument!). Still, meaning and application are not the same thing, and it might well be objected that the transition performed is not legitimate, since the different contexts of the impetus theory, on the one hand, and of Newton's theory, on the other, confer different meanings upon one and the same word "force". This being the case, our last argument is based upon a *quaternio terminorum* and is, therefore, invalid. In order to meet this objection, we may repeat our argument using the word "cause" instead of the word "force" (the latter has a somewhat more specific meaning). But if someone again retorts that "cause" has a different meaning in Newton's theory from what it has in the impetus theory, then all I can say is that a consistent continuation of that kind of objection will in the end establish what I wanted to show in a more simple manner, viz., the impossibility of defining the notion of an impetus in terms of the descriptive terms of Newton's theory. To sum up: the concept *impetus* is not "explicable in terms of the theoretical primitives of the primary science."[14] And this is exactly as it should be, considering the inconsistency between some very basic principles of these two theories.

However, explication in terms of the primitives of the primary science is not the only method which was considered by Nagel in his discussion of the process of reduction. Another way to achieve reduction, which he mentions immediately after the above quotation, "is to adopt a material, or physical hypothesis according to which the occurrence of the properties designated by some expression in the premises of the primary science is a sufficient, or a necessary and sufficient condition for the occurrence of the properties designated by the expressions of the secondary discipline." Both procedures are in accordance with (4), or with (2), or at least Nagel thinks that they are: ". . . in this case" he says, referring to the procedure just outlined, "the meaning of the expressions of the secondary science *as fixed by the established usage of the latter*, is not declared to be analytically related to the meanings of the corresponding expressions of the primary science."[15] Let us now see what this second method achieves in the present case.

To start with, this method amounts to introducing a hypothesis of the form

(9) impetus = momentum

where each side retains the meaning it possesses in its respective discipline. The hypothesis then simply asserts that wherever momentum is present, impetus will also be present (see the above quotation of Nagel's), and it also asserts that the measure will be the same in both cases. Now this hypothesis, although acceptable within the impetus theory (after all, this theory permits the incorporation of the concept of momentum), is incompatible with Newton's theory. It is therefore not possible to achieve reduction and explanation by the second method.

To sum up: a law such as (8) which, as I have argued, is empirically adequate, and in quantitative agreement with Newton's first law, is yet incapable of reduction to Newton's theory and therefore incapable of explanation in terms of the latter. Whereas the reasons we have so far found for irreducibility were of a quantitative nature, this time we met a qualitative reason, as it were, i.e., the incommensurable character of the conceptual apparatus of (8), on the one side, with that of Newton's theory, on the other.

Taking together the quantitative as well as the qualitative argument, we are now presented with the following situation: there exist pairs of theories, T and T′, which overlap in a domain D′ and which are incompatible (though experimentally indistinguishable) in this domain. Outside D′, T has been confirmed, and it is also more coherent, more general, and less *ad hoc* than T′. The conceptual apparatus of T and T′ is such that it is possible neither to define the primitive descriptive terms of T′ on the basis of the primitive descriptive terms of T′ nor to establish correct empirical relations involving both these terms (correct, that is, from the point of view of T). This being the case, explanation of T′ on the basis of T or reduction of T′ to T is clearly impossible if both explanation and reduction are to satisfy (A) and (B). Altogether, *the use of T will necessitate the elimination both of the conceptual apparatus of T′* and of the laws of T′. The conceptual apparatus will have to be eliminated because its use involves principles, such as (7) in the example above, which are inconsistent with the principles of T; and the laws will have to be eliminated because they are inconsistent with what follows from T for events inside D′. . . . This being the case the demand for explanation and reduction clearly cannot arise if this demand is interpreted as the demand for the explanation, or reduction, of T′, rather than of a set of laws that is in some respect similar to T′ but in other respects (meanings of fundamental terms included) very different from it. For such a demand would imply the demand to derive, from correct premises, what is false, and to incorporate what is incommensurable.

The effect of the transition from T′ and T is rather to be described in the manner indicated in the introductory remarks of the present paper: where I said: What happens when transition is made from a restricted theory T′ to a wider theory T (which is capable of covering all the phenomena which have been covered by T′) is something much more radical than incorporation of the *unchanged* theory T′ into the wider context of T. What happens is rather a *complete replacement* of the ontology of T′ by the ontology of T, and a corresponding change in the meanings of all descriptive terms of T′ (provided these terms are still employed). Let me add here that the not-too-well-known example of the impetus theory versus Newton's mechanical theory is not the only instance where this assertion holds. As I shall show a little later, more recent theories also correspond to it. Indeed, it will turn out that the principle correctly describes the relation between the elements of *any* pair of noninstantial theories satisfying the conditions which I have just enumerated.

This finishes step one of the argument against the assumption that reduction and explanation are by derivation. What I have shown . . . is that some very important cases which have been, or could be used as examples of reduction (and explanation) are not in agreement with the condition of derivability. It will be left to the reader to verify that this holds in almost all cases of explanation by theories: assumption (A) does not give a correct account of actual scientific practice. It has also been shown that in this respect the thesis formulated in the beginning of this paper is much more adequate.

. . .

6. Criticism of the Assumption of Meaning Invariance

In Section 5 it was shown that the "inertial law" (8) of the impetus theory is incommensurable with Newtonian physics in the sense that the main concept of the former, viz., the concept of impetus, can neither be defined on the basis of the primitive descriptive terms of the latter, nor related to them via a correct empirical statement. The reason for this incommensurability was also exhibited: although (8), *taken by itself*, is in quantitative agreement both with experience and with Newton's theory, the "rules of usage" to which we must refer in order to explain the meanings of its main descriptive terms contain the law (7) and, more especially, the law that constant forces bring about constant velocities. Both of these laws are inconsistent with Newton's theory. Seen from the point of view of this theory, any concept of a force whose content is dependent upon the two laws just mentioned will possess zero magnitude, or zero denotation, and will therefore be incapable of expressing features of actually existing situations. Conversely, it will be capable of being used in such a manner only if all connections with Newton's theory have first been severed. It is clear that this example refutes (B) if we interpret that thesis as the description of how science actually proceeds.

We may generalize this result in the following fashion: consider two theories, T′ and T, which are both empirically adequate inside D′, but which differ widely outside D′. In this case the demand may arise to explain T′ on the basis of T, i.e., to derive T′ from T and suitable initial conditions (for D′). Assuming T and T′ to be in quantitative agreement inside D′, such derivation will still be impossible if T′ is part of a theoretical context whose "rules of usage" involve laws inconsistent with T.[16]

It is my contention that the conditions just enumerated apply to many pairs of theories which have been used as instances of explanation and reduction. Many (if not all) of such pairs on closer inspection turn out to consist of elements which are incommensurable and therefore incapable of mutual reduction and explanation. However, the above conditions admit of still wider application and then lead to very important consequences with regard to the structure and development both of our knowledge and of the language used for the expression of it. After all, the principles of the context of which T′ is a part need not be explicitly formulated, and as a matter of fact they rarely are. To bring about the situation described above (sets of mutually incommensurable concepts), it is sufficient that they govern the use of the main terms of T′. In such a case T′ is formulated in an idiom some of whose implicit rules of usage are inconsistent with T (or with some consequences of T in the domain where T′ is successful). Such inconsistency will not be obvious at a glance; it will take considerable time before the incommensurability of T and T′ can be demonstrated. However, as soon as this demonstration has been carried out, in the very same moment, the idiom of T′ must be given up and must be replaced by the idiom of T. Of course, one need not go through the laborious and very uninteresting task of analyzing the context of which T′ is part. All that is needed is the adoption of the terminology and the "grammar" of the most detailed and most successful theory throughout the domain of its application. This automatically takes care of whatever incommensurabilities may arise, and it does so without any linguistic detective work (which therefore turns out to be entirely unnecessary for the progress of knowledge).

What has just been said applies most emphatically to the relation between (theories formulated in) some commonly understood language and more abstract theories. That is, I assert that languages such as the "everyday language," this notorious abstraction of contemporary linguistic philosophy, frequently contain (not explicitly formulated, that is, but implicit in the way in which its terms are used) principles which are inconsistent with newly introduced theories, and that they must therefore be either abandoned and replaced by the language of the new and better theories even in the most common situations, or they must be completely separated from these theories (which would lead to a situation where it is possible to believe in various kinds of "truth"): it is far from correct to assume that the everyday languages are so widely conceived, so tolerant, indefinite, and vague that they will be compatible with any scientific theory, that science can at most fill in details, and that a scientific theory will never run against the principles implicitly contained in them. *The very opposite is the case.* As will be shown later, even everyday languages, like languages of highly theoretical systems, have been introduced in order to give expression to some theory or point of view, and they therefore contain a well-developed and sometimes very abstract ontology. It is very surprising that the champions of the "ordinary language" should have such a low opinion of its descriptive power.

However, before turning to this part of the argument, I shall briefly discuss another example where the questionable principles of T′ have been explicitly formulated, or can at least be easily unearthed.

The example which is dealt with by Nagel is the relation between phenomenological thermodynamics and the kinetic theory. Employing his own theory of reduction . . . Nagel claims that the terms of the statements which have been derived from the kinetic theory (with the help of correlating hypotheses similar to (9)) will have the meanings they originally possessed within the phenomenological theory, and he repeatedly emphasizes that these meanings are fixed by "its own procedures" (i.e., by the procedures of the phenomenological theory) "whether or not [this theory] has been, or will be, reduced to some other discipline."[17]

As in the case of the impetus theory, we shall begin our study of the correctness of this assertion with an examination of these "procedures" and "usages"; more especially, we shall start with an examination of the usage of the term "temperature," "as fixed by the established procedures" of thermodynamics.

Within thermodynamics proper,[18] temperature ratios are defined by reference to reversible processes of operating between two levels, L′ and L″, each of these levels being characterized by one and the same temperature throughout. The definition, viz.,

(10) $T':T'' = Q':Q''$

identifies (after a certain arbitrary choice of units) the ratio of the temperature with the ratio between the amount of heat absorbed at the higher level and the amount of heat rejected at the lower level. Closer inspection of the "established usage" of the temperature thus defined shows that it is supposed to be

(11) independent of the material of the substance chosen for the cycle, and unique.

This property can be inferred from the extension of the concept of temperature thus defined to radiation fields and from the fact that the constants of the main laws in this domain are universal, rather than dependent upon either the thermometric substance or the substance of the system investigated.

Now, it can be shown by an argument not to be presented here that (10) and (11) taken together imply the second law of thermodynamics in its strict (phenomenological) form: the concept of temperature as "fixed by the established usages" of thermodynamics is such that its application to concrete situations entails the strict (i.e., nonstatistical) second law.

Now whatever procedure is adopted, the kinetic theory does *not* give us such a concept. First of all, there does not exist any dynamical concept that possesses the required property. The statistical account, on the other hand, allows for fluctuations of heat back and forth between two levels of temperature and, therefore, again contradicts one of the laws implicit in the "established usage" of the thermodynamic temperature. The relation between the thermodynamic concept of temperature and what can be defined in the kinetic theory, therefore, can be seen to conform to the pattern that has been described at the beginning of the present section: we are again dealing with two incommensurable concepts. The same applies to the relation between the purely thermodynamic entropy and its statistical counterpart; whereas the latter admits of very general application, the former can be measured by infinitely slow reversible processes only. Taking all this into consideration we must admit that it is impossible to relate the kinetic theory and the phenomenological theory in the manner described by Nagel, or to explain all the laws of the phenomenological theory in the manner demanded by Hempel and Oppenheim on the basis of the statistical theory. Again replacement rather than incorporation, or derivation (with the help, perhaps, of premises containing statistical as well as phenomenological concepts), is seen to be the process that characterizes the transition from a less general theory to a more general one.

It ought to be pointed out that the discussion is very idealized. The reason is that a purely kinetic account of the phenomena of heat does not yet seem to exist. What exists is a curious mixture of phenomenological and statistical elements, and it is this mixture which has received the name "statistical thermodynamics." However, even if this is admitted, it remains that the concept of temperature as it is used in this new and mixed theory is different from the original, purely phenomenological concept. To our point of view, according to which terms change their meanings with the progress of science, Nagel raises the following objection: "The redefinition of expressions with the development of inquiry [so it is noted], is a recurrent feature in the history of science. Accordingly, though it must be admitted that in an earlier use the word 'temperature' had a meaning specified exclusively by the rules and procedures of thermometry and classical thermodynamics, it is *now* so used that temperature is 'identical by definition' with molecular energy. The deduction of Boyle-Charles' law does not therefore require the introduction of a further postulate, whether in the form of a coordinating definition or a special empirical hypothesis, but simply makes use of this definitional identity. This objection illustrates the unwitting double talk into which it is so easy to fall. It is certainly possible to redefine the word 'temperature' so that it becomes synonymous with 'mean kinetic energy.' But it is equally certain that on this redefined usage the word has a different meaning from the one associated with it in the classical science of heat, and therefore a meaning different from the one associated with the word in the statement of the Boyle-Charles law. However, if thermodynamics is to be reduced to mechanics, it is temperature in the sense of the term in the classical science of heat which must be asserted to be proportional to the mean kinetic energy of gas molecules. Accordingly, if the word 'temperature' is redefined as suggested by the objection, the hypothesis must be invoked that the state of bodies described as 'temperature' (in the classical thermodynamic sense) is also characterized by 'temperature' in the redefined sense of the term. This hypothesis, however, will then be one that does not hold as a matter of definition . . . Unless this hypothesis is adopted, it is not the Boyle-Charles law which can be derived from the assumptions of the kinetic theory of gases. What is derivable without the hypothesis is a sentence similar in syntactical structure to the standard formulation of the law, but possessing a sense that is unmistakably different from what the law asserts."[19] So far Nagel.

Commencing my criticism, I shall at once admit the correctness of the last assertion. After all, it has been my contention all through this paper that extension of knowledge leads to a decisive modification of the previous theories both as regards the quantitative assertions made and as regards the meanings of the main descriptive terms used. Applying this to the present case I shall therefore at once admit that incorporation into the context of the statistical theory is bound to change the meanings of the main descriptive terms of the phenomenological theory. The difference between Nagel and myself lies in the following. For me, such a change to new meanings and new quantitative assertions is a natural occurrence which is also desirable for methodological reasons (the last point will be established later in the present section). For Nagel such a change is an indication that reduction has not been achieved, for reduction in Nagel's sense is supposed to leave untouched the meanings of the main descriptive terms of the discipline to be reduced (cf. his "if thermodynamics is to be reduced to mechanics, it is temperature in the sense of the term in the classical science of heat which must be asserted to be proportional to the mean kinetic energy of gas-molecules"). "Accordingly," he continues, quite obviously assuming that reduction in his sense can be carried through, "if the word 'temperature' is redefined as suggested by the objection, the hypothesis must be invoked that the state of bodies described as 'temperature' (in the classical thermodynamic sense) is also characterized by 'temperature' in the redefined sense of the term. *This* hypothesis . . . will then be one that does not hold as a matter of definition." It will also be a false hypothesis because the conditions for the definition of the phenomenological temperature are *never* satisfied in nature (see the arguments above in the text and compare also the arguments in connection with formula (9)), which is only another sign of the fact that reduction, in the sense of Nagel, of the phenomenological theory to the statistical theory is not possible (obviously the additional premises used in the reduction are not supposed to be false). Once more arguments of meaning have led to quite unnecessary complications.

. . .

Our argument against meaning invariance is simple and clear. It proceeds from the fact that usually some of the principles involved in the determination of the meanings of older theories or points of view are inconsistent with the new, and better, theories. It points out that it is natural to resolve this contradiction by eliminating the troublesome and unsatisfactory older principles and to replace them by principles, or theorems, of the new and better theory. And it concludes by showing that such a procedure will also lead to the elimination of the old meanings and thereby to the violation of meaning invariance.

. . .

7. Summary and Conclusion

Two basic assumptions of the orthodox theory of reduction and explanation have been found to be in disagreement with actual scientific practice and with reasonable methodology. The first assumption was that the explanandum is *derivable* from the explanans. The second assumption was that *meanings are invariant* with respect to the process of reduction and explanation. We may sum up the results of our investigation in the following manner:

Let us assume that T and T′ are two theories satisfying the conditions outlined at the beginning of Section 3. Then, from the point of view of scientific method, T will be most satisfactory if it is (α) *inconsistent* with T′ in the domain where they both overlap,[20] and if it is (β) *incommensurable* with T′.

Now it is clear that a theory which is satisfactory according to the criterion just pronounced will not be capable of functioning as an explanans in any explanation or reduction that satisfies the principles put forth by Hempel and Oppenheim or Nagel. Paradoxically speaking: *Hempel–Oppenheim explanations cannot use satisfactory theories as explanantia. And satisfactory theories cannot function as explanantia in Hempel–Oppenheim explanations*. How is the theory of explanation and reduction to be changed in order to eliminate this very undesirable paradox?

It seems to me that the changes that are necessary will make it impossible to retain a formal theory of explanation, because these changes will introduce pragmatic or "subjective" considerations into the theory of explanation. This being the case, it seems perhaps advisable to eliminate altogether considerations of explanation from the domain of scientific method and to concentrate upon those rules which enable us to compare two theories with respect to their formal character and their predictive success and which guarantee the constant modification of our theories in the direction of greater generality, coherence, and comprehensiveness. I shall now give a more detailed outline of the reasons which have prompted me to adopt this pragmatic point of view.

Consider again T′ and T′ as described above. Under these circumstances, the set of laws T″ following from T inside D′ will either be inconsistent with T′ or incommensurable with it. In what sense, then, can T be said to explain T′? This question has been answered by Popper for the case of the inconsistency of T′ and T″. "Newton's theory," he says, "unifies Galileo's and Kepler's. But far from being a mere conjunction of these two theories—which play the part of *explicanda* for Newton—*it corrects them while explaining them*. The original explanatory task was the deduction of the earlier results. It is solved, not by deducing them, but by deducing something better in their place: new results which, under the special conditions of the older results, come numerically very close to these older results, and at the same time correct them. Thus the empirical success of the old theory may be said to corroborate the new theory; and in addition, the corrections may be tested in their turn ... What is brought out strongly by [this] ... situation ... is the fact that the new theory cannot possibly be *ad hoc* ... Far from repeating its *explicandum*, the new theory contradicts it and corrects it. In this way, even the evidence of the *explicandum* itself becomes independent evidence for the new theory."[21]

In a letter to me, J. W. N. Watkins has suggested that this theory may be summarized as follows: Explanation consists of two steps. The first step is derivation, from T, of those laws which obtain under the conditions characterizing D′. The second step is comparison of T″ and T′ and realization that both are empirically adequate, i.e., fall within the domain of uncertainty of the observational results. Or, to express it in a more concise manner: T explains T′ satisfactorily only if T is true and there exists a consequence T″ of T for the conditions of validity of T′ such that T″ and T′ are at least equally strong and also experimentally indistinguishable.

The first question that arises in connection with Dr. Watkins' formulation is this: experimentally indistinguishable on the basis of which observations? T′ and T″ may be indistinguishable by the crude methods used at the time when T was first suggested, but they may well be distinguishable on the basis of later and more refined methods. Reference to a certain observational method will therefore have to be included in the clause of experimental indistinguishability. The notion of explanation will be relative to this observational material. It will not make sense any longer to ask whether or not T explains T′. The proper question will be whether T explains T′ *given the observational material, or the observational methods* O. Using this new mode of speech we are forced to deny that Kepler's laws are explained by Newton's theory *relative to the present observations*—and this is perfectly in order; for these present observations in fact refute Kepler's laws and thereby eliminate the demand for explanation. It seems to me that this theory can well deal with all the problems that arise when T and T′ are commensurable, but inconsistent inside D′. It does not seem to me that it can deal with the case where T′ and T are incommensurable. The reason is as follows.

As soon as reference to certain observational material has been included in the characterization of what counts as a satisfactory explanation, in the very same moment the question arises as to how this observational material is to be presented. If it is correct, as has been argued all the way through the present paper, that the meanings of observational terms depend on the theory on behalf of which the observations have been made, then the observational material referred to in this modified sketch of explanation must be presented in terms of this theory also. Now incommensurable theories may not possess any comparable consequences, observational or otherwise. Hence, there may not exist any possibility of finding a characterization of the observations which are supposed to confirm two incommensurable theories. How, then, is the above account of explanation to be modified to cover the case of incommensurable theories also?[22]

It seems to me that the only possible way lies in closest adherence to the pragmatic theory of observation. According to this theory . . . we must carefully distinguish between the *causes* of the production of a certain observational sentence, or the features of the process of production, on the one side, and the *meaning* of the sentence produced in this manner on the other. More especially, a sentient being must distinguish between the fact that he possesses a certain sensation, or disposition to verbal behavior, and the interpretation of the sentence being uttered in the presence of this sensation, or terminating this verbal behavior. Now our theories, apart from being pictures of the world, are also instruments of prediction. And they are good instruments if the information they provide, taken together with information about initial conditions characterizing a certain observational domain D_0, would enable a robot, who has no sense organs, but who has this information built into himself (or herself), to react in this domain in exactly the same manner as sentient beings who, without knowledge of the theory, have been trained to find their way about D_0 and who are able to answer, "on the basis of observation," many questions concerning their surroundings.[23] This is the criterion of predictive success, and it is seen not at all to involve reference to the *meanings* of the reactions carried out either by the robot or by the sentient beings (which latter need not be humans, but can also be other robots). All it involves is agreement of behavior.

Now this criterion involves "subjective" elements. Agreement is demanded between the behavior of (nonsentient, but theory-fed) robots and that of sentient beings, and it is thereby assumed that the latter possesses a privileged position. Considering that perceptions are influenced by belief in theories and that behavior, too, is influenced by belief in theories, this criterion would seem to be somewhat arbitrary. It is easily seen, however, that it cannot be replaced by a less arbitrary and more "objective" criterion. What would such an objective criterion be? It would be a criterion which is either based upon behavior that is not connected with any theoretical element—and this is impossible (cf. my criticism of the theory of sense data above)—or it would be behavior that is tied up with an irrefutable and firmly established theory—which is equally impossible. We have to conclude, therefore, that a formal and "objective" account of explanation cannot be given.

Notes

1 [5], p. 301. A more elaborate form of this condition is called the "condition of derivability" on p. 354 of [11].

2 [5], p. 301. My italics. See also [11], pp. 345, 352.

3 [14], Sec. 12.

4 [8], p. 321.

5 An exception is Nagel who, in [11], p. 338, defines reduction as "the explanation of a theory or a set of experimental laws established in one area of inquiry by a theory usually, though not invariably, formulated for some other domain." This implies that the condition of meaning invariance formulated by him for the process of reduction is supposed to be valid in the case of explanation also. On pp. 86–87, meaning invariance for observational terms is stated quite explicitly: an experimental law "retains a meaning that can be formulated independently of [any] theory . . . [It] has . . . a life of its own, not contingent on the continued life of any particular theory that may explain the law."

6 [7], p. 105, condition (8.3). It was J. W. N. Watkins who drew my attention to this property of Hempel's theory.

7 [5], p. 291. I am aware that, from a historical point of view, the discussion to follow is not adequate. However, I am here interested in the systematic aspect, and I have therefore allowed myself what could only be regarded as great liberties if the main interest were historical.

8 For a more detailed account of the theories mentioned in this section, see M. Clagett [4]. Concerning the first part of the present section, see J. Burnet [2], as well as Clagett [3] and Popper [13].

9 Clagett [4], p. 425.

10 The parentheses I have added because of the absence from the earlier forms of the impetus theory of an explicit consideration of direction.

11 See Nagel [5], p. 301.

12 See Clagett [4], p. 523.

13 For an elaborate discussion of the difference between momentum and impetus, see Anneliese Maier [9]. For what follows, see also M. Bunge [1], Ch. 4.4.

14 Nagel [5], p. 302.

15 *Ibid.* My italics.

16 Since this difficulty can arise even in the domain of empirical generalizations, the orthodox account may be inappropriate for them as well.

17 Nagel [5], p. 301.

18 See Fermi [6], Sec. 9.

19 [11], pp. 357–358.

20 This condition has been discussed with great clarity in [12]. It was this

discussion (as well as dissatisfaction with [10]) that was the starting point of the present analysis of the problem of explanation.

21 Popper [12], p. 33.

22 As Professor Feigl has pointed out to me, this difficulty also arises in the case of crucial experiments.

23 Of course, the *motivations* of the robot and of the sentient being must also be the same.

References

1 Bunge, M. *Causality*. Cambridge, Mass.: Harvard University Press, 1959.

2 Burnet, J. *Early Greek Philosophy*. London: Adam and Charles Black, 1930.

3 Clagett, M. *Greek Science in Antiquity*. London: Abelard-Schuman, 1957.

4 Clagett, M. *The Science of Mechanics in the Middle Ages*. Madison: University of Wisconsin Press, 1959.

5 Danto, A., and S. Morgenbesser, eds. *Philosophy of Science*. New York: Meridian Books, 1960.

6 Fermi, E. *Thermodynamics*. New York: Dover Publications, 1956.

7 Hempel, C. G. "Studies in the Logic of Confirmation," *Mind*, 54:1–26, 97–121 (1945).

8 Hempel, C. G., and P. Oppenheim. "Studies in the Logic of Explanation," *Philosophy of Science*, 15:135–175 (1948).

9 Maier, A. *Die Vorlaeufer Galilei's im 14, Jahrhundert*. Rome: Edizioni di Storiae Litteratura, 1949.

10 Nagel, E. "The Meaning of Reduction in the Natural Sciences," in *Science and Civilization*, R. C. Stauffer, ed. Madison: University of Wisconsin Press, 1949, pp. 99–145.

11 Nagel, E. *The Structure of Science*. New York: Harcourt, Brace, and Company, 1961.

12 Popper, K. R. "The Aim of Science," Ratio, 1:24–35 (1957).

13 Popper, K. R. "Back to the Pre-Socratics," *Proceedings of the Aristotelian Society*, New Series, 54:1–24 (1959).

14 Popper, K. R. *The Logic of Scientific Discovery*. New York: Basic Books, 1959.

11

Philip Kitcher, "Theories, Theorists and Theoretical Change"

Philip Kitcher's article responds to the challenge of conceptual relativism by exploring the connection between the philosophy of science and the philosophy of language, in particular by drawing on the resources of novel theories of reference.

Joseph Priestley, the most famous advocate of the phlogiston theory, claimed, on numerous occasions, that when a metal is obtained from its calx it absorbs phlogiston from the air, and that the dephlogisticated air which remains is better than ordinary air. Historians of science are interested in discovering what Priestley was talking about, and how much of what he said is true. Their researches are relevant to philosophy. For it is a commonplace of modern philosophy that our views about the nature and development of science can be illuminated by studying the processes through which the great dead theories were replaced by their more modern counterparts; and, since our only access to the great dead theories is through the writings of the great dead theorists, we can only engage in such study if we can decide what those theorists were talking about.

Paradoxically enough, the writers who have contended most vigorously that history of science is relevant to philosophy of science have also argued for theses which imply that the task of the historian of science cannot be successfully completed. After beginning *The Structure of Scientific Revolutions* with an exhortation to philosophers to take the history of science seriously, Thomas Kuhn eventually concludes that the content of past theories resists expression in modern terms.[1] In a similar vein, Paul Feyerabend uses historical examples to challenge philosophical theses, while repeatedly claiming that different theories share no common statements.[2] If Kuhn and Feyerabend are right, then one important part of the historian's

P. Kitcher, "Theories, Theorists and Theoretical Change," *Philosophical Review*, 1978, 87: 519–47.

enterprise is impossible; we cannot formulate past theories in contemporary language.

My aim in this paper is to show that a sensitive reading of some episodes in the history of science combined with a crude approach to semantical issues will indeed yield the theses which Kuhn and Feyerabend champion. I shall then propose a strategy for understanding the semantical aspects of theoretical change.

I. Conceptual Relativism

Whatever disagreements they may have on points of detail, Kuhn and Feyerabend concur in maintaining that most of traditional philosophy of science is bankrupt, and, in particular, that the logical empiricist account of the resolution of intertheoretical debates is hopelessly misguided. During the major upheavals in the history of science, those episodes which Kuhn calls "scientific revolutions," scientists of different persuasions do not have recourse to a common body of observational evidence or to a shared set of methodological rules. More fundamentally, they are unable to communicate. Both sides lack the ability to express, within their own language, the assertions of the rival theory. Where logical empiricists have seen orderly debate between rational men of good will, Kuhn and Feyerabend depict situations in which the presuppositions of debate break down.

Conceptual relativism is the doctrine that the language used in a field of science changes so radically during a revolution in that field that the old language and the new language are not intertranslatable. The examples are familiar. The languages of Ptolemaic and Copernican astronomy are each allegedly inadequate to the expression of the opposing view—as are the languages of phlogiston theory and Lavoisier's theory of combustion, of Newtonian dynamics and the special theory of relativity. But the careful attention which Kuhn and Feyerabend have given to these examples does not make up for the unclarity of the main thesis. At the heart of conceptual relativism is the idea of a radical difference in meaning across the revolutionary divide. Conceptual relativism inherits the philosophical difficulties of the notion of meaning.

Ever since Frege distinguished two semantic functions of linguistic expressions—the functions of expressing a sense and of referring—many philosophers have felt uncomfortable with the notion of sense, and have urged the benefits of doing as much semantics as possible within the theory of reference. Perhaps there is a legitimate heir to the Fregean notion of sense, but it is not to be found in the writings of Kuhn and Feyerabend. Indeed, many writers have exposed widespread confusions in the remarks which Kuhn and Feyerabend make about meaning.[3]

If we are to clarify the doctrine of conceptual relativism, it is natural to turn to the notion of reference, reformulating the doctrine as the thesis that, for any two languages used in the same scientific field at times separated by a revolution, there are some expressions in each language whose referents are not specifiable in the other language.[4]

The idea that conceptual relativism is a thesis about reference has been cogently presented by Israel Scheffler.[5] However, Scheffler fails to distinguish conceptual relativism from another position defended by Kuhn and Feyerabend, the position that, in a scientific revolution, the referents of some expressions are changed. As a result, he tries to combat conceptual relativism by claiming stability of reference through revolutions. But, strictly speaking, referential change is neither necessary nor sufficient for conceptual relativism. Trivially, conceptual relativism can occur without referential change if the languages involved contain completely different expressions. More importantly, even if some (or all) terms were to change in reference, this would not imply that there are some expressions of one language whose referents cannot be specified in the other language. Hence Scheffler's attempt to defend referential stability may commit him to a position which is unnecessarily strong.

To see this, we need only imagine a number of cases. Suppose first that a perverse scientist decides to interchange some of his terms in a systematic fashion. Doubtless this practice would cause some confusion, but, given enough time and interest on the part of his colleagues, his utterances could be translated into the normal idiom. (The translation would be adequate in the sense of preserving reference.) Even if the practice were extended to all terms, there is no reason to suppose that the deviant utterances would inevitably remain incomprehensible. Finally, if the systematic interchange were to be carried out simultaneously with a change in belief, the task of the interpreters might become more complicated. Yet we should not concede that the task is impossible, and that the parties to the dispute are necessarily prevented from formulating their differences. What *would* block understanding—and what is of interest to Feyerabend and Kuhn—is a situation involving a special type of referential change, namely change which culminates in a mutual inability to specify the referents of terms used in presenting the rival position. Cases of *this* kind do appear to threaten the possibility of an objective comparison between the rival theories and hence to subvert traditional accounts of intertheoretic debate.

Kuhn and Feyerabend often confuse the issue by citing examples of referential change in defense of conceptual relativism.[6] In examining these examples we should not be sidetracked into arguing for referential stability, but should attend to the possibility of formulating the opposing theories in the rival languages. If we proceed without a more refined approach to the reference of scientific terms than is usual in discussions in the philosophy of science, the historical evidence will indeed support conceptual relativism. The remedy is to begin with the notion of reference.

II. Theories of Reference: Some Basic Distinctions

Imagine that we are setting out to give a theory of reference for the language of Aristotelian physics. What is our goal?

There is a simple answer. We aim to correlate expression-*types* of Aristotelian language with expression-*types* of contemporary English, to complete matrices of the form ⌜In the language of Aristotelian physics, *e* refers to . . .⌝, where *e* is an expression-type and where the blanks are filled with an expression of English which is coreferential with *e*. Not any way of filling in the blanks will do. For example, it will not help to insert the expression "the entity which Aristotelians referred to when they produced tokens of *e*." Rather, what we require is an expression which is informative, in the sense of enabling speakers of contemporary English to pick out the Aristotelian referents. (The task of saying in a precise way what conditions must be met by the expressions inserted is tricky; I shall not try to complete it here, but shall rely on an intuitive grasp of what kinds of expressions are appropriate.)

A full theory of reference for the Aristotelian language is a set of completed matrices such that a name of each primitive expression of Aristotelian language occurs in the place of *e* in exactly one matrix. A full theory of reference would help us to understand the sentence-*tokens* produced by Aristotelians. Confronted with an Aristotelian expression-token, we consult the theory of reference to discover which English expression-type is correlated with the Aristotelian expression-type of which the token in question is a token. We then replace the Aristotelian token with a token of the appropriate English expression-type. Proceeding in this way, we can turn Aristotelian sentence-tokens into English sentence-tokens.

This rather pedantic exposition spells out some of the details of the task which philosophers of science usually envisage when they think about constructing a theory of reference for Aristotelian language. They implicitly assume what I have made explicit, namely that we are looking for a mapping of expression-types onto expression-types and that all tokens of the same Aristotelian expression-type can be treated in the same way. Let us call a theory of reference of the kind just described a *context-insensitive* theory of reference (a *CIT*). *CIT*s are well known to be inadequate for coping with natural languages. In natural languages, different tokens of some types—such as demonstratives, personal pronouns, proper names and ambiguous expressions—refer to different entities in virtue of their production in different contexts. Since the languages used in mathematics and the natural sciences do not appear to contain such expressions, it is tempting to suppose that *CIT*s will be able to cope with them.

Even when we are investigating a language which contains context-sensitive expressions we might hope to find general ways of specifying the dependence of tokens of those expressions on the context in which they are produced. Ideally, we try to frame clauses for context-sensitive terms which allow for identification of the referents of their tokens given minimal information about context. Simple paradigms are easy to find. When we give a theory of reference for German in English, we naturally propose that a token of "ich" refers to the person who produced it, and that a token of "jetzt" refers to the time at which it was produced. (The specifications allow for the assignment of referents even if we are ignorant concerning many features of the context of utterance.)

I want to show that the languages used in presenting scientific theories demand a theory of reference which is different from *CIT*s and from theories which supplement *CIT*s with the simple kinds of clauses just mentioned. I shall explain what I have in mind by attending to a different aspect of the enterprise of constructing theories of reference.

Theories of reference for particular languages must meet standards of adequacy which are laid down by what I shall call the *general theory of reference*. The general theory of reference provides us with universal principles for the determination of reference, principles which we accept independently of our views about the referents of expressions in particular languages and to which we appeal to evaluate such views.[7] Although the general theory of reference thus serves as a standard when we attempt to construct a theory of reference for a particular language, we do not expect the principles themselves to appear in particular theories of reference. We endeavor to replace them with reference-specifying clauses which are specific to the expressions of the language under study. In some cases, this expectation may be defeated. We may find that, for some expressions of the language under study, the only way to specify the referents of tokens of those expressions is to appeal directly to part of the general theory. If there are indeed such languages, then adequate theories of reference for them will be *context-sensitive theories* (*CST*s), which specify the referents of tokens of some expressions of the language (the context-sensitive expressions) by invoking general principles about reference.

We can make the discussion more concrete by trying to sketch the general theory of reference and by looking at a particular example. Following a number of writers (in particular, Keith Donnellan),[8] I shall suppose that the general theory of reference is an "historical explanation" theory. The central principle of the theory is the thesis that the referent of a token of an expression is the entity which figures in the appropriate way in the correct historical explanation of the production of that token. The thesis is obviously not precise, and an articulation of the general theory of reference should explain what is meant by "figuring in an appropriate way in the correct historical explanation," but we need not worry about this imprecision just yet. Roughly, the idea is that the production of the expression-token is the terminal event in a sequence of events which would be described in detail by the correct (and complete) explanation of that terminal event. This sequence links the expression-token produced to an entity singled out in the first event of the sequence, and that entity is the referent of the token.

Some examples may help to explain this terminology. Consider our current use of the word "Socrates." Behind (most of our utterances of "Socrates" stand sequences of events with a common first member, an event in which a particular Greek baby was singled out and given a name. (The name was probably not "Socrates," but that does not matter.) Socrates was causally involved in the event. His presence led to the production of a token of the name. Contemporary uses of "Socrates" derive from the event, and they refer to Socrates through his causal involvement in it.

The explanations of current utterances of other terms are slightly different. Many of our present uses of "Neptune" are backed by sequences of events whose common first member is an event in which Neptune was picked out by description. Adams and Leverrier decided to give the name "Neptune" to the planet responsible for the perturbation of Uranus. Their decision determines the referent of many of our tokens of "Neptune"—even if we do not know that Neptune fits the description they gave.

Armed with some understanding of the general theory of reference, we can now illustrate the idea of a *CST* by considering the following example. Eustacia Evergreen, a famous and eccentric millionairess, has become tired of the publicity which she has received for many years. Determined to secure privacy, she hatches a plot, calling for an impersonator to lead her public life. The impersonator moves into a new community and quickly becomes friends with many of her neighbors. Prior to their first encounter with the impostor, the neighbors had known about Ms. Evergreen from newspaper and television reports, and they had sometimes talked about her. After meeting the impostor they continue to produce tokens of "Eustacia Evergreen" (or of related abbreviatory expressions such as "Eustacia"). I shall assume that they are not snobbish, and that they are interested in discussing such issues as whether to accept the impostor's kind invitations or how to reciprocate her lavish hospitality.

To whom do the unwitting friends of the impostor refer when they produce their tokens of "Eustacia Evergreen"? If we merely consider their utterances prior to the first meeting with the impersonator, the answer is easy: all tokens of "Eustacia Evergreen" refer to the millionairess. But, once they become deceived, the task of assigning referents becomes more difficult. *Some* of their later tokens refer to the real Eustacia. When one member of the circle promises to introduce a recently arrived guest to Eustacia Evergreen, he is referring to the millionairess—he is promising to acquaint his guest with a famous person, and he is unable to fulfil his promise. Other tokens refer to the impostor. When George announces to Alice that Eustacia has invited them to go skiing he refers to the impostor and to an invitation which she has issued.

The reference of a token of "Eustacia Evergreen" varies according to which of two candidates, the millionairess and the impostor, figures appropriately in the explanation of the production of the token. An omniscient observer would be able to spell out the details of the explanation. He would see how an event involving one of the candidates produced a state in the utterer, and how that state gave rise to the utterance. Using the general theory of reference, this observer could usually pick out the referent of the token.[9]

I claim that any theory of reference we can provide for the language of the community will be a *CST*. Our clause for assigning referents to tokens of "Eustacia Evergreen" will have to appeal to the idea that the referent of each token is the object (the person) figuring appropriately in the explanation of the production of the token. The language under study contains the context-sensitive expression "Eustacia Evergreen" and we need a *CST* to accommodate it.

However, we may not be as lucky as my envisaged omniscient observer. The evidence available to us may not enable us to construct explanations of the productions of all tokens of "Eustacia Evergreen" in sufficient detail, or the versions of the general theory of reference at our command may not be sufficiently precise, to enable us to specify the referent of each token, even in cases where, with greater knowledge, such specification would be possible. Yet our predicament is not hopeless. We can specify a set of entities (the pair set of the millionairess and the impostor) such that each token of "Eustacia Evergreen" refers to one member of the set, even if, in the case of some tokens, we are unable to decide which member is the referent.

Using this example and the distinctions developed so far we can gain a clearer view of the issue of conceptual relativism and of the semantical aspects of theoretical change. Consider the enterprise of providing a theory of reference for the language of Aristotelian physics, or more generally, for a language used in presenting some past scientific theory. In general, there are four possible outcomes for our enterprise.

(1) We can find a *CIT* adequate for the language under study.
(2) We cannot find a *CIT* which is adequate for the language under study.

We can find an adequate *CST*, and, using the *CST* and the available evidence, we can specify the referent of each token produced by a speaker of the language.

(3) We cannot find an adequate *CIT*. We can only find a *CST* and, in the light of the *CST* and the available evidence, there are some tokens produced by speakers of the language whose referents we cannot specify. However, for each expression-type of the language we can specify a set of entities such that the referent of any token of that type belongs to the set.

(4) We can only find a *CST*, and, for some expression types we are unable even to specify a set of entities such that the referent of any token of that type belongs to the set.

As I remarked above, conceptual relativism is intended to be the first part of an attack on the thesis that scientific theories can objectively be compared. I suggest that the central claim of conceptual relativism is the thesis that, in general, scientists working in the same field but separated by a revolution, find, when they attend to one another's languages, that they are confronted by cases of type (4). For, if he can establish this central claim, the conceptual relativist can plausibly contend that, lacking any language in which to formulate their disagreements, scientists separated by a revolution are unable to compare their theories. Conversely, if *any* of the other cases obtains, the scientists in question will be able to *formulate* their disagreements. (This does not mean, of course, that they will be able to *resolve* them. For there may be difficulties in reaching agreement on evidence or on methodological principles. However, my concern here is only with the entering wedge of the attack on the objectivity of scientific decision, the thesis that the preconditions for debate are not satisfied.) In cases of type (3), of course, the theory of reference which a scientist proposes for the language of his rival may allow for a number of different formulations of his opponent's hypothesis, but each of these may be compared with his own position.

To put the point another way, the idea of the incomparability ("incommensurability") of two scientific theories presupposes that there is *no* adequate translation of the language used in presenting one of those theories in the language of the other. The type of situation described in (4) articulates this idea. *Underdetermination* of translation of the kind described in (3) does not preclude the possibility of comparing the two theories; rather, it demands that each theorist consider a number of different ways of formulating his opponent's position.

Typically, in responding to the historical claims made by Kuhn and Feyerabend, philosophers of science have tacitly supposed that the only alternative to conceptual relativism is to claim that all scientific language is amenable to the treatment described in (1). The terms used in presenting scientific theories have been assumed to be context-insensitive. I shall try to show that neither conceptual relativism nor the traditional approaches offer a satisfactory account of the languages of past science. I shall argue that apparently problematic parts of the languages used by past theorists can be treated as cases of type (2), and my discussion will suggest that, at worst, such languages provide cases of type (3).

III. The Language of the Phlogiston Theory

I shall undertake a detailed study of one example, the language used by defenders of the phlogiston theory. Several key terms employed in formulating the theory are context-sensitive.

The phlogiston theory[10] attempted to give an account of a number of chemical reactions, and, in particular, it offered an explanation of processes of combustion. Substances which burn are rich in a "principle," *phlogiston*, which is imparted to the air in combustion. So, for example, when we burn wood, phlogiston is given to the air leaving ash as a residue. Similarly, when a metal is heated, phlogiston is emitted, and we obtain the *calx* of the metal.

Champions of the phlogiston theory knew that, after a while, combustion in an enclosed space will cease. They explained this phenomenon by supposing that air has a limited capacity for absorbing phlogiston. By heating the red calx of mercury on its own, Priestley found that he could obtain the metal mercury, and a new kind of "air," which he called *dephlogisticated air*. (According to the phlogiston theory, the calx of mercury has been turned into the metal mercury by taking up phlogiston; since the phlogiston must have been taken from the air, the resultant air is dephlogisticated.) Dephlogisticated air supports combustion (and respiration) better than ordinary air—but this is only to be expected, since the removal of phlogiston from the air leaves the air with a greater capacity for absorbing phlogiston.

In the last decades of the eighteenth century, the phlogiston theorists became interested in the properties of a gas, which they obtained by pouring a strong acid (concentrated sulphuric acid, for example) over a metal or by passing steam over heated iron. They called the gas *inflammable air*.

It may be useful to compare the descriptions of these familiar reactions given by the phlogiston theory, and by modern elementary chemistry:

Phlogiston theory	Modern theory
Metal + air $\xrightarrow[\text{heat}]{}$ Calx of metal + phlogisticated air	Metal + air $\xrightarrow[\text{heat}]{}$ Metal oxide + air which is poor in oxygen

Red calx of mercury $\xrightarrow[\text{heat}]{}$ Mercury + dephlogisticated air	Oxide of mercury $\xrightarrow[\text{heat}]{}$ Mercury + oxygen
Metal + acid → Salt + inflammable air	Metal + acid → Salt + hydrogen
Iron + steam $\xrightarrow[\text{heat}]{}$ Calx of iron + inflammable air	Iron + steam $\xrightarrow[\text{heat}]{}$ Iron oxide + hydrogen.

A glance at this table suggests some obvious identifications. We might naturally suppose that dephlogisticated air is oxygen and that inflammable air is hydrogen. We shall consider the merits of these identifications shortly.

Historians of science want to say that the phlogiston theory is incorrect, but they do not want to deny that some of its proponents discovered important things, or that some of what they said is true.[11] However, there is an obvious problem in reconstructing the content of the phlogiston theory, a problem which arises because a false presupposition, the idea that something is *emitted* in combustion, infects most of the terminology. This is the problem on which Kuhn and Feyerabend fasten: the "theory-ladenness" of the key terms of the phlogiston theory presents an insuperable bar to the formulation of that theory in language which is laden with a contrary theory.

The view that phlogiston is a substance emitted in combustion is central to the phlogiston theory, and is the doctrine from which the theory develops. Hence, it is quite natural to assume that the reference of "phlogiston" is fixed by this view, so that "phlogiston" refers to that which is emitted in all cases of combustion. But there is *nothing* which is emitted in all cases of combustion. So it seems that we must conclude that "phlogiston" fails to refer.

This conclusion results from two very plausible assumptions. Firstly, the phlogiston theory descends from the work of Stahl[12] who takes phlogiston as being, *by definition*, that which is emitted in combustion; the phlogistonian tradition repeats Stahl's definition and honors his usage. Secondly, in construing the definition, we suppose that some terms employed by proponents of the phlogiston theory are coreferential with homonymous terms of modern English. For example, we assume that the expression "① is emitted from ②" has the same referent as a term of the language of the phlogiston theory and as a term of modern English. In general, there are a large number of expressions—including "① is absorbed by ②," "① is the substance which results from adding ② to ③ till no more of ② can be absorbed," "① is the substance which results from removing ② from ③" (and many more similar expressions)—which we naturally take to have preserved their reference.

Let us assume, for the moment, that the proposal that "phlogiston" fails to refer is correct, and that the ideas on which it is anchored are also sound. Problems will result when we try to square this proposal with the claim that champions of the phlogiston theory made some important, true statements. Consider the status of some of the complex terms used in presenting the phlogiston theory. The terms "dephlogisticated air" and "phlogisticated air" are abbreviations for the expressions "the substance which results from removing phlogiston completely from air" and "the substance which results from adding phlogiston to the air until no more phlogiston can be absorbed," respectively.[13] In denying that phlogiston exists we seem to be forced to deny that these complex terms which contain "phlogiston" refer. How can there be a substance which remains when phlogiston is removed from air if there is no such substance as phlogiston?[14]

However, we are also tempted to suppose that Priestley and Cavendish used the terms "dephlogisticated air" and "phlogisticated air" to refer, and that they made some true utterances using these terms. Consider Priestley's account of his first experience of breathing oxygen.

> My reader will not wonder, that, after having ascertained the superior goodness of dephlogisticated air by mice living in it, and the other tests above mentioned, I should have had the curiosity to taste it myself. I have gratified that curiosity, by breathing it, . . . The feeling of it to my lungs was not sensibly different from that of common air; but I fancied that my breast felt peculiarly light and easy for some time afterwards.[15]

Priestley's token of "dephlogisticated air" refers to the substance which he and the mice breathed—namely, oxygen.

Similarly, it seems that Cavendish's uses of the terms refer, when he describes the formation of water by synthesis of hydrogen and oxygen:

> From the foregoing experiments it appears, that when a mixture of inflammable and dephlogisticated air is exploded in such proportion that the burnt air is not much phlogisticated, the condensed liquor contains a little acid which is always of the nitrous kind, whatever substance the dephlogisticated air is procured from; but if the proportion be such that the burnt air is almost entirely phlogisticated, the condensed liquor is not at all acid, but seems pure water, without any addition whatever; . . .[16]

We readily understand what prompted this description. Cavendish had performed a series of experiments in which samples of hydrogen and oxygen were "exploded" together; in some cases, the oxygen obtained was not entirely pure, and the small amount of nitrogen mixed in it, participated in a reaction to form a small amount of nitric acid; when the sample of oxygen was pure there was no formation of nitric acid, and the only reaction was the combination of hydrogen and oxygen to form water. In reporting the experiments he had done, Cavendish used the terms of the phlogiston theory to refer to the substances on which he had performed the experiments: that is, he used "inflammable air" to refer to hydrogen, "dephlogisticated air" to refer to oxygen, and so forth.

Our difficulties in deciding whether or not the terms used by phlogiston theorists refer are not confined to "dephlogisticated air." From the time of Stahl onwards, phlogistonians ordered the metals according to the ease with which they burned. They formulated their results using the relational expression "① is richer in phlogiston than ②," and, in seeking to understand their reports, we naturally take this expression to refer to the relation to which we refer by using "① has a greater affinity for oxygen than ②." Even the expression "phlogiston" seems sometimes to have been used in acts of successful reference. For a time, Priestley and Kirwan believed that the inflammable air which they had isolated was phlogiston. Once they had made this identification, they went on to record the properties of inflammable air (hydrogen) using the term "phlogiston."[17]

We encounter problems in assigning referents to such terms as "dephlogisticated air" because we attempt to combine uniform semantic treatment for all tokens of the term with the demands of a legitimate constraint on translation. Successful translation should accord with a principle which Richard Grandy has aptly dubbed the "principle of humanity."[18] This principle enjoins us to impute to the speaker whom we are trying to translate a "pattern of relations among beliefs, desires and the world [which is] as similar to ours as possible."[19] Historians standardly employ a principle of this type, and I have tacitly adopted it in my ascriptions of reference above. If we attempt to satisfy this principle, and if we treat all tokens of the same type in the same way, then we shall be led to the position defended by Kuhn and Feyerabend: there is no term of contemporary English which specifies the referent of "dephlogisticated air," so that a term which is central to the presentation of the phlogiston theory resists translation into contemporary language.

The remedy is not to opt for an inferior translation which renders some of Priestley's arguments or assertions inexplicable. We should abandon the search for a *CIT*, and allow that different tokens of "dephlogisticated air" refer differently. Some tokens of "dephlogisticated air" refer to oxygen, others fail to refer—and, in assigning referents to tokens we can do no better than to appeal to such general principles about reference as the principle of humanity. To decide on the referent of a token, we must construct an explanation of its production. Our explanation, and the hypothesis about reference which we choose, should enable us to trace familiar connections among Priestley's beliefs and between his beliefs and entities in the world. So, for example, when we find Priestley concluding that whatever gas remains after heating the red calx of mercury will be dephlogisticated air, we hypothesize that the referent of his token of "dephlogisticated air" is fixed as that which remains when phlogiston is removed from the air. This hypothesis enables us to reconstruct Priestley's reasoning and to render his judgment explicable: believing that the liberation of mercury involves absorption of phlogiston, Priestley reasonably infers that the residual air will be poor in phlogiston. At other times, when he concludes that dephlogisticated air supports combustion better than ordinary air, we recognize a basis for his judgment by supposing that it proceeds from an experiment in which oxygen was evolved, that his beliefs concern the gas evolved in that experiment, and that the tokens of "dephlogisticated air" which he uses in recording his beliefs refer to oxygen. In both cases, our ascriptions of reference are guided by the principle of humanity.[20]

Hence the case of the phlogiston theory does not force us to accept conceptual relativism.[21] The evidence that shows us that we cannot provide an adequate *CIT* for the language of the phlogiston theory shows us how to construct an adequate *CST*. Different tokens of the same type can be linked to the world in different ways. As I shall suggest in the next section, we may view the linkage between scientific terms and the world as being constantly renewed.

IV. Reference Potential

If the conclusions of the last section are correct, then a number of questions naturally arise. How is it possible for scientists to use different tokens of the same expression-type to refer to different entities? Does this possibility endanger scientific communication? Does it reflect carelessness on the part of the scientists involved? I shall try to show how the "historical explanation" account of reference can be developed to provide answers to these questions and to illuminate the phenomenon of conceptual change in science.

Let us look more closely at the circumstances under which Priestley came to use "dephlogisticated air" to refer to oxygen. Priestley began his discussion by talking about various attempts he had made to remove phlogiston from the air. He then recorded the details of his experiments on the red calx of mercury from which he had liberated a "new air." After a number of mistaken efforts to identify the gas, he finally managed to describe it in the terminology of the phlogiston theory.

> Being now fully satisfied with respect to the *nature* of this new species of air, viz. that, being capable of taking more phlogiston from nitrous air, it therefore originally contains less of this principle [i.e. phlogiston]; my next inquiry was, by what means it comes to be so pure, or philosophically speaking, to be so much dephlogisticated; . . .[22]

From our perspective Priestley has misdescribed the new gas. His remarks on this occasion identify the gas obtained by heating the red calx of mercury as dephlogisticated air, *and the token of "dephlogisticated air" produced here has its referent fixed through Stahl's definition (that is, it fails to refer).* Yet it is plausible to suppose that Priestley's mistake, the false statement made using a token of "dephlogisticated air" whose referent is fixed in the *old* way, set the stage for the subsequent production of tokens of "dephlogisticated air" whose referents are fixed differently. Priestley attached an old name to the new gas, and when he later went on to record the properties of dephlogisticated air he was talking about oxygen.

The general theory of reference which I espoused above proposes that the referent of a token is the entity which figures in an appropriate way in a historical explanation of the production of that token. An explanation of the production of a token will consist in a description of a sequence of events whose final member is the production of the token and whose first member is either an event in which the referent of the token is causally involved, or an event which involves the singling out, by description, of the referent of the token. Let us call the first member of the sequence the *initiating event* for the production of the token.[23] Priestley's early utterances of "dephlogisticated air" were initiated by an event in which Stahl specified phlogiston as the substance emitted in combustion. After Priestley had isolated oxygen and misidentified it, things changed. His later utterances could be initiated either by the event in which Stahl fixed the referent of "phlogiston" or by events of quite a different sort, to wit, encounters with oxygen.[24] Thus we can answer the question of how different tokens of a scientific term can refer to different entities by supposing that the production of different tokens can be initiated by different events.

The discussion of the last section points the way to develop further the "historical explanation" account of reference for scientific terms. When we construct an explanation of the production of a token, we attempt to link that token to an entity in the world through an initiating event, and, as we saw, our construction aims to make comprehensible the judgments and inferences of our subject. To say that an entity "figures in an appropriate way" in the explanation of the production of a token is, I suggest, to claim that the hypothesis that the token was initiated by an event in which that entity was causally involved or singled out by description best explains why our subject makes the assertions and arguments he does. We might say that the initiating event is that event whose effects are manifested in the subject's linguistic behavior when he produces the token. Thus, for example, Priestley's report of how the mice flourished on dephlogisticated air results from his perception of them, and we understand why he made his utterances by taking his experiment as the initiating event for his tokens of "dephlogisticated air."

To summarize, we identify the initiating event as the event such that the hypothesis that our subject is referring to the entity involved in that event best explains why he says the things he does. In some cases, our search for the initiating event may lead us back through events involving other speakers to some primal act of baptism by the first user of the term. However, when we are concerned with the utterances of scientists who are developing particular theories and using the vocabulary peculiar to those theories, I think that different explanations will often be appropriate. We may find, for example, that a scientist's argument presupposes that the referent of a term satisfies a particular description, and best explain his utterances by hypothesizing that they are initiated by an event in which the referent of the term is fixed by that description. Or, as in the case of Priestley, we may find reports which are prompted by particular observations and account for the scientist's judgments by taking the tokens he produces to be initiated by those observations. "Historical explanation" accounts of reference often paint the picture of a single chain connecting tokens of a term to the object singled out on the first occasion of its use. I am suggesting a rival picture in which the connections of terms to the world are often extended in subsequent uses. This picture appears to accord better with the continued reapplication and redefinition which is typical of scientific usage, and of which the case of Priestley furnishes a striking example.[25]

It would plainly be wrong to pretend that my development of the "historical explanation" account is entirely precise, but it does solve one apparent problem. There is an obvious connection between the event in which Stahl originally fixed the referent of "phlogiston" and the later event in which Priestley misdescribed oxygen. Because of this, we might be tempted to conclude that the historical explanations of *all* Priestley's tokens of "dephlogisticated air" will trace them back to Stahl's original act of reference-fixing. My criterion for identifying the initiating event enables us to avoid this conclusion. Consider, first, the occasion of Priestley's original misidentification of the gas he had isolated as dephlogisticated air. We best understand his linguistic behavior on this occasion by adopting the hypothesis that the referent of his token of "dephlogisticated air" was fixed as the substance obtained when phlogiston (that is, the substance emitted in combustion) is

Variation on Historical Explanation

removed from the air. However, when we consider his later reports of the antics of the mice, this hypothesis fails to explain why Priestley says what he does. Instead, we understand his assertions by viewing him as referring to the substance which inspired the mice, namely oxygen. On my version of the "historical explanation" account, different tokens can therefore be initiated by different events, even when those events are themselves causally connected.[26]

This picture of the ways in which scientific terms refer may appear to threaten the possibility of a common scientific language, for I have proposed that the links between words and the world vary with the utterer and the occasion of his utterance. Anxiety can be quelled by attending to further aspects of my example. Priestley was the first to use "dephlogisticated air" to refer to the new gas, but his practice was quickly followed by his fellow phlogistonians. The reason is obvious. Like Priestley, other phlogistonians believed that the gas evolved from the red oxide of mercury was the result of removing phlogiston from the air. In reporting on the properties of the gas, their tokens, too, were initiated by their observations, and thus referred to oxygen.

I suggest that an expression-type used by a scientific community is associated with a set of events such that productions of tokens of that type by members of the community are normally initiated by an event in the associated set. The set which is associated with a particular expression-type (in a particular community) will be called the *reference potential* of the expression (for that community). Terms which have a heterogeneous reference potential, that is, terms whose reference potential contains two or more different initiating events, may reasonably be called *theory-laden*. For their use depends upon hypotheses to the effect that the same entity is involved in the appropriate way in the different events which belong to the same reference potential. (These hypotheses may or may not have been explicitly formulated by the community which uses the terms.) The nature of the dependence is straightforward: if one of the hypotheses were to be seriously questioned then the use of the term which depends on it would have to be revised.

"Dephlogisticated air" is a good example of a theory-laden term. Once Priestley's practice became common, the reference potential of the term contained events of two different kinds. Later tokens could either be initiated by events in which dephlogisticated air was specified as the substance resulting when phlogiston (the substance emitted in combustion) is removed from the air, or by events in which the gas obtained by heating the red calx of mercury was isolated and investigated.

Does the use of theory-laden terms pose problems for scientific communication? Can it be avoided? I believe that the answer to both questions is

"No." Consider first, communication among the defenders of the phlogiston theory. Phlogistonians accept the hypothesis:

> *H*. There is a substance which is emitted in combustion and which is normally present in the air. The result of removing this substance from the air is a gas which can also be produced by heating the red calx of mercury.

Because they accept *H* they are willing to use tokens of "dephlogisticated air" to refer to the gas obtained by heating the red calx of mercury. When they come to identify the referents of one another's tokens they will either use the description "the substance obtained by removing from the air the substance which is emitted in combustion" or the description "the gas obtained by heating the red calx of mercury." They will use these descriptions interchangeably and this may lead them, on occasion, to misidentify the referent of a speaker's token. However, since the speaker, too, is a phlogistonian who accepts *H*, he will agree that either description identifies the referent of his token. To put the point another way, the mistakes which an audience will make in identifying the referent of an utterer's token will be mistakes which the utterer himself would have made.

Let us now consider the predicament of those who do *not* share the theoretical assumptions on which the use of a term with a heterogeneous reference potential depends. Their attempts to understand the language of their rival theorists will be governed by the principles mentioned at the end of the last section, that is they will endeavor to formulate hypotheses about the referents of their rivals' tokens which will explain their rivals' linguistic behavior. This enterprise can be advanced if the hypotheses on which theory-laden terms depend are formulated explicitly. Because of Priestley's straightforward claim that the gas produced by heating the red calx of mercury is the result of removing phlogiston from the air, Lavoisier and other French chemists were easily able to attribute to him a belief in *H*. Having done so, they were prepared to find Priestley identifying oxygen as dephlogisticated air, and so, with relative ease, the phlogistonians and their opponents were able to pinpoint their areas of agreement and disagreement.[27] In cases where the hypotheses on which theory-laden terms depend are not presented explicitly, the process of translation may be more difficult, but, even here, there is no reason to suspect that it is impossible. (Indeed, there are interesting examples from the history of science where the presuppositions of a theory-laden term are exposed and challenged.)

The idea that theory-ladenness stems from scientific irresponsibility is vulnerable to familiar objections. Hempel's lucid critique of operationalism[28] shows how and why the scientist inevitably courts ambiguity. We may be tempted to suppose that Priestley could have been more scrupulous. We imagine him, confronting the new gas and carefully refraining from giving it an old name. Let us suppose that he had called it "oxygen." But now we imagine that Priestley needs more samples. He wants to investigate thoroughly the properties of the new gas, and so he repeats the process of heating red calx of mercury. If Priestley continues to be scrupulous he must not call the new sample "oxygen," but must give it another new name—for the application of "oxygen" to the new sample would presuppose the hypothesis that the method of preparing a gas through heating the red calx of mercury always yields the *same gas*. Plainly, the demand for caution in applying old words quickly reduces to absurdity.

Our discussion so far suggests a framework within which we can elaborate the Hempelian dictum that "concept formation and theory formation go hand in hand." The notion of reference potential seems to capture something of Frege's nonreferential dimension of meaning. As several writers have pointed out,[29] Frege specifies the notion of sense in two different ways: the sense of an expression is that which is grasped by someone who understands the expression, and it is also the way in which the reference is determined. The reference potential is akin to the second idea of sense as "the manner in which the reference is presented." If we identify scientific concepts with reference potentials (thus explicating the ordinary, *non*-Fregean, notion of "concept"), we can clarify the idea that theoretical concepts must absorb theoretical hypotheses and so enhance our understanding of conceptual change in science.

In some cases, when the referent is singled out by *description* in the initiating event, it will be convenient to represent the referent as determined by the description, and, as a result, we may sometimes use a set of associated descriptions, instead of a set of events, to exhibit the ways in which the referents of tokens of an expression-type are usually determined.[30] By doing so, we may appear to make concessions to accounts of reference within the Fregean tradition which are quite different from the "historical explanation" account advocated above. The concessions are only apparent. For the description which determines the referent of a speaker's token will not necessarily be a description which he would provide, but rather a description used to single out an entity in the event which initiated the production of the token.

As the example of Priestley shows, acceptance of a hypothesis or a set of hypotheses may lead to the absorption within the reference potential of a term of a new event or class of events. If the members of a scientific community come to believe that circumstances of different types involve them in encounters with the same entity, or if they become convinced that the entity which satisfies one description satisfies another, then a term whose previous usage was initiated by *one* type of encounter or by *one* of the descriptions may undergo an expansion of its reference potential. Conversely, disconfirmation of previously accepted hypotheses of these kinds can lead to contraction of reference potentials. Putting the two types of change together we can allow for radical conceptual revision without conceptual discontinuity. The simplest way for this to occur is as follows: imagine that the reference potential of a term is first extended through the addition of a new class of events and that theoretical progress then leads to a contraction of the reference potential through deletion of all except the newly added class. Hence, our approach promises to solve a problem which has bedevilled most accounts of conceptual change in science, the problem of expressing the idea that while scientific concepts can change radically, they also change continuously.

At this point, someone may object that the promise is illusory, and that the notion of reference potential is less interesting than I have taken it to be. It may seem that I have stacked the deck by choosing examples from a badly mistaken theory, and, specifically, by focussing on terms which are either used to refer to different entities on different occasions, or else sometimes used to refer and, at other times, used in ways which fail to refer. I think that this objection is incorrect. Obviously the most *prominent* cases in which a term has a heterogeneous reference potential are cases in which the initiating events relate the term to different entities. However, even when the term is always used to refer to the same entity, it may, nonetheless, be used to refer to that entity in different ways; that is, its reference, on different occasions of utterance, may be effected *via* quite different initiating events. Appreciation of this point lends support to the view that the notion of reference potential is akin to Frege's idea of sense.[31]

Priestley and Cavendish again provide us with an example. They were luckier with hydrogen than with oxygen, and their assumption that both methods for preparing inflammable air produce the same gas is correct. However, we can suppose that things did not go so smoothly. Imagine a counterfactual situation in which water does not occur naturally, but in which there is a substance which resembles water in its most obvious properties. When this substance is boiled and is passed over heated iron, a gas, pseudohydrogen, which is very similar to hydrogen, is evolved. Cavendish prepares what he calls "inflammable air" by pouring sulphuric acid over iron *and* by passing the waterlike substance over heated iron. Sometimes he reports on the properties of samples of gas prepared in one way; sometimes on the properties of samples of gas prepared in the other.

I suggest that, in the situation envisaged, Cavendish sometimes refers to hydrogen and sometimes to pseudohydrogen when he uses the term "inflammable air." His utterances are initiated by two different types of event—just as some utterances of Priestley's were seen to be initiated by two different types of event in the case we considered above. But Cavendish's actual life and his life in the imagined situation are very similar. Sometimes Cavendish *actually* reported on the properties of a gas evolved from passing acid over metals, sometimes he reported on a gas evolved when steam is passed over heated iron. He always referred to hydrogen. But the similarity of his life in the actual and the imagined situation should lead us to construct parallel explanations of his various utterances, and since, in the imagined situation, Cavendish's shifting references compel us to regard his utterances as initiated by events of two distinct types, so, in the actual situation we must also see his utterances as initiated by two distinct types of events. (The difference is, of course, that, in actuality, the same entity, hydrogen, is appropriately involved in events of both types.) To put the point another way, if we were to suppose that Cavendish's actual utterances were initiated by only one type of event (for example, encounters with a gas obtained by pouring acid over metal), then we would only be able to account for Cavendish's shifts of reference, in the imagined situation, by supposing that utterances produced in a very similar way are to be explained quite differently. Since this last supposition is extremely implausible, we should conclude that in Cavendish's actual utterances he referred to hydrogen via two different kinds of initiating events.

I do not wish to pretend that we can always decide which event initiated the production of a given token, or that we can always identify the reference potential of a given expression-type. But I do suggest that we can often determine changes in the reference potential of a scientific term, and that the claims of historians about conceptual change are best understood as changes of reference potential. The notion of reference potential is a tool for exposing the fine grain of the history of science.[32,33]

V. Conclusions

To scotch the thesis of conceptual relativism it is not enough to point out that the thesis leads to absurd conclusions or that the arguments advanced in its support are self-defeating.[34] What is needed is to show how to accommodate the historical evidence which Kuhn and Feyerabend cite. I propose that we should abandon a traditional assumption of the philosophy of science, the assumption that we can reconstruct the language of a theorist by reconstructing the language of his theory. Instead, we should recognize that scientific expressions are associated with a complex apparatus—their reference potential—which changes as science develops. Claims about the development of scientific concepts are, I suggest, best understood as claims concerning the changes of reference potentials.

Because the reference potential of scientific expressions is frequently heterogeneous and presupposes the generalizations of a particular theory, it will often be the case that there are some crucial expressions in the languages of theorists, separated by a large change in theory, whose reference potential cannot be matched by any expression of the rival theory. Nevertheless, successful communication can continue, even when reference potential has changed, because each theorist can specify the referents of his rival's individual tokens.

This conclusion should not be surprising. Trivially, there are just the entities there are. When we succeed in talking about anything at all, these entities are the things we talk about, even though our ways of talking about them may be radically different. However variable the connections we draw among its constituents, the world supplies a common content for our references.[35]

Notes

1 For Kuhn's scepticism about expressing the content of one theory in the language of another, see *The Structure of Scientific Revolutions* (Chicago, 2nd edition, 1970) pp. 101–103, 132, 149–150. Kuhn appears to retract his scepticism a little in the *Postscript* (ibid. pp. 199–202), but scepticism soon re-emerges once again (pp. 203–204).

2 This thesis has survived radical changes in Feyerabend's methodological views. See, for example, "Explanation, Reduction, and Empiricism" (in H. Feigl and G. Maxwell (eds.) *Minnesota Studies in the Philosophy of Science*, Volume III, Minneapolis, 1962, pp. 28–97), especially pp. 52–62, 68–70, 74–85; "Problems of Empiricism" (in R. Colodny (ed.) *Beyond the Edge of Certainty*, Prentice-Hall, Englewood Cliffs, 1965, pp. 145–260), especially pp. 168–173, 179–181; "Reply to Criticism" (in R.S. Cohen and M. Wartofsky (eds.), *Boston Studies in the Philosophy of Science*, Volume II, Humanities Press, New York, 1965, pp. 223–261), especially pp. 230–235; "On the 'meaning' of Scientific Terms," (*Journal of Philosophy*, 61, 1964, pp. 497–509); "Against Method," (in M. Radner and S. Winokur (eds.) *Minnesota Studies in the Philosophy of Science*, Volume IV, Minneapolis, 1970, pp. 17–130), especially pp. 81–90.

3 See, for example, Dudley Shapere, "Meaning and Scientific Change," (in R. Colodny (ed.) *Mind and Cosmos*, Pittsburgh, 1966, pp. 41–85); Peter Achinstein, *Concepts of Science*, (Johns Hopkins, 1968) pp. 91–98; Donald Davidson, "On the Very Idea of a Conceptual Scheme," (*Proceedings and Addresses of the American Philosophical Association*, 1973–4, pp. 5–20; Carl Kordig, *The Justification of Scientific Change* (D. Reidel, 1971), Chapters II and III.

4 There is some difficulty in saying what exactly must be done to specify the referent of a term. See §II below.

5 In *Science and Subjectivity* (Bobbs-Merrill, Indianapolis, 1967) pp. 57–64.

6 See, for example, Kuhn's "Reflections on my Critics" in I. Lakatos and A. Musgrave (eds.) *Criticism and the Growth of Knowledge* (Cambridge, 1970, pp. 231–278) and Feyerabend's "On the 'meaning' of Scientific Terms."

7 Of course, we do not have a precise, explicit version of this theory. Rather we have some tacit views about the determination of reference which it is the task of the general theory of reference to elaborate and codify.

8 See Keith Donnellan, "Speaking of Nothing," (*Philosophical Review*, LXXXIII, 1974, pp. 3–31). Similar views can be found in Saul Kripke, "Naming and Necessity," (in D. Davidson and G. Harman (eds.) *Semantics of Natural Language* (D. Reidel, 1972), pp. 253–355), in Hilary Putnam "The Meaning of 'Meaning'," (in K. Gunderson (ed.) *Language, Mind, and Knowledge* (Minneapolis, 1975) pp. 131–193), and in Michael Devitt, "Singular Terms," (*Journal of Philosophy*, LXXI, 1974, pp. 183–205), and "Semantics and the Ambiguity of Proper Names," (*The Monist*, 59, (1976), pp. 404–424).

9 I add the qualification to allow for the possibility that, in some cases, the facts of reference may be genuinely indeterminate.

10 The account I shall give derives chiefly from the following sources: Joseph Priestley *Experiments and Observations on Different Kinds of Air* (3 Volumes, Kraus Reprint, New York, 1970; R. E. Schofield (ed.) *A Scientific Autobiography of Joseph Priestley, 1733–1804* (MIT, 1966); Henry Cavendish, *Experiments on Air* (Alembic Club Reprints No. 3, Edinburgh, 1961); R. Kirwan *An Essay on Phlogiston* (Frank Cass Reprint, London, 1968); J. R. Partington, *A Short History of Chemistry* (London, 1937), Chapters V–VII; J.B. Conant "The Overthrow of the Phlogiston Theory" (in *Harvard Case Histories in Experimental Science*, Harvard, 1948, pp. 67–115).

11 Kuhn and Feyerabend are, of course, sceptical of talk of theory-independent truth. Hence it might appear that here and in the following pages I am begging the question against them. However, we should remember that their discussion of historical examples is meant to wean us away from the natural way of describing the claims of past theorists as true or false. So it suffices to show that we can account for the historical material while continuing to talk about the claims of past theorists as true or false. My aim is to trace the way in which apparent problems arise and to indicate a way of overcoming them.

12 This is historically controversial. Stahl's predecessor, Becher, might be given the credit of founding the phlogiston theory. However, accurate resolution of the controversy is beside the point here.

13 The semantic connections between "phlogiston," "dephlogisticated air" and "phlogisticated air" are obvious from the inferences which the

phlogiston theorists make. See, for example, the opening sentence of Priestley, *Experiments* Vol. II, Part I, Section II (quoted below in §IV) and Cavendish, op.cit. pp. 5–9.

14 The intuitive idea that, if "phlogiston" fails to refer, then "dephlogisticated air" will fail to refer, can be defended either by adopting a Fregean principle to the effect that a complex expression formed by extensionally embedding a nonreferring expression fails to refer, or by accepting an analogue of a principle of substitutability for nonreferring names. On the latter approach, we would suppose that the referent of a complex expression formed by extensionally embedding a nonreferring expression preserves its reference under substitution of nonreferring expressions for nonreferring expressions of the same syntactic category. Hence, the expressions, "the result of removing phlogiston from air," "the result of removing the philosopher's stone from air," "the result of removing the substance which is both lighter and heavier than gold from air," would all be coreferential. Since the latter expressions clearly fail to refer, we must suppose that the former does too. (The general idea of substitutability principles for nonreferring expressions is due to Hartry Field.)

15 Priestley, *Experiments* Vol. II, pp. 161–162.

16 Cavendish, op. cit., p. 19.

17 For the identification, see Priestley's letters of 1782–4 in Schofield, op. cit., especially numbers 93 (to Josiah Wedgwood), 95 (to Benjamin Franklin), 98 (to Wedgwood), 108 (to Arthur Young), and 109 (to Wedgwood). The subsequent use of "phlogiston" to refer to hydrogen can be found in letters to Sir Joseph Banks, number 114, and to Jean André DeLuc, number 115.

18 See Richard Grandy "Reference, Meaning and Belief" (*Journal of Philosophy*, LXX, [1973], pp. 439–452). A similar principle is advocated by Kathryn Pyne Parsons in "A Criterion for Meaning Change" (*Philosophical Studies*, 28, [1975], pp. 367–396). Parsons' discussion is particularly useful in showing how the principle can be used in studying the history of science (see pp. 378–388).

19 Grandy, op. cit. p. 443.

20 Perhaps it is worth pointing out explicitly that, in some cases, two (or more) hypotheses about the referents of a speaker's tokens may provide equally good accounts of his linguistic behavior. In such cases, the principle of humanity will not decide between the hypotheses. But even if this principle (and other canons of translation) leave ascriptions of reference undetermined, this will not support the conclusions favored by Kuhn and Feyerabend (see above pp. 170–171).

21 The case of the phlogiston theory is one of Kuhn's favorite examples. For reasons of space, I have not tackled directly Feyerabend's most detailed argument which rests on the difficulties of specifying the referent of the

medieval term "impetus" within the language of Newtonian theory (see "Explanation, Reduction, and Empiricism" pp. 52–62). A close study of Feyerabend's reasoning will show that it invites the type of approach I have recommended. In brief, Feyerabend correctly denies that we can take Buridan's term (-type) "impetus" to refer either to momentum or to inertia, because, in Buridan's usage, impetus is fixed as a force which sustains bodies in motion. Feyerabend wishes to contend, however, that "impetus" refers to *something*—presumably on the grounds that some of Buridan's remarks seem to anticipate important ideas of modern dynamics—even though Newtonian dynamics denies the existence of such forces. Hence, he concludes that the referent of "impetus" is not specifiable in Newtonian terms. But we are not compelled to accept this conclusion. We may reasonably suppose that *most* of Buridan's tokens of "impetus" fail to refer, while *some* refer to momentum and others to inertia. That is, we can construct a *CST* for Buridan's language.

22 Priestley, *Experiments*, Vol. II p. 120.

23 This notion is akin to Hilary Putnam's idea of an *introducing event*. (See his essay "Explanation and Reference" in Glenn Pearce and Patrick Maynard (eds.) *Conceptual Change* [D. Reidel, 1973] pp. 199–221, especially p. 203.) My notion differs from Putnam's in that I am explicitly concerned with cases in which *various* events are associated with one expression-*type*.

24 More exactly, events in which oxygen is liberated from the red oxide of mercury.

25 In fact, a similar picture of reference may also work well for proper names. It seems more plausible to suppose that those who have met me and use my name refer to me via their encounters with me rather than via a chain which extends back to my baptism. This supposition can be defended by considering cases of misidentification, but, since the issue of the reference of proper names is secondary to my main point, I shall not pursue it at length.

26 Tokens of proper names can also be initiated by different initiating events which are themselves causally connected. The naming of Albert Herring Sr. may play a causal role in the naming of Albert Herring Jr., but it would be folly to suggest that the father's baptism initiates all tokens of the name which he shares with his son. Here, too, the approach which I have recommended will enable us to identify the initiating event and arrive at the correct ascription of reference.

27 Clearly I am assuming that some expressions (such as those which figure in *H*) pose no problems for translation. It is important to notice that the Kuhn–Feyerabend thesis that such expressions cannot be translated homophonically depends on the prior conclusion that key expressions of the theory cannot be translated at all. Put starkly, the argument would

run as follows: (*a*) the referent of "phlogiston" cannot be specified in contemporary English; (*b*) Stahl's sentence "Phlogiston is the substance which is emitted in combustion" fixes the referent of "phlogiston"; therefore, (*c*) some (or all) of the terms "substance," "emission," "combustion," as used by Stahl, cannot be translated into contemporary English. (The clearest presentation of an argument like this is Feyerabend's discussion of "impetus," referred to in note 21 above.) I have tried to show that the evidence for (*a*) can be accommodated without abandoning the assumption that terms like "substance," "emission," "combustion" can be translated homophonically. If this is correct, we have no reason to accept (*a*), and hence no reason to accept the idea that the central terms of a theory infect all the language which is used in presenting it. Ironically, the historical evidence gives overwhelming support to the claim that translations of the kind I envisage are possible. The interchanges between Kirwan and the French chemists are free from problems of communication (see Kirwan op. cit., especially pp. 56–57, 104–105, 115–117, 176–177, 201–203, 281–283, 314–316). Moreover, Cavendish suggests a clear procedure for translating between his own language and that of Lavoisier (Cavendish op. cit. pp. 35–38).

28 C. G. Hempel, *Philosophy of Natural Science* (Prentice-Hall, 1966) Chapter 7, and "A Logical Appraisal of Operationism" (in *Aspects of Scientific Explanation*, New York, Free Press, pp. 123–133).

29 See, for example, Kripke, op. cit., pp. 277 and note 28.

30 Berent Enç has argued cogently that some scientific terms have their referents fixed by description. (See his "Reference of Theoretical Terms," *Noûs*, X, 1976, pp. 261–282.) It is possible that the referents of different tokens of some scientific terms are fixed via events in which different descriptions are used, and we might naturally represent the reference potentials of such terms as sets of descriptions.

31 In particular, the notion of reference potential has a property analogous to a prominent feature of Fregean senses: changes in reference potential need not bring any change in the set of entities referred to via events in the reference potential.

32 Feyerabend's example of the impetus theory mentioned in note 21 shows the need for a tool of this kind. Historians of medieval physics (such as Marshall Clagett, Alexandre Koyré, and Analiese Maier) would like to view the impetus theory as a transitional phase between Aristotelian dynamics and Newtonian dynamics. Their idea that there is a development of the concept of inertia from Buridan through the work of Benedetti and Galileo to Newton can readily be understood in terms of developments in the reference potential of "impetus." By contrast, Feyerabend's thesis of conceptual relativism fails to allow for any type of conceptual continuity. (For a

penetrating study of the impetus theory, see Marshall Clagett, *The Science of Mechanics in the Middle Ages* [Wisconsin, 1959], chapters 8–11 and, for Clagett's assessment of the theory, pp. 669–671.)

33 A rather different approach to the question of how to identify the referents of terms in the languages of past scientists has been suggested by Hartry Field ("Theory Change and the Indeterminacy of Reference," *Journal of Philosophy*, LXX, 1973, pp. 462–481). Field's strategy is interesting and important, and might prove complementary to that which I have described above. However, Field's major example does not seem to be a case of the type for which his approach is needed. (See John Earman and Arthur Fine "Against Indeterminacy," *Journal of Philosophy*, LXXIV, 1977 pp. 535–538.) Moreover, like Earman and Fine, I have been unable to find a convincing example from the history of science which would demand the use of Field's apparatus of partial reference.

34 Thus while the arguments of Achinstein, Scheffler, Shapere, Davidson and Kordig show that something is amiss with the Kuhn–Feyerabend position they fail to account for the evidence on which the position rests. Unfortunately, sceptical positions cannot be satisfactorily dismissed with a quick *reductio*. We need a diagnosis of the reasons leading to scepticism. So, for example, Davidson's assurances that we will always be able to translate any alien *language* fail to show us what is especially difficult about reconstructing the languages used by past scientists, and how the difficulties can be overcome.

35 I would like to thank Dale Kent, David Lewis and George Sher for helpful comments. I am particularly grateful to Patricia Kitcher and to the editors of *The Philosophical Review* for forcing me to clarify my thinking on these issues.

QUESTIONS

1 According to the "received view," theoretical terms, such as "electron" or "gene", derive their entire meaning from their partial observational interpretation, which proceeds by connecting such terms with observable states of affairs by rules of correspondence. Does this view provide a satisfactory analysis of the meaning of theoretical terms?

2 Some philosophers of science debated the virtues of the so-called Correspondence Principle stating that an older theory of a particular domain of phenomena must be a *limiting case* of a newer theory superseding it. For example, classical mechanics is a limiting case of relativistic mechanics for velocities much smaller than the speed of light or, alternatively, in the limit $c \to \infty$; the same classical mechanics is a limiting case of quantum

mechanics when h (Planck's constant) goes to zero. Defend or criticize this principle, drawing on the readings in this part and your knowledge of the history of science. (You may want to return to this question after you have read Larry Laudan's chapter in the next set of readings.)

3 Reflect on the Kuhn–Feyerabend thesis of "conceptual relativism" or "incommensurability." Kitcher argues that despite conceptual change (which he takes to be primarily change of "reference potentials"), communication across theoretic boundaries is possible and, in many cases, unproblematic. Do you agree with his conclusion? Support or challenge it by appealing to other examples from the history of science. (You may want to return to this question after studying Kuhn's and Shapere's chapters in Part VI.)

FURTHER READING

Two classic sources for the "received view" of scientific theories are Nagel (1961) and Hempel (1966). An important paper by Hempel, "The Theoretician's Dilemma" (reprinted in Hempel 1965), focuses on the tension between the empiricist demand that all theoretical terms receive observational (if only partial) interpretation and the indispensability of such terms to the development of science. Suppe (1977) is a seminal volume containing a very detailed discussion of the received view, as well as its critique.

For Nagel's theory of intertheoretic reduction, see Chapter 11 of his (1961). The literature on reduction (and reductionism) is extensive. The interested reader could begin with Chapter 4 of Rosenberg (1985), where the issue is discussed with applications to biology.

Kuhn's *The Structure of Scientific Revolutions* (1996) is a must for everyone studying philosophy of science and a locus classicus for the incommensurability thesis. Kuhn's later thoughts on the matter are collected in his (1977).

The "historical explanation" theory of reference for scientific terms mentioned in Kitcher's article was put forward in the 1970s by Kripke (1972) and Putnam (1975).

PART IV

SCIENTIFIC REALISM

INTRODUCTION

Many scientists and some philosophers reject the dilemma created by the explanatory indispensability of theoretical terms and the epistemological impossibility of theoretical knowledge. They hold that though we may not be able to hear, taste, smell, touch, or see electrons, genes, quasars, and neutron stars, or their properties, we have every reason to think that they exist. For our scientific theories tell us that they do, and these theories have great predictive and explanatory power. If the most well-confirmed theory of the nature of matter includes the laws about molecules, atoms, leptons, bosons, and quarks, then surely such things exist. If our most well-confirmed theories attribute charge, angular momentum, spin, or van der Waals forces to these things, then surely such properties exist. On this view theories must be interpreted literally, not as making claims whose meaning is entirely grounded in observations, but as telling us about things and their properties, where the meaning of the names for these things and their properties is no more or less problematical than the meaning of terms that name observable things and their properties. And if this conclusion is incompatible with the view which makes observational terms the basement level of language and requires all other terms to be built out of them, then so much the worse for that view. And so much the worse for the empiricist epistemology that goes along with it.

This approach to the problem of theoretical terms is widely known as Scientific Realism, since it takes the theoretical commitments of scientific theories seriously, as pertaining to reality, and not just as (disguised) abbreviations for a set of observational claims. Scientific Realists start with a manifestly obvious fact about science: its great and ever-increasing predictive power. Over time our theories have improved both in the range and the precision of their predictions. Not only can we predict the occurrence of more and more different kinds of phenomena, but over time we have been able to increase the precision of our predictions—the number of decimal places or significant digits to which our scientifically derived expectations match up with our actual meter readings. These long-term improvements translate themselves into technological applications on which we increasingly rely, indeed on which we literally stake our lives every day. This so-called "instrumental success" of science cries out for explanation. Or at least the Realist insists that it does. How can it be explained? What is the best explanation for the fact that science "works"? The answer seems evident to the Realist: science works so well because it is (approximately) true. It would a miracle of cosmic proportions if science's predictive success and its technological applications were just lucky guesses, if science worked, as it were, by accident.

Despite its attractions to scientists, the difficulty of reconciling Realism with Empiricism has made Realism a controversial thesis in philosophy of science. Nagel's selection "The Cognitive Status of Theories" recapitulates the alternative arguments for Realism and its traditional opponent, Instrumentalism, as of about 1960. Instrumentalism names the view that scientific theories are useful instruments, heuristic devices, tools we employ for organizing our experience, but not literal claims about it that are either true or false. This philosophy of science goes back at least to the eighteenth-century British empiricist philosopher Berkeley, and is also attributed to leading figures of the Inquisition who sought to reconcile Galileo's heretical claims about the motion of the earth round the sun with Holy Writ and Papal pronouncements. According to some versions of the history these learned churchmen recognized that the heliocentric hypothesis was at least as powerful in prediction as Ptolemaic theories, according to which the sun and the planets moved around the Earth; they accepted that it might be simpler to use in calculations of the apparent positions of the planets in the night sky. But the alleged motion of the Earth was observationally undetectable; it does not feel to us that the earth is moving. Galileo's theory required that we disregard the evidence of observation, or heavily reinterpret it. Therefore, these officers of the Inquisition urged Galileo to advocate his improved theory not as literally true, but as more useful, convenient, and effective an instrument for astronomical expectations than the traditional theory. Were he so to treat his theory, and remain silent on whether he believed it was true, Galileo was promised that he would escape the wrath of the Papal Inquisition. Although at first he recanted, Galileo eventually declined to adopt an instrumentalist view of the heliocentric hypothesis and spent the rest of his life under house arrest. Subsequent instrumentalist philosophers and historians of science have suggested that the Church's view was more reasonable than Galileo's. And although Berkeley did not take sides in this matter, his empiricist arguments against the intelligibility of Realism (and of realistic interpretations of parts of Newton's theories) made Instrumentalism more attractive. Berkeley went on to insist that the function of scientific theorizing was not to explain but simply to organize our experiences in convenient packages. On this view, theoretical terms are not abbreviations for observational ones; they are more like mnemonic devices, acronyms, uninterpreted symbols without empirical or literal meaning. And the aim of science is constantly to improve the reliability of its instruments, without worrying about whether reality corresponds to these instruments when they are interpreted literally.

Nagel's selection treats the dispute between Realists and Instrumentalists as a stand-off and concludes that the dispute about how we are to interpret theories is largely a verbal one. Since Nagel's time the controversy has taken on greater urgency, in some measure because Realism is seen as a strong bulwark against the attack on the objectivity of science mounted by sociologists and historians, as we shall see in Part VI below.

In "A Confutation of Convergent Realism" Larry Laudan, no friend to subjectivism or for that matter Instrumentalism, nevertheless mounts a strong challenge to Realism's claims that (1) the history of science is a history of convergence on the truth or successive approximation to it, and (2) the suggestion that only Realism can explain such convergence. He shows how many basic problems in the philosophy of logic and language need to be solved fully even to understand these two theses, let alone to substantiate them. What is more, he argues, the history of science teaches us that many successful scientific theories have completely failed to substantiate the Scientific Realist's picture of why theories succeed. Well before Kepler, and certainly since his time, successful scientific theories have not only been false and improvable, but if current science is any guide, they have sometimes been radically false in their claims about what exists and what the properties of things are, even as their predictive power has been persistently improved. One classical example is the eighteenth-century phlogiston theory, which embodied significant predictive improvements over prior theories of combustion, but whose central explanatory entity, *phlogiston*, is nowadays cited with ridicule. Still another example is the classical theory of light as a wave phenomenon. This theory managed to increase substantially our predictive (and our explanatory) grasp on light and its properties. Yet the theory claims that light moves through a medium of propagation, an ether. Without such a medium, light would turn out to be a mysterious phenomenon. Subsequent physics revealed that despite its great predictive improvements, the central theoretical postulate of the ether theory does not exist. It is not required by more adequate accounts of the behavior of light. Postulating the ether contributed to the "unrealism" of the classical theory of light. This at least must be the judgment of contemporary scientific theory. But by a "pessimistic induction" from the falsity—sometimes radical falsity—of predictively successful theories in the past, it would be unsafe to assume that our current "best-estimate" theories are immune to a similar fate. Since science is fallible, one might expect that such stories can be multiplied to show that over the long term, as science progresses in predictive power and technological application, the posits of its theories vary so greatly as to undermine any straightforward inference to Scientific Realism's interpretation of its claims. Laudan offers no competing explanation for the success of science, but his article must reduce confidence in the explanation which Scientific Realism can offer.

What is more, Scientific Realism is silent on how to reconcile the knowledge it claims we have about the (approximate) truth of our theories about unobservable entities with the empiricist epistemology that makes observation indispensable for knowledge. In a sense, Scientific Realism is part of the problem of how scientific knowledge is possible, not part of the solution.

The stand-off between Realism and Instrumentalism has spawned at least one attempt at a compromise, van Fraassen's Constructive Empiricism. This view suggests that, like Realists, we interpret the claims of science literally as claims about reality, and not figuratively as instruments for organizing our experience; but like Instrumentalists we restrict the goals of science to successful prediction of our experiences, to "empirical adequacy," in van Fraassen's terms. We can thus honor the literal meaning of the indispensable theoretical claims of science, without any unscientific commitment to their (increasing approximation to the) truth. The distinction between observable and unobservable entities plays a central role in this account. But how is one supposed to draw it? After all, the methods and techniques of observation and experimentation are constantly improving: thus what was unobservable yesterday is observable today, and what is unobservable today may become observable tomorrow. Gary Gutting's "Scientific Realism versus Constructive Empiricism" illuminates the strength and weakness of van Fraassen's purported compromise in the form of dialog. In "A Case for Scientific Realism" Ernan McMullin advances a sophisticated defense of Scientific Realism that is cognizant of the philosophical and historical difficulties it faces. The reader is invited to reflect critically on some of the important lessons McMullin wants to draw from his analysis. (1) There are many varieties of "realisms" and "antirealisms" and they should be carefully distinguished. (2) Laudan's "pessimistic induction" may be based on a rather superficial interpretation of the historical cases he considers; when read properly, most of them are consistent with Realism. (3) Scientific Realism is not a blanket claim to the effect that any empirically successful theory should automatically be regarded as a true story about the world, but as a more cautious claim that empirical success manifested by a theory over a considerable period of time gives one strong reason to believe in the existence of its theoretical posits, especially when "empirical adequacy" (in van Fraassen's sense, involving successful predictions of new phenomena) is accompanied by a steady progress in "structural explanation."

12

Ernest Nagel, "The Cognitive Status of Theories"

Although Ernest Nagel's (1901–1985) classic book (1961) was written long before the current debate about Scientific Realism took shape and bears many marks of the Logical Positivist agenda, the selection below provides both a clear statement and a useful discussion of the two rival views of the cognitive status of theories, Instrumentalism and Realism.

I. The Descriptive View of Theories

The cognitive status of universal statements in general, and of scientific theories in particular, has been the subject of a long and inconclusive debate. . . . According to the first and historically oldest account, a theory is literally either true or false; and, although a theory can at best be established only as "probable," it is as significant to ask whether a theory is true or false as it is to ask a similar question about a statement concerning some individual matter of fact, such as the statement "Krakatoa was destroyed by a volcanic eruption in 1883." A corollary often drawn from this view is that when a theory is well supported by empirical evidence, the objects ostensibly postulated by the theory (e.g., atoms, in the case of an atomic theory) must be regarded as possessing a physical reality at least on par with the physical reality commonly ascribed to familiar objects such as sticks and stones.

A second (and historically the youngest) position on the cognitive status of theories maintains that theories are primarily logical instruments for organizing our experience and for ordering experimental laws. Although some theories are more effective than others for attaining these ends, theories are not statements, and belong to a different category of linguistic expressions than do statements. For theories function as rules or principles in

E. Nagel, *The Structure of Science: Problems in the Logic of Scientific Explanation*, 1961, pp. 106–52 (excerpts). New York and Burlingame: Harcourt, Brace & World, Inc.

accordance with which empirical materials are analyzed or inferences drawn, rather than as premises from which factual conclusions are deduced; and they cannot therefore be usefully characterized as either true or false, or even as probably true or probably false. However, those who adopt this position do not always agree in their answers to the question whether physical reality is to be assigned to such theoretical entities as atoms.

Finally, the third stand on the cognitive status of theories is a sort of halfway position between the other two. According to it, a theory is a compendious but elliptic formulation of relations of dependence between observable events and properties. Although the assertions of a theory cannot be properly characterized as either true or false when they are taken at face value, a theory can nevertheless be so characterized insofar as it is translatable into statements about matters of observation. Proponents of this position usually maintain, therefore, that in the sense that a theory (such as an atomic theory) can be said to be true, theoretical terms like "atom" are simply a shorthand notation for a complex of observable events and traits, and do not signify some observationally inaccessible physical reality.

This third view, which we shall consider first, is associated with the historically influential conception that the sciences never "explain" anything, but merely "describe" in a "simple" or "economical" fashion the succession and concomitance of events. . . . The conception was vigorously espoused by many nineteenth-century scientists in reaction to the development of atomistic theories in physics and chemistry, since these theories appeared to them not only to be unnecessary for systematizing the experimental facts but also to assign an unwarranted absolute priority to Newtonian mechanics. Moreover, the descriptive account of science was espoused by many thinkers who rejected the assumptions of classical rationalism and sought to emancipate science from any dependence on unverifiable "metaphysical" commitments. In its inception, at any rate, the descriptive thesis was regarded both as an accurate analysis of the nature of physical science and as a weapon in the struggle against philosophies that were felt to hinder the development of science.

. . .

The most radical form of the descriptive thesis is simply the consistent extension of the phenomenalist theory of knowledge to the materials of the sciences. According to this theory, the psychologically primitive and indubitable objects of knowledge are the immediate "impressions" or "sense contents" of introspective and sensory experience. Moreover, if the postulation of inherently unknowable (because observationally inaccessible) things is to be avoided, all expressions ostensibly referring to such hypothetical objects (which include the physical objects of common sense) must be *defined* in terms of these immediate data. In consequence, every empirical statement containing expressions other than those designating such data (or complexes of such data) must in principle be translatable without loss of verifiable meaning into statements about the succession or coexistence of the allegedly immediate objects of experience. . . .

An allied but in some ways less radical form of the descriptive view of science . . . accepts the common-sense notion that normally we directly observe sticks and stones and animals, the motions of bodies and the actions of men, and the like. It therefore takes ordinary "gross experience" as the starting point for its analyses, even though it recognizes that judgments based on such experience are frequently erroneous and must be corrected in the light of further reflection. The thesis which this version of the doctrine maintains is that all theoretical statements are in principle translatable, again without loss of meaningful content, into statements . . . about the observable events, things, properties, and relations of common-sense and gross experience. Accordingly, on this conception of the doctrine also, the claim that theories are simply conveniently compendious descriptions is once more a thesis concerning the translability of theoretical statements, though this time into the familiar language that formulates the materials of publicly verifiable experience.

However, both versions of the descriptive view as here interpreted encounter serious problems.

1. The first version is beset by the standing difficulty of phenomenalism: that, although it is a thesis about the translatability of theoretical statements into the "language" of sense data, an autonomous language of bare sense contents actually does not exist, nor is the prospect bright for constructing one. As a matter of psychological fact, elementary sense data are not the primitive materials of experience out of which all our ideas are built like houses out of initially isolated bricks. On the contrary, sense experience normally is a response to complex though unanalyzed patterns of qualities and relations; and the response usually involves the exercise of habits of interpretation and recognition based on tacit beliefs and inferences, which cannot be warranted by any single momentary experience. Accordingly, the language we normally use to describe even our immediate experiences is the common language of social communication, embodying distinctions and assumptions grounded in a large and collective experience, and not a language whose meaning is supposedly fixed by reference to conceptually uninterpreted atoms of sensation.

. . . In short, the "language" of sense data is not an autonomous language, and no one has yet succeeded in constructing such a language. However, if there is indeed no such language, the thesis that all theoretical statements are in principle translatable into the language of pure sense contents is questionable from the outset.

2. But however this may be, further difficulties emerge in connection with the notion of translatability. In the familiar sense of the word "translatable," a statement in one language is translatable into another language only if there is a statement (or a finite conjunction of statements) in the latter equivalent in meaning (or logically) to the given statement. In this sense translations from one natural language into another are plentiful, despite occasional disagreements on the adequacy of proposed translations. For example, no one who understands French and English will seriously question that the English statement "At constant temperature, the volume of a given mass of a gas is inversely proportional to its pressure" is a translation of the French statement "*A une même température, les volumes occupés par une même masse de gaz sont en raison inverse des pressions qu'elle supporte.*"

Is there any evidence that every statement in science, and in particular every theoretical statement, is translatable in this sense either into a phenomenalistic language or into the language of gross experience? The evidence would be conclusive, if each subject-matter term employed in the sciences were actually introduced by way of an *explicit* definition (or by way of some other variant of substitutive definitions) whose subject-matter expressions all belong to the language of observation. For in that case, all terms in the sciences not occurring in this language would be eliminable in favor of those occurring in it. But in point of fact, as has already been noted, theoretical notions are not introduced in this way, so that neither version of the descriptive view of science is immediately warranted by considerations of actual scientific practice. The question remains whether, despite the facts of actual procedure, theoretical terms cannot in principle be eliminated in consonance with the descriptive thesis.

. . .

3. There is indeed a general consensus that the outlook for establishing the thesis is dim when the word "translatable" is understood in its customary sense. In current discussions, at any rate, that thesis has been considerably weakened. The thesis is asserted not in the form mentioned above, but in the sense that for every theoretical statement there is a *class* of observation statements which is logically equivalent to the given statement, thus leaving it open whether the class is finite or not. The point of this emendation, and the import of its consequences, will be evident from an example. Let us assume that the expression "electric current" is a theoretical term, for which appropriate rules of correspondence have been established. It would then be

generally acknowledged that the statement "There is now an electric current in this wire" (asserted at a given time for a given wire) is not equivalent in content to, say, the conditional observation statement "If the galvanometer on that shelf were introduced into this circuit, the pointer of the instrument would be deflected from its present position." The equivalence does not obtain for at least two reasons. On the assumption that the theoretical statement implies anything at all about the behavior of any galvanometer, it implies not only a single statement about a particular galvanometer but an indefinitely large class of similar statements about all other such instruments. Accordingly, if the original statement about a wire is at all equivalent to statements about the behavior of galvanometers, the statement must be equivalent to an indefinitely large (perhaps infinitely large) class of them.

In the second place, the presence of an electric current in the wire is associated with observable phenomena other than the behavior of galvanometers. As is well known, optical, thermal, chemical, and other magnetic phenomena could also be used as evidence upon which to decide whether or not the wire is carrying a current. In consequence, the class of statements which is supposedly equivalent to the theoretical statement must also include statements about these additional phenomena as well. On the other hand, it is difficult to fix the membership of that supposed class, and it is certainly not possible to specify that membership once and for all and in detail. For we cannot foresee the experimental discoveries that may be made in the future, some of which may provide still further ways (at present unsuspected) of detecting the presence of a current in a wire. In consequence, statements about these still unknown but hypothetically relevant phenomena must also be included in the class equivalent to the theoretical statement, so that the variety and number of such member statements may be greater than those we can specify at any given time. Accordingly, the indicated emendation of the translatability thesis is consonant with the possibility that this supposed class is not only infinitely numerous but is also incapable of being definitely specified.

. . .

II. The Instrumentalist View of Theories

The position which we shall call, for the sake of brevity, the "instrumentalist" view of the status of scientific theory has received a variety of formulations. . . .

The central claim of the instrumentalist view is that a theory is neither a summary description nor a generalized statement of relations between observable data. On the contrary, a theory is held to be a rule or a principle for analyzing and symbolically representing certain materials of gross experience, and at the same time an instrument in a technique for inferring observation statements from other such statements. For example, the theory that a gas is a system of rapidly moving molecules is not a description of anything that has been or can be observed. The theory is rather a rule which prescribes a way of symbolically representing, for certain purposes, such matters as the observable pressure and temperature of a gas; and the theory shows among other things how, when certain empirical data about a gas are supplied and incorporated into that representation, we can calculate the quantity of heat required for raising the temperature of the gas by some designated number of degrees (i.e., we can calculate the specific heat of a gas). The molecular theory of gases is thus neither logically implied by nor (according to some proponents of the instrumentalist view) does it logically imply any statements about matters of observation. The *raison d'être* of the theory is to serve as a rule or guide for making logical transitions from one set of experimental data to another set. More generally, a theory functions as a "leading principle" or "inference ticket" *in accordance with which* conclusions about observable facts may be drawn from given factual premises, not as a premise *from which* such conclusions are obtained.

Several consequences follow directly from this account.

1. The view that a theory is a "convenient shorthand" for a class of observation statements (whether finite or infinite in number), and the correlative claim that a theory must be translatable into the language of observation are both irrelevant and misleading approaches to understanding the role of theories. The value of a theory for the conduct of inquiry would not be enhanced if perchance it could be shown to be logically equivalent to some class of observation statements; and failure to establish such an equivalence for any of the theories in physics does not diminish their importance as instruments for analyzing the materials of experience with a view to solving concrete experimental problems and systematically relating experimental laws. . . .

2. It is common if not normal for a theory to be formulated in terms of ideal concepts such as the geometrical ones of straight line and circle, or the more specifically physical ones of instantaneous velocity, perfect vacuum, infinitely slow expansion, perfect elasticity, and the like. Although such "ideal" or "limiting" notions may be suggested by empirical subject matter, for the most part they are not descriptive of anything experimentally observable. Indeed, in the case of some of them it seems quite impossible that when they are understood in a literal sense they could be used to characterize any existing thing. For example, we can attribute a velocity to a physical body only if the body moves through a finite, nonvanishing distance

during a finite, nonvanishing interval of time. But instantaneous velocity is defined as the *limit* of the ratios of the distance and time as the time interval diminishes toward zero. In consequence, it is difficult to see how the numerical value of this limit could possibly be the measure of any actual velocity.

There is nevertheless a rationale for using such limiting concepts in constructing a theory. With their help a theory may lend itself to a relatively simple formulation—simple enough, at any rate, to render it amenable to treatment by available methods of mathematical analysis. To be sure, standards of simplicity are vague, they are controlled in part by intellectual fashions and the general climate of opinion, and they vary with improvements in mathematical techniques. But in any event, considerations of simplicity undoubtedly enter into the formulation of theories. Despite the fact that a theory may employ simplifying concepts, it will in general be preferred to another theory using more "realistic" notions if the former answers to the purposes of a given inquiry and can be handled more conveniently than the latter.

On the other hand, the use of such limiting concepts in the formulation of a theory presents difficulties to the view that factual truth or falsity can be significantly predicated of the theory. For a factual statement is normally said to be true if it formulates some indicated relation either between existing things and events (in the omnitemporal sense of "exist") or between properties of existing things and events. However, if a theory formulates relations between properties that ostensibly do not (or cannot) characterize existing things, it is not clear in what sense the theory can be said to be factually true or false.

Analogous difficulties for this view are raised by the circumstance that in general a theory contains terms for which no rules of correspondence are given, whether or not an interpretation is provided for the theory on the basis of some model. In consequence, no experimental notions are associated with such terms, so that in effect those terms have the status of *variables*. . . . The point can be illustrated by examples from actual physical theories. We have already noted that in the molecular theory of gases there is no correspondence rule for the expression "the velocity of an individual molecule," though there is such a rule for the expression "the average value of the velocities of all the molecules." Similarly, the expression $\psi(x, t)$ is employed in the Schrödinger equation in quantum mechanics for characterizing the state of an electron. There is in effect a correspondence rule for the expression $\psi(x,t)\psi^*(x,t)$ (where ψ^* is the complex conjugate of ψ), but no such rule for $\psi(x,t)$ itself. On the face of it, therefore, theories containing such terms are statement-forms, and cannot be said to be true or false.

These and similar difficulties do not arise for the instrumentalist view of theories, since on this view the pertinent question about theories is not whether they are true or false but whether they are effective techniques for representing and inferring experimental phenomena. The fact that theories contain expressions which describe or designate nothing in actual existence, or which are not associated with experimental notions is indeed taken as confirmation for the claim that theories must be construed in terms of their intermediary, instrumental function in inquiry, rather than in terms of their adequacy as objective accounts of some subject matter. From the perspective of this standpoint, it is not a flaw in the molecular theory of gases, for example, that it employs limiting concepts such as the notions of point-particle, instantaneous velocity, or perfect elasticity. For the task of the theory is not to give a faithful portrayal of what transpires within a gas but to provide a method for analyzing and symbolizing certain properties of the gas, so that when information is available about some of these properties in concrete experimental situations the theory makes it possible to infer information having a required degree of precision about other properties.

Similarly, it is not a source of embarrassment to the instrumentalist position that in inquiries into the thermal properties of a gas we use a theory which analyzes a gas as an aggregation of discrete particles, although when we study acoustic phenomena in connection with gases we employ a theory that represents the gas as a continuous medium. Construed as statements that are either true or false, the two theories are on the face of it mutually incompatible. But construed as techniques or leading principles of inference, the theories are simply different though complementary instruments, each of which is an effective intellectual tool for dealing with a special range of questions. In any event, physicists show no noticeable compunction in using one theory for dealing with one class of problems and an apparently discordant theory for handling another class. They employ the inclusive wave theory of light, according to which optical phenomena are represented in terms of periodic wave motion, when dealing with questions of diffraction and polarization; but they continue to use the relatively simpler theory of geometrical optics, according to which light is analyzed as a rectilinear propagation, when handling problems in reflection and refraction. They introduce considerations based on the theory of relativity in applying quantum mechanics to the analysis of the fine structure of spectral lines; they ignore such considerations when quantum theory is exploited for analyzing the nature of chemical bonds. Such examples can be multiplied; and if they prove nothing else, they show at least that the literal truth of theories is not the object of primary concern when theories are used in experimental inquiry.

It does not follow, however, that on the instrumentalist view theories are "fictions," except in the quite innocent sense that theories are human creations. For in the pejorative sense of the word, to say that a theory is a fiction is to claim that the theory is not true to the facts; and this is not a claim which is consistent with the instrumentalist position that truth and falsity are inappropriate characterizations for theories. It is indeed possible to maintain, consistent with that position, that many of the models in terms of which theories have been interpreted are fictions (in some cases even explicitly introduced as fictions, as were some of Lord Kelvin's mechanical models of the ether). In maintaining this much, one is merely asserting either that there simply is no empirical evidence satisfying some assumed criterion for the physical reality of those models, or that in terms of this criterion the available evidence is negative. On the other hand, it is also consistent with the instrumentalist view to recognize that some theories are superior to others—either because one theory serves as an effective leading principle for a more inclusive range of inquiries than does another, or because one theory supplies a method of analysis and representation that makes possible more precise and more detailed inferences than does the other. However, a theory is an effective tool in inquiry only if things and events are actually so related that the conclusions the theory enables us to infer from given experimental data are generally in good agreement with further matters of observed fact. As in the case of other instruments, the effectiveness of a theory as an instrument, or its superiority to some other theory, is thus contingent on objective features of a subject matter and depends on something other than personal whim or preference.

. . .

3. But it is time for noting some limitations in the instrumentalist standpoint. Proponents of this view often seem to believe that, if the instrumental role of theories is once established, theories are thereby shown to be improper subjects for the characterizations "true" and "false." There is, however, no necessary incompatibility between saying that a theory is true and maintaining that the theory performs important functions in inquiry. Few will deny that statements such as "The distance between New York and Washington, D.C., is approximately 225 miles" may be true, and yet play valuable roles in the plans of men. Indeed, most statements that by common consent can be significantly affirmed as true or false can also be studied for the use that is made of them. In brief, it does not follow that theories cannot be regarded as "genuine statements" and cannot therefore be investigated for their truth or falsity, merely because theories have indispensable functions in inquiry.

. . .

. . . There is little doubt, for example, that in many cases the wave theory of light is used, or can be construed, as a leading principle or technique for inferring statements about experimentally identifiable data from other such data. Nor is it disputable that this way of viewing the theory brings out a role it plays in inquiry which might otherwise be overlooked, or that this perspective on theories is a salutary antidote to dogmatic affirmations that some particular theory is the final truth about the "ultimate nature" of things. It nevertheless does not follow that theories do not or cannot also serve as premises in scientific explanations and predictions, as bona fide statements concerning which it therefore seems quite proper to raise questions of truth and falsity.

... Some of the most eminent scientists, both living and dead, certainly have viewed theories as statements about the constitution and structure of a given subject matter; and they have conducted their investigations on the assumption that a theory is a *projected map* of some domain of nature, rather than a set of *principles of mapping*. Much experimental research is undoubtedly inspired by a desire to ascertain whether or not various hypothetical entities and processes postulated by a theory (e.g., neutrons, mesons, and neutrinos of current atomic physics) do indeed occur in circumstances and relations stated by the theory. But research which is ostensibly directed to testing a theory proceeds on the *prima facie* assumption that the theory is affirming some things and denying others. ...

One final comment on the instrumentalist view must be made. It has already been briefly noted that proponents of this view supply no uniform account of the various "scientific objects" (such as electrons or light waves) which are ostensibly postulated by microscopic theories. But the further point can also be made that it is far from clear how, on this view, such "scientific objects" can be said to be physically existing things. For if a theory is just a leading principle—a technique for drawing inferences based upon a method of representing phenomena—terms like "electron" and "light wave" presumably function only as conceptual links in rules of representation and inference. On the face of it, therefore, the meaning of such terms is exhausted by the roles they play in guiding inquiries and ordering the materials of observation; and in this perspective the supposition that such terms might refer to physically existing things and processes that are not phenomena in the strict sense seems to be excluded. Proponents of the instrumentalist view have indeed sometimes flatly contradicted themselves on this issue. Thus, while maintaining that the atomic theory of matter is simply a technique of inference, some writers have nevertheless seriously discussed the question whether atoms exist and have argued that the evidence is sufficient to show that atoms really do exist. Others have explicitly asserted that atoms and other "scientific objects" are generalized statements of relations between sets of changes, and cannot be individual existing things; but they have also declared that atoms are in motion, and possess a mass. Such inconsistencies suggest that those who are guilty of them are not really prepared to exclude, as improper, questions of truth and falsity concerning a theory. In any event, it is clearly not inconsistent to admit the logical propriety of such questions, and also to recognize the important instrumental function of theories.

III. The Realist View of Theories

Are theories then "really" statements, of which truth and falsity are meaningfully predicable, despite the difficulties that have been noted in this view? Enough has already been said to suggest that, whether the question is answered affirmatively or negatively, the answer given may not be the exclusively reasonable one. Indeed, those who differ in their answers to it frequently disagree neither on matters falling into the province of experimental inquiry nor on points of formal logic nor on the facts of scientific procedure. What often divides them are, in part, loyalties to different intellectual traditions, in part inarbitrable preferences concerning the appropriate way of accommodating our language to the generally admitted facts. It is a matter of historical record that, while many distinguished figures in both science and philosophy have adopted as uniquely adequate the characterization of theories as true or false statements, a no less distinguished group of other scientists and philosophers has made a similar claim for the description of theories as instruments of inquiry. However, a defender of either view cannot only cite eminent authority to support his position; with a little dialectical ingenuity he can usually remove the sting from apparently grave objections to his position. In consequence, the already long controversy as to which of the two is the proper way of construing theories can be prolonged indefinitely. The obvious moral to be drawn from such a debate is that once both positions are so stated that each can meet the *prima facie* difficulties it faces, the question as to which of them is the "correct position" has only terminological interest.

Let us consider the chief obstacles to each of the two views under discussion, beginning with those facing the conception of theories as true or false statements.

. . .

In the [first] place, there is the objection previously mentioned that theories are commonly formulated in terms of limiting concepts which characterize nothing actually in existence, so that at any rate non-vacuous factual truth cannot be claimed for such theories. This objection can be turned in a number of ways. A familiar gambit is to challenge the contention that limiting concepts do not apply to existing things. To be sure, we cannot, for example, ascertain by overt measurement the value of an instantaneous velocity of the magnitude of some length whose theoretical value is stipulated to be equal to the square root of 2. But unless accessibility to overt measurement (or more generally to observation) is made the criterion of physical existence, so it is sometimes said, this does not show that bodies cannot have instantaneous velocities or lengths with real number magnitudes. On the contrary, if a theory postulating such values is supported by competent evidence, then according to the rejoinder under discussion there is good reason to maintain that these limiting concepts do designate certain phases of things and processes. Since in testing a theory we test the totality of assumptions it makes, so the rejoinder continues, if a theory is regarded as well established on the available evidence, all its component assumptions must also be regarded. Accordingly, unless we introduce quite arbitrary distinctions, we cannot pick and choose between the component assumptions, counting some as descriptions of what exists and others as not.

There is another way in which the objection under discussion is sometimes countered. The rejoinder then consists in admitting that limiting concepts are simplifying devices, and that a theory employing them does not in general assert anything for which literal truth can reasonably be claimed. Nevertheless, existing things possess traits that often are either indistinguishable from the "ideal" traits mentioned in a theory or differ from such "ideal" traits by a negligible factor. In consequence, on this rejoinder to the objection, a theory is said to be true in the sense that the discrepancy between what a theory asserts and what even ultrarefined observation can discover is small enough to be counted as arising from experimental error.

A [second] type of difficulty for the conception of theories as true or false statements is created by the fact, to which attention has already been directed, that apparently incompatible theories are sometimes employed for the same subject matter. Thus, a liquid cannot be both a system of discrete particles and also a continuous medium, though theories dealing with the properties of liquids adopt one assumption in some cases and the opposing assumption in others.

The usual reply to this objection consists of two parts. One of them is essentially a repetition of the rejoinder mentioned in the preceding paragraph. A theory may be employed in a given area of inquiry, even though it is apparently incompatible with some other theory that is also used, because the former is simpler than the latter and because for the problems under discussion the more complex theory does not yield conclusions in better agreement with the facts than are the conclusions of the simpler theory. Accordingly, the simpler theory can be regarded as in a sense a special case of the more complex one, rather than as a contrary of the latter.

The second part of the reply is that, though incompatible theories may be used for a time, their use is but a temporary makeshift, to be abandoned as soon as an internally consistent theory is developed, more comprehensive than either of the previous ones. Thus, although there were serious discrepancies between the atomic theories employed at the turn of the present century to account for many facts of both physics and chemistry, these conflicting theories have been replaced by a single theory of atomic structure currently used in both these sciences. Indeed, inconsistencies between theories, each of which is nevertheless useful in some limited domain of inquiry, are often a powerful incentive for the construction of a more inclusive but consistent theoretical structure. Accordingly, a proponent of the view that theories are true or false statements can escape any embarrassment for his position from the circumstance that incompatible theories are sometimes employed in the sciences; he can insist on the corrigible character of every theory and refuse to claim final truth for any theory. He can freely admit that even a false theory may be quite useful for handling many problems; and he can join this admission with the claim that the succession of theories in some branch of science is a series of progressively better approximations to the unattainable but valid ideal of a finally true theory.

And finally, there is the objection currently raised against the position under discussion because of the difficulties encountered in interpreting quantum mechanics in terms of some familiar model. For example, theoretical as well as experimental considerations have led physicists to ascribe to electrons (and to other entities postulated by quantum theory) apparently incompatible and in any case puzzling characteristics. Thus, electrons are construed to have features which make it appropriate to think of them as a system of waves; on the other hand, electrons also have traits which lead us to think of them as particles, each having a spatial location and a velocity, though no determinate position and velocity can in principle be assigned simultaneously to any of them. Many physicists have therefore concluded that quantum theory cannot be viewed as a statement about an "objectively existing" domain of things and processes, as a map that outlines even approximately the microscopic constitution of matter. On the contrary, the theory must be regarded simply as a conceptual schema or a policy for guiding and coordinating experiments.

The rejoinder to this objection follows a familiar pattern. The fact that a visualizable model embodying the laws of classical physics cannot be given for quantum theory, so runs the reply, is not an adequate ground for denying that the quantum theory does formulate the structural properties of subatomic processes. It is doubtless desirable to have a satisfactory model for the theory. But the type of model that is regarded as satisfactory at any given time is a function of the prevailing intellectual climate. Even though current models for quantum theory may strike us as strange and even "unintelligible," there are no compelling reasons for assuming that the strangeness will not wear away with increased familiarity, or that a more satisfactory interpretation for the theory will not be eventually found. . . .

This sample of objections to the view that theories are true or false statements suffices to show that the view has dialectical resources for maintaining itself in the face of severe criticism. Undoubtedly the rejoinders to these criticisms can be met with counterrejoinders, though none to which defenders of the view under attack cannot offer at least a *prima facie* suitable reply. . . .

13

Larry Laudan,* "A Confutation of Convergent Realism"

Larry Laudan builds his case against "convergent realism" around traditional problems about the notion of reference (here the relation between theoretical terms of mature science and the world) and by appeal to the historical record, which suggests that many theories have been empirically successful without being true or securing reference for their terms.

The positive argument for realism is that it is the only philosophy that doesn't make the success of science a miracle.

H. Putnam (1975)

1. The Problem

It is becoming increasingly common to suggest that epistemological realism is an empirical hypothesis, grounded in, and to be authenticated by, its ability to explain the workings of science. A growing number of philosophers (including Boyd, Newton-Smith, Shimony, Putnam, Friedman and Niiniluoto) have argued that the theses of epistemic realism are open to empirical test. The suggestion that epistemological doctrines have much the same empirical status as the sciences is a welcome one: for, whether it stands up to detailed scrutiny or not, it marks a significant facing-up by the philosophical community to one of the most neglected (and most notorious) problems of philosophy: the status of epistemological claims.

But there are potential hazards as well as advantages associated with the "scientizing" of epistemology. Specifically, once one concedes that epistemic doctrines are to be tested in the court of experience, it is possible that one's favorite epistemic theories may be refuted rather than confirmed. It is

L. Laudan, "A Confutation of Convergent Realism," *Philosophy of Science*, 1981, 48: 19–38, 45–9.

the thesis of this paper that precisely such a fate afflicts a form of realism advocated by those who have been in the vanguard of the move to show that realism is supported by an empirical study of the development of science. Specifically, I shall show that epistemic realism, at least in certain of its extant forms, is neither supported by, nor has it made sense of, much of the available historical evidence.

2. Convergent Realism

Like other philosophical *-isms*, the term "realism" covers a variety of sins. Many of these will not be at issue here. For instance, "semantic realism" (in brief, the claim that all theories have truth values and that some theories—we know not which—are true) is not in dispute. Nor shall I discuss what one might call "intentional realism" (i.e., the view that theories are generally intended by their proponents to assert the existence of entities corresponding to the terms in those theories). What I shall focus on instead are certain forms of *epistemological* realism. As Hilary Putnam has pointed out, although such realism has become increasingly fashionable, "very little is said about what realism *is*" (1978). The lack of specificity about what realism asserts makes it difficult to evaluate its claims, since many formulations are too vague and sketchy to get a grip on. At the same time, any efforts to formulate the realist position with greater precision lay the critic open to charges of attacking a straw man. In the course of this paper, I shall attribute several theses to the realists. Although there is probably no realist who subscribes to all of them, most of them have been defended by some self-avowed realist or other; taken together, they are perhaps closest to that version of realism advocated by Putnam, Boyd and Newton-Smith. Although I believe the views I shall be discussing can be legitimately attributed to certain contemporary philosophers (and will frequently cite the textual evidence for such attributions), it is not crucial to my case that such attributions can be made. Nor will I claim to do justice to the complex epistemologies of those whose work I will criticize. My aim, rather, is to explore certain epistemic claims which those who are realists might be tempted (and in some cases have been tempted) to embrace. If my arguments are sound, we will discover that some of the most intuitively tempting versions of realism prove to be chimeras.

The form of realism I shall discuss involves variants of the following claims:

R1) Scientific theories (at least in the "mature" sciences) are typically approximately true and more recent theories are closer to the truth than older theories in the same domain;

True

R2) The observational and theoretical terms within the theories of a mature science genuinely refer (roughly, there are substances in the world that correspond to the ontologies presumed by our best theories);

R3) Successive theories in any mature science will be such that they "preserve" the theoretical relations and the apparent referents of earlier theories (i.e., earlier theories will be "limiting cases" of later theories).[1]

R4) Acceptable new theories do and should explain why their predecessors were successful insofar as they were successful.

To these semantic, methodological and epistemic theses is conjoined an important meta-philosophical claim about how realism is to be evaluated and assessed. Specifically, it is maintained that:

R5) Theses (R1)–(R4) entail that ("mature") scientific theories should be successful; indeed, these theses constitute the best, if not the only, explanation for the success of science. The empirical success of science (in the sense of giving detailed explanations and accurate predictions) accordingly provides striking empirical confirmation for realism.

I shall call the position delineated by (R1) to (R5) *convergent epistemological realism*, or CER for short. Many recent proponents of CER maintain that (R1), (R2), (R3), and (R4) are empirical hypotheses which, via the linkages postulated in (R5), can be tested by an investigation of science itself. They propose two elaborate abductive arguments. The structure of the first, which is germane to (R1) and (R2), is something like this:

I
1. If scientific theories are approximately true, they will typically be empirically successful;
2. If the central terms in scientific theories genuinely refer, those theories will generally be empirically successful;
3. Scientific theories are empirically successful.
4. (Probably) Theories are approximately true and their terms genuinely refer.

The argument relevant to (R3) is of slightly different form, specifically:

II
1. If the earlier theories in a "mature" science are approximately true and if the central terms of those theories genuinely refer, then later

more successful theories in the same science will preserve the earlier theories as limiting cases;
2. Scientists seek to preserve earlier theories as limiting cases and generally succeed.
3. (Probably) Earlier theories in a "mature" science are approximately true and genuinely referential.

Taking the success of present and past theories as givens, proponents of CER claim that *if* CER were true, it would follow that the success and the progressive success of science would be a matter of course. Equally, they allege that if CER were false, the success of science would be "miraculous" and without explanation.[2] Because (on their view) CER explains the fact that science is successful, the theses of CER are thereby confirmed by the success of science and non-realist epistemologies are discredited by the latter's alleged inability to explain both the success of current theories and the progress which science historically exhibits.

As Putnam and certain others (e.g., Newton-Smith) see it, the fact that statements about reference (R2, R3) or about approximate truth (R1, R3) function in the explanation of a contingent state of affairs, establishes that "the notions of 'truth' and 'reference' have a causal explanatory role in epistemology" (Putnam 1978, p. 21).[3] In one fell swoop, both epistemology and semantics are "naturalized" and, to top it all off, we get an explanation of the success of science into the bargain!

The central question before us is whether the realist's assertions about the interrelations between truth, reference and success are sound. It will be the burden of this paper to raise doubts about both I and II. Specifically, I shall argue that *four* of the five premises of those abductions are either false or too ambiguous to be acceptable. I shall also seek to show that, even if the premises were true, they would not warrant the conclusions which realists draw from them. . . .

3. Reference and Success

The specifically referential side of the "empirical" argument for realism has been developed chiefly by Putnam, who talks explicitly of reference rather more than most realists. On the other hand, reference is usually implicitly smuggled in, since most realists subscribe to the (ultimately referential) thesis that "the world probably contains entities very like those postulated by our most successful theories."

If (R2) is to fulfill Putnam's ambition that reference can explain the success of science, and that the success of science establishes the presumptive truth of (R2), it seems he must subscribe to claims similar to these:

S1) The theories in the advanced or mature sciences are successful;
S2) A theory whose central terms genuinely refer will be a successful theory;
S3) If a theory is successful, we can reasonably infer that its central terms genuinely refer;
S4) All the central terms in theories in the mature sciences do refer.

There are complex interconnections here. (S2) and (S4) explain (S1), while (S1) and (S3) provide the warrant for (S4). Reference explains success and success warrants a presumption of reference. The arguments are plausible, given the premises. But there is the rub, for with the possible exception of (S1), none of the premises is acceptable.

The first and toughest nut to crack involves getting clearer about the nature of that "success" which realists are concerned to explain. Although Putnam, Sellars and Boyd all take the success of certain sciences as a given, they say little about what this success amounts to. So far as I can see, they are working with a largely *pragmatic* notion to be cashed out in terms of a theory's workability or applicability. On this account, we would say that a theory is successful if it makes substantially correct predictions, if it leads to efficacious interventions in the natural order, if it passes a battery of standard tests. One would like to be able to be more specific about what success amounts to, but the lack of a coherent theory of confirmation makes further specificity very difficult.

Moreover, the realist must be wary—at least for these purposes—of adopting too strict a notion of success, for a highly robust and stringent construal of "success" would defeat the realist's purposes. What he wants to explain, after all, is why science in general has worked so well. If he were to adopt a very demanding characterization of success (such as those advocated by inductive logicians or Popperians) then it would probably turn out that science has been largely "unsuccessful" (because it does not have high confirmation) and the realist's avowed explanandum would thus be a non-problem. Accordingly, I shall assume that a theory is "successful" so long as it has worked well, i.e., so long as it has functioned in a variety of explanatory contexts, has led to confirmed predictions and has been of broad explanatory scope. As I understand the realist's position, his concern is to explain why certain theories have enjoyed this kind of success.

If we construe "success" in this way, (S1) can be conceded. Whether one's criterion of success is broad explanatory scope, possession of a large number of confirming instances, or conferring manipulative or predictive control, it is clear that science is, by and large, a successful activity.

What about (S2)? I am not certain that any realist would or should endorse it, although it is a perfectly natural construal of the realist's claim that "reference explains success." The notion of reference that is involved here is highly complex and unsatisfactory in significant respects. Without endorsing it, I shall use it frequently in the ensuing discussion. The realist sense of reference is a rather liberal one, according to which the terms in a theory may be genuinely referring even if many of the claims the theory makes about the entities to which it refers are false. Provided that there are entities which "approximately fit" a theory's description of them, Putnam's charitable account of reference allows us to say that the terms of a theory genuinely refer.[4] On this account (and these are Putnam's examples), Bohr's "electron," Newton's "mass," Mendel's "gene," and Dalton's "atom" are all referring terms, while "phlogiston" and "aether" are not (Putnam 1978, pp. 20–22).

Are genuinely referential theories (i.e., theories whose central terms genuinely refer) invariably or even generally successful at the empirical level, as (S2) states? There is ample evidence that they are not. The chemical atomic theory in the 18th century was so remarkably unsuccessful that most chemists abandoned it in favor of a more phenomenological, elective affinity chemistry. The Proutian theory that the atoms of heavy elements are composed of hydrogen atoms had, through most of the 19th century, a strikingly unsuccessful career, confronted by a long string of apparent refutations. The Wegenerian theory that the continents are carried by large subterranean objects moving laterally across the earth's surface was, for some thirty years in the recent history of geology, a strikingly unsuccessful theory until, after major modifications, it became the geological orthodoxy of the 1960s and 1970s. Yet all of these theories postulated basic entities which (according to Putnam's "principle of charity") genuinely exist.

The realist's claim that we should expect referring theories to be empirically successful is simply false. And, with a little reflection, we can see good reasons why it should be. To have a genuinely referring theory is to have a theory which "cuts the world at its joints," a theory which postulates entities of a kind that really exist. But a genuinely referring theory need not be such that all—or even most—of the specific claims it makes about the properties of those entities and their modes of interaction are true. Thus, Dalton's theory makes many claims about atoms which are false; Bohr's early theory of the electron was similarly flawed in important respects. Contra-(S2), genuinely referential theories need not be strikingly successful, since such theories may be "massively false" (i.e., have far greater falsity content than truth content).

(S2) is so patently false that it is difficult to imagine that the realist need be committed to it. But what else will do? The (Putnamian) realist wants attributions of reference to a theory's terms to function in an explanation of that theory's success. The simplest and crudest way of doing that involves a claim like (S2). A less outrageous way of achieving the same end would involve the weaker

(S2′) A theory whose terms refer will usually (but not always) be successful.

Isolated instances of referring but unsuccessful theories, sufficient to refute (S2), leave (S2′) unscathed. But, if we were to find a broad range of referring but unsuccessful theories, that would be evidence against (S2′). Such theories can be generated at will. For instance, take any set of terms which one believes to be genuinely referring. In any language rich enough to contain negation, it will be possible to construct indefinitely many unsuccessful theories, all of whose substantive terms are genuinely referring. Now, it is always open to the realist to claim that such "theories" are not really theories at all, but mere conjunctions of isolated statements—lacking that sort of conceptual integration we associate with "real" theories. Sadly a parallel argument can be made for genuine theories. Consider, for instance, how many inadequate versions of the atomic theory there were in the 2000 years of atomic "speculating," before a genuinely successful theory emerged. Consider how many unsuccessful versions there were of the wave theory of light before the 1820s, when a successful wave theory first emerged. Kinetic theories of heat in the seventeenth and eighteenth century, developmental theories of embryology before the late nineteenth century sustain a similar story. (S2′), every bit as much as (S2), seems hard to reconcile with the historical record.

As Richard Burian has pointed out to me (in personal communication), a realist might attempt to dispense with both of those theses and simply rest content with (S3) alone. Unlike (S2) and (S2′), (S3) is not open to the objection that referring theories are often unsuccessful, for it makes no claim that referring theories are always or generally successful. But (S3) has difficulties of its own. In the first place, it seems hard to square with the fact that the central terms of many relatively successful theories (e.g., aether theories, phlogistic theories) are evidently non-referring. I shall discuss this tension in detail below. More crucial for our purposes here is that (S3) is *not strong enough* to permit the realist to utilize reference to explain success. Unless genuineness of reference entails that all or most referring theories will be successful, then the fact that a theory's terms refer scarcely provides a convincing explanation of that theory's success. If, as (S3) allows, many (or even most) referring theories can be unsuccessful, how can the fact that a successful theory's terms refer be taken to explain why it is successful? (S3) may or may not be true; but in either case it arguably gives the realist no explanatory access to scientific success.

A more plausible construal of Putnam's claim that reference plays a role in explaining the success of science involves a rather more indirect argument. It might be said (and Putnam does say this much) that we can explain why a theory is successful by assuming that the theory is true or approximately true. Since a theory can only be true or nearly true (in any sense of those terms open to the realist) if its terms genuinely refer, it might be argued that reference gets into the act willy-nilly when we explain a theory's success in terms of its truth(like) status. On this account, reference is piggy-backed on approximate truth. The viability of this indirect approach is treated at length in section 4 below so I shall not discuss it here except to observe that if the only contact point between reference and success is provided through the medium of approximate truth, then the link between reference and success is extremely tenuous.

What about (S3), the realist's claim that success creates a rational presumption of reference? We have already seen that (S3) provides no explanation of the success of science, but does it have independent merits? The question specifically is whether the success of a theory provides a warrant for concluding that its central terms refer. Insofar as this is—as certain realists suggest—an empirical question, it requires us to inquire whether past theories which have been successful are ones whose central terms genuinely referred (according to the realist's own account of reference).

A proper empirical test of this hypothesis would require extensive sifting of the historical record of a kind that is not possible to perform here. What I can do is to mention a range of once successful, but (by present lights) non-referring, theories. A fuller list will come later (see section 5), but for now we shall focus on a whole family of related theories, namely, the subtle fluids and aethers of 18th and 19th century physics and chemistry.

Consider specifically the state of aetherial theories in the 1830s and 1840s. The electrical fluid, a substance which was generally assumed to accumulate on the surface rather than permeate the interstices of bodies, had been utilized to explain *inter alia* the attraction of oppositely charged bodies, the behavior of the Leyden jar, the similarities between atmospheric and static electricity and many phenomena of current electricity. Within chemistry and heat theory, the caloric aether had been widely utilized since Boerhaave (by, among others, Lavoisier, Laplace, Black, Rumford, Hutton, and Cavendish) to explain everything from the role of heat in chemical reactions to the conduction and radiation of heat and several standard problems of thermometry. Within the theory of light, the optical aether functioned centrally in explanations of reflection, refraction, interference, double refraction, diffraction and polarization. (Of more than passing interest, optical aether theories had also made some very startling predictions, e.g., Fresnel's prediction of a bright spot at the center of the shadow of a circular disc; a surprising prediction which, when tested, proved correct. If that does not count as empirical success, nothing does!) There were also gravitational (e.g., LeSage's) and physiological (e.g., Hartley's) aethers which enjoyed some measure of empirical success. It would be difficult to find a family of theories in this period which were as successful as aether theories; compared to them, 19th century atomism (for instance), a genuinely referring theory (on realist accounts), was a dismal failure. Indeed, on any account of empirical success which I can conceive of, non-referring 19th-century aether theories were more successful than contemporary, referring atomic theories. In this connection, it is worth recalling the remark of the great theoretical physicist, J. C. Maxwell, to the effect that the aether was better confirmed than any other theoretical entity in natural philosophy!

What we are confronted by in 19th-century aether theories, then, is a wide variety of once successful theories, whose central explanatory concept Putnam singles out as a prime example of a non-referring one (Putnam 1978, p. 22). What are (referential) realists to make of this historical case? On the face of it, this case poses two rather different kinds of challenges to realism: (1) it suggests that (S3) is a dubious piece of advice in that *there can be* (and have been) *highly successful theories some central terms of which are non-referring*; and (2) it suggests that *the realist's claim that he can explain why science is successful is false at least insofar as a part of the historical success of science has been success exhibited by theories whose central terms did not refer*.

But perhaps I am being less than fair when I suggest that the realist is committed to the claim that *all* the central terms in a successful theory refer. It is possible that when Putnam, for instance, says that "terms in a mature [or successful] science typically refer" (Putnam 1978, p. 20), he only means to suggest that *some* terms in a successful theory or science genuinely refer. Such a claim is fully consistent with the fact that certain other terms (e.g., "aether") in certain successful, mature sciences (e.g., 19th-century physics) are nonetheless non-referring. Put differently, the realist might argue that the success of a theory warrants the claim that at least some (but not necessarily all) of its central concepts refer.

Unfortunately, such a weakening of (S3) entails a theory of evidential support which can scarcely give comfort to the realist. After all, part of what separates the realist from the positivist is the former's belief that the evidence for a theory is evidence for *everything* which the theory asserts. Where the stereotypical positivist argues that the evidence selectively confirms only the more "observable" parts of a theory, the realist generally asserts (in the language of Boyd) that:

> the sort of evidence which ordinarily counts in favor of the acceptance

> of a scientific law or theory is, ordinarily, evidence for the (at least approximate) truth of the law or theory as an account of the causal relations obtaining between the entities ["observation or theoretical"] quantified over in the law or theory in question. (Boyd 1973, p. 1)[5]

For realists such as Boyd, either all parts of a theory (both observational and non-observational) are confirmed by successful tests or none are. In general, realists have been able to utilize various holistic arguments to insist that it is not merely the lower-level claims of a well-tested theory which are confirmed but its deep-structural assumptions as well. This tactic has been used to good effect by realists in establishing that inductive support "flows upward" so as to authenticate the most "theoretical" parts of our theories. Certain latter-day realists (e.g., Glymour) want to break out of this holist web and argue that certain components of theories can be "directly" tested. This approach runs the very grave risk of undercutting what the realist desires most: a rationale for taking our deepest-structure theories seriously, and a justification for linking reference and success. After all, if the tests to which we subject our theories only test *portions* of those theories, then even highly successful theories may well have central terms which are non-referring and central tenets which, because untested, we have no grounds for believing to be approximately true. Under those circumstances, a theory might be highly successful and yet contain important constituents which were patently false. Such a state of affairs would wreak havoc with the realist's presumption (R1) that success betokens approximate truth. In short, to be less than a holist about theory testing is to put at risk precisely that predilection for deep-structure claims which motivates much of the realist enterprise.

There is, however, a rather more serious obstacle to this weakening of referential realism. It is true that by weakening (S3) to only certain terms in a theory, one would immunize it from certain obvious counter-examples. But such a maneuver has debilitating consequences for other central realist theses. Consider the realist's thesis (R3) about the retentive character of inter-theory relations (discussed below in detail). The realist both recommends as a matter of policy and claims as a matter of fact that successful theories are (and should be) rationally replaced only by theories which preserve reference for the central terms of their successful predecessors. The rationale for the normative version of this retentionist doctrine is that the terms in the earlier theory, *because it was successful, must* have been referential and thus a constraint on any successor to that theory is that reference should be retained for such terms. This makes sense just in case success provides a blanket warrant for presumption of reference. But if (S3) were weakened so as to say merely that it is reasonable to assume that *some* of the terms in a successful theory genuinely refer, then the realist would have no rationale for his retentive theses (variants of R3), which have been a central pillar of realism for several decades.[6]

Something apparently has to give. A version of (S3) strong enough to license (R3) seems incompatible with the fact that many successful theories contain non-referring central terms. But any weakening of (S3) dilutes the force of, and removes the rationale for, the realist's claims about convergence, retention and correspondence in inter-theory relations.[7] If the realist once concedes that some unspecified set of the terms of a successful theory may well not refer, then his proposals for restricting "the class of candidate theories" to those which retain reference for the *prima facie* referring terms in earlier theories is without foundation (Putnam 1975, p. 22).

More generally, we seem forced to say that such linkages as there are between reference and success are rather murkier than Putnam's and Boyd's discussions would lead us to believe. If the realist is going to make his case for CER, it seems that it will have to hinge on approximate truth, (R1), rather than reference, (R2).

4. Approximate Truth and Success: the "Downward Path"

Ignoring the referential turn among certain recent realists, most realists continue to argue that, at bottom, epistemic realism is committed to the view that successful scientific theories, even if strictly false, are nonetheless "approximately true" or "close to the truth" or "verisimilar."[8] The claim generally amounts to this pair:

(T1) if a theory is approximately true, then it will be explanatorily successful; and

(T2) if a theory is explanatorily successful, then it is probably approximately true.

What the realist would *like* to be able to say, of course, is:

(T1′) if a theory is true, then it will be successful.

(T1′) is attractive because self-evident. But most realists balk at invoking (T1′) because they are (rightly) reluctant to believe that we can reasonably presume of any given scientific theory that it is true. If all the realist could explain was the success of theories which were true *simpliciter*, his explanatory repertoire would be acutely limited. As an attractive move in the direction of broader explanatory scope, (T1) is rather more appealing. After all, presumably many theories which we believe to be false (e.g., Newtonian mechanics, thermodynamics, wave optics) were—and still are—highly successful across a broad range of applications.

Perhaps, the realist evidently conjectures, we can find an *epistemic* account of that pragmatic success by assuming such theories to be "approximately true." But we must be wary of this potential sleight of hand. It may be that there is a connection between success and approximate truth; *but if there is such a connection it must be independently argued for*. The acknowledgedly uncontroversial character of (T1′) must not be surreptitiously invoked—as it sometimes seems to be—in order to establish (T1). When (T1′)'s antecedent is appropriately weakened by speaking of approximate truth, it is by no means clear that (T1) is sound.

Virtually all the proponents of epistemic realism take it as unproblematic that if a theory were approximately true, it would deductively follow that the theory would be a relatively successful predictor and explainer of observable phenomena. Unfortunately, few of the writers of whom I am aware have defined what it means for a statement or theory to be "approximately true." Accordingly, it is impossible to say whether the alleged entailment is genuine. This reservation is more than perfunctory. Indeed, on the best known account of what it means for a theory to be approximately true, it does *not* follow that an approximately true theory will be explanatorily successful.

Suppose, for instance, that we were to say in a Popperian vein that a theory, T_1, is approximately true if its truth content is greater than its falsity content, i.e.,

$$Ct_T(T_1) >> Ct_F(T_1).^9$$

(Where $Ct_T(T_1)$ is the cardinality of the set of true sentences entailed by T_1 and $Ct_F(T_1)$ is the cardinality of the set of false sentences entailed by T_1.) When approximate truth is so construed, it does *not* logically follow that an arbitrarily selected class of a theory's entailments (namely, some of its observable consequences) will be true. Indeed, it is entirely conceivable that a theory might be approximately true in the indicated sense and yet be such that *all* of its thus far tested consequences are *false*.[10]

Some realists concede their failure to articulate a coherent notion of approximate truth or verisimilitude, but insist that this failure in no way compromises the viability of (T1). Newton-Smith, for instance, grants that "no one has given a satisfactory analysis of the notion of verisimilitude" (1981, p. 197), but insists that the concept can be legitimately invoked "even if one cannot at the time give a philosophically satisfactory analysis of it." He quite rightly points out that many scientific concepts were explanatorily useful long before a philosophically coherent analysis was given for them.

But the analogy is unseemly, for what is being challenged is not whether the concept of approximate truth is philosophically rigorous but rather whether it is even clear enough for us to ascertain whether it entails what it purportedly explains. Until someone one provides a clearer analysis of approximate truth than is now available, it is not even clear whether truthlikeness would explain success, let alone whether, as Newton-Smith insists, "the concept of verisimilitude is *required* in order to give a satisfactory theoretical explanation of an aspect of the scientific enterprise." If the realist would de-mystify the "miraculousness" (Putnam) or the "mysteriousness" (Newton-Smith)[11] of the success of science, he needs more than a promissory note that somehow, someday, someone will show that approximately true theories must be successful theories.[12]

Whether there is some definition of approximate truth which does indeed entail that approximately true theories will be predictively successful (and yet still probably false) is not clear.[13] What can be said is that, promises to the contrary notwithstanding, *none* of the proponents of realism has yet articulated a coherent account of approximate truth which entails that approximately true theories will, across the range where we can test them, be successful predictors. Further difficulties abound. Even if the realist had a semantically adequate characterization of approximate or partial truth, and even if that semantics entailed that most of the consequences of an approximately true theory would be true, he would still be without any criterion that would *epistemically* warrant the ascription of approximate truth to a theory. As it is, the realist seems to be long on intuitions and short on either a semantics or an epistemology of approximate truth.

These should be urgent items on the realists' agenda since, until we have a coherent account of what approximate truth is, central realist theses like (R1), (T1) and (T2) are just so much mumbo-jumbo.

5. Approximate Truth and Success: the "Upward Path"

Despite the doubts voiced in section 4, let us grant for the sake of argument that if a theory is approximately true, then it will be successful. Even granting (T1), is there any plausibility to the suggestion of (T2) that explanatory success can be taken as a rational warrant for a judgment of approximate truth? The answer seems to be "no."

To see why, we need to explore briefly one of the connections between "genuinely referring" and being "approximately true." However the latter is understood, I take it that *a realist would never want to say that a theory was approximately true if its central theoretical terms failed to refer*. If there were nothing like genes, then a genetic theory, no matter how well confirmed it was, would not be approximately true. If there were no entities similar to atoms, no atomic theory could be approximately true; if there were no sub-atomic particles, then no quantum theory of chemistry could be approximately true. In short, a necessary condition—especially for a scientific realist—for a theory being close to the truth is that its central explanatory terms genuinely refer. (An *instrumentalist*, of course, could countenance the weaker claim that a theory was approximately true so *long* as its directly testable consequences were close to the observable values. But as I argued above, the realist must take claims about approximate truth to refer alike to the observable and the deep-structural dimensions of a theory.)

Now, what the history of science offers us is a plethora of theories which were both successful and (so far as we can judge) non-referential with respect to many of their central explanatory concepts. I discussed earlier one specific family of theories which fits this description. Let me add a few more prominent examples to the list:

the crystalline spheres of ancient and medieval astronomy;
the humoral theory of medicine;
the effluvial theory of static electricity;
"catastrophist" geology, with its commitment to a universal (Noachian) deluge;
the phlogiston theory of chemistry;
the caloric theory of heat;
the vibratory theory of heat;
the vital force theories of physiology;
the electromagnetic aether;
the optical aether;
the theory of circular inertia;
theories of spontaneous generation.

This list, which could be extended *ad nauseam*, involves in every case a theory which was once successful and well confirmed, but which contained central terms which (we now believe) were non-referring. Anyone who imagines that the theories which have been successful in the history of science have also been, with respect to their central concepts, genuinely referring theories has studied only the more "whiggish" versions of the history of science (i.e., the ones which recount only those past theories which are referentially similar to currently prevailing ones).

It is true that proponents of CER sometimes hedge their bets by suggesting that their analysis applies exclusively to "the mature sciences" (e.g., Putnam and Krajewski). This distinction between mature and immature sciences proves convenient to the realist since he can use it to dismiss any *prima facie* counter-example to the empirical claims of CER on the grounds that the example is drawn from an "immature" science. But this insulating maneuver is unsatisfactory in two respects. In the first place, it runs the risk of making CER vacuous since these authors generally define a mature science as one in which correspondence or limiting case relations obtain invariably between any successive theories in the science once it has passed "the threshold of maturity." Krajewski grants the tautological character of this view when he notes that "the thesis that there is [correspondence] among successive theories becomes, indeed, analytical" (1977, p. 91). Nonetheless, he believes that there is a version of the maturity thesis which "may be and must be tested by the history of science." That version is that "every branch of science crosses at some period the threshold of maturity." But the testability of this hypothesis is dubious at best. There is no historical observation which could conceivably *refute* it since, even if we discovered that no sciences yet possessed "corresponding" theories, it could be maintained that eventually every science will become corresponding. It is equally difficult to *confirm* it since, even if we found a science in which corresponding relations existed between the latest theory and its predecessor, we would have no way of knowing whether that relation will continue to apply to subsequent changes of theory in that science. In other words, the much-vaunted empirical testability of realism is seriously compromised by limiting it to the mature sciences.

But there is a second unsavory dimension to the restriction of CER to the "mature" sciences. The realists' avowed aim, after all, is to explain why science is successful: that is the "miracle" which they allege the non-realists leave unaccounted for. The fact of the matter is that parts of science, including many "immature" sciences, have been successful for a very long time; indeed, many of the theories I alluded to above were empirically successful by any criterion I can conceive of (including fertility, intuitively high confirmation, successful prediction, etc.). If the realist restricts himself to explaining only how the "mature" sciences work (and recall that very few sciences indeed are yet "mature" as the realist sees it), then he will have completely failed in his ambition to explain why science in general is successful. Moreover, several of the examples I have cited above come from the history of mathematical physics in the last century (e.g., the electromagnetic and optical aethers) and, as Putnam himself concedes, "*physics* surely counts as a 'mature' science if any science does" (1978, p. 21). Since realists would presumably insist that many of the central terms of the theories enumerated above do not genuinely refer, it follows that none of those theories could be approximately true (recalling that the former is a necessary condition for the latter). Accordingly, cases of this kind cast very grave doubts on the plausibility of (T2), i.e., the claim that nothing succeeds like approximate truth.

I daresay that for every highly successful theory in the past of science which we now believe to be a genuinely referring theory, one could find half a dozen once successful theories which we now regard as substantially non-referring. If the proponents of CER are the empiricists they profess to be about matters epistemological, cases of this kind and this frequency should give them pause about the well-foundedness of (T2).

But we need not limit our counter-examples to non-referring theories. There were many theories in the past which (so far as we can tell) were both genuinely referring and empirically successful which we are nonetheless loath to regard as approximately true. Consider, for instance, virtually all those geological theories prior to the 1960s which denied any lateral motion to the continents. Such theories were, by any standard, highly successful (and apparently referential); but would anyone today be prepared to say that their constituent theoretical claims—committed as they were to laterally stable continents—are almost true? Is it not the fact of the matter that structural geology was a successful science between (say) 1920 and 1960, even though geologists were fundamentally mistaken about many—perhaps even most—of the basic mechanisms of tectonic construction? Or what about the chemical theories of the 1920s which assumed that the atomic nucleus was structurally homogenous? Or those chemical and physical theories of the late 19th century which explicitly assumed that matter was neither created nor destroyed? I am aware of no sense of approximate truth (available to the realist) according to which such highly successful, but evidently false, theoretical assumptions could be regarded as "truthlike."

More generally, the realist needs a riposte to the *prima facie* plausible claim that there is no necessary connection between increasing the accuracy of our deep-structural characterizations of nature and improvements at the level of phenomenological explanations, predictions and manipulations. It *seems* entirely conceivable intuitively that the theoretical mechanisms of a new theory, T_2, might be closer to the mark than those of a rival T_1 and yet T_1 might be more accurate at the level of testable predictions. In the absence of an argument that greater correspondence at the level of unobservable claims is more likely than not to reveal itself in greater accuracy at the experimental level, one is obliged to say that the realist's hunch that increasing deep-structural fidelity must manifest itself pragmatically in the form of heightened experimental accuracy has yet to be made cogent. (Equally problematic, of course, is the inverse argument to the effect that increasing experimental accuracy betokens greater truthlikeness at the level of theoretical, i.e., deep-structural, commitments.)

. . .

6. The Realists' Ultimate "Petitio Principii"

It is time to step back a moment from the details of the realists' argument to look at its general strategy. Fundamentally, the realist is utilizing, as we have seen, an abductive inference which proceeds from the success of science to the conclusion that science is approximately true, verisimilar, or referential (or any combination of these). This argument is meant to show the sceptic that theories are not ill-gotten, the positivist that theories are not reducible to their observational consequences, and the pragmatist that classical epistemic categories (e.g., "truth", "falsehood") are a relevant part of metascientific discourse.

It is little short of remarkable that realists would imagine that their critics would find the argument compelling. As I have shown elsewhere (1978), ever since antiquity critics of epistemic realism have based their scepticism upon a deep-rooted conviction that the fallacy of affirming the consequent is indeed fallacious. When Sextus or Bellarmine or Hume doubted that certain theories which saved the phenomena were warrantable as true, their doubts were based on a belief that the exhibition that a theory had some true consequences left entirely open the truth-status of the theory. Indeed, many non-realists have been non-realists precisely because they believed that false theories, as well as true ones, could have true consequences.

Now enters the new breed of realist (e.g., Putnam, Boyd and Newton-Smith) who wants to argue that epistemic realism can reasonably be presumed to be true by virtue of the fact that it has true consequences. But this is a monumental case of begging the question. The non-realist refuses to admit that a *scientific* theory can be warrantedly judged to be true simply because it has some true consequences. Such non-realists are not likely to be impressed by the claim that a *philosophical* theory like realism can be warranted as true because it arguably has some true consequences. If non-realists are chary about first-order abductions to avowedly true conclusions, they are not likely to be impressed by second-order abductions, particularly when, as I have tried to show above, the premises and conclusions are so indeterminate.

But, it might be argued, the realist is not out to convert the intransigent sceptic or the determined instrumentalist.[14] He is perhaps seeking, rather, to show that realism can be tested like any other scientific hypothesis, and that realism is at least as well confirmed as some of our best scientific theories. Such an analysis, however plausible initially, will not stand up to scrutiny. I am aware of no realist who is willing to say that a *scientific* theory can be reasonably presumed to be true or even regarded as well confirmed just on the strength of the fact that its thus far tested consequences are true. Realists have long been in the forefront of those opposed to *ad hoc* and *post hoc* theories. Before a realist accepts a scientific hypothesis, he generally wants to know whether it has explained or predicted more than it was devised to explain; he wants to know whether it has been subjected to a battery of controlled tests; whether it has successfully made novel predictions; whether there is independent evidence for it.

What, then, of realism itself as a "scientific" hypothesis?[15] Even if we grant (contrary to what I argued in section 4) that realism entails and thus explains the success of science, ought that (hypothetical) success warrant, by the realist's own construal of scientific acceptability, the acceptance of realism? Since realism was devised in order to explain the success of science, it remains purely *ad hoc* with respect to that success. If realism has made some novel predictions or been subjected to carefully controlled tests, one does not learn about it from the literature of contemporary realism. At the risk of apparent inconsistency, the realist repudiates the instrumentalist's view that saving the phenomena is a significant form of evidential support while endorsing realism itself on the transparently instrumentalist grounds that it is confirmed by those very facts it was invented to explain. No proponent of realism has sought to show that realism satisfies those stringent empirical demands which the realist himself minimally insists on when appraising scientific theories. The latter-day realist often calls realism a "scientific" or "well-tested" hypothesis, but seems curiously reluctant to subject it to those controls which he otherwise takes to be a *sine qua non* for empirical well-foundedness.

7. Conclusion

The arguments and cases discussed above seem to warrant the following conclusions:

1. The fact that a theory's central terms refer does not entail that it will be successful; and a theory's success is no warrant for the claim that all or most of its central terms refer.
2. The notion of approximate truth is presently too vague to permit one to judge whether a theory consisting entirely of approximately true laws would be empirically successful; what is clear is that a theory may be empirically successful even if it is not approximately true.
3. Realists have no explanation whatever for the fact that many theories which are not approximately true and whose "theoretical" terms seemingly do not refer are nonetheless often successful.
4. The convergentist's assertion that scientists in a "mature" discipline usually preserve, or seek to preserve, the laws and mechanisms of earlier theories in later ones is probably false; his assertion that when such laws

are preserved in a successful successor, we can explain the success of the latter by virtue of the truthlikeness of the preserved laws and mechanisms, suffers from all the defects noted above confronting approximate truth.

5. Even if it could be shown that referring theories and approximately true theories would be successful, the realists' argument that successful theories are approximately true and genuinely referential takes for granted precisely what the non-realist denies (namely, that explanatory success betokens truth).
6. It is not clear that acceptable theories either *do* or *should* explain why their predecessors succeeded or failed. If a theory is better supported than its rivals and predecessors, then it is not epistemically decisive whether it explains why its rivals worked.
7. If a theory has once been falsified, it is unreasonable to expect that a successor should retain either all of its content *or* its confirmed consequences *or* its theoretical mechanisms.
8. Nowhere has the realist established—except by fiat—that non-realist epistemologists lack the resources to explain the success of science.

With these specific conclusions in mind, we can proceed to a more global one: it is not yet established—Putnam, Newton-Smith and Boyd notwithstanding—that realism can explain *any* part of the success of science. What is very clear is that realism *cannot*, even by its own lights, explain the success of those many theories whose central terms have evidently not referred and whose theoretical laws and mechanisms were not approximately true. The inescapable conclusion is that insofar as many realists are concerned with explaining how science works and with assessing the adequacy of their epistemology by that standard, they have thus far failed to explain very much. Their epistemology is confronted by anomalies which seem beyond its resources to grapple with.

It is important to guard against a possible misinterpretation of this essay. *Nothing* I have said here refutes the possibility in principle of a realistic epistemology of science. To conclude as much would be to fall prey to the same inferential prematurity with which many realists have rejected in principle the possibility of explaining science in a non-realist way. My task here is, rather, that of reminding ourselves that there *is* a difference between wanting to believe something and having good reasons for believing it. All of us would like realism to be true; we would like to think that science works because it has got a grip on how things really are. But such claims have yet to be made out. Given the *present* state of the art, it can only be wish fulfilment that gives rise to the claim that realism, and realism alone, explains why science works.

Notes

* I am indebted to all of the following for clarifying my ideas on these issues and for saving me from some serious errors: Peter Achinstein, Richard Burian, Clark Glymour, Adolf Grünbaum, Gary Gutting, Allen Janis, Lorenz Krüger, James Lennox, Andrew Lugg, Peter Machamer, Nancy Maull, Ernan McMullin, Ilkka Niiniluoto, Nicholas Rescher, Ken Schaffner, John Worrall, Steven Wykstra.

1 Putnam, evidently following Boyd, sums up (R1) to (R3) in these words:

> 1) Terms in a mature science typically *refer*.
> 2) The laws of a theory belonging to a mature science are typically approximately true . . . I will only consider [new] theories . . . which have this property—[they] contain the [theoretical] laws of [their predecessors] as a limiting case (1978, pp. 20–21).

2 Putnam insists, for instance, that if the realist is wrong about theories being referential, then "the success of science is a miracle" (Putnam 1975, p. 69).

3 Boyd remarks: "scientific realism offers an *explanation* for the legitimacy of ontological commitment to theoretical entities" (Putnam 1978, Note 10, p. 2). It allegedly does so by explaining why theories containing theoretical entities work so well: because such entities genuinely exist.

4 Whether one utilizes Putnam's earlier or later versions of realism is irrelevant for the central arguments of this essay.

5 See also p. 3: "experimental evidence for a theory is evidence for the truth of even its non-observational laws." See also Sellars (1963, p. 97).

6 A caveat is in order here. *Even if* all the central terms in some theory refer, it is not obvious that every rational successor to that theory must preserve all the referring terms of its predecessor. One can easily imagine circumstances when the new theory is preferable to the old one even though the range of application of the new theory is less broad than the old. When the range is so restricted, it may well be entirely appropriate to drop reference to some of the entities which figured in the earlier theory.

7 For Putnam and Boyd both "it will be a constraint on T_2 [i.e., any new theory in a domain] . . . that T_2 must have this property, the property that *from its standpoint* one can assign referents to the terms of T_1 [i.e., an earlier theory in the same domain]" (Putnam 1978, p. 22). For Boyd, see (1973, p. 8): "new theories should, *prima facie*, resemble current theories with respect to their accounts of causal relations among theoretical entities."

8 For just a small sampling of this view, consider the following: "The claim of a realist ontology of science is that the only way of explaining why the models

of science function so successfully . . . is that they approximate in some way the structure of the object" (McMullin 1970, pp. 63–64); "the continued success [of confirmed theories] can be *explained* by the hypothesis that they are in fact close to the truth . . . " (Niiniluoto 1980, p. 448); the claim that "the laws of a theory belonging to a mature science are typically approximately *true* . . . [provides] an *explanation* of the behavior of scientists and the success of science" (Putnam 1978, pp. 20–21). Smart, Sellars, and Newton-Smith, among others, share a similar view.

9 Although Popper is generally careful not to assert that actual historical theories exhibit ever-increasing truth content (for an exception, see his (1963, p. 220)), other writers have been more bold. Thus, Newton-Smith writes that "the historically generated sequence of theories of a mature science is a sequence in which succeeding theories are increasing in truth content without increasing in falsity content." [See Newton-Smith 1981, p. 184. Laudan is quoting from a manuscript of Newton-Smith's book. The wording of the quotation differs slightly from the published version.]

10 On the more technical side, Niiniluoto has shown that a theory's degree of corroboration co-varies with its "estimated verisimilitude" (1977, pp. 121–147 and 1980). Roughly speaking, "estimated truthlikeness" is a measure of how closely (the content of) a theory corresponds to *what we take to be* the best conceptual systems that we so far have been able to find (1980, pp. 443ff.). If Niiniluoto's measures work it follows from the abovementioned co-variance that an empirically successful theory will have a high degree of estimated truthlikeness. But because estimated truthlikeness and genuine verisimilitude are not necessarily related (the former being parasitic on existing evidence and available conceptual systems), it is an open question whether—as Niiniluoto asserts—the continued success of highly confirmed theories can be *explained* by the hypothesis that they in fact are close to the truth at least in the relevant respects. Unless I am mistaken, this remark of his betrays a confusion between "true verisimilitude" (to which we have no epistemic access) and "estimated verisimilitude" (which is accessible but non-epistemic).

11 Newton-Smith claims that the increasing predictive success of science through time "would be totally mystifying . . . if it were not for the fact that theories are capturing more and more truth about the world" (1981, p. 196).

12 I must stress again that I am *not* denying that there *may* be a connection between approximate truth and predictive success. I am only observing that until the realists show us what that connection is, they should be more reticent than they are about claiming that realism can explain the success of science.

13 A *non-realist* might argue that a theory is approximately true just in case all its *observable* consequences are true or within a specified interval from the true value. Theories that were "approximately true" in this sense would indeed be demonstrably successful. But, the realist's (otherwise commendable) commitment to taking seriously the theoretical claims of a theory precludes him from utilizing any such construal of approximate truth, since he wants to say that the theoretical as well as the observational consequences are approximately true.

14 I owe the suggestion of this realist response to Andrew Lugg.

15 I find Putnam's views on the "empirical" or "scientific" character of realism rather perplexing. At some points, he seems to suggest that realism is both empirical and scientific. Thus, he writes: "If realism is an explanation of this fact [namely, that science is successful], realism must itself be an overarching scientific *hypothesis*" (1978, p. 19). Since Putnam clearly maintains the antecedent, he seems committed to the consequent. Elsewhere he refers to certain realist tenets as being "our highest level empirical generalizations about knowledge" (p. 37). He says moreover that realism "could be false," and that "facts are relevant to its support (or to criticizing it)" (pp. 78–79). Nonetheless, for reasons he has not made clear, Putnam wants to deny that realism is either scientific or a hypothesis (p. 79). How realism can consist of doctrines which 1) explain facts about the world, 2) are empirical generalizations about knowledge, and 3) can be confirmed or falsified by evidence and yet be neither scientific nor hypothetical is left opaque.

References

Boyd, R. (1973), "Realism, Underdetermination, and a Causal Theory of Evidence," *Noûs* 7: 1–12.

Krajewski, W. (1977), *Correspondence Principle and Growth of Science*. Dordrecht: Reidel.

Laudan, L. (1978), "Ex-Huming Hacking," *Erkenntnis 13*: 417–435.

McMullin, Ernan (1970), "The History and Philosophy of Science: A Taxonomy," Stuewer, R. (ed.), *Minnesota Studies in the Philosophy of Science V*: 12–67. Minneapolis: University of Minnesota Press.

Newton-Smith, W. (1981), *The Rationality of Science*. London: Routledge & Kegan Paul.

Niiniluoto, Ilkka (1977), "On the Truthlikeness of Generalizations," Butts, R. and Hintikka, J. (eds.), *Basic Problems in Methodology and Linguistics*: 121–147. Dordrecht: Reidel.

Niiniluoto, Ilkka (1980), "Scientific Progress," *Synthese 45*: 427–462.

Popper, K. (1963), *Conjectures and Refutations*. London: Routledge & Kegan Paul.

Putnam, H. (1975), *Mathematics, Matter and Method, Vol. 1*. Cambridge: Cambridge University Press.

Putnam, H. (1978), *Meaning and the Moral Sciences*. London: Routledge & Kegan Paul.

Sellars W. (1963), *Science, Perception and Reality*. New York: The Humanities Press.

14

Gary Gutting, "Scientific Realism versus Constructive Empiricism: A Dialogue"

Following Plato, many philosophers have used the dialog form to present their views. The debate about Realism naturally invites such a format. Gary Gutting's engaging dialog between a Scientific Realist and a Constructive Empiricist aims at clarifying both van Fraassen's position and its critique.

Note: *The following is a discussion between a scientific realist* (SR) *who has been strongly influenced by the work of Wilfrid Sellars and a constructive empiricist* (CE) *who has been equally influenced by the work of Bas van Fraassen. Indeed, the influence is so great that the interlocuters occasionally lapse into direct quotation of their masters. I do not, however, want anyone to* identify *the views of my two characters with those of Sellars and van Fraassen. What they say merely represents the dialectic of my own mind as I think through the issues raised by the debate between Sellars and van Fraassen on scientific realism.*

SR: Realism is encapsulated in the claim that "to have good reason for holding a theory is *ipso facto* to have good reason for holding that the entities postulated by the theory exist."[1] For an appropriate scientific theory (say atomic theory), this claim allows us to argue as follows:

(1) If we have good reason for holding atomic theory, we have good reasons for holding that atoms exist;
(2) We do have good reason for holding atomic theory (it's highly confirmed, fruitful, simple, etc.);
(3) Therefore, we have good reason for holding that atoms exist.

CE: Everything depends on what we mean by "holding a theory." If it

G. Gutting, "Scientific Realism versus Constructive Empiricism: A Dialogue," *The Monist*, 1982, 65: 336–49.

means "believing that the theory is true," then premise (1) of your argument is obvious but premise (2) strikes me as false. There's a lot to be said for atomic theory but none of it constitutes a cogent case for its truth. At the most, the evidence shows that atomic theory is empirically adequate, by which I mean that there may be reason to think that all its *observable* consequences are true. But the evidence does not support the existence of the particular unobservable mechanisms and entities the theory postulates. On the other hand, if you take "holding a theory," as I do, to mean "believing it to be empirically adequate," then there's no problem with premise (2) but, for the reasons I've just been urging, there is no basis for premise (1).

SR: I'm willing to stand by the argument even if we take "holding a theory" to mean "believing it to be empirically adequate"; but we need to get clear on just what's involved in empirical adequacy. For example, it won't do to take empirical adequacy in the minimal sense of "accurately describing all the observable phenomena." With this sense of "empirically adequate" the argument will fall to the old problem of the underdetermination of theory by data. Specifically, with this meaning of "empirically adequate," premise (1) says: "If we have good reason for believing that atomic theory accurately describes all the observable phenomena, then we have good reason for believing that atoms exist." But this isn't so since, first, there are an infinity of other sorts of theoretical entities that would produce the same observable phenomena and, second, we could just as well believe only in the phenomena and forget about underlying entities.[2]

CE: Your second point is just my view. I'm not saying theoretical entities don't exist or that talk about them is meaningless. I don't even say there's anything wrong with believing in them if you want to. My point is simply that there's no evidence that makes it irrational to withhold judgment about their existence. I'm defending my right to be an agnostic on the issue. I suspect however that, just like theists who deny the rationality of religious agnosticism, you're going to invoke the explanatory power of your postulations to support their existence.

SR: Of course, though the case for scientific realism can avoid the pitfalls of "theological realism." The point is that the empirical adequacy of a scientific theory needs to be taken broadly enough to include the theory's explanatory power and, specifically, its explanatory superiority to the physical thing language used in the observation framework. Atomic theory and its associated ontology is needed precisely because of the explanatory failures of the observation framework. So, roughly, premise (1) needs to be understood as saying this: If we have good reasons to believe that atomic theory is needed to explain the observable data, then we have good reason to believe that atoms exist.

CE: What I want to question is the move you're trying to make from an explanatory need for a theory to the existence of its entities. Consider, for example, Sellars's fictional case of observationally identical samples of gold that dissolve at different rates in *aqua regia*. I agree that available microtheory might explain the empirical fact of the different dissolution rates. "The microtheory of chemical reactions might admit of a simple modification to the effect that there are two structures of microentities each of which 'corresponds' to gold as an observational construct, but such that pure samples of one dissolve, under given conditions of pressure, temperature, etc., at a different rate from samples of the other. Such a modification of the theory would explain the observationally unpredicted variation in the rate of dissolution of gold by saying that samples of observational gold are mixtures of these two theoretical structures in various proportions, and have a rate of dissolution which varies with the proportion."[3] Of course, I'd expect the realist to admit in his turn that it might also be possible to sue the correspondence rules of our microtheory to "derive observational criteria for distinguishing between observational golds of differing theoretical compositions."[4] If so, we could formulate two empirical laws (in the observation framework), one for each variety of gold, that would explain the differing dissolution rates.

SR: But remember that it might also happen that a good theory does not allow the formulation of any such empirical generalizations. It might simply itself directly explain the singular observable fact of differing dissolution rates. In such a case, the theory would be necessary to give any explanation of the singular observed facts and so would have to be accepted as true if we were to have any explanation at all of these facts. And, of course, this is precisely my claim regarding postulational scientific theories that have actually been developed. If, for example, we do not accept the existence of atoms there are numerous singular empirical facts for which we simply have no explanation. This discussion lets me formulate more precisely premise (1) of the argument for realism. We should take it to say: If we have good reason to believe that atomic theory provides the only way of explaining some singular observed facts, then we have good reason to believe in the existence of atoms.

CE: Your more accurate formulation only serves to pinpoint the weakness of your position. Just what is the "explanatory failure" of the observation language? You agree that it can sustain inductive generalizations but insist that it cannot sustain "enough to explain all the singular facts that require explanation. In the case of the gold, one might have achieved a very precise statistical generalization: for each number r, the probability that a random sample of gold dissolves at rate r equal p(r). . . . But faced with the question why a given sample of gold dissolves at rate r, the physical thing language provides us with no property X such that we could say: this is an X-sample, and all X-samples dissolve at that rate."[5] You conclude from this that we need a theory to explain what the observation language cannot. But the claim is far too strong. On your principles, an exactly parallel argument could be made for the existence of hidden variables underlying quantum phenomena. As is well known, the laws of quantum mechanics are irreducibly statistical; that is, they can explain why certain events occur a certain percentage of times over a given period but they cannot explain why one particular event occurs rather than another. For example, quantum mechanics has no explanation of the fact that a particular radium atom decays at a particular time, though it can explain why, over a period of time, a given fraction of the atoms in a sample of radium will decay. Using your principles, a quantum physicist would have to accept the suggestion, made by some physicists, that there are "hidden" entities and processes, not taken account of by quantum theory, that are responsible for the occurrence of the singular facts that theory cannot explain. But this conclusion, generated by an a priori demand for explanation, conflicts with the fact that the irreducibly statistical character of its laws does not, in the mind of the scientific community, constitute a case for the explanatory inadequacy of quantum theory and the need for the acceptance of hidden variables. So, just like the theist with his cosmological argument, you make your case by insisting on the need to explain something that there is no reason to think has to be explained. If the singular facts not explained by quantum theory don't need explanation, neither do the singular facts not explained by the observation framework.

SR: I entirely agree with you about quantum mechanics, and in arguing for realism I do not mean to "demand that all singular matters of fact be capable of explanation."[6] Perhaps the fictional gold example is misleading. The inadequacy of the observation framework does not consist in its inability to explain some singular empirical facts but in its inability to explain without relying on theoretical concepts. "It is not that the 'physical thing framework' doesn't sustain *enough* inductive generalizations, but rather that what inductive generalizations it *does* sustain, it sustains by a covert introduction of the framework of theory into the physical thing framework itself."[7]

CE: I take it, then, that you're revising even your most recent statement of premise (1) in your argument for realism. You now seem to be taking it to mean something like this: If we have reason to believe that all explanations of singular empirical facts (in a given empirical domain) must rely on the concepts of atomic theory, then we have good reason to believe in the existence of atoms.

SR: That's about what I have in mind.

CE: But then we need some clarification of what you mean by an explanation's "relying on the concepts of a theory." You're obviously assuming that explanation is at least partly a matter of subsuming singular empirical facts under generalizations. So one possible meaning of the claim that all of a set of observation-framework explanations "rely on the concepts of a theory" is that all the generalizations that accurately subsume the singular empirical facts must be expressed in theoretical terms. But taken this way the claim is clearly false. Even when theoretical corrections are provided for empirical generalizations, the results are typically expressible by a new empirical generalization in the observation language. For example, the correction of Boyle's Law by kinetic theory results in van der Waals Law, which is still entirely observational.

SR: But my claim about the explanatory reliance of the observation framework on theory is not about the *content* of the generalizations used to explain singular empirical facts but about the way in which these generalizations are *inductively justified*. To a theoretical correction of an empirical generalization there will, I agree, typically correspond a revised empirical generalization. Further, this revised generalization will be compatible with the observational evidence. But, I claim, it will not typically be the law that would be accepted "on purely inductive grounds—i.e., in the absence of theoretical considerations."[8] Rather, lacking theoretical guidance, purely inductive reasoning in the observation framework alone would lead us to accept an empirical generalization that is shown by theoretical considerations to be false. So explanations of singular empirical facts rely on theory in the following sense: The empirical generalizations that explain these facts must be justified, ultimately, by theoretical considerations. This can be expressed by putting premise (1) in the following way: If we have good reason to believe that the empirical generalizations that explain singular empirical facts require an appeal to atomic theory for their justification, then we have good reason for believing in the existence of atoms.

CE: I have a number of reservations about your factual claim that the empirical data alone wouldn't lead us to the same law that theoretical considerations yield. We'd have to look at some examples to probe that thesis. But suppose I agree that theory plays the role you say it does, that it's essential for the development of adequate empirical generalizations. I still don't think there's any reason to accept such a theory as a true (or approximately true) description of the world. In other words, I might be prepared to accept the antecedent of your last formulation of premise (1), but I'll still deny the consequent. Here my position is like Duhem's. He shared your conviction that "postulation of unobservable entities is indispensable to science." But since he also held that improved description of observable phenomena is the only basic role for theoretical postulation, he maintained that there is no need for us to accept the existence of postulated entities. "If that is how one sees it, then truth of the postulates becomes quite irrelevant.

When a scientific theory plays [its] role well, we shall have reason to use the theory whether we do or do not believe it to be true; and we may do well to reserve judgment on the question of truth. The only thing we need to believe here is that the theory is empirically adequate, which means that in its round-about way it has latched on to actual regularities in the observable phenomena. Acceptance of the theory need involve no further beliefs."[9]

SR: Don't we at least need an explanation of why the theory is empirically adequate; that is, of why what we observe is just as it would be if the theory were true? And isn't the best explanation just that the theory is in fact true (or anyway near the truth)? Don't we ultimately need to invoke realism to explain the success of science?

CE: It seems to me that you're slipping back into the theological demand for explanation for its own sake. Why are you so sure that we need an explanation for science's success? But let me agree, at least for the moment, that we do need such an explanation. Even so, I think the appropriate explanation is quite different from the one you've proposed. After all, "science is a biological phenomenon, an activity by one kind of organism which facilitates its interaction with its environment."[10] Just as, from a Darwinian viewpoint, the only species that survive are those that are successful in coping with their environment, so too only successful scientific theories have survived. "Any scientific theory is born into a life of fierce competition, a jungle red in tooth and claw. Only the successful theories survive—the ones which in fact latched on to actual regularities in nature."[11] But this process of selection on the basis of empirical success need have nothing to do with the *truth* of the theories selected. Your argument is no different from that of an antievolutionist who holds that the survival of a species can be explained only by some design that has preadapted the species to its environment. But we don't need the hypothesis that theories are successful because truth has preadapted them to the world. We need only the hypothesis that theories that are not empirically successful have not survived.

SR: It seems to me you're ignoring the amazing *rate* at which empirically successful theories have emerged. Perhaps an extended process of trial and error would eventually lead to an empirically adequate scientific description of what we observe. But the use of theoretical postulations has led to success far more quickly than we could reasonably expect from mere trial and error selection. So even from your Darwinian viewpoint I think we need the realist hypothesis. But this is taking us off the track. The defense of realism that I'm interested in proposing need not be based on putting it forward as an explanation of science's success. Rather I have been arguing from the indispensable role of theoretical postulation in the formulation of empirically adequate laws. Think about it this way: Imagine that we are doing science initially only in the observation framework. We are aware of various singular observational facts and are trying to explain them by subsuming them under empirical generalizations—that is, under inductive generalizations that employ only the concepts of the observation framework.

CE: Excuse me a moment. Just how are you understanding the notion of *observation* when you speak of the observation framework? Some realists, you know, have maintained that anything—even electrons and other postulated submicroscopic entities—are in principle observable.

SR: I sympathize with some of the epistemological motives behind such claims. It is important to reject the idea that the realm of entities and properties we in fact observe functions as an unchangeable given. But in this context we must avoid trivializing the distinction between what is observable and what is not. When someone says that everything is observable, he is envisaging a situation in which concepts from the theoretical framework of science have ingressed into the observation framework. I'm speaking of the observation framework prior to any theoretical ingressions.

CE: All right. So what we observe are the ordinary objects of everyday life and their properties.

SR: Yes, but we need to distinguish two sorts of properties that we attribute to observable things. On the one hand, there are occurrent (nondispositional) properties that are, strictly speaking, *what we perceive of an object* that we perceive. There is, for example, "the occurrent sensuous redness of the facing side" of a brick.[12] On the other hand, there are dispositional or, more broadly, causal properties that correspond to *what we perceive an object as* (e.g., the brick seen as made of baked clay). Within the observation framework properties of the first sort are a constant factor. They correspond to the way that, for physiological reasons, we must perceive the world. The second sort of properties corresponds to our conceptual resources for classifying objects into kinds with distinctive causal features. These kinds are not constant but can change as "our classification of physical objects . . . becomes more complex and sophisticated."[13] An essential feature of scientific inquiry is its *revision* of the causal concepts of the observation framework in order to arrive at maximally accurate empirical generalizations. In many cases, these revisions take place entirely within the observation framework. In such cases, the causal properties are always built out of concepts expressing the occurrent properties of physical objects. We can imagine that science never required any conceptual revision other than this sort. In that case, the observation framework would be conceptually autonomous; that is, its conceptual resources would suffice for formulating and justifying entirely accurate empirical generalizations about singular observable facts. If this were the case, then the framework of postulational science would be "in principle otiose,"[14] and what Sellars calls the "manifest image" would provide a correct ontology for the world. But, as we have learned from the development of science, the observation framework is not autonomous. We cannot do the job of science using only its conceptual resources. Rather, we can arrive at justified empirical generalizations in the observation framework only by appealing to theories that employ concepts that cannot be built out of the conceptual resources of the observation framework.

CE: You seem to be missing the point of my antirealism. I'm willing to admit everything you've said about the *indispensability* of theories. But why should we also have to accept the *truth* of theories? As I see it, the highest virtue we need attribute to a theory needed for the successful practice of science is empirical adequacy. In other words, we need only agree that all of a successful theory's *observable consequences* are true. If we regard a theory as empirically adequate—and of course we are entitled to so regard highly successful theories—then from that alone we have sufficient justification for accepting the empirical generalizations that theory entails. The further assertion of the theory's truth is a gratuitous addition, entirely unnecessary for the fulfillment of science's fundamental aim; namely, an exact account of observable phenomena. The enterprise of science can be entirely successful without ever accepting the truth of its theories. Consider two scientists. One accepts atomic theory in the sense that he thinks it is empirically adequate: he knows that it fits all observations to date and expects that it will continue to fit all further observations. Accordingly, he thinks in terms of atomic theory and uses its conceptual resources to solve relevant scientific problems. However, he remains agnostic on the question of whether atoms really exist. A second scientist shares the first's views about the empirical adequacy of atomic theory, but he also holds that atoms really do exist. But what sort of work is done by this latter belief? It makes absolutely no difference for what the second scientist expects to observe or for how he proceeds in his scientific work. His expectations and procedures are exactly the same as those of his agnostic colleague. So, while I agree with you that theories are not "otiose in principle," I do maintain that a realistic interpretation of theories is. There is nothing in the aims of scientific inquiry that is in the least affected by the acceptance or the rejection of the existence of theoretical entities.

SR: It seems to me that it's you who are missing the point of my realism. First of all, you misrepresent my view by taking it as a thesis about the aims of science. I am entirely content with the view that science aims only at empirical adequacy. Indeed, I agree that this aim might have (in some other possible world) been attained without the postulation of theoretical entities. My thesis is rather that empirical adequacy in fact requires theories that postulate unobservable entities and that this fact provides good reason for thinking such entities exist. My realism is not a thesis about the aim of science but rather a thesis about the philosophical (specifically, metaphysical) significance of the means that scientists have had to use in fulfilling this aim.

CE: I'm happy to accept this clarification of your views, but it does not affect the central point: that you haven't offered an argument from the indispensability of theories to the existence of the entities they postulate. Even Sellars, who develops the indispensability thesis much more cogently than you do, seems to ignore the need for such an argument. He seems just to assume that, once theories have been shown to be indispensable, the reality of the entities they postulate has been established. But in fact there is a gap that needs bridging. Sellars himself has emphasized this very point in parallel contexts. For example, he agrees that semantic concepts such as *meaning* and moral concepts such as *person* are indispensable, but nonetheless insists this doesn't entail that meanings and persons exist. Similarly, I think you should admit that the indispensability of theoretical language does not entail the existence of theoretical entities. Further, I submit that the only way of bridging the gap between indispensability and existence is by an act of faith that may be permissible for those who want to believe in atoms and similar things but is in no way required by the evidence.

SR: I agree that the gap needs to be bridged and I even agree that it cannot be bridged by a deductive argument. I admit that indispensability does not entail existence; theories could be indispensable and yet theoretical entities not exist. But I insist that there is a good *inductive* case for realism; reasons that make it highly probable that theoretical entities exist.

CE: I take it then that you're about to follow Putnam in presenting realism as a quasi-scientific hypothesis that is the best explanation of the success of science. I've already suggested my criticism of this sort of approach.

SR: On the contrary. I'm not entirely happy myself with this sort of "empirical" approach. Besides the difficulties you raised earlier about possible alternative hypotheses, it seems to me that realism lacks the fruitfulness we require of a good scientific hypothesis. From a purely scientific viewpoint, it looks a lot like an *ad hoc* explanation of one fact with no other explanatory significance. I rather see realism as a *philosophical* thesis based on an analysis of the nature of theoretical explanation in science. I have in mind the following strategy of argument: First find a generally valid type of argumentation from the explanatory power of a hypothesis to the reality of the entities the hypothesis refers to; then show that, in some specific cases, the results of theoretical science enable us to construct an argument of just this type for the existence of theoretical entities. Such a case is inductive because the argument type employed is inductive. But it's philosophical rather than scientific because it is not postulating an explanatory hypothesis but rather pointing out the essential similarity of two ways of arguing.

CE: I need to hear the type of argumentation you have in mind.

SR: The point is really quite simple. There's a standard way of arguing—in both everyday and scientific contexts—for the existence of unobserved entities. The mode of argument is this: from the ability of a hypothesis to (a) subsume all known facts and to (b) predict new and even unexpected facts, we infer the reality of the entities the hypothesis postulates. There's no doubt that we all accept this mode of argument in many cases that involve unobserved though observable entities. For example, this is just the way we proceed in arguing for the past existence of dinosaurs or for the present existence of stars, conceived as huge, tremendously hot, gaseous masses far distant from us. But the very same mode of argumentation used in these non-controversial cases can be employed to argue for the existence of the unobservable entities postulated by microphysics. Just as we accepted the existence of dinosaurs because the hypothesis of their existence subsumed the known facts and successfully predicted new ones, so too we ought to accept the existence of electrons, neutrinos, etc. for precisely the same sort of reasons. The case for the existence of electrons and neutrinos is logically identical to the case for the existence of dinosaurs and stars. Since you can hardly reject the latter, I submit that you cannot consistently reject the former.

CE: You yourself have mentioned but ignored the crucial point: the non-observability of the entities postulated by microphysics in contrast to the observability of stars and dinosaurs. This difference undermines your claim that the mode of argument for the two sorts of entities is the same. As I see it, the mode of inference at work in the examples you've given is not from a theory's explanatory power to its truth but from its explanatory power to its empirical adequacy. At any rate, the uncontroversial uses of the mode of argumentation—to the existence of stars and dinosaurs—cannot decide between the realist and the antirealist interpretations of it. For these are cases of inference to *observable* entities; and, for such cases, the claim that a postulation is empirically adequate is equivalent to the claim that it is true. For example, "Stars (as described by modern astronomy) exist" is equivalent to "All observable phenomena are as if stars exist," since the existence of stars is itself (in principle) observable. So I can accept the inference to stars and reject the inference to electrons, etc. on the grounds that, in both cases, we are only entitled to infer the empirical adequacy of a hypothesis from its explanatory and predictive success. In the case of stars, this is equivalent to inferring their existence; but in the case of electrons, which are not even in principle observable, it is not.

SR: This response keeps your position consistent but at the price of arbitrariness. By maintaining that explanatory and predictive success supports only the empirical adequacy of a theory, you are implicitly committing yourself to a sharp epistemic distinction between the observable and the unobservable. You say the explanatory success of a hypothesis is evidence of its truth only if the hypothesis is about observable entities. But why should observability matter in this context?

CE: The answer depends, of course, on what you mean by "observable." As we noted above, in one sense everything is observable: there might be some creature with sense organs appropriate for perceiving it. But, as we agreed when discussing this point, here a more restricted sense of "observable" is appropriate. Specifically, observability must be taken as a function of certain empirical limitations of human beings. "The human organism is, from the point of view of physics, a certain kind of measuring apparatus. As such it has certain inherent limitations—which will be described in detail in the final physics and biology. It is these limitations to which the 'able' in 'observable' refers—our limitations, *qua* human beings."[15]

SR: I agree with this construal of "observable." But my question is, What does observability in this sense have to do with the existence or non-existence of an entity? You're surely not so much of a positivist as to deny the possibility of the existence of what's unobservable?

CE: You're misconstruing my point. Of course observability in the sense we're taking it has nothing to do with the existence or the nonexistence of an entity. But it has a great deal to do with what we have reason to believe exists. "The question is . . . how much we shall believe when we accept a scientific theory. What is the proper form of acceptance: belief that the theory, as a whole, is true; or something else? To this question, what is observable by us seems eminently relevant." And my answer to the question is this: "to accept a theory is (for us) to believe that it is empirically adequate—that what the theory says *about what is observable* (by us) is true."[16]

SR: Since you refer to what is observable "by us," I take it you make observability relative to an epistemic community, not to individuals?

CE: Of course. The dimmer component of the double star in the Big Dipper's handle is observable because some sharp-eyed people can see it, even if most of us can't.

SR: But then I have a problem. You surely must admit the possibility that our epistemic community might be enlarged, say by the inclusion of animals or extraterrestrials capable of observing things that we now can't observe. For example, we might encounter space travellers who, when we tell them our theories about electrons, say, "Of course, we see them all the time." If this happened, your principles would require that we then, for the first time, accept the existence of electrons. But this seems absurd. Why should the testimony of these aliens be decisive when the overwhelming evidence of our science was not? More generally, isn't it absurd to say that, just because our epistemic community has been enlarged in this way, our beliefs about what there is should change?

CE: Not at all. Your objection has weight only if we believe that "our epistemic policies should give the same results independent of our beliefs about the range of evidence accessible to us."[17] But I see absolutely no reason to believe this. On the contrary, it seems to me that to deny that what evidence is accessible is relevant to what we should believe is to open the door to scepticism or irrationalism.

SR: It seems to me you're equivocating on the expression "evidence accessible to us." Of course such evidence is relevant to our beliefs if it means "the evidence that we are in fact aware of." But this isn't what you mean here. Rather, you're saying that our beliefs ought to depend on the range of evidence that we *might* have even if we don't. Specifically, you're saying that believing that an entity exists ought to depend on whether or not there *could be* direct observations of its existence. Of course, actual observations of an entity are relevant to belief in its existence. And it's also true that, since evidence of actual observations of unobservable entities is not available, it's often harder to produce an adequate case for the existence of such things. But what reason could you have for thinking that the mere question of whether or not an entity could in principle be observed by us is decisive for the question of whether we ought to believe that it exists?

CE: What reason do you have for thinking this isn't the case?

SR: Well, consider an example. Suppose astronomical theory postulates the existence of a far distant star that has not been observed but which we have every reason to think we could observe if we were close enough. It might be, for example, that the star has been postulated as the much smaller double of a known star to explain certain anomalies in its motion. I suppose that if the evidence supporting this postulation is strong enough, you will agree that we have good reason to believe in the existence of this star.

CE: Of course, since it is in principle observable.

SR: All right. But suppose further that, entirely independent of astronomical investigations, physiological studies subsequently show that there are previously unknown limits on human powers of observation that make the postulated star unobservable. We may, for example, have assumed that the star was observable because it emitted light in the visible spectrum and so could be seen if we got close enough to it. But physiologists might discover that the visible spectrum is not continuous, that there are small "holes" of invisibility corresponding to specific wave lengths, one of which is that emitted by the star. On your principles, such a physiological result would require us to abandon our conclusion that the star exists, even though all the empirical evidence that led us to postulate it remains the same. But surely such a move would be unreasonable;

whether or not the star is observable does not alter the evidence in favor of its existence.

Notice that I'm not saying that observations we in fact have made are not relevant to our beliefs about what exists. But the mere fact that something is observable does not give us any reason to think that it ever has or will in fact be observed. The issue between us is whether mere observability—as distinct from actual observation—is relevant to our beliefs about what exists. I submit that it is not.

Another difficulty for your view derives from the fact that an observable entity may have unobservable properties. The sun, for example, is observable but the temperature of its interior is not. What then is your attitude toward the claim that the temperature at the sun's center is about 20 million degrees Centigrade? It would be odd not to accept it: After all, the claim is very well supported by a calculation based on observed facts (the average temperature of the earth, etc.) If these observed facts were appropriately different, the calculation would yield a temperature of the center of the sun that is observable (e.g., about 10 degrees Centigrade). Since you would accept the truth of the result of the calculation in the latter case, it's hard to see how you could coherently not accept its truth in the former case. But, if you accept the claim that the temperature of the sun's center is about 20 million degrees, then you've implicitly given up your principle that observability is relevant to the justification of existence claims.

CE: Not necessarily. The principle might distinguish between unobservable entities and unobservable properties of observable entities.

SR: Possibly. But then we'd need an explanation of why such a distinction is epistemically relevant.

CE: Of course, but you can see where this would lead. To respond to this objection—and your other one about the star that turns out to be unobservable—in a convincing way would require a very elaborate excursion into the theory of knowledge. "But we cannot settle the major questions of epistemology *en passant* in philosophy of science."[18] I'll just acknowledge the relevance of your objections but maintain that more careful epistemological analysis would disarm them. Furthermore, even if I can't answer your objections and your argument stands, remember that the argument is only inductive. This means that, even if I can't directly refute it, I might be able to blunt its force by pointing to overriding considerations that make realism implausible. There is, for example, the fact that, for any theory whose ontology you propose to accept, we can always formulate another theory with a different ontology that is just as well supported by the evidence as the theory you favor. Also, there's the strong historical evidence that scientific postulations are not converging to any single picture of what the unobservable world is like. Until you've dealt with these historical and logical objections to realism, you can't be content with your case for realism.

SR: I agree that the issue isn't fully settled, but your remarks strike me as a strategic retreat that, at least for the present, leaves me in control of the battlefield.

Notes

1 Wilfrid Sellars, *Science, Perception, and Reality* (New York: Humanities, 1963), p. 91.

2 Cf. Bas van Fraassen, *The Scientific Image* (Oxford: The Clarendon Press, 1980), p. 12.

3 Sellars, *Science, Perception, and Reality*, cited in n1, above, pp. 121–22.

4 Ibid., p. 122.

5 Bas van Fraassen, "Wilfrid Sellars and Scientific Realism," *Dialogue* 14 (1975): 611.

6 Wilfrid Sellars, "Is Scientific Realism Tenable?," *PSA 1976*, ed. F. Suppe and P. Asquith (East Lansing, MI: Philosophy of Science Association, 1977), p. 315.

7 Ibid.

8 Ibid., p. 319.

9 Bas van Fraassen, "On the Radical Incompleteness of the Manifest Image," *PSA 1976*, ed. F. Suppe and P. Asquith (East Lansing, MI: Philosophy of Science Association, 1977), p. 325.

10 van Fraassen, *The Scientific Image*, p. 39, cited in n2, above.

11 Ibid., p. 40.

12 Sellars, "Is Scientific Realism Tenable?," cited in n6, above, p. 316.

13 Ibid., p. 318.

14 Sellars, *Science, Perception, and Reality*, p. 118, cited in n1, above.

15 van Fraassen, *The Scientific Image*, p. 17.

16 Ibid., p. 18.

17 Ibid.

18 Ibid., p. 19.

15

Ernan McMullin, "A Case for Scientific Realism"

The realist view of scientific theories has been attacked from different directions. Some popular criticisms capitalize on the difficulties in the traditional philosophical accounts of reference and truth as correspondence with the external world. Other writers (e.g., Laudan) argue that the evidence from the history of science makes Realism untenable. Ernan McMullin takes these challenges in turn.

When Galileo argued that the familiar patterns of light and shade on the face of the full moon could best be accounted for by supposing the moon to possess mountains and seas like those of earth, he was employing a joint mode of inference and explanation that was by no means new to natural science but which since then has come to be recognized as central to scientific explanation. In a retroduction, the scientist proposes a model whose properties allow it to account for the phenomena singled out for explanation. Appraisal of the model is a complex affair, involving criteria such as coherence and fertility, as well as adequacy in accounting for the data. The theoretical constructs employed in the model may be of a kind already familiar (such as "mountain" and "sea" in Galileo's moon model) or they may be created by the scientist specifically for the case at hand (such as "galaxy," "gene," or "molecule").

Does a successful retroduction permit an inference to the existence of the entities postulated in the model? The instincts of the working scientist are to respond with a strong affirmative. Galaxies, genes, and molecules exist (he would say) in the straightforward sense in which the mountains and seas of the earth exist. The immense and continuing success of the retroductions employing these constructs is (in the scientist's eyes) a sufficient testimony to this. Scientists are likely to treat with incredulity the suggestion that

E. McMullin, "A Case for Scientific Realism," in *Scientific Realism*, ed. J. Leplin, 1984, pp. 8–40. Berkeley: University of California Press.

constructs such as these are no more than convenient ways of organizing the data obtained from sophisticated instruments, or that their enduring success ought not lead us to believe that the world actually contains entities corresponding to them. The near-invincible belief of scientists is that we come to discover more and more of the entities of which the world is composed through the constructs around which scientific theory is built.[1]

But how reliable *is* this belief? And how is it to be formulated? This is the issue of scientific realism that has once again come to be vigorously debated among philosophers, after a period of relative neglect. The "Kuhnian revolution" in the philosophy of science has had two quite opposite effects in this regard. On the one hand, the new emphasis on the *practice* of science as the proper basis for the philosophy of science led to a more sensitive appreciation of the role played by theoretical constructs in guiding and defining the work of science. The restrictive empiricism of the logical positivists had earlier shown itself in their repeated attempts to "reduce" theoretical terms to the safer language of observation. The abandonment of this program was due not so much to the failure of the reduction techniques as to a growing realization that theoretical terms have a distinctive and indispensable part to play in science.[2] It was only a step from this realization to an acknowledgment that these terms carry with them an ontology, though admittedly an incomplete and tentative one. For a time, it seemed as though realism was coming into its own again.

But there were also new influences in the opposite direction. The focus of attention in the philosophy of science was now on scientific change rather than on the traditional topic of justification, and so the instability of scientific concepts became a problem with which the realist had to wrestle. For the first time, philosophers of language were joining the fray, and puzzles about truth and reference began to build into another challenge for realism. And so antirealism has reemerged, this time, however, much more sophisticated than it was in its earlier positivist dress.

When I say "antirealism," I make it sound like a single coherent position. But of course, antirealism is at least as far from a single coherent position as realism itself is. Though my concern is to construct a case for realism, it will be helpful first to survey the sources and varieties of antirealism. I will comment on these as I go, noting ambiguities and occasional misunderstandings. This will help to clarify the sort of scientific realism that in the end can be defended.

Sources of Antirealism: Science

Classical Mechanics

It is important to recall that scientists themselves have often been dubious about some of their own theoretical constructs, not because of some general antirealist sentiment, but because of some special features of the particular constructs themselves. Such constructs may seem like extra baggage—additional interpretations imposed on the theories themselves—much as the crystalline spheres seemed to many of the astronomers of the period between Ptolemy and Copernicus. Or it may be very difficult to characterize them in a consistent way, a problem that frequently bedeviled the proponents of ethers and fluids in nineteenth-century mechanics.

The most striking example of this sort of hesitation is surely that of Newton in regard to his primary explanatory construct, *attraction*. Despite the success of the mechanics of the *Principia*, Newton was never comfortable with the implications of the notion of attraction and the more general notion of force. Part of his uneasiness stemmed from his theology; he could not conceive that matter might of itself be active and thus in some sense independent of God's directing power. The apparent implication of action at a distance also distressed him. But then, how were these forces to be understood ontologically? *Where* are they, in what do they reside, and does the postulating of an inverse-square law of force between sun and planet say anything more than that each tends to move in a certain way in the proximity of the other?

The Cartesians, Leibniz, and later Berkeley, charged that the new mechanics did not really *explain* motion, since its central notion, *force*, could not be given an acceptable interpretation. Newton was sensitive to this charge and, in the decades following the publication of *Principia*, kept trying to find an ontology that might satisfy his critics.[3] He tried "active principles" that would somehow operate outside bodies. He even tried to reintroduce an ether with an extraordinary combination of properties—this despite his convincing refutation of mechanical ethers in *Principia*.[4] None of these ideas, however, were satisfactory. There were either problems of coherence and fit (the ether) or of specification (the active principles). After Newton's death, the predictive successes of his mechanics gradually stilled the doubts about the explanatory credentials of its central concept. But these doubts did not entirely vanish; Mach's *Science of Mechanics* (1881) would give them enduring form.

What are the implications of this often-told story for the realist thesis? It might seem that the failure of the attempts to interpret the concept of force in terms of previously familiar causal categories was a failure for realism also, and that the gradual laying aside in mechanics of questions about the underlying ontology was, in effect, an endorsement of antirealism. This would be so, however, only if one were to suppose the realist to be committed to theories that permit interpretation in familiar categories or, at the very least, in categories that are immediately interpretable. Naive realism of *this* sort is, indeed, easily undermined. But this is not the view that scientific realists ordinarily defend, as will be seen.

How should Newton's attempts at "interpretation" be regarded, after the fact? Were they an improper intrusion of "metaphysics," the sort of thing that science today would bar? The term "underlying ontology" that I have used might mislead here. A scientist *can* properly attempt to specify the mechanisms that underlie his equations. Newton's ether *might* have worked out; it was a potentially testable hypothesis, prompted by analogies with the basic explanatory paradigm of an earlier mechanical tradition. The metaphor of "active principle" proved a fruitful one; it was the ancestor of the notion of field, which would much later show its worth.[5]

In one of his critiques of "metaphysical realism," Putnam argues that "the whole history of science has been antimetaphysical from the seventeenth century on."[6] Where different "metaphysical" interpretations can be given of the same set of equations (e.g., the action-at-a-distance and the field interpretations of Newtonian gravitation theory), Putnam claims that competent physicists have focused on the equations and have left to philosophers the discussion of which of the empirically equivalent interpretations is "right." But this is not a good reading of the complicated history of Newtonian physics. First and foremost, it does not apply to Newton himself nor to many of his most illustrious successors, such as Faraday and Maxwell.[7]

Scientists have never thought themselves disqualified from pursuing one of a number of physical models that, for the moment, appear empirically equivalent. As metaphors, these models may give rise to quite different lines of inquiry, leading eventually to their empirical separation. Or it may be that one of the alternative models appears undesirable on other grounds than immediate empirical adequacy (as action at a distance did to Newton). If prolonged efforts to separate the models empirically are unsuccessful, or if it comes to be shown that the models are in principle empirically equivalent, scientists will, of course, turn to other matters. But this is not a rejection of realism. It is, rather, an admission that no decision can be made in this case as to what the theory, on a realist reading, commits us to.

What makes mechanics unique (and therefore an improper paradigm for the discussion of realism with regard to the theoretical entities of science generally) is that this kind of barrier occurs so frequently there. This would seem to derive from its status as the "ultimate" natural science, the basic mode of explanation of motions. The realist can afford to be insouciant about his inability to construe, for example, "a force of attraction between sun and earth . . . [as] responsible for the elliptical shape of the earth's orbit" in ontological terms, as long as he *can* construe astrophysics to give at least tentative warrant to his claim that the sun is a sphere of gas emitting light through a process of nuclear fusion. There was no way for Newton to know that attempts to interpret force in terms of the simple ontological alternatives he posed would ultimately fail, whereas the ontology of "insensible corpuscles," which he proposes in *Opticks*, would prosper. Each of these ventures was "metaphysical" in the sense that no evidence then available could determine the likelihood of its ever becoming an empirically decidable issue. But it is of such ventures that science is made.

Quantum Mechanics

In the debates between realists and antirealists, one claim that antirealists constantly make is that quantum mechanics has decided matters in their favor. In particular, the outcome of the famous controversy involving Bohr and Einstein, leading to the defeat (in most physicists' eyes) of Einstein, is taken to be a defeat for realism also. Once again, I want to show that this inference cannot be directed against the realist position proper.

Was the Copenhagen interpretation of quantum mechanics antirealist in its thrust?[8] Did Bohr's "complementary principle" imply that the theoretical entities of the new mechanics do not license any sort of existence claims about the structures of the world? It would seem not, for Bohr argues that the world is much more complex than classical physics supposed, and that the debate as to whether the basic entities of optics and mechanics are waves or particles cannot be resolved because its terms are inadequate. Bohr believes that the wave picture and the particle picture are *both* applicable, that *both* are needed, each in its own proper context. He is not holding that from his interpretation of quantum mechanics nothing can be inferred about the entities of which the world is composed; quite the reverse. He is arguing that what can be inferred is entirely at odds with what the classical world view would have led one to expect.

Of course, Einstein was a realist in regard to science. But he was also much more than a realist. He maintained a quite specific view about the nature of the world and about its relationship to observation; namely, that dynamic variables have unique real values at all times, that measurement reveals (or should reveal) these values as they exist prior to the measurement, and that there is a deterministic relationship between successive sets of these values. It was this further specification of realism that Bohr disputed.[9]

It is important to note that Einstein *might* have been right here. There is nothing about the nature of science per se that, in retrospect, allows us to say that Bohr *had* to be right. There could well be a world of which Einstein's version of realism would hold true. And in the 1930s, it was not yet clear that it might not just be our world. We now know that it is not and, furthermore, that this was implicit from the beginning in certain features of the quantum formalism itself, once this formalism was shown to predict correctly. (J. S. Bell's theorem could, in principle, have been proved as easily in 1934 as in 1964; no new empirical results were needed for it.)

What we have discovered as a result of this controversy is, in the first instance, something about the kind of world we live in.[10] The dynamical variables associated with its macro- and microconstituents are measurement-dependent in an unexpected way. (E. Wigner tried to show more specifically that they are *observer*-dependent, in the sense of being affected by the consciousness of the observer, but few have followed him in this direction.) Does the fact that quantum systems are partially indeterminate in this way affect the realist thesis? Not as far as I can see, unless a confusion is first made between scientific realism and the "realism" that is opposed to idealism, and then the measurement-dependence is somehow read as idealist in its implications. It *does* mean, of course, that the quantum formalism is incomplete by the standards of classical mechanics and that a quantum system lacks some kinds of ontological determinacy that classical systems possessed.

This was what Einstein objected to. This was why he sought an "underlying reality" (specifiable ultimately in terms of "hidden parameters" or the like) which would restore determinism of the classical sort. But to search for a completeness of the classical sort was no more "realist" than to maintain (as Bohr did) that the old completeness could never be regained. Recall that realism has to do with the existence-implications of the theoretical entities of successful theories. Einstein's ideal of physics would have the world entirely determinate against the mapping of variables of a broadly Newtonian type; Bohr's would not. The implications for the realist of Bohr's science are, it is true, more difficult to grasp. But why should we have expected the ontology of the microworld to be like that of the macroworld? Newton's third rule of philosophizing (which decreed that the macroworld should resemble the microworld in all essential details) was never more than a pious hope.

Elementary-Particle Physics

And this dissimilarity of the macrolevel and microlevel is even plainer when one turns from dynamic variables to the entities which these variables characterize. In the plate tectonic model that has had such striking success in recent geology, the continents are postulated to be carried on large plates of rocky material which underlie the continents as well as the oceans and which move very slowly relative to one another. There is no problem as to what an existence-claim means in this case. But problems do arise when we consider such microentities as electrons. For one thing, these are not particles strictly speaking, though custom dies hard and the label "elementary-particle physics" is still widely used. Electrons do not obey classical (Boltzman) statistics, as the familiar enduring individuals of our middle-sized world do.

The use of namelike terms, such as "electron," and the apparent causal simplicity of oil-drop or cloud-track experiments, could easily mislead one into supposing that electrons are very small localized individual entities with the standard mechanical properties of mass and momentum. Yet a bound electron might more accurately be thought of as a state of the system in which it is bound than as a separate discriminable entity. It is only because the charge it carries (which is a measure of the proton coupling to the electron) happens to be small that the free electron can be represented as a independent entity. When the coupling strength is greater, as it is between such nuclear entities as protons and neutrons, the matter becomes even more problematic. According to relativistic quantum theory, the forces between these entities are produced by the exchange of mesons. What is meant by "particle" in this instance reduces to the expression of a force characteristic of a particular field, a far cry from the hard massy points of classical mechanics. And the situation is still more complicated if one turns to the quark hypothesis in quantum field theory. Though quarks are supposed to "constitute" such entities as protons, they cannot be regarded as "constituents" in the ordinary physical sense; that is, they cannot be dissociated nor can they exist in the free state.

The moral is not that elementary-particle physics makes no sort of realist claim, but that the claim it makes must be construed with caution. The denizens of the microworld with their "strangeness" and "charm" can hardly be said to be imaginable in the ordinary sense. At that level, we have lost many of the familiar bearings (such as individuality, sharp location, and measurement-independent properties) that allow us to anchor the reference of existence-claims in such macrotheories as geology or astrophysics. But imaginability must not be made the test for ontology. The realist claim is that the scientist is discovering the structures of the world; it is not required in addition that these structures be imaginable in the categories of the macroworld.

The form of the successful retroductive argument is the same at the micro- as at the macrolevel. If the success of the argument at the macrolevel is to be explained by postulating that something like the entities of the theory exist, the same ought to be true of arguments at the microlevel. Are there electrons? Yes, there are, just as there are stars and slowly moving geological plates bearing the continents of earth. What are electrons? Just what the theory of electrons says they are, no more, no less, always allowing for the likelihood that the theory is open to further refinement. If we cannot quite imagine what they are, this is due to the distance of the microworld from the world in which our imaginations were formed, not to the existential shortcomings of electrons (if I may so express the doubts of the antirealist).

A Strategy for Scientists?

Some of the critics of realism assume that defenders of realism are prescribing a strategy for scientists, a kind of regulative principle that will separate the good from the bad among proposed explanatory models. Since the critics believe this strategy to be defective, they have an additional argument against realism. In their view, nonrealist strategies as often as not work out. Indeed, two such episodes might be said to be foundational for modern science: Einstein's laying aside of ontological scruples in his rejection of classical space and time when formulating his general theory of relativity, and Heisenberg's restriction of matrix mechanics to observable quantities only.[11]

A contemporary example of a similarly non-realist strategy can be found among the proponents of S-matrix theory. Geoffrey Chew defends this approach against its rival, quantum field theory with its horde of theoretical entities, by claiming as an advantage that it has no "implication of physical meaning" and that its ability to dispense with an equation of motion allows it also to dispense with any sort of fundamental entities, such as particles or fields.[12] In some of his later essays, Heisenberg (the original proponent of the S-matrix formalism) dwelt on the choice facing quantum physicists of whether to opt for the Democritean approach, utilizing constituent entities, which has been canonical since the seventeenth century, or the Pythagorean approach, which relies on the resources of pure mathematics alone.[13] Heisenberg argued that the Pythagorean approach is now coming into its own, as the resources of the Democritean physical models are close to exhaustion at the quantum level.[14]

It is important to see why a realist could have supported Chew's effort and why the success of Heisenberg's early matrix mechanics must not be credited to antirealism. The realism/antirealism debate has to do with the assessment of the existential implications of successful theories *already in place*. It is not directed to strategies for *further* development, for deciding among alternative formalisms with respect to their likely future potential. A scientist who is persuaded of the truth of realism *might* very well decide that a fresh start is needed when he cannot find a coherent physical model around which to build a new theory. Positivism of this sort may well be called for in some situations, and the realist need not oppose it.

A realist might even decide that at some point the program of Heisenberg and Chew offers more promise, without repudiating his confidence in constructs that have been validated by earlier work. It is true, of course, that a realist will be less likely to turn in this direction than a non-realist would; the extended successes of the Democritean approach and the knowledge of physical structure it has made possible might weigh more heavily, as a sort of inductive argument, with the realist.

Nevertheless, there is no necessary connection; the defender of realism must not be saddled with a normative doctrine of the kind attributed here. One reason, perhaps, why this sort of confusion occurs is that Einstein's stand against Bohr is so often taken to be the paradigm of realism. And it did, indeed, involve a strongly normative doctrine in regard to the proper strategy for quantum physics. But Einstein's world view included, as I have shown, much more than realism; where it failed was not in its realistic component, but in the conservative constraints on future inquiry that Einstein felt the success of classical physics warranted.

As a footnote to this discussion, it may be worth emphasizing that the realist of whom I speak here is, in the first instance, a philosopher. The qualifier "scientific" in front of "realist" should not be allowed to mislead. It is used to distinguish the realism I am discussing from the many others that dot the history of philosophy. The realisms that philosophers in the past opposed to nominalism and to idealism are very different doctrines, and neither is connected, in any straightforward way at least, with the realism being referred to here. In the past, the realism I am speaking of has been most often contrasted with fictionalism or with "instrumentalism"; but at this point the term is almost hopelessly equivocal.

"Scientific realism" is scientific because it proposes a thesis *in regard to* science. Though the case to be made for it may employ the inference-to-best-explanation technique also used in science, the doctrine itself is still a philosophic one. The scientist qua scientist is not called on to take a stand on it one way or another. Most scientists *do* have views on the issue, sometimes on the basis of much reflection but more often of a spontaneous kind. Indeed, it could be argued that worrying about whether or not their constructs approximate the real is more apt to hinder than to help their work as scientists!

Sources of Antirealism: History of Science

The most obvious source of antirealism in recent decades is the new concern for the history of science on the part of philosophers of science. Thomas Kuhn's emphasis on the discontinuity that, according to him, characterizes the "revolutionary" transitions in the history of science also led him to a rejection of realism: "I can see [in the systems of Aristotle, Newton and Einstein] no coherent direction of ontological development."[15] Kuhn is willing to attribute a cumulative character to the low-level empirical laws of science. But he denies any cumulative character to theory: theories come and go, and many leave little of themselves behind.

Among the critics of realism, Larry Laudan is perhaps the one who sets most store in considerations drawn from the history of science. He displays an impressive list of once-respected theories that now have been discarded, and guesses that "for every highly successful theory in the past of science which we now believe to be a genuinely referring theory, one could find half a dozen once successful theories which we now regard as substantially non-referring."[16]

To meet this challenge adequately, it would be necessary to look closely at Laudan's list of discarded theories, and that would require an essay in its own right. But a few remarks are in order. The sort of theory on which the realist grounds his argument is one in which an increasingly finer specification of internal structure has been obtained over a long period, in which the theoretical entities function *essentially* in the argument and are not simply intuitive postulations of an "underlying reality," and in which the original metaphor has proved continuously fertile and capable of increasingly further extension. (More on this will follow.)

This excludes most of Laudan's examples right away. The crystalline spheres of ancient astronomy, the universal Deluge of catastrophist geology, theories of spontaneous generation—none of these would qualify. That is not to say that the entities or events they postulated were not firmly believed in by their proponents. But realism is not a blanket approval for all the entities postulated by long-supported theories of the past. Ethers and fluids are a special category, and one which Laudan stresses. I would argue that these were often, though not always, interpretive additions, that is, attempts to specify what "underlay" the equations of the scientist in a way which the equations (as we now see) did not really sanction. The optical ether, for example, in whose existence Maxwell had such confidence, was no more than a carrier for variations in the electromagnetic potentials. It seemed obvious that a vehicle of some sort *was* necessary; undulations cannot occur (as it was often pointed out) unless there is something to undulate! Yet nothing could be inferred about the carrier itself; it was an "I-know-not-what," precisely the sort of unknowable "underlying reality" that the antirealist so rightly distrusts.

The theory of circular inertia and the effluvial theory of static electricity were first approximations, crude it is true, but effective in that the metaphors they suggested gradually were winnowed through, and something of the original was retained. Phlogiston left its antiself, oxygen, behind. The view that the continents were static, which preceded the plate-tectonic model of contemporary geology, was not a theory; it was simply an assumption, one that is correct to a fairly high approximation. The early theories of the nucleus, which assumed it to be homogeneous, were simply idealizations; it was not known whether the nucleus was homogeneous or not, but a decision on that could be put off until first the notion of the nuclear atom itself could be fully explored. These are all examples given by Laudan. Clearly, they need more scrutiny than I have given them. But equally clearly, Laudan's examples may not be taken without further examination to count on the antirealist side. The value of this sort of reminder, however, is that it warns the realist that the ontological claim he makes is at best tentative, for surprising reversals *have* happened in the history of science. But the nonreversals (and a long list is easy to construct here also) still require some form of (philosophic) explanation, or so I shall argue.

Sources of Antirealism: Philosophy

According to the classic ideal of science as demonstration which dominated Western thought from Aristotle down to Descartes, hypothesis can be no more than a temporary device in science. Of course, one can find an abundance of retroductive reasoning in Aristotle's science as in Descartes', a tentative working back from observed effect to unobserved cause. But there was an elaborate attempt to ensure that *real* science, *scientia propter quid*, would not contain theoretical constructs of a hypothetical kind. And there was a tendency to treat these latter constructs as fictions, in particular the constructs of mathematical astronomy. Duhem has left us a chronicle of the antirealism with which the medieval philosophers regarded the epicycles and eccentrics of the Ptolemaic astronomer.

Empiricism

As the bar to hypothesis gradually came to be dropped in the seventeenth century, another source of opposition to theoretical constructs began to appear. The new empiricism was distrustful of unobserved entities, particularly those that were unobservable in principle. One finds this sort of skepticism already foreshadowed in some well-known chapters of Locke's *Essay Concerning Human Understanding*. Locke concluded there (Book IV) that a "science of bodies" may well be forever out of reach because there is no way to reason securely from the observed secondary qualities of things to the primary qualities of the minute parts on which those secondary qualities are supposed to depend. Hume went much further and restricted science to the patterning of sense impressions. He simply rejects the notion of cause according to which one could try to infer from these impressions to the unobserved entities causing them.

Kant tried to counter this challenge to the realistic understanding of Newtonian physics. He argued that entities such as the "magnetic matter pervading all bodies" need not be perceivable by the unaided senses in order to qualify as real.[17] He established a notion of cause sufficiently large to warrant causal inference from sense-knowledge to such unobservables as the "magnetic matter." Even though the transcendental deductions of the first *Critique* bear on the prerequisites of possible experience, "experience" must be interpreted here as extending to all spatiotemporal entities that can be causally connected with the deliverances of our senses.[18]

Despite Kant's efforts, the skeptical empiricism of Hume has continued to find admirers. The logical positivists were attracted by it but were sufficiently impressed by the central role of theoretical constructs in science not to be quite so emphatic in their rejection of the reality of unobservable theoretical entities. The issue itself tended to be pushed aside and to be treated by them as undecidable; E. Nagel's *The Structure of Science* gives classical expression to this view. This sort of agnosticism alternated with a more definitely skeptical view in logical positivist writings. If one takes empiricism as a starting point, it is tempting to push it (as Hume did) to yield the demand not just that every claim about the world must ultimately rest on sense experience but that every admissible entity must be directly certifiable by sense experience.

This is the position taken by Bas van Fraassen. His antirealism is restricted to those theoretical entities that are in principle unobservable. He has no objection to allowing the reality of such theoretical entities as stars (interpreted as large glowing masses of gas) because these are, in his view, observable in principle since we could approach them by spaceship, for example. It is part of what he calls the "empirical adequacy" of a stellar theory that it should predict what we would observe should we come to a star. This criterion, which he makes the single aim of science, is sufficiently broad, therefore, to allow reality-claims for any theoretical entity that, though at present unobserved, is at least in principle directly observable by us. His antirealism has more than a tinge of old-fashioned nominalism about it, the rejection of what he calls an "inflationary metaphysics" of redundant entities.[19] Since neither of the two main arguments he lists for realism, inference to the best explanation and the common cause argument, are (in his view) logically compelling, this is taken to justify his application of Occam's razor.

One immediate difficulty with this position is, of course, the distinction drawn between the observable and the unobservable. Since entities on one side of the line are ontologically respectable and those on the other are not, it is altogether crucial that there be some way not only to draw the distinction but also to confer on it the significance that van Fraassen attributes to it. In one of the classic papers in defense of scientific realism, Grover Maxwell argued in 1962 that there is a continuum in the spectrum of observation from ordinary unaided seeing down to the operation of a high-power microscope.[20] Van Fraassen concedes that the distinction is not a sharp one, that "observe" is a vague predicate, but insists that it is sufficient if the ends of the spectrum be clearly distinct, that is, that there be at least some clear cases of supposed interaction with theoretical entities which would not count as "observing."[21] He takes the operation of a cloud chamber, with its ionized tracks allegedly indicating the presence of charged entities such as electrons, to be a case where "observe" clearly ought not be used. One must not say, on noting such a track: I observed an electron.

To lay as much weight as this on the contingencies of the human sense organs is obviously problematic, as van Fraassen recognizes. There are organisms with sense-organs very different from ours that can perceive phenomena such as ultraviolet light or the direction of optical polarization. Why could there not, in principle, be organisms much smaller than we, able to perceive microentities that for us are theoretical and able also to communicate with us? Is not the notion "observable in principle" hopelessly vague in the face of this sort of objection? How can it be used to draw a usable distinction between theoretical entities that do have ontological status and those that do not? Van Fraassen's response is cautious:

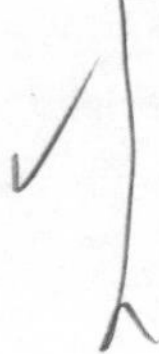

> It is, on the face of it, not irrational to commit oneself only to a search for theories that are empirically adequate, ones whose models fit the observable phenomena, while recognizing that what counts as an observable phenomenon is a function of what the epistemic community is (that *observable* is *observable-to-us*).[22]

So "observable" means here "observable in principle by us with the sense organs we presently have." But once again, why would "unobservable" in this sense be allowed the implications for epistemology and ontology that van Fraassen wants to attach to it?[23] The question is not whether the aim of science ought to be broadened to include the search for unobservable but real entities, though something could be said in favor of such a proposal. It is sufficient for the purposes of the realist to ask whether theories that are in van Fraassen's sense empirically adequate can also be shown under certain circumstances to have likely ontological implications.

Van Fraassen allows that the moons of Jupiter can be observed through a telescope; this counts as observation proper "since astronauts will no doubt be able to see them as well from close up."[24] But one cannot be said to "observe" by means of a high-power microscope (he alleges) because no such direct alternative is available to us in this case. What matters here is not so much the way the instrument works, the precise physical or theoretical principles involved. It is whether there is also, in principle, a direct unmediated alternative mode of observation available to us. The entity need not be observable *in practice*. The iron core that geologists tell us lies at the center of the earth is certainly not observable in practice; it is a theoretical entity since its existence is known only through a successful theory, but it may nonetheless be regarded as real, van Fraassen would say, because *in principle* we could go down there and check it out.

The quality of the evidence for this geological entity might, however, seem no better than that available for the chromosome viewed by microscope. Van Fraassen rests his case on an analysis of the aims of science, in an abstract sense of the term "aim," on the "epistemic attitude" (as he calls it) proper to science as an activity. And he thinks that reality-claims in the case of the chromosomes, but not the iron core, lie outside the permissible aims of science. Is there any way to make this distinction more plausible?

Reference

Some theoretical entities (such as the iron core or the star) are of a kind that is relatively familiar from other contexts. We do not need a theory to tell us that iron exists or how it may be distinguished. But electrons are what quantum theory says they are, and our only warrant for knowing that they exist is the success of that theory. So there is a special class of theoretical entities whose *entire* warrant lies in the theory built around them. They correspond more or less to the unobservables of van Fraassen.

What makes them vulnerable is that the theory postulating them may itself change or even be dropped. This is where the problems of meaning change and of theory replacement so much discussed in recent philosophy of science become relevant. The antirealist might object to a reality-claim for electrons or genes not so much because they are unobservable but because the reference of the term "electron" may shift as theory changes. To counter this objection, it sounds as though the realist will have to provide a theory of reference that is able to secure a constancy of reference in regard to such theoretical terms. R. Rorty puts it this way:

> The need to pick out objects without the help of definitions, essences, and meanings of terms, produced (philosophers thought) a need for a "theory of reference" which would not employ the Fregean machinery which Quine had rendered dubious. This call for a theory of reference became assimilated to the demand for a "realistic" philosophy of

science which would reinstate the pre-Kuhnian and pre-Feyerabendian notion that scientific inquiry made progress by finding out more and more about the same objects.[25]

Rorty is, of course, skeptical of theories of reference generally, and derides the idea that the problems of realism could be handled by such a theory. He chides Putnam, in particular, for leading philosophers to believe that they could be. Recall the celebrated realist's nightmare conjured up by Putnam:

> What if all the theoretical entities postulated by one generation (molecules, genes, etc. as well as electrons) invariably "don't exist" from the standpoint of later science? . . . One reason this is a serious worry is that eventually the following meta-induction becomes compelling: just as no term used in the science of more than 50 (or whatever) years ago referred, so it will turn out that no term used now (except maybe observation-terms if there are such) refers.[26]

This is the "disastrous meta-induction" which at that time Putnam felt had to be blocked at all costs. But, of course, if the theoretical entities of one generation really did *not* have any existential claim on the next, realism simply would be false. It is, in part at least, because the history of science testifies to a substantial continuity in theoretical structures that we are led to the doctrine of scientific realism at all. Were the history of science *not* to do so, then we would have no logical or metaphysical grounds for believing in scientific realism in the first place. But this is to get ahead of the story. I introduced the issue of reference here not to argue its relevance one way or the other, but to note that one form of antirealism can be directed against the subset of theoretical entities which derive their definition entirely from a particular theory.

One way for a realist to evade objections of this kind is to focus on the manner in which theoretical entities can be causally connected with our measurement apparatus. An electron may be defined as the entity that is causally responsible for, among other things, certain kinds of cloud tracks. A small number of parameters, such as mass and charge, can be associated with it. Such an entity will be said to exist, that is, not to be an artifact of the apparatus, if a number of convergent sorts of causal lines lead to it. There would still have to be a theory of some sort to enable the causal tracking to be carried out. But the reason to affirm the entity's existence lies not in the success of the theory in which it plays an explanatory role, but in the operation of traceable causal lines. Ian Hacking urges that this defense of realism, which relies on experiential interactions, avoids the problems of meaning-change that beset arguments based on inference to the best explanation.[27]

Truth as Correspondence

The most energetic criticisms of realism, of late, have been coming from those who see it as the embodiment of an old-fashioned, and now (in their view) thoroughly discredited, attachment to the notion of truth as some sort of "correspondence" with an "external world." These criticisms take quite different forms, and it is impossible to do them justice in a short space. The rejected doctrine is one that would hold that even in the ideal limit, the best scientific theory, one that has all the proper methodological virtues, could be false. This embodies what the critics have come to call the "God's eye view," the view that there may be more to the world than our language and our sciences can, even in principle, express. They concede that the doctrine has been a persuasive one ("it is impossible to find a philosopher before Kant who was *not* a metaphysical realist");[28] its denial seems, indeed, shockingly anthropomorphic. But they are in agreement that no philosophic sense can be made of the central metaphor of correspondence: "To single out a correspondence between two domains, one needs some independent access to both domains."[29] And, of course, an independent "access to the noumenal objects" is impossible.

The two main protagonists of this view are, perhaps, Rorty and Putnam. Rorty is the more emphatic of the two. He defends a form of pragmatism that discounts the traditional preoccupations of the philosopher with such Platonic notions as truth and goodness. He sees the Greek attempt to separate *doxa* and *epistēmē* as misguided; he equally refuses the modern trap of trying to analyze the meaning of "true," because it would involve an "impossible attempt to step outside our skins."[30] The pragmatist

> drops the notion of truth as correspondence with reality altogether, and says that modern science does not enable us to cope because it corresponds, it just plain enables us to cope. His argument for the view is that several hundred years of effort have failed to make interesting sense of the notion of "correspondence," either of thoughts to things or of words to things.[31]

Does Rorty deny scientific realism, that is, the view that the long-term success of a scientific theory gives us a warrant to believe that the entities it postulates do exist? It is not clear. What is clear, first, is that he rejects any kind of argument for scientific realism that would explain the success of a theory in terms of a correspondence with the real. And second, he denies that scientific claims have a privileged status, that the scientists' table (in Eddington's famous story) is the only real table. Science, he retorts, is just "one genre of literature," a way "to cope with various bits of the universe," just as ethics helps us cope with other bits.[32]

Putnam, in contrast, is willing to ask the traditional philosophic questions. His patron is Kant rather than James.[33] "Truth" he defines as "an idealization of rational acceptability."[34] He has more specific objections to urge against the offending metaphysical version of realism than does Rorty, whose argument amounts to claiming that it has failed to make "interesting sense."[35] Does he link this rejection with a rejection of scientific realism? Certainly not in his *Meaning and the Moral Sciences* (1978), where he defends scientific realism by urging that it permits the best explanation of the success of science. It is somewhat more difficult to be sure where his allegiances lie in his more recent pieces; his earlier enthusiasm for scientific realism seems, at the least, to be waning.[36] He attacks materialism with its assumption of mind-independent things,[37] as well as reductionism.

> We are too realistic about physics ... [because] we see physics (or some hypothetical future physics) as the One True Theory, and not simply as a rationally acceptable description suited for certain problems and purposes.[38]

This does not sound like scientific realism. Be this as it may, however, it seems clear that scientific realism is not the main target in this debate. The target is a set of metaphysical views, views (it is true) that scientific realists have in the past usually taken for granted. I suspect that Rorty would allow that genes exist and that dinosaurs once roamed the earth, as long as these claims are not given a status that is denied to more mundane statements about chairs and goldfish. But can we allow him this position so easily?

Recall that the original motivation for the doctrine of scientific realism was not a perverse philosopher's desire to inquire into the unknowable or to show that only the scientist's entities are "really real." It was a response to the challenges of fictionalism and instrumentalism, which over and over again in the history of science asserted that the entities of the scientist are fictional, that they do not exist in the everyday sense in which chairs and goldfish do. Now, how does Rorty respond to this? Has he an argument to offer? If he has, it would be an argument for scientific realism. It would also (as far as I can see) be a return to philosophy in the "old style" that he thinks we ought to have outgrown.

My own inclinations are to defend a form of metaphysical realism, though not necessarily under all the diverse specifications Putnam offers of it.[39] But that is not to the point here. What is to the point is that scientific realism is not immediately undermined by the rejection of metaphysical realism, though the character of the claim scientific realism makes obviously depends on whether or not it is joined to a concept of truth in which the embattled notion of "correspondence" plays a part. Further, the type of argument most often alleged in its support *does* use the language of correspondence: it is the approximate correspondence between the physical structure of the world and postulated theoretical entities that is held to explain why a theory succeeds as well as it does.[40] Readers will have to decide for themselves whether my argument below does "make interesting sense" or not.

Varieties of Antirealism

It may be worthwhile at this point, looking back at the territory we have traversed, to draw two rough distinctions between types of antirealism. *General antirealism* denies ontological status to theoretical entities of science generally, while *limited antirealism* denies it only to certain classes of theoretical entities, such as those that are said to be unobservable in principle. Thus, the arguments of Laudan, based as they are on a supposedly general review of the history of scientific theories, would lead him to a *general* form of antirealism, one that would exclude existence status to *any* theoretical entity whose existence is warranted only by the success of the theory in which it occurs. In contrast, van Fraassen is claiming, as I have shown, only a *limited* form of antirealism.

Second, we might distinguish between *strong antirealism*, which denies any kind of ontological status to all (or part) of the theoretical entities of science, and *weak antirealism* which allows theoretical entities existence of an everyday "chairs and goldfish" kind,[41] but insists that there is some further sense of "really really there," which realists purportedly have in mind, that is to be rejected. Classical instrumentalism would be of the former kind (strong antirealists), whereas many of the more recent critics of scientific realism appear to fall in the latter category (weak antirealists). These (weak antirealist) critics are often, as I have shown, hard to place. They reject any attempt to justify scientific realism as involving dubious metaphysics, but appear to accept a weak (realist) claim of the "everyday" kind without any form of supporting argument.[42] Their rhetoric is antirealist in tone, but their position often seems compatible with the most basic claim of scientific realism, namely that there is reason to believe that the theoretical terms of successful theories refer. This gives the weak antirealists' position a puzzling sort of undeclared status where they appear to have the best of both worlds. I am inclined to think that their effort to have it both ways must in the end fail.

The Convergences of Structural Explanation

The basic claim made by scientific realism, once again, is that the long-term success of a scientific theory gives reason to believe that something like the entities and structure postulated by the theory actually exists. There are four important qualifications built into this: (1) the theory must be successful over a significant period of time; (2) the explanatory success of the theory gives some reason, though not a conclusive warrant, to believe it; (3) what is believed is that the theoretical structures are *something like* the structure of the real world; (4) no claim is made for a special, more basic, privileged, form of existence for the postulated entities.[43] These qualifications: "significant period," "some reason," "something like," sound very vague, of course, and vagueness is a challenge to the philosopher. Can they not be made more precise? I am not sure that they can; efforts to strengthen the thesis of scientific realism have, as I have shown, left it open to easy refutation.

The case for scientific realism can be made in a variety of ways. Maxwell, Salmon, Newton-Smith, Boyd, Putnam, and others have argued it in well-known essays. I am not going to comment on their arguments here since my aim is to outline what I think to be the best case for scientific realism. My argument will, of course, bear many resemblances to theirs. What may be the most distinctive feature of my argument is my stress on structural types of explanation, and on the role played by the criterion of fertility in such explanations.

Stage one of the argument will be directed especially against general anti-realism. I want to argue that in many parts of natural science there has been, over the last two centuries, a progressive discovery of *structure*. Scientists construct theories which explain the observed features of the physical world by postulating models of the hidden structure of the entities being studied. This structure is taken to account causally for the observable phenomena, and the theoretical model provides an approximation of the phenomena from which the explanatory power of the model derives. This is the standard account of structural explanation, the type of explanation that first began to show its promise in the eighteenth and early nineteenth centuries in such sciences as geology and chemistry.[44]

I want to consider some of the areas where the growth in our knowledge of structure has been relatively steady. Let me begin with geology, a good place for a realist to begin. The visible strata and their fossil contents came to be interpreted as the evidence for an immense stretch of time past in which various processes such as sedimentation and volcanic activity occurred. There was a lively debate about the mechanisms of mountain building and the like, but gradually a more secure knowledge of the past aeons built up. The Carboniferous period succeeded the Devonian and was, in turn, succeeded by the Permian. The length of the periods, the climatic changes, and the dominant life forms were gradually established with increasing accuracy. It should be stressed that a geological period, such as the Devonian, is a theoretical entity. Further, it is, in principle, inaccessible to our direct observation. Yet our theories have allowed us to set up certain temporal boundaries, in this case (the Devonian period) roughly 400 to 350 million years ago, when the dominant life form on earth was fish and a number of important developments in the vertebrate line occurred.

The long-vanished species of the Devonian are theoretical entities about which we have come to know more and more in a relatively steady way. Of course, there have been controversies, particularly over the sudden extinction of life forms such as occurred at the end of the Cretaceous period and over the precise evolutionary relationships among given species. But the very considerable theory changes that have occurred since Hutton's day do not alter the fact that the growth in our knowledge of the sorts of life forms that inhabited the earth aeons ago has been pretty cumulative. The realist would say that the success of this synthesis of geological, physical, and biological theories gives us good reason to believe that species of these kinds did exist at the times and in the conditions proposed. Most antirealists (I suspect) would agree. But if they do, they must concede that this mode of retroductive argument can warrant, at least in some circumstances, a realist implication.

Geologists have also come to know (in the scientists' sense of the term "know") a good deal about the interior of the earth. There is a discontinuity between the material of the crust and the much denser mantle, the "Moho" as it is called after its Yugoslavian discoverer, about 5 kilometers under the ocean bed and much deeper, around 30 or 40 kilometers, under the continents. There is a further discontinuity between the solid mantle and the molten core at a depth of 2,900 kilometers. All this is inferred from the characteristics of seismic waves at the surface. Does this structural model of the earth simply serve as a device to enable the scientist to predict the seismic findings more accurately, or does it enable an additional ontological claim to be made about the actual hidden structures of earth? The realist would argue that the explanatory power of the geologist's hypothesis, its steadily improving accuracy, gives good ground to suppose that something can be inferred about real structures that lie far beneath us.

An elegant example of a quite different sort would come from cell biology. Here, the techniques of microscopy have interwoven with the theories of genetics to produce an ever more detailed picture of what goes on inside the cell. The chromosome first appeared under a microscope; only gradually was the gene, the theoretical unit of hereditary transmission, linked to it.

Later the gene came to be associated with a particular locus on the chromosome. The unraveling by Crick and Watson of the biochemical structure of the chromosome made it possible to define the structure of the gene in a relatively simple way and has allowed at least the beginnings of an understanding of how the gene operates to direct the growth of the organism. In his book, *The Matter of Life*, Michael Simon has traced this story in some detail, and has argued that its progressive character can best be understood in terms of a realist philosophy of science.[45]

One further example of this sort of progression can be found in chemistry. The complex molecules of both inorganic and organic chemistry have been more accurately charted over the past century. The atomic constituents and the spatial relations among them can be specified on the basis both of measurement, using X-ray diffraction patterns, for example, and on the basis of a theory that specifies where each kind of atom *ought* to fit. Indeed, this knowledge has enabled a computer program to be designed that can "invent" molecules, can suggest that certain configurations would yield a new type of complex molecule and can even predict what some of the molecule's properties are likely to be.

To give a realist construal to the molecular models of the chemist is not to imply that the nature of the constituent atoms and of the bonding between them is exhaustively known. It is only to suppose that the elements and spatial relationships of the model disclose, in a partial and tentative way, real structures within complex molecules. These structures are coming to be more exactly charted, using a variety of techniques both experimental and theoretical. The coherence of the outcome of these widely different techniques, and the reliability of the chemist's intuitions as he decides which atom must fit a particular spot in the lattice, are most easily understood in terms of the realist thesis.

These examples may serve to make two points. The first is that the discontinuous replacement account of the history of theories favored by antirealists is seen to be one-sided. If one focuses on global explanatory theories, particularly in mechanics, it can come to seem that theoretical entities are modified beyond recognition as theories change. Dirac's electron has little in common with the original Thomson electron; Einstein's concept of time is a long way from Newton's, and so on. These conventional examples of conceptual change could themselves be scrutinized to see whether they will bear the weight the antirealist gives them. But it may be more effective to turn from explanatory elements such as electrons to explanatory structures such as those of the organic chemist, and note, as a historical fact, the high degree of continuity in the relevant history.

Second, one could note the sort of confidence that scientists have in structural explanations of this sort. It is not merely a confidence in the empirical adequacy of the predictions these models enable them to make. It is a confidence in the model itself as an analysis of complex real structure. Look at any textbook of polymer chemistry to verify this. Of course, the chemists could be wrong to build this sort of realist expectation into their work, but the arguments of philosophers are not likely to convince them of it.

A third consequence one might draw from the history of the structural sciences is that there is a single form of retroductive inference involved throughout. As C. S. Peirce stressed in his discussion of retroduction, it is the degree of success of the retroductive hypothesis that warrants the degree of its acceptance as truth. The point is a simple one, and indeed is already implicit in Aristotle's *Posterior Analytics*. Aristotle indicates that what certifies as *demonstrative* a piece of reasoning about the relation between the nearness of planets and the fact that they do not twinkle, is the degree to which the reasoning *explains*. This connection between the explanatory and the epistemic character of scientific reasoning is constantly stressed in Renaissance and early modern discussions of hypothetical reasoning.[46]

What the history of recent science has taught us is not that retroductive inference yields a plausible knowledge of causes. We already knew this on *logical* grounds. What we have learned is that retroductive inference *works* in the world we have and with the senses we have for investigating that world. This is a contingent fact, as far as I can see. This is why realism as I have defined it is in part an empirical thesis. There could well be a universe in which observable regularities would *not* be explainable in terms of hidden structures, that is, a world in which retroduction would not work. Indeed, until the eighteenth century, there was no strong empirical case to be made against that being *our* universe. Scientific realism is not a logical doctrine about the implications of successful retroductive inference. Nor is it a metaphysical claim about how any world *must* be. It has both logical and metaphysical components. It is a quite limited claim that purports to explain why certain ways of proceeding in science have worked out as well as they (contingently) have.

That they have worked out well in such structural sciences as geology, astrophysics, and molecular biology, is apparent. And the presumption in these sciences is that the model-structures provide an increasingly accurate insight into the real structures that are causally responsible for the phenomena being explained. This may be thought to give a reliable presumption in favor of the realist implications of retroductive inference in natural science generally. But one has to be wary here. Much depends on the sort of theoretical entity one is dealing with; I have already noted, for instance, some of the perplexities posed by quantum-mechanical entities. Much depends too on how *well* the theoretical entity has served to explain: How important a part of the theory has it been? Has it been a sort of optional extra feature like the solid spheres of Ptolemaic astronomy? Or has it guided research in the way the Bohr model of the hydrogen atom did? What kind of fertility has the theoretical entity shown?

Fertility and Metaphor

Kuhn lists five values that scientists look for when evaluating a scientific theory: predictive accuracy, consistency, breadth of scope, simplicity, fertility.[47] It is the last of these that bears most directly on the problem of realism. Fertility is usually equated with the ability to make novel predictions. A good theory is expected to predict novel phenomena, that is, phenomena that were not part of the set to be explained. The further in kind these novel phenomena are from the original set, and thus the more unexpected they are, the better the model is said to be. The display of this sort of fertility reduces the likelihood of the theory's being an ad hoc one, one invented just for the original occasion but with no further scope to it.

There has been much debate about the significance of this notion of ad hoc. Clearly, it will appeal to the realist and will seem arbitrary to the antirealist. The realist takes an ad hoc hypothesis not to be a genuine theory, that is, not to give any insight into real structure and therefore to have no ground for further extension. The fact that it accounts for the original data is accidental and testifies to the ingenuity of the inventor rather than to any deeper fit. When the theory is first proposed, it is often difficult to tell whether or not it is ad hoc on the basis of the other criteria of theory appraisal. This is why fertility is so important a criterion from the realist standpoint.

The antirealist will insist that the novel facts predicted by the theory simply increase its scope and thus make it more acceptable. They will say that there is no significance to the time order in which predictions are made; if they are successful, they count as evidence whether or not they pertain to the data originally to be explained. A straightforward application of Bayes's theorem shows this, assuming of course the antirealist standpoint. Yet scientists seem to set a lot of store in the notion of ad hoc. Are scientific intuitions sufficiently captured by a translation into antirealist language? Is an ad hoc hypothesis one that just happens not to be further generalizable, or is it one that does not give sufficient insight into real structure to permit any further extension?

Rather than debate this already much-debated issue further, let me turn to a second aspect of fertility which is less often noted but which may be more significant for our problem.[48] The first aspect of fertility, novelty, had to do with what could logically be inferred from the theory, its logical resources, one might put it. But a good model has more resources than these. If an anomaly is encountered or if the theory is unable to predict one way or the other in a domain where it seems it *should* be able to do so, the model itself may serve to suggest possible modifications or extensions. These are *suggested*, not implied. Therefore, a creative move on the part of the scientist is required.

In this case, the model functions somewhat as a metaphor does in language. The poet uses a metaphor not just as decoration but as a means of expressing a complex thought. A good metaphor has its own sort of precision, as any poet will tell you. It can lead the mind in ways that literal language cannot. The poet who is developing a metaphor is led by suggestion, not by implication; the reader of the poem queries the metaphor and searches among its many resonances for the ones that seem best to bear insight. The simplistic "man is a wolf" examples of metaphor have misled philosophers into supposing that what is going on in metaphor is a comparison between two already partly understood things. The only challenge then would be to decide in what respects the analogy holds. In the more complex metaphors of modern poetry, something much more interesting is happening. The metaphor is helping to illuminate something that is not well understood in advance, perhaps, some aspect of human life that we find genuinely puzzling or frightening or mysterious. The manner in which such metaphors work is by tentative suggestion. The minds of poet and reader alike are actively engaged in creating. Obviously, much more would need be said about this, but it would lead me too far afield at this point.[49]

The good model has something of this metaphoric power.[50] Let me recall another one here, from geology once again. It had long been known that the west coast of Africa and the east coast of South America show striking similarities in terms of strata and their fossil contents. In 1915, Alfred Wegener put forward a hypothesis to explain these and other similarities, such as those between the major systems of folds in Europe and North America. The continental drift notion that he developed in *The Origins of Continents and Oceans* was not at first accepted, although it admittedly did explain a great deal. There were too many anomalies: How could the continents cut through the ocean floor, for example, since the material of the ocean floor is considerably harder than that of the continents? In the 1960s, new evidence of seafloor spreading led H. Hess and others to a modification of the original model. The moving elements are not the continents but rather vast plates on which the continents as well as the seafloor are carried. And so the continental drift hypothesis developed into the plate tectonic model.

The story has been developed so ably from the methodological standpoint by Rachel Laudan[51] and Henry Frankel[52] that I can be very brief, and simply refer you to their writings. The original theoretical entity, a floating continent, did not logically entail the plates of the new model. But in the context of anomalies and new evidence, it did *suggest* them. And these plates in turn suggested new modifications. What happens when the plates pull apart are seafloor rifts, with quite specific properties. The upwelling lava will have magnetic directional properties that will depend on its orientation relative to the earth's magnetic field at the time. This allows the lava to be dated, and the gradual pulling apart of the plates to be charted. It was the discovery of such dated strips paralleling the midocean rifts that proved decisive in swinging geologists over to the new model in the mid-1960s. What happens when the plates collide? One is carried down under (subduction); the other may be upthrust to form a mountain ridge. One can see here how the original metaphor is gradually extended and made more specific.

In a recent critical discussion of my views on fertility and metaphor,[53] Michael Bradie has urged as a weakness of my argument that one needs to give a sufficiently precise account of metaphor to allow one to understand what would count as a metaphorical extension, so as to know when two theory stages can be identified as different stages of the same theory. My response is simple and, perhaps, simplistic. If the original model (say, continental drift) suggested the later modification as a plausible way of meeting the known anomalies and of incorporating the new evidence, then I would call this a metaphorical extension. Are continental drift and the plate tectonic model two stages of the same theory or two different theories? It all depends on how "theory" is defined and how sharply theories are individuated. I do not see that very much hangs on this decision, one way or the other. The important thing to note is that there are structural continuities from one stage to the next, even though there are also important structural modifications. What provides the continuity is the underlying metaphor of moving continents that had been in contact a long time ago and had very gradually developed over the course of time. One feature of the original theory, that the continents are the units, is eventually dropped; other features, such as what happens when the floating plates collide, are thought through and made specific in ways that allow a whole mass of new data to fall into place.

How does all this bear on the argument for realism? The answer should be obvious. This kind of fertility is a persistent feature of structural explanations in the natural sciences over the last three centuries and especially during the last century. How can it best be understood? It appears to be a contingent feature of the history of science. There seems to be no a priori reason why it *had* to work out that way, as I have already shown. What best explains it is the supposition that the model approximates sufficiently well the structures of the world that are causally responsible for the phenomena to be explained to make it profitable for the scientist to take the model's metaphoric extensions seriously. It is because there is something like a floating plate under our feet that it is proper to ask: What happens when plates collide, and what mechanisms would suffice to keep them in motion? These questions do not arise from the original theory if it is taken as no more than a formalism able to give a reasonably accurate predictive account of the data then at hand. If the continental drift hypothesis had no implications for what is really going on beneath us, for the hidden structures responsible for the phenomena of the earth's surface, then the subsequent history of that hypothesis would be unintelligible. The antirealist cannot, it seems to me, make sense of such sequences, which are pretty numerous in the recent history of all the natural sciences, basic mechanics, as always, constituting a special case.

One further point is worth stressing in regard to our geological story. Some theoretical features of the model, such as the midocean rifts, could be checked directly and their existence observationally shown. Here, as so often in science, theoretical entities previously unobserved, or in some cases even thought to be unobservable, are in fact observed and the expectations of theory are borne out, to no one's surprise. The separation between observable and unobservable postulated by many antirealists in regard to ontological status does not seem to stand up. The same mode of argument is used in each case; it is not clear why in one case expectations of real existence are accorded to the theoretical entity whereas in other cases, logically similar in explanatory character, these expectations are denied. The ontological inference, let me insist again, must be far more hesitant in some cases than in others. There is no question of according the same ontological status to *all* theoretical entities by virtue of a similar degree of fertility evinced over a significant period of time. Nonetheless, such fertility finds its best explanation in a broadly realist account of science.

Does this form of argument commit the realist to holding that every regularity in the world must be explained in terms of ontological structure? This turns out to be van Fraassen's main line of attack against realism. He takes it that the realist is committed to finding hidden variables in quantum mechanics. Since the odds against this are now quite high, and since, in any event, this would commit the realist to one possible world where the other looks just as possible, van Fraassen takes this to refute realism. But as I have shown, realism is not a regulative principle, and it does not lay down a strategy for scientists. Realism would not be refuted if the decay of individual radioactive atoms turns out to be genuinely undetermined. It does not look to the future; much more modestly, realism looks to quite specific past historical sequences and asks what best explains them. Realism does not look at *all* science, nor at all future science, just at a good deal of past science which (let me say it again) might not have worked out to support realism the way it did. The realist seeks an explanation for the regularities he finds in science, just as the scientist seeks an explanation for regularities he finds in the world. But if in particular cases he cannot find an explanation or cannot even show that there is no explanation, this in no sense shows that his original aim has somehow been discredited.

Thus, what van Fraassen describes as the "nominalist response" of the antirealist must in the end be rejected. He characterizes it in this way:

> That the observable phenomena exhibit these regularities, because of which they fit the theory, is merely a brute fact, and may or may not have an explanation in terms of unobservable facts "behind the phenomena"—it really does not matter to the goodness of the theory, nor to our understanding of the world.[54]

I hope I have shown that the nominalist resolve to leave such regularities as the extraordinary fertility of our scientific theories at the level of brute fact is unphilosophical. Furthermore, I hope I have shown that it makes a very great deal of difference to the explanatory power or goodness of a theory whether it can call on effective metaphors of hidden structure. And I doubt whether it is really necessary to prove that such metaphors are important to our understanding of the world and of the role of science in achieving such understanding.

Epilogue

Finally, I return to the weighty issues of reference and truth which are so dear to the heart of the philosopher. Clearly, my views on metaphor would lead me to reject the premise on which so much of the recent debate on realism has been based. Van Fraassen puts it thus:

> Science aims to give us, in its theories, a literally true story of what the world is like; and acceptance of a scientific theory involves the belief that it is true. This is the correct statement of scientific realism.[55]

I do not think that acceptance of a scientific theory involves the belief that it is true. Science aims at fruitful metaphor and at ever more detailed structure. To suppose that a theory is literally true would imply, among other things, that no further anomaly could, in principle, arise from any quarter in regard to it. At best, it is hard to see this as anything more than an idealized "horizon-claim," which would be quite misleading if applied to the actual work of the scientist. The point is that the resources of metaphor are essential to the work of science and that the construction and retention of metaphor must be seen as part of the aim of science.

Scientists in general accept the quantum theory of radiation. Do they believe it to be true? Scientists are very uncomfortable at this use of the word "true," because it suggests that the theory is definitive in its formulation. As has often been pointed out, the notion of *acceptance* is very complex, indeed ambiguous. It is basically a pragmatic notion: one accepts an explanation as the best one available; one accepts a theory as a good basis for further research, and so forth. In no case would it be correct to say that acceptance of a theory entails belief in its truth.

The realist would not use the term "true" to describe a good theory. He would suppose that the structures of the theory give some insight into the structures of the world. But he could not, in general, say how good the insight is. He has no independent access to the world, as the antirealist constantly reminds him. His assurance that there is a fit, however rough, between the structures of the theory and the structures of the world comes not from a comparison between them but from the sort of argument I sketched above, which concludes that only this sort of reasoning would explain certain contingent features of the history of recent science. The term "approximate truth," which has sometimes been used in this debate, is risky because it immediately invites questions such as: *how* approximate, and how is the degree of approximation to be measured? If I am right in my presentation of realism, these questions are unanswerable because they are inappropriate.

The language of theoretical explanation is of a quite special sort. It is open-ended and ever capable of further development. It is metaphoric in the sense in which the poetry of the symbolists is metaphoric, not because it uses explicit analogy or because it is imprecise, but because it has resources of suggestion that are the most immediate testimony of its ontological worth. Thus, the M. Dummett-Putnam claim that a realist is committed to holding with respect to any given theory, that the sentences of the theory are either true or false,[56] quite misses the mark where scientific realism is concerned. Indeed, I am tempted to say (though this would be a bit too strong) that if they are literally true or false, they are not of much use as the basis for a research program.

Ought the realist be apologetic, as his pragmatist critic thinks he should be, about such vague-sounding formulations as these: that a good model gives an insight into real structure and that the long-term success of a theory, in most cases, gives reason to believe that something like the theoretical entities of that theory actually exist? I do not think so. The temptation to try for a sharper formulation must be resisted by the realist, since it would almost certainly compromise the sources from which his case derives its basic strength. And the antirealist must beware of the opposite temptation to suppose that whatever cannot be said in a semantically definitive way is not worth saying.

Notes

The first version of this essay was delivered as an invited paper at the Western Division meeting of the American Philosophical Association in April 1981. I am indebted to Larry Laudan for his incisive commentary on that occasion, and to the numerous discussions we have had on this topic.

1 It was the confidence that, as a student of physics, I had developed in this belief that led me, in my first published paper in philosophy, to formulate a defense of scientific realism against the instrumentalism prevalent at the time among philosophers of science. (See "Realism in Modern Cosmology," *Proceedings of the American Catholic Philosophical Association* 29 [1955]: 137–150.) Much has changed in philosophy of science since that time; a different sort of defense is (as we shall see) now called for.

2 This is the theme of C. G. Hempel's classic essay. "The Theoretician's Dilemma," *Minnesota Studies in the Philosophy of Science* 3 (1958): 37–98.

3 For the details of this story, see E. McMullin, *Newton on Matter and Activity* (Notre Dame: University of Notre Dame Press, 1978), especially chap. 4: "How is Matter Moved?"

4 In a recent critique of "metaphysical realism," Hilary Putnam has Newton defending the view that particles act at a distance across empty space. *Reason, Truth and History* (Cambridge: Cambridge University Press, 1981), 73. Though the *Principia* has often been made to yield that claim, this view is, in fact, the one alternative that Newton at all times steadfastly rejected.

5 Newton's other suggestion, briefly explored in the 1690s, that forces might be nothing other than the manifestations of God's direct involvement in the governance of the universe, *could*, however, be properly described as "metaphysical": this is not, of course, to say that it was illegitimate.

6 H. Putnam, "Why There Isn't a Ready-Made World," *Synthese* 51 (1982): 141–168; see 163. Also available in volume 3 of Putnam's Philosophical Papers Series, *Realism and Reason* (Cambridge: Cambridge University Press, 1983).

7 According to Putnam, Newton, though no positivist, "strongly rejected the idea that his theory of universal gravitation could or should be read as a description of metaphysically ultimate fact. '*Hypotheses non fingo*' was a rejection of metaphysical hypotheses, not of scientific ones" (*Reason, Truth and History*, 163). This supposed rejection of metaphysics would, however, place Newton much closer to positivism than he really was. In the *Principia*, Newton shows himself well aware that different interpretations (he calls them "physical," not "metaphysical") can be given of attraction, and he tries to deflect anticipated criticism of this ambiguity by intimating that one

can prescind such interpretation by remaining at the "mathematical" level. But he knew perfectly well that he could not *remain* at this level and still claim to have "explained" the planetary motions. In his own later writing, much of it unpublished in his lifetime, he constantly tried out different hypotheses, as I have already noted. He knew, of course, that these were speculative, that none of them was "metaphysically ultimate fact." But I can find nothing in his writing to suggest that he believed that in principle a decision between these alternatives could not be reached. The task of the natural philosopher (he would have said) was to try to adjudicate between them.

8 As Fine argues in "The Natural Ontological Attitude," this volume [J. Leplin (ed.), *Scientific Realism* (Berkeley: University of California Press, 1984), 83–107].

9 Richard Healey calls it "naive realism"; "naive" not in a deprecatory sense, but as connoting the "natural attitude." See "Quantum Realism: Naiveté Is No Excuse," *Synthese* 42 (1979): 121–144.

10 Especially owing to the developments in recent years of the original quantum formalism, associated not only with physicists (Bell, Kochen, Specker, Wigner) but also with philosophers of science (Cartwright, Fine, Gibbins, Glymour, Putnam, Redhead, Shimony, van Fraassen, and others).

11 This argument may be found, for example, in Fine, "Natural Ontological Attitude," sec. II.

12 G. Chew, "Impasse for the Elementary-Particle Concept," *Great Ideas Today* (Chicago: Encyclopedia Britannica, 1973), 367–389; see 387–389. In his more recent, and very speculative combinatorial topology, Chew has managed to construct a formalism in which the various elementary "particles" are replaced by combinations of triangles (shades of the *Timaeus*!). Though quarks do not appear in his formalism, Chew has hopes of obtaining all the results that quantum field theory does and perhaps even more.

13 See, for example, W. Heisenberg, "Tradition in Science," in *The Nature of Scientific Discovery*, ed. O. Gingerich (Washington: Smithsonian, 1975), 219–236.

14 In the last few years, this claim has come to seem a lot less plausible, in the short run at least, since quantum field theory has been scoring notable successes, while work on the S-matrix formalism has been all but abandoned.

15 T. Kuhn, *The Structure of Scientific Revolutions*, 2nd ed. (Chicago: University of Chicago Press, 1970), 206.

16 See, in particular, L. Laudan, "A Confutation of Convergent Realism," this volume [see note 8]. The quotation is from p. 232.

17 I. Kant, *Critique of Pure Reason*, A226/B273.

18 See G. G. Brittan, *Kant's Theory of Science* (Princeton: Princeton University Press, 1978), chap. 5.

19 B. C. van Fraassen, *The Scientific Image* (Oxford: Clarendon Press, 1980), 73.

20 G. Maxwell, "The Ontological Status of Theoretical Entities," *Minnesota Studies in Philosophy of Science* 3 (1962): 3–27.

21 Van Fraassen, *The Scientific Image*, 16.

22 Ibid., 19.

23 Van Fraassen complicates the picture further by also allowing the sense of "observable" to depend on the theory being tested. "To find the limits of what is observable in the world described by theory T, we must inquire into T itself, and the theories used as auxiliaries in the testing and application of T." Ibid., 57.

24 Ibid., 16.

25 R. Rorty, *Philosophy and the Mirror of Nature* (Princeton: Princeton University Press, 1979), 274–275.

26 H. Putnam, "What is Realism?" this volume [see note 8] p. 145.

27 See I. Hacking, "Experimentation and Scientific Realism," this volume [see note 8]. It is not clear to me whether one comes up with the same list of entities using Hacking's way as one does with the more usual form of argument relying on explanatory efficacy.

28 Putnam, *Reason, Truth and History*, 57.

29 Ibid., 74.

30 R. Rorty, *Consequences of Pragmatism* (Minneapolis: University of Minnesota Press, 1982), xix.

31 Ibid., xvii.

32 Ibid., xliii.

33 I must say that I have difficulties in seeing that Kant "all but says that he is giving up the correspondence theory of truth" (Putnam, *Reason, Truth and History*, 63), and that he "is best read as proposing for the first time what I have called the 'internalist' or 'internal realist' view of truth" (ibid., 60).

34 Ibid., 55. This puts him close to Dummett's camp in a different philosophical battle.

35 These are briefly sketched in "Realism and Reason," final chapter of H. Putnam's *Meaning and the Moral Sciences* (London: Routledge, 1978). See also Putnam, "Why There Isn't a Ready-Made World." His main argument is that even if the world did have a "built-in structure" (which he denies), this could not single out *one* correspondence between signs and objects.

36 "Scientific realism" does not occur in the topic index of Putnam's *Reason, Truth and History*, even though other "realisms" are discussed extensively.

37 See Putnam, "Why There Isn't a Ready-Made World."

38 Putnam, *Reason, Truth and History*, 143. It is curious that both he and Rorty (*Consequences of Pragmatism*, xxvi) criticize the realistic tendency to suppose that physics can reach the "one true theory." But they both define the offending sort of realism precisely as the view that supposes that even in the ideal limit such a theory may not be reached. In fact, according to Putnam's own definition, the "one true theory" is, by definition, what physics *does* reach!

39 These become less and less sympathetic as times goes on. I do not see, for example, why a metaphysical realist should defend the claim that "the world consists of some fixed totality of mind-independent objects," or that "there is exactly one true and complete description of the way 'the world is'" (Putnam, *Reason, Truth and History*, 49). Paul Horwich, in an attempt to pin down Putnam's notion, makes it follow from "a more general and fundamental aspect of metaphysical realism," namely, "the view according to which truth is so inexorably separated from our practice of confirmation that we can have no reasonable expectation that our methods of justification are even remotely correct." Horwich claims that Putnam's notion is "committed to an uncomfortable extent to the possibility of unverifiable truth: no truths are verifiable or even inconclusively confirmable" (P. Horwich, "Three Forms of Realism," *Synthese* 51 [1982]: 181–201; see 188, 189). Not only does this go a long way, in my opinion, beyond what Putnam believes metaphysical realism amounts to, but it also makes a straw man of the position. In fact, I know of no philosopher who would defend it in the form in which Horwich states it.

40 Since this was the type of argument that Putnam endorsed in his earlier work, citing Boyd, one can see why he might now have backed away not only from the supporting argument but also from the thesis itself.

41 This is what Horwich calls "epistemological realism." P. Horwich, "Three Forms of Realism," 181. I am not as convinced as he is that this position is "opposed only by the rare skeptic."

42 Fine's essay in this volume [see note 8] appears to fall into this category. The first section of it is devoted to a critique of all the arguments normally brought in support of scientific realism; the second section argues that instrumentalism had a much more salutary influence than realism did on the growth of modern science. But the final section proposes, as the consequence of a "natural ontological attitude," that "there really are molecules and atoms" and rejects the instrumentalist assertion that they are just fictions. But some argument is needed for this, beyond calling this attitude "natural." And to say that the realist adds to this acceptable "core position" an unacceptable "foot-stamping shout of 'Really,'" an "emphasis that all this is really so," leaves me puzzled as to what this difference is supposed to amount to.

43 The issues as to whether these entities *ought* to be attributed privileged status (as materialism and various forms of reductionism maintain) will not be discussed here.

44 I traced the history and main features of this form of explanation in "Structural Explanation," *American Philosophical Quarterly* 15 (1978): 139–147.

45 M. Simon, *The Matter of Life* (New Haven: Yale University Press, 1971).

46 See the discussion of this in E. McMullin, "The Conception of Science in Galileo's Work," *New Perspectives on Galileo*, ed. R. Butts and J. Pitt (Dordrecht: Reidel, 1978), 209–257.

47 T. Kuhn, *The Essential Tension* (Chicago: University of Chicago Press, 1977), 321–322. See also E. McMullin, "Values in Science," PSA Presidential Address 1982, in *PSA* 1982, vol. 2.

48 For a fuller discussion of the criterion of fertility, see E. McMullin, "The Fertility of Theory and the Unit for Appraisal in Science," *Boston Studies in the Philosophy of Science*, ed. R. S. Cohen et al., 39 (1976): 395–432.

49 See, for instance, P. Wheelwright, *Metaphor and Reality* (Bloomington: Indiana University Press, 1962), esp. chap. 4, "Two Ways of Metaphor"; and E. McMullin, "The Motive for Metaphor," *Proceedings of the American Catholic Philosophical Association* 55 (1982): 27–39.

50 I have elsewhere developed one instance of this in some detail, the Bohr model of the H-atom as it guided research from 1911 to 1926. See E. McMullin, "What Do Physical Models Tell Us?" in *Logic, Methodology and Philosophy of Science*, Proceedings Third International Congress, ed. B. van Rootselaar (Amsterdam, 1968), 3: 389–396.

51 See, for example, R. Laudan, "The Recent Revolution in Geology and Kuhn's Theory of Scientific Change," in *Paradigms and Revolutions*, ed. G. Gutting (Notre Dame: University of Notre Dame Press, 1980), 284–296; R. Laudan, "The Method of Multiple Working Hypotheses and the Development of Plate-Tectonic Theory," in press.

52 H. Frankel, "The Reception and Acceptance of Continental Drift Theory as a Rational Episode in the History of Science," in *The Reception of Unconventional Science*, ed. S. Mauskopf (Boulder: Westview Press, 1978), 51–89; H. Frankel, "The Career of Continental Drift Theory," *Studies in the History and Philosophy of Science* 10 (1979): 21–66.

53 M. Bradie, "Models, Metaphors and Scientific Realism," *Nature and System* 2 (1980): 3–20.

54 Van Fraassen, *The Scientific Image*, 24.

55 Ibid., 8.

56 H. Putnam, "What is Mathematic Truth?", *Mathematics, Matter and Method* (Cambridge: Cambridge University Press), 69–70.

QUESTIONS

1 What is "convergent realism"? How is it distinct from other forms of Realism? How does it bear on the commonsense thesis that science has progressed over its history? State and discuss critically Laudan's thesis that the history of science undermines this form of Realism. Take into account McMullin's response to Laudan's claim.

2 Gutting's dialog focuses on Constructive Empiricism. Situate this position in the spectrum of Realist and Antirealist views. Be sure to indicate how Constructive Empiricism is different (if it is) from both Realism and Instrumentalism. Do you think Constructive Empiricism is coherent? What is, in your view, the strongest argument against it?

3 Reflect on the relationship between two major issues in contemporary philosophy of science, Explanation and Realism. Does a particular view of the nature of Explanation dispose or at least incline one to take a stance in the debate about Realism?

FURTHER READING

Maxwell (1962) is a classic early discussion of the ontological status of theoretical entities. Many significant contributions to the Realism-Antirealism debate are collected in Leplin (1984). Leplin (1997) is his own recent defense of Realism. Hardin and Rosenberg (1982) is another response to Laudan's critique of "convergent realism."

Van Fraassen (1980) introduces and defends Constructive Empiricism. His influential book was followed by a volume containing contributions by a number of authors assessing his position and van Fraassen's reply to his critics (Churchland and Hooker 1985).

Important developments in the debate about Realism not covered in the present collection include: (1) Fine's (1984, 1986, 1991) idea of the "Natural Ontological Attitude," a proposed neutral stance as between Realism and Instrumentalism (see McMullin 1991 for a critique); (2) Putnam's evolution from a robust, run-of-the-mill Realism to what he later called "internal realism" (Putnam 1981, 1983, 1987); (3) Cartwright's (1983) "entity realism," as opposed to "theory realism," and Hacking's defense of a version of the former based on the analysis of experimentation (1983); and (4) very specific (and technical) difficulties for Realism raised by the development of quantum theory (see, in this connection, Fine 1996 and Cushing 1994).

PART V

TESTING AND CONFIRMATION OF THEORIES

INTRODUCTION

Suppose the dispute between Realism and Instrumentalism can be resolved. The problem still remains of exactly how observation and evidence, the collection of data, etc., enable us to choose among scientific theories. On the one hand, that they do so has been taken for granted across several centuries of science and its philosophy. On the other hand, no one has explained exactly how they do so, and over time the challenges facing the explanation of exactly how evidence controls theory have increased.

The enormity of the task has been evident at least since the eighteenth century, when Hume first broached the *problem of induction*. Hume's argument is often reconstructed as follows: there are two and only two ways to justify a conclusion: deductive argument, in which the conclusion follows logically from the premises, and inductive argument, in which the premises support the conclusion but do not guarantee it. A deductive argument is colloquially described as one in which the premises "contain" the conclusion, whereas an inductive argument is often described as one that moves from the particular to the general, as when we infer from observation of a hundred white swans to the conclusion that all swans are white. Now, if we are challenged to justify the claim that inductive arguments—arguments from the particular to the general, or from the past to the future—will be reliable in the future, we can do so only by employing a deductive argument or an inductive argument. The trouble with any deductive argument to this conclusion is that at least one of the premises will itself require the reliability of induction. For example, consider the deductive argument below:

1 If a practice has been reliable in the past, it will be reliable in the future.
2 In the past inductive arguments have been reliable.
   Therefore:
3 Inductive arguments will be reliable in the future.

This argument is deductively valid, but its first premise requires justification and the only satisfactory justification for the premise would be the reliability of induction, which is what the argument is supposed to establish. Any deductive argument for the reliability of induction will include at least one question-begging premise. This leaves only inductive arguments to justify induction. But clearly, no inductive argument for induction will support its reliability, for such arguments are also question begging. An inductive argument for the reliability of induction is like underwriting your promise to pay back a loan by promising that you always keep your promises. If your reliability as a promise keeper is what is in question, offering a second promise to assure the first one is pointless. Hume's argument has for 250 years been treated as an argument for skepticism about empirical science, for it suggests that all inductive conclusions about scientific laws, and all prediction science makes about future events, are unwarranted, until we justify our reliance on induction. Hume's own conclusion was quite different. He noted that as a person who acts in the world, he was satisfied that inductive arguments were reasonable; what he thought the argument shows is that we have not yet found the right justification for induction, not that there is no justification for it. Bertrand Russell's paper "On Induction" provides an updated account of Hume's problem.

Even in the absence of a solution to the problem of induction philosophers and scientists have held up testability, if not complete verification, as a hallmark of scientific hypotheses. It may not be evident how scientific hypotheses are established but what is clear is that only observational data and experimental evidence will test them, and that it is the role of testing in our decision whether to accept or reject a hypothesis that makes it scientific. Among scientists testability has often come to be understood as "falsifiability" following a dictum of Karl Popper's. Since laws make claims about indefinite numbers of objects and events, they could never be verified by a finite number of observations. But they can be falsified by just one. Moreover, the ability and willingness of a theory to "stick its neck out"—to make bold predictions that can actually be falsified—may be employed in a criterion or *principle of demarcation* allowing one to distinguish genuinely scientific theories from pseudoscience or superstition. For these reasons, Popper advanced the claim in "Science: Conjectures and Refutations" that to be scientific a hypothesis must be falsifiable. Popper applies his dictum to argue that Charles Darwin's theory of natural selection is not falsifiable and therefore not properly speaking a scientific theory. Popper's selection is followed by a selection from the famous chapter from Darwin's *On the Origin of Species*, "Difficulties of the Theory," in which Darwin explicitly recognizes the importance of exposing his theory to potentially falsifying evidence if it is to have real explanatory content.

The real philosophical problems of theory testing are, however, much more fundamental. As Peter Achinstein's paper notes, there is another difficulty that faces us even beyond Hume's problem. This is the famous "new riddle of induction," first revealed by Nelson Goodman, that we do not even have a good grasp on what counts as a piece of positive observational evidence for a theory, let alone decisive evidence that it is true or false. Indeed, the whole empiricist idea that there is a level of observation, free from theory, that provides a neutral basis on which to comparatively test scientific theories, has been increasingly challenged in the course of the twentieth century. Russell Hanson mounted one of the earliest and most vigorous challenges to this doctrine. "Seeing and Seeing As" not only undermines the Empiricist conception of theory testing, but gives substantial support to the attack on scientific objectivity explored in Part VI below.

The challenge became even more severe as philosophers began to see that falsification of scientific hypotheses is as complex a matter as their confirmation. To see the problem with falsifying hypotheses notice that nothing follows from a general law alone. From "All swans are white" it does not follow that there are any swans, still less that there are white ones. Testing even the simplest hypothesis requires "auxiliary assumptions"—further statements about the conditions under which the hypothesis is tested. Consider a test of the ideal gas law, $PV = RT$. To subject this law to test we measure two of the three variables, say the volume of the gas container and temperature, use the law to predict a pressure, and then compare the predicted gas pressure to its actual value. If the predicted value is identical to the observed value, the evidence supports the hypothesis. If it does not, then presumably the hypothesis is falsified. But in this test of the ideal gas law we needed to measure the volume of the gas and its temperature. Measuring its temperature requires a thermometer, and employing a thermometer requires us to accept one or more rather complex hypotheses about how thermometers measure heat, for example the scientific law that mercury in an enclosed glass tube expands as it is heated, and does so uniformly. But this is another general hypothesis—an auxiliary we need to invoke in order to put the ideal gas law to the test. If the predicted value of the gas pressure diverges from the observed value, the problem may be that our thermometer was defective, or that our hypothesis about how expansion of mercury in an enclosed tube measures temperature change is false. But to show that a thermometer was defective—because, say, the glass tube was broken—presupposes another general hypothesis: thermometers with broken tubes do not measure temperature accurately. Now in many cases of testing the auxiliary hypotheses are, of course, among the most basic generalizations of a discipline, which no one would seriously challenge. But the logical possibility that they might be mistaken, a possibility which cannot be denied, means that any hypothesis which is tested under the assumption that the auxiliary assumptions are true can in principle be preserved from falsification, by giving up the auxiliary assumptions and attributing the falsity to these auxiliary assumptions. And sometimes, hypotheses are in practice preserved from falsification. Here is a classic example in which the falsification of a test is rightly attributed to the falsity of auxiliary hypotheses and not the theory under test. In the nineteenth century, predictions of the location in the night sky of Uranus derived from Newtonian mechanics were falsified as telescopic observation improved. But instead of blaming the falsification on Newton's laws of motion, astronomers challenged the auxiliary assumption that there were no other forces, beyond those due to the known planets, acting on Uranus. By calculating how much additional gravitational force was necessary and from what direction to render Newton's laws consistent with the data apparently falsifying them, astronomers were led to the discovery of Neptune.

As a matter of logic, a scientific law can neither be completely established by available evidence nor conclusively falsified by a finite body of evidence. What this means, according to W.V.O. Quine ("Two Dogmas of Empiricism"), is that scientific claims do not meet experience for testing one sentence at a time. Rather it is the totality of our beliefs that are tested against observations. This claim is often referred to as the *Duhem–Quine Thesis*, Pierre Duhem (1861–1916) being the first to articulate it early in the twentieth century (see Duhem 1954). If science meets experience in large blocks or even *en masse*, we need to face the prospect of "underdetermination," the thesis that the sum total of observational data is not capable of discriminating between competing theories. For negative data cannot point uniquely to those components of theory that require revision, and often two or more changes in a large theory will be equivalent in their effects on its acceptability. Which one should we choose? If observation affords no guidance, then it underdetermines theory. This threat, widely trumpeted as a latter-day version of Hume's problem, is rejected by Laudan and Leplin in "Empirical Equivalence and Underdetermination."

Philosophers still wedded to the Empiricist program of explaining the rationality and objectivity of scientific beliefs as a reflection of empirical evidence have in recent years taken a certain radical interpretation of the theory of probability as not only a descriptive account of theory testing in the history of science but also a prescriptive guide to how theories ought in fact to be tested by observations. Wesley Salmon's paper "Bayes's Theorem and the History of Science" provides a critical introduction to this strategy.

16

Bertrand Russell, "On Induction"

Bertrand Russell (1872–1970) gives a succinct, even graphic, statement of the classical problem of induction in his book *The Problems of Philosophy* (first published in 1912), from which this excerpt is taken.

Experience has shown us that, hitherto, the frequent repetition of some uniform succession or coexistence has been a *cause* of our expecting the same succession or coexistence on the next occasion. Food that has a certain appearance generally has a certain taste, and it is a severe shock to our expectations when the familiar appearance is found to be associated with an unusual taste. Things which we see become associated, by habit, with certain tactile sensations which we expect if we touch them; one of the horrors of a ghost (in many ghost-stories) is that it fails to give us any sensations of touch. Uneducated people who go abroad for the first time are so surprised as to be incredulous when they find their native language not understood.

And this kind of association is not confined to men; in animals also it is very strong. A horse which has been often driven along a certain road resists the attempt to drive him in a different direction. Domestic animals expect food when they see the person who usually feeds them. We know that all these rather crude expectations of uniformity are liable to be misleading. The man who has fed the chicken every day throughout its life at last wrings its neck instead, showing that more refined views as to the uniformity of nature would have been useful to the chicken.

But in spite of the misleadingness of such expectations, they nevertheless exist. The mere fact that something has happened a certain number of times causes animals and men to expect that it will happen again. Thus our instincts certainly cause us to believe that the sun will rise to-morrow, but we

B. Russell, *The Problems of Philosophy*, 1959 (first published 1912), pp. 60–9 (with some cuts). New York: Oxford University Press.

may be in no better a position than the chicken which unexpectedly has its neck wrung. We have therefore to distinguish the fact that past uniformities *cause* expectations as to the future, from the question whether there is any reasonable ground for giving weight to such expectations after the question of their validity has been raised.

The problem we have to discuss is whether there is any reason for believing in what is called "the uniformity of nature." The belief in the uniformity of nature is the belief that everything that has happened or will happen is an instance of some general law to which there are *no* exceptions. The crude expectations which we have been considering are all subject to exceptions, and therefore liable to disappoint those who entertain them. But science habitually assumes, at least as a working hypothesis, that general rules which have exceptions can be replaced by general rules which have no exceptions. "Unsupported bodies in air fall" is a general rule to which balloons and aeroplanes are exceptions. But the laws of motion and the law of gravitation, which account for the fact that most bodies fall, also account for the fact that balloons and aeroplanes can rise; thus the laws of motion and the law of gravitation are not subject to these exceptions.

The belief that the sun will rise to-morrow might be falsified if the earth came suddenly into contact with a large body which destroyed its rotation; but the laws of motion and the law of gravitation would not be infringed by such an event. The business of science is to find uniformities, such as the laws of motion and the law of gravitation, to which, so far as our experience extends, there are no exceptions. In this search science has been remarkably successful, and it may be conceded that such uniformities have held hitherto. This brings us back to the question: Have we any reason, assuming that they have always held in the past, to suppose that they will hold in the future?

It has been argued that we have reason to know that the future will resemble the past, because what was the future has constantly become the past, and has always been found to resemble the past, so that we really have experience of the future, namely of times which were formerly future, which we may call past futures. But such an argument really begs the very question at issue. We have experience of past futures, but not of future futures, and the question is: Will future futures resemble past futures? This question is not to be answered by an argument which starts from past futures alone. We have therefore still to seek for some principle which shall enable us to know that the future will follow the same laws as the past.

The reference to the future in this question is not essential. The same question arises when we apply the laws that work in our experience to past things of which we have no experience—as, for example, in geology, or in theories as to the origin of the Solar System. The question we really have to ask is: "When two things have been found to be often associated, and no instance is known of the one occurring without the other, does the occurrence of one of the two, in a fresh instance, give any good ground for expecting the other?" On our answer to this question must depend the validity of the whole of our expectations as to the future, the whole of the results obtained by induction, and in fact practically all the beliefs upon which our daily life is based.

It must be conceded, to begin with, that the fact that two things have been found often together and never apart does not, by itself, suffice to *prove* demonstratively that they will be found together in the next case we examine. The most we can hope is that the oftener things are found together, the more probable it becomes that they will be found together another time, and that, if they have been found together often enough, the probability will amount *almost* to certainty. It can never quite reach certainty, because we know that in spite of frequent repetitions there sometimes is a failure at the last, as in the case of the chicken whose neck is wrung. Thus probability is all we ought to seek.

It might be urged, as against the view we are advocating, that we know all natural phenomena to be subject to the reign of law, and that sometimes, on the basis of observation, we can see that only one law can possibly fit the facts of the case. Now to this view there are two answers. The first is that, even if *some* law which has no exceptions applies to our case, we can never, in practice, be sure that we have discovered that law and not one to which there are exceptions. The second is that the reign of law would seem to be itself only probable, and that our belief that it will hold in the future, or in unexamined cases in the past, is itself based upon the very principle we are examining.

The principle we are examining may be called the *principle of induction*, and its two parts may be stated as follows:

(a) When a thing of a certain sort A has been found to be associated with a thing of a certain other sort B, and has never been found dissociated from a thing of the sort B, the greater the number of cases in which A and B have been associated, the greater is the probability that they will be associated in a fresh case in which one of them is known to be present;

(b) Under the same circumstances, a sufficient number of cases of association will make the probability of a fresh association nearly a certainty, and will make it approach certainty without limit.

As just stated, the principle applies only to the verification of our expectation in a single fresh instance. But we want also to know that there is a probability in favour of the general law that things of the sort A are *always* associated with things of the sort B, provided a sufficient number of cases of association are known, and no cases of failure of association are known. The probability of the general law is obviously less than the probability of the particular case, since if the general law is true, the particular case must also be true, whereas the particular case may be true without the general law being true. Nevertheless the probability of the general law is increased by repetitions, just as the probability of the particular case is. We may therefore repeat the two parts of our principle as regards the general law, thus:

(a) The greater the number of cases in which a thing of the sort A has been found associated with a thing of the sort B, the more probable it is (if no cases of failure of association are known) that A is always associated with B;
(b) Under the same circumstances, a sufficient number of cases of the association of A with B will make it nearly certain that A is always associated with B, and will make this general law approach certainty without limit.

It should be noted that probability is always relative to certain data. In our case, the data are merely the known cases of coexistence of A and B. There may be other data, which *might* be taken into account, which would gravely alter the probability. For example, a man who had seen a great many white swans might argue, by our principle, that on the data it was *probable* that all swans were white, and this might be a perfectly sound argument. The argument is not disproved by the fact that some swans are black, because a thing may very well happen in spite of the fact that some data render it improbable. In the case of the swans, a man might know that colour is a very variable characteristic in many species of animals, and that, therefore, an induction as to colour is peculiarly liable to error. But this knowledge would be a fresh datum, by no means proving that the probability relatively to our previous data had been wrongly estimated. The fact, therefore, that things often fail to fulfil our expectations is no evidence that our expectations will not *probably* be fulfilled in a given case or a given class of cases. Thus our inductive principle is at any rate not capable of being *disproved* by an appeal to experience.

The inductive principle, however, is equally incapable of being *proved* by an appeal to experience. Experience might conceivably confirm the inductive principle as regards the cases that have been already examined; but as regards unexamined cases, it is the inductive principle alone that can justify any inference from what has been examined to what has not been examined. All arguments which, on the basis of experience, argue as to the future or the unexperienced parts of the past or present, assume the inductive principle; hence we can never use experience to prove the inductive principle without begging the question. Thus we must either accept the inductive principle on the ground of its intrinsic evidence, or forgo all justification of our expectations about the future. If the principle is unsound, we have no reason to expect the sun to rise to-morrow, to expect bread to be more nourishing than a stone, or to expect that if we throw ourselves off the roof we shall fall. When we see what looks like our best friend approaching us, we shall have no reason to suppose that his body is not inhabited by the mind of our worst enemy or of some total stranger. All our conduct is based upon associations which have worked in the past, and which we therefore regard as likely to work in the future; and this likelihood is dependent for its validity upon the inductive principle.

The general principles of science, such as the belief in the reign of law, and the belief that every event must have a cause, are as completely dependent upon the inductive principle as are the beliefs of daily life. All such general principles are believed because mankind have found innumerable instances of their truth and no instances of their falsehood. But this affords no evidence for their truth in the future, unless the inductive principle is assumed.

Thus all knowledge which, on a basis of experience tells us something about what is not experienced, is based upon a belief which experience can neither confirm nor confute, yet which, at least in its more concrete applications, appears to be as firmly rooted in us as many of the facts of experience. The existence and justification of such beliefs—for the inductive principle, as we shall see, is not the only example—raises some of the most difficult and most debated problems of philosophy. . . .

17

Karl Popper, "Science: Conjectures and Refutations"*

Sir Karl Raimund Popper's (1902–1994) methodology of science based on the principle of falsifiability has exerted great influence on his contemporaries among scientists. Philosophers have appreciated the importance of his principle of falsifiability but have resisted turning it into the single overarching criterion demarcating science from pseudoscience. This passage vividly describes Popper's motivations to adopt the doctrine of "falsificationism" in the first place.

> Mr. Turnbull had predicted evil consequences, . . . and was now doing the best in his power to bring about the verification of his own prophecies.
>
> (Anthony Trollope)

I

When I received the list of participants in this course and realized that I had been asked to speak to philosophical colleagues I thought, after some hesitation and consultation, that you would probably prefer me to speak about those problems which interest me most, and about those developments with which I am most intimately acquainted. I therefore decided to do what I have never done before: to give you a report on my own work in the philosophy of science, since the autumn of 1919 when I first began to grapple with the problem, "*When should a theory be ranked as scientific?*" or "*Is there a criterion for the scientific character or status of a theory?*"

The problem which troubled me at the time was neither, "When is a theory true?" nor, "When is a theory acceptable?" My problem was different. I *wished to distinguish between science and pseudo-science*; knowing

K. Popper, *Conjectures and Refutations*, 1963, pp. 33–9. London: Routledge and Kegan Paul.

very well that science often errs, and that pseudo-science may happen to stumble on the truth.

I knew, of course, the most widely accepted answer to my problem: that science is distinguished from pseudo-science—or from "metaphysics"—by its *empirical method*, which is essentially *inductive*, proceeding from observation or experiment. But this did not satisfy me. On the contrary, I often formulated my problem as one of distinguishing between a genuinely empirical method and a non-empirical or even a pseudo-empirical method—that is to say, a method which, although it appeals to observation and experiment, nevertheless does not come up to scientific standards. The latter method may be exemplified by astrology, with its stupendous mass of empirical evidence based on observation—on horoscopes and on biographies.

But as it was not the example of astrology which led me to my problem I should perhaps briefly describe the atmosphere in which my problem arose and the examples by which it was stimulated. After the collapse of the Austrian Empire there had been a revolution in Austria: the air was full of revolutionary slogans and ideas, and new and often wild theories. Among the theories which interested me, Einstein's theory of relativity was no doubt by far the most important. Three others were Marx's theory of history, Freud's psycho-analysis, and Alfred Adler's so-called "individual psychology."

There was a lot of popular nonsense talked about these theories, and especially about relativity (as still happens even today), but I was fortunate in those who introduced me to the study of this theory. We all—the small circle of students to which I belonged—were thrilled with the result of Eddington's eclipse observations which in 1919 brought the first important confirmation of Einstein's theory of gravitation. It was a great experience for us, and one which had a lasting influence on my intellectual development.

The three other theories I have mentioned were also widely discussed among students at that time. I myself happened to come into personal contact with Alfred Adler, and even to co-operate with him in his social work among the children and young people in the working-class districts of Vienna where he had established social guidance clinics.

It was during the summer of 1919 that I began to feel more and more dissatisfied with these three theories—the Marxist theory of history, psycho-analysis, and individual psychology; and I began to feel dubious about their claims to scientific status. My problem perhaps first took the simple form, "What is wrong with Marxism, psycho-analysis, and individual psychology? Why are they so different from physical theories, from Newton's theory, and especially from the theory of relativity?"

To make this contrast clear I should explain that few of us at the time would have said that we believed in the *truth* of Einstein's theory of gravitation. This shows that it was not my doubting the *truth* of those other three theories which bothered me, but something else. Yet neither was it that I merely felt mathematical physics to be more *exact* than the sociological or psychological type of theory. Thus what worried me was neither the problem of truth, at that stage at least, nor the problem of exactness or measurability. It was rather that I felt that these other three theories, though posing as sciences, had in fact more in common with primitive myths than with science; that they resembled astrology rather than astronomy.

I found that those of my friends who were admirers of Marx, Freud, and Adler were impressed by a number of points common to these theories, and especially by their apparent *explanatory power*. These theories appeared to be able to explain practically everything that happened within the fields to which they referred. The study of any of them seemed to have the effect of an intellectual conversion or revelation, opening your eyes to a new truth hidden from those not yet initiated. Once your eyes were thus opened you saw confirming instances everywhere: the world was full of *verifications* of the theory. Whatever happened always confirmed it. Thus its truth appeared manifest; and unbelievers were clearly people who did not want to see the manifest truth; who refused to see it, either because it was against their class interest, or because of their repressions which were still "un-analysed" and crying aloud for treatment.

The most characteristic element in this situation seemed to me the incessant stream of confirmations, of observations which "verified" the theories in question; and this point was constantly emphasized by their adherents. A Marxist could not open a newspaper without finding on every page confirming evidence for his interpretation of history; not only in the news, but also in its presentation—which revealed the class bias of the paper—and especially of course in what the paper did *not* say. The Freudian analysts emphasized that their theories were constantly verified by their "clinical observations." As for Adler, I was much impressed by a personal experience. Once, in 1919, I reported to him a case which to me did not seem particularly Adlerian, but which he found no difficulty in analysing in terms of his theory of inferiority feelings, although he had not even seen the child. Slightly shocked, I asked him how he could be so sure. "Because of my thousandfold experience," he replied; whereupon I could not help saying: "And with this new case, I suppose, your experience has become thousand-and-one-fold."

What I had in mind was that his previous observations may not have been much sounder than this new one; that each in its turn had been interpreted in the light of "previous experience," and at the same time counted as additional confirmation. What, I asked myself, did it confirm? No more than that a case could be interpreted in the light of the theory. But this meant very little, I reflected, since every conceivable case could be interpreted in the light of Adler's theory, or equally of Freud's. I may illustrate this by two very different examples of human behaviour: that of a man who pushes a child into the water with the intention of drowning it; and that of a man who sacrifices his life in an attempt to save the child. Each of these two cases can be explained with equal ease in Freudian and in Adlerian terms. According to Freud the first man suffered from repression (say, of some component of his Oedipus complex), while the second man had achieved sublimation. According to Adler the first man suffered from feelings of inferiority (producing perhaps the need to prove to himself that he dared to commit some crime), and so did the second man (whose need was to prove to himself that he dared to rescue the child). I could not think of any human behaviour which could not be interpreted in terms of either theory. It was precisely this fact—that they always fitted, that they were always confirmed—which in the eyes of their admirers constituted the strongest argument in favour of these theories. It began to dawn on me that this apparent strength was in fact their weakness.

With Einstein's theory the situation was strikingly different. Take one typical instance—Einstein's prediction, just then confirmed by the findings of Eddington's expedition. Einstein's gravitational theory had led to the result that light must be attracted by heavy bodies (such as the sun), precisely as material bodies were attracted. As a consequence it could be calculated that light from a distant fixed star whose apparent position was close to the sun would reach the earth from such a direction that the star would seem to be slightly shifted away from the sun; or, in other words, that stars close to the sun would look as if they had moved a little away from the sun, and from one another. This is a thing which cannot normally be observed since such stars are rendered invisible in daytime by the sun's overwhelming brightness; but during an eclipse it is possible to take photographs of them. If the same constellation is photographed at night one can measure the distances on the two photographs, and check the predicted effect.

Now the impressive thing about this case is the *risk* involved in a prediction of this kind. If observation shows that the predicted effect is definitely absent, then the theory is simply refuted. The theory is *incompatible with certain possible results of observation*—in fact with results which everybody before Einstein would have expected.[1] This is quite different from the situation I have previously described, when it turned out that the theories in question were compatible with the most divergent human behaviour, so that it was practically impossible to describe any human behaviour that might not be claimed to be a verification of these theories.

These considerations led me in the winter of 1919–20 to conclusions which I may now reformulate as follows.

(1) It is easy to obtain confirmations, or verifications, for nearly every theory—if we look for confirmations.
(2) Confirmations should count only if they are the result of *risky predictions*; that is to say, if, unenlightened by the theory in question, we should have expected an event which was incompatible with the theory—an event which would have refuted the theory.
(3) Every "good" scientific theory is a prohibition: it forbids certain things to happen. The more a theory forbids, the better it is.
(4) A theory which is not refutable by any conceivable event is non-scientific. Irrefutability is not a virtue of a theory (as people often think) but a vice.
(5) Every genuine *test* of a theory is an attempt to falsify it, or to refute it. Testability is falsifiability; but there are degrees of testability: some theories are more testable, more exposed to refutation, than others; they take, as it were, greater risks.
(6) Confirming evidence should not count *except when it is the result of a genuine test of the theory*; and this means that it can be presented as a serious but unsuccessful attempt to falsify the theory. (I now speak in such cases of "corroborating evidence.")
(7) Some genuinely testable theories, when found to be false, are still upheld by their admirers—for example by introducing *ad hoc* some auxiliary assumption, or by re-interpreting the theory *ad hoc* in such a way that it escapes refutation. Such a procedure is always possible, but it rescues the theory from refutation only at the price of destroying, or at least lowering, its scientific status. (I later described such a rescuing operation as a "*conventionalist twist*" or a "*conventionalist stratagem.*")

One can sum up all this by saying that *the criterion of the scientific status of a theory is its falsifiability, or refutability, or testability*.

II

I may perhaps exemplify this with the help of the various theories so far mentioned. Einstein's theory of gravitation clearly satisfied the criterion of falsifiability. Even if our measuring instruments at the time did not allow us to pronounce on the results of the tests with complete assurance, there was clearly a possibility of refuting the theory.

Astrology did not pass the test. Astrologers were greatly impressed, and misled, by what they believed to be confirming evidence—so much so that they were quite unimpressed by any unfavourable evidence. Moreover, by making their interpretations and prophecies sufficiently vague they were able to explain away anything that might have been a refutation of the theory had the theory and the prophecies been more precise. In order to escape falsification they destroyed the testability of their theory. It is a typical soothsayer's trick to predict things so vaguely that the predictions can hardly fail: that they become irrefutable.

The Marxist theory of history, in spite of the serious efforts of some of its founders and followers, ultimately adopted this soothsaying practice. In some of its earlier formulations (for example in Marx's analysis of the character of the "coming social revolution") their predictions were testable, and in fact falsified.[2] Yet instead of accepting the refutations the followers of Marx re-interpreted both the theory and the evidence in order to make them agree. In this way they rescued the theory from refutation; but they did so at the price of adopting a device which made it irrefutable. They thus gave a "conventionalist twist" to the theory; and by this stratagem they destroyed its much advertised claim to scientific status.

The two psycho-analytic theories were in a different class. They were simply non-testable, irrefutable. There was no conceivable human behaviour which could contradict them. This does not mean that Freud and Adler were not seeing certain things correctly: I personally do not doubt that much of what they say is of considerable importance, and may well play its part one day in a psychological science which is testable. But it does mean that those "clinical observations" which analysts naïvely believe confirm their theory cannot do this any more than the daily confirmations which astrologers find in their practice.[3] And as for Freud's epic of the Ego, the Super-ego, and the Id, no substantially stronger claim to scientific status can be made for it than for Homer's collected stories from Olympus. These theories describe some facts, but in the manner of myths. They contain most interesting psychological suggestions, but not in a testable form.

At the same time I realized that such myths may be developed, and become testable; that historically speaking all—or very nearly all—scientific theories originate from myths, and that a myth may contain important anticipations of scientific theories. Examples are Empedocles' theory of evolution by trial and error, or Parmenides' myth of the unchanging block universe in which nothing ever happens and which, if we add another dimension, becomes Einstein's block universe (in which, too, nothing ever happens, since everything is, four-dimensionally speaking, determined and laid down from the beginning). I thus felt that if a theory is found to be non-scientific, or "metaphysical" (as we might say), it is not thereby found to be unimportant, or insignificant, or "meaningless," or "nonsensical."[4] But it cannot claim to be backed by empirical evidence in the scientific sense—although it may easily be, in some genetic sense, the "result of observation."

(There were a great many other theories of this pre-scientific or pseudo-scientific character, some of them, unfortunately, as influential as the Marxist interpretation of history; for example, the racialist interpretation of history—another of those impressive and all-explanatory theories which act upon weak minds like revelations.)

Thus the problem which I tried to solve by proposing the criterion of falsifiability was neither a problem of meaningfulness or significance, nor a problem of truth or acceptability. It was the problem of drawing a line (as well as this can be done) between the statements, or systems of statements, of the empirical sciences, and all other statements—whether they are of a religious or of a metaphysical character, or simply pseudo-scientific. Years later—it must have been in 1928 or 1929—I called this first problem of mine the "*problem of demarcation.*" The criterion of falsifiability is a solution to this problem of demarcation, for it says that statements or systems of statements, in order to be ranked as scientific, must be capable of conflicting with possible, or conceivable, observations.

Notes

* A lecture given at Peterhouse, Cambridge, in Summer 1953, as part of a course on developments and trends in contemporary British philosophy, organized by the British Council; originally published under the title "Philosophy of Science: a Personal Report" in *British Philosophy in Mid-Century*, ed. C. A. Mace, 1957.

1 This is a slight oversimplification, for about half of the Einstein effect may be derived from the classical theory, provided we assume a ballistic theory of light.

2 See, for example, my *Open Society and Its Enemies* [Princeton University Press, 1945], ch. 15, section iii, and notes 13–14.

3 "Clinical observations," like all other observations, are *interpretations in the light of theories* . . . and for this reason alone they are apt to seem to support those theories in the light of which they were interpreted. But real support can be obtained only from observations undertaken as tests (by "attempted refutations"); and for this purpose *criteria of refutation* have to be laid down beforehand: it must be agreed which observable situations, if actually observed, mean that the theory is refuted. But what kind of clinical responses would refute to the satisfaction of the analyst not merely a particular analytic diagnosis but psycho-analysis itself? And have such criteria ever been discussed or agreed upon by analysts? Is there not, on the contrary, a whole family of analytic concepts, such as "ambivalence" (I do

not suggest that there is no such thing as ambivalence), which would make it difficult, if not impossible, to agree upon such criteria? Moreover, how much headway has been made in investigating the question of the extent to which the (conscious or unconscious) expectations and theories held by the analyst influence the "clinical responses" of the patient? (To say nothing about the conscious attempts to influence the patient by proposing interpretations to him, etc.) Years ago I introduced the term "*Oedipus effect*" to describe the influence of a theory or expectation or prediction *upon the event which it predicts* or describes; it will be remembered that the causal chain leading to Oedipus' parricide was started by the oracle's prediction of this event. This is a characteristic and recurrent theme of such myths, but one which seems to have failed to attract the interest of the analysts, perhaps not accidentally. (The problem of confirmatory dreams suggested by the analyst is discussed by Freud, for example in *Gesammelte Schriften*, III, 1925, where he says on p. 314: "If anybody asserts that most of the dreams which can be utilized in an analysis . . . owe their origin to [the analyst's] suggestion, then no objection can be made from the point of view of analytic theory. Yet there is nothing in this fact," he surprisingly adds, "which would detract from the reliability of our results.")

4 The case of astrology, nowadays a typical pseudo-science, may illustrate this point. It was attacked, by Aristotelians and other rationalists, down to Newton's day, for the wrong reason—for its now accepted assertion that the planets had an "influence" upon terrestrial ("sublunar") events. In fact Newton's theory of gravity, and especially the lunar theory of the tides, was historically speaking an offspring of astrological lore. Newton, it seems, was most reluctant to adopt a theory which came from the same stable as for example the theory that "influenza" epidemics are due to an astral "influence." And Galileo, no doubt for the same reason, actually rejected the lunar theory of the tides; and his misgivings about Kepler may easily be explained by his misgivings about astrology.

18

Karl Popper, "Darwinism as a Metaphysical Research Programme"

Despite his great respect for Darwin's theory of evolution, Popper thought its greatness lay not in its being a scientific theory capable of passing the test of his principle of falsifiability, but in its being a "metaphysical research program."

I have come to the conclusion that Darwinism is not a testable scientific theory, but a *metaphysical research programme*—a possible framework for testable scientific theories.[1] . . . One might say that it "almost predicts" a great variety of forms of life.[2] In other fields, its predictive or explanatory power is still more disappointing. Take "adaptation." At first sight natural selection appears to explain it, and in a way it does, but it is hardly a scientific way. To say that a species now living is adapted to its environment is, in fact, almost tautological. Indeed we use the terms "adaptation" and "selection" in such a way that we can say that, if the species were not adapted, it would have been eliminated by natural selection. Similarly, if a species has been eliminated it must have been ill adapted to the conditions. Adaptation or fitness is *defined* by modern evolutionists as survival value, and can be measured by actual success in survival: there is hardly any possibility of testing a theory as feeble as this.[3]

And yet, the theory is invaluable. I do not see how, without it, our knowledge could have grown as it has done since Darwin. In trying to explain experiments with bacteria which become adapted to, say, penicillin, it is quite clear that we are greatly helped by the theory of natural selection. Although it is metaphysical, it sheds much light upon very concrete and very practical researches. It allows us to study adaptation to a new environment (such as a penicillin-infested environment) in a rational way: it suggests the

K. Popper, "Darwinism as a Metaphysical Research Programme," in *The Philosophy of Karl Popper*, vol. 1, ed. P.A. Schilpp, 1974, pp. 134, 136–8. La Salle, IL: Open Court.

existence of a mechanism of adaptation, and it allows us even to study in detail the mechanism at work. And it is the only theory so far which does all that.

This is, of course, the reason why Darwinism has been almost universally accepted. Its theory of adaptation was the first nontheistic one; and theism was worse than an open admission of failure, for it created the impression that an incontrovertible explanation had been reached.

Now to the degree that Darwinism creates the same impression, it is not so very much better than the theistic view of adaptation; it is therefore important to show that Darwinism is not a scientific theory, but metaphysical. But its value for science as a metaphysical research programme is very great, especially if it is admitted that it may be criticized and improved upon.

Let us now look a little more deeply into the research programme of Darwinism.

First, though Darwin's theory of evolution does not have sufficient explanatory power to *explain* the terrestrial evolution of a great variety of forms of life, it certainly *suggests* it, and thereby draws attention to it. And it certainly does *predict* that *if* such an evolution takes place, it will be *gradual*.

. . .

Gradualness is thus, from a logical point of view, the central prediction of the theory. (It seems to me that it is its only prediction.) Moreover, as long as changes in the genetic base of the living forms are gradual, they are—at least "in principle"—explained by the theory; for the theory does predict the occurrence of small changes, each due to mutation. However, "explanation in principle"[4] is something very different from the type of explanation which we demand in physics. While we can explain a particular eclipse by predicting it, we cannot predict or explain any particular evolutionary change (except perhaps certain changes in the gene population *within* one species); all we can say is that if it is not a small change, there must have been some intermediate steps—an important suggestion for research: a research programme.

Notes

1 The term "metaphysical research programme" was used in my lectures from about 1949 on, if not earlier; but it did not get into print until 1958, though clearly in evidence in the last chapter of the *Postscript* (in galley proofs since 1957). I made the *Postscript* available to my colleagues, and Professor Lakatos acknowledges that what he calls "scientific research programmes" are in the tradition of what I described as "metaphysical research programmes" ("metaphysical" because nonfalsifiable). See

p. 183 of his paper "Falsification and the Methodology of Scientific Research Programmes," in *Criticism and the Growth of Knowledge*, ed. by Imre Lakatos and Alan Musgrave (Cambridge: Cambridge University Press, 1970).

2 For the problem of "degrees of prediction" see F. A. Hayek, "Degrees of Explanation," first published in 1955 and now Chap. 1 of his *Studies in Philosophy, Politics and Economics* (London: Routledge & Kegan Paul, 1967); see esp. n. 4 on p. 9. For Darwinism and the production of "a great variety of structures," and for its irrefutability, see esp. p. 32.

3 Darwin's theory of sexual selection is partly an attempt to explain falsifying instances of this theory; such things, for example, as the peacock's tail, or the stag's antlers. . . .

4 For the problem of "explanation in principle" (or "of the principle") in contrast to "explanation in detail," see Hayek, *Philosophy, Politics and Economics*, esp. section VI, pp. 11–14.

19

Charles Darwin, "Difficulties of the Theory"

This short excerpt from *The Origin of Species* demonstrates that Darwin was aware of the significance of putting his theory to test, as well as of the empirical difficulties confronting it.

As natural selection acts solely by the preservation of profitable modifications, each new form will tend in a fully-stocked country to take the place of, and finally to exterminate, its own less improved parent-form and other less favoured forms with which it comes into competition. Thus extinction and natural selection go hand in hand. Hence, if we look at each species as descended from some unknown form, both the parent and all the transitional varieties will generally have been exterminated by the very process of the formation and perfection of the new form.

But, as by this theory innumerable transitional forms must have existed, why do we not find them embedded in countless numbers in the crust of the earth? It will be more convenient to discuss this question in the chapter on the Imperfection of the Geological Record; and I will here only state that I believe the answer mainly lies in the record being incomparably less perfect than is generally supposed. The crust of the earth is a vast museum; but the natural collections have been imperfectly made, and only at long intervals of time.

. . .

To sum up, I believe that species come to be tolerably well-defined objects, and do not at any one period present an inextricable chaos of varying and intermediate links; first, because new varieties are very slowly formed, for variation is a slow process, and natural selection can do nothing until favourable individual differences or variations occur, and until a place in the natural polity of the country can be better filled by some modification of some one or more of its inhabitants. And such new places will depend on

C. Darwin, *The Origin of Species*, Chapter 6 (extract), first published in 1859.

slow changes of climate, or on the occasional immigration of new inhabitants, and, probably, in a still more important degree, on some of the old inhabitants becoming slowly modified, with the new forms thus produced, and the old ones acting and reacting on each other. So that, in any one region and at any one time, we ought to see only a few species presenting slight modifications of structure in some degree permanent; and this assuredly we do see.

Secondly, areas now continuous must often have existed within the recent period as isolated portions, in which many forms, more especially amongst the classes which unite for each birth and wander much, may have separately been rendered sufficiently distinct to rank as representative species. In this case, intermediate varieties between the several representative species and their common parent must formerly have existed within each isolated portion of the land, but these links during the process of natural selection will have been supplanted and exterminated, so that they will no longer be found in a living state.

Thirdly, when two or more varieties have been formed in different portions of a strictly continuous area, intermediate varieties will, it is probable, at first have been formed in the intermediate zones, but they will generally have had a short duration. For these intermediate varieties will, from reasons already assigned (namely from what we know of the actual distribution of closely allied or representative species, and likewise of acknowledged varieties), exist in the intermediate zones in lesser numbers than the varieties which they tend to connect. From this cause alone the intermediate varieties will be liable to accidental extermination; and during the process of further modification through natural selection, they will almost certainly be beaten and supplanted by the forms which they connect; for these from existing in greater numbers will, in the aggregate, present more varieties, and thus be further improved through natural selection and gain further advantages.

Lastly, looking not to any one time, but to all time, if my theory be true, numberless intermediate varieties, linking closely together all the species of the same group, must assuredly have existed; but the very process of natural selection constantly tends, as has been so often remarked, to exterminate the parent-forms and the intermediate links. Consequently evidence of their former existence could be found only amongst fossil remains, which are preserved, as we shall attempt to show in a future chapter, in an extremely imperfect and intermittent record.

20

Peter Achinstein, "The Grue Paradox"[1]

Since its introduction by Nelson Goodman in the 1950s, the "new riddle of induction," or the "grue paradox," has been extensively debated in the literature. In a paper written for this volume Peter Achinstein suggests a fresh look at the paradox and a novel way to meet its challenge.

1. Goodman's New Riddle of Induction

Nelson Goodman's great paradox[2] begins with the fact that

e: All the emeralds so far examined are green.

From this fact, by inductive generalization, we ought to be able to conclude that

h: All emeralds are always green.

Now define "grue" as follows:

Definition: x is grue at time t if and only if t is prior to A.D. 2500 and x is green at t, or t is A.D. 2500 or later and x is blue at t.[3]

Since e is true, and since the emeralds so far examined have all been examined prior to 2500, the following is also true:

e′: All the emeralds so far examined are grue.

So by parity of reasoning, from the fact that e′ is true, we ought to be able to conclude that

P. Achinstein, *The Book of Evidence*, 2001, Chapter 9 (with modifications). New York: Oxford University Press.

h′: All emeralds are always grue.

But h′ says that while emeralds before 2500 are green, emeralds beginning in 2500 are blue. And even though it is true that all emeralds observed so far are grue (because they are green and it is prior to 2500), this fact does not warrant the inference that all emeralds are always grue, that is, green before 2500 and blue thereafter. That would be absurd! Can this conclusion be avoided?

One approach, suggested first by Carnap, and defended later by Barker and me,[4] is that grue is a temporal property in the sense that a specific time, namely 2500, is invoked in characterizing the property, whereas this is not so in the case of green. The claim, then, is that induction works only for non-temporal properties, and not for temporal ones. In the present case this means that since h′ attributes a temporal property (grue) to all emeralds, we cannot "project" grue (to use Goodman's term). We cannot make an inductive generalization from e′ to h′. By contrast, since h attributes a non-temporal property (green) to all emeralds, we can make an inductive generalization from e to h.

This solution (which I no longer believe to be adequate) raises two important questions: What is a temporal property? And why should such properties not be projected? My answer to these questions, and my solution to the paradox, will be developed step by step in what follows.

2. A Solution to Grue: The First Step

To begin with, it is untrue that every property that mentions a specific time cannot be projected. Suppose, for example, that there is a certain necktie produced by Harvard University emblazoned with the letters MCMLVI, and that all the owners of such ties whom we have interviewed were graduated from Harvard in 1956. We might then legitimately infer that all owners of this type of tie were graduated from Harvard in 1956, despite the fact that "were graduated from Harvard in 1956" expresses a temporal property.

We need to be more selective with the temporal properties we say cannot be projected. Grue is a very special type of temporal property. It is a disjunctive one having this form:

(1) x has property P at time t if and only if x has property Q_1 at t and t is prior to a specific time T or x has property Q_2 at t and t is T or later.

This is not yet a sufficient characterization. Two provisos must be added. First, the properties Q_1 and Q_2 must be incompatible, e.g., green and blue (something cannot have both at once). Second (this will become important later), the properties Q_1 and Q_2 (e.g., green and blue) must not be thought of as disjunctive properties satisfying (1).[5]

Grue, and indeed any property of type (1), is a property of an even more general type that is not necessarily temporal. Consider disjunctive properties with this form:

(2) x has P if and only if x has Q_1 and condition C obtains or x has Q_2 and condition C does not obtain.

Again the properties Q_1 and Q_2 must be incompatible. And they must not be disjunctive properties satisfying (2). In the grue case, Q_1 is green and Q_2 is blue. Condition C is that the time at which x has whatever color it has is before 2500. Goodman's paradox can be generated with respect to any property of type (2), whether temporal or not, if all the Ps examined have been Q_1 and condition C obtains. For example, again let Q_1 be green, let Q_2 be blue, but let condition C be that x's temperature is less than some fixed value M. Suppose that all emeralds so far examined have been green and have been at temperatures below M degrees. Then if we project P with respect to emeralds, we generate a conclusion that entails that at temperatures reaching or exceeding M emeralds are blue. We generate the paradox even when the property P in question is non-temporal.[6]

How can Goodman's puzzle be resolved? Suppose that information e says that all P_1s so far examined are P_2, and hypothesis h states that all P_1s are P_2. Under what conditions can we project P_2 relative to P_1? Let us look at the grue case first.

We *can* project the property grue relative to the property of being an emerald, but only if evidence e reports on times both before and after T (2500), that is, only if e reports that the emeralds examined before T are grue (and hence green) and that emeralds examined at T or later are grue (and hence blue).

More generally, if P_2 (e.g., grue) is a disjunctive property of types (1) or (2), with Q_1 and Q_2 (e.g., green and blue) as disjuncts, and if P_1 (e.g., being an emerald) is not a disjunctive property of types (1) or (2), and if e reports that all the P_1s examined are P_2 in virtue of being Q_1 and none in virtue of being Q_2, then from e we cannot conclude that all P_1s are P_2.

I will speak of a *selection procedure* as a rule for determining how to test, or obtain evidence for or against, a hypothesis. Consider the following two types of selection procedures for a hypothesis h of the form "All P_1s are P_2":

SP_1: Select P_1s to observe at times that are both before and after T. (Or more generally, for properties of form (2), select P_1s to observe at times that satisfy condition C and also ones that fail to satisfy C.)

SP_2: Select P_1s to observe at times that are only before T. (Select only P_1s to observe at times that satisfy C.)

Where P_1 and P_2 are properties of the sort described in the previous paragraph, from e (the fact that all observed P_1s are P_2) we can infer h (all P_1s are P_2) only if the selection procedure SP_1 is employed, not SP_2.

The basic idea derives from an injunction to "vary the instances." A disjunctive property P of the type depicted in (1) and (2) applies to two different sorts of cases: ones in which an item that is P (e.g., grue) has property Q_1 (green) before time T (condition C is satisfied) and ones in which an item that is P has an incompatible property Q_2 (blue) at or after T (condition C is not satisfied). Since property P, when projected, is supposed to apply to items of both types, where these types are incompatible, items of both types need to be obtained as instances of the generalization. That is, SP_1 is to be followed, not SP_2.

For example, projecting the property grue, in the case of emeralds, requires that some emeralds be examined before 2500 to determine whether they are then green, and that some emeralds be examined after 2500 to determine whether they are then blue. Only if both of these determinations are made, and the emeralds examined before 2500 are green and those examined later are blue, can the resulting information e warrant a generalization to the hypothesis that all emeralds are grue.

Contrast this case with one in which the property green is projected with respect to emeralds. This property is not being construed as one that applies to two different sorts of cases: ones in which a green item has some non-disjunctive property Q_1 before time T and ones in which a green item has some incompatible non-disjunctive property Q_2 after T. So projecting the property green, in the case of emeralds, does not require that some emeralds be examined before T to determine whether they have such a property Q_1 and that some emeralds be examined after T to determine whether they have Q_2. Accordingly, it is not the case that only if both determinations are made and the emeralds examined before T have Q_1 while those examined after T have Q_2 can the resulting information (that all the examined emeralds are green) warrant the conclusion that all emeralds are green. In this case selection procedure SP_2 (as well as SP_1) can be used. One can select emeralds to observe at times that are only before some T, or at times that are before T and after.

The important point is not that grue, unlike green, is a temporal property. That is not enough to prevent grue from being projected. The important point is that grue is a certain type of disjunctive property, while green is not. To project this type of disjunctive property P with respect to some type of item, one needs to vary the instances observed by examining items that satisfy the condition C and items that do not.

3. "I Object. Don't Forget Bleen!" (Nelson Goodman)

All of this is subject to what seems like a devastating objection. In characterizing a disjunctive property such as grue as one satisfying condition (1), or, more generally, (2), I indicated that the properties Q_1 and Q_2 must not be disjunctive properties satisfying (1) and (2). But there lies the rub, as Goodman gleefully points out. To illustrate the problem we define "bleen" as follows:

Definition: x is bleen at time t if and only if t is prior to A.D. 2500 and x is blue at t or t is A.D. 2500 or later and x is green at t.

Now, thinking of grue and bleen as our basic properties, we can characterize the properties green and blue in a way that satisfies conditions (1) and (2):

(3) x is green at t if and only if t is prior to A.D. 2500 and x is grue at t or t is A.D. 2500 or later and x is bleen at t.
(4) x is blue at t if and only if t is prior to A.D. 2500 and x is bleen at t or t is A.D. 2500 or later and x is grue at t.

Looking at the properties green and blue this way and treating grue and bleen as our basic non-disjunctive properties, green and blue become disjunctive properties satisfying conditions (1) and (2). Accordingly, to project green with respect to emeralds we need to examine emeralds both before and after 2500. We must use selection procedure SP_1 and not SP_2. This directly contradicts what was said earlier.

So where do we stand? Can the property grue be projected with respect to emeralds by examining only emeralds before 2500? In gathering information that will warrant the hypothesis that all emeralds are grue can we use SP_2 and select emeralds to observe at times that are only before 2500? Similarly, can the property green be projected with respect to emeralds only by examining emeralds both before and after 2500, i.e., by following only SP_1?

My answer is that *for us*, that is, for normal human beings, green and blue are not disjunctive properties of types (1) and (2) subject to a temporal condition, while grue and bleen are. What I mean by this is explained as follows:

(a) For us, the properties green and blue are not defined in the disjunctive way given above. Our dictionaries do not define the terms "green" and "blue" in terms of "grue" and "bleen" and a specific time. Nor do dictionaries in other languages with words for the properties blue and green. Indeed, the dictionaries I own do not even contain the words "grue" and "bleen."

(b) When we attempt to ascertain whether something we are examining is green (or blue) at a certain time t we do not, and do not need to, ascertain whether it is grue at t and t is before 2500 or whether it is bleen at t and t is 2500 or later. For example, if it is within five minutes of midnight, one way or the other, December 31, 2499, but we do not know which, and we are presented with a colored object, we could examine it and determine whether it is then green (or blue) without knowing whether midnight has passed.

By contrast,

(c) For us, the properties grue and bleen are defined disjunctively in the manner of (1) and (2) and are subject to a temporal condition. We understand these properties only by reference to such definitions.

(d) When we attempt to ascertain whether something is grue (or bleen) at a certain time t we need to ascertain whether it is green at t and t is before 2500 or whether it is blue at t and t is 2500 or later. For example, if it is within five minutes of midnight, one way or the other, December 31, 2499, but we do not know which, and we are presented with a colored object, by examining it we could not determine whether it is then grue (or bleen) without knowing whether midnight has passed.

We might, however, imagine some extraordinary group of individuals very different from us in the following respects:

(a′) For members of this group the properties grue and bleen are not defined disjunctively in the manner of (1) and (2). Their dictionaries do not define "grue" and "bleen" in terms of "green" and "blue" and a specific time. Indeed, their dictionaries do not even contain the words "green" and "blue."

(b′) When members of this extraordinary group attempt to ascertain whether something they are examining is grue (or bleen) at a certain time t, they do not, and do not need to, ascertain whether it is green at t and t is before 2500 or whether it is blue at t and t is 2500 or later. For example, if it is five minutes before or after midnight, December 31, 2499, but they do not know which, if they are presented with a colored object they could determine whether it is then grue (or bleen) without knowing whether midnight has passed.

(c′) For them, the properties green and blue are defined in the manner of (3) and (4). They understand these properties only by reference to such definitions.

(d′) When they attempt to ascertain whether something is green (or blue) at a certain time t they need to ascertain whether it is grue at t and t is before 2500 or whether it is bleen at t and t is 2500 or later. If it is five minutes before or after midnight, December 31, 2499, but they do not know which, and they are presented with a colored object, by examining it they could not determine whether it is then green (or blue) without knowing whether midnight has passed.

It may be useful to draw an analogy with a different sort of case involving a disjunction that is non-temporal but is different from ones that can spawn Goodman's paradox. Suppose there is an extraordinary group of persons who have a word in their language for male robins and a different word for female robins, but no word for robins. (Perhaps they regard male robins and female ones as belonging to different species.) Using their words for "male robin" and "female robin" we can then define our word "robin" for them, as follows:

x is a robin if and only if x is a male robin or a female robin.

This is how they will understand the word "robin" which is new for them. Moreover, when members of this group attempt to ascertain whether something is a robin they will determine whether or not it is a male robin, and if it is not, whether or not it is a female robin. If it is one or the other it is a robin; if it is neither it is not a robin. For them, but not us, "robin" is a sex-linked term.

In the case of grue, what we are imagining is that for members of the extraordinary group the properties green and blue are disjunctive ones subject to a temporal condition, while grue and bleen are not. We have no idea how they do what they do, in particular how they determine whether something is grue at a certain time t without knowing whether t is before 2500 or later. Nor do we have any idea why, in order to determine whether something is green at a certain time t they need to know whether t is before or after 2500. We are imagining simply that these things are so.

My claim is that if there were (or could be) such extraordinary people, they would be justified in projecting the property grue with respect to emeralds after examining emeralds before 2500; they would not need to wait until 2500 to examine emeralds then as well. They would be justified in using selection procedure SP_2. And if there were such extraordinary people they would be justified in projecting the property green with respect to emeralds only by examining emeralds both before and after 2500, that is, by using SP_1.

However, the claim is not that there *are* people such as the extraordinary ones being imagined. Nor is it that there *could be*, in some robust sense of "could be." It may not be physically possible. The claim is only that it is *logically* possible.[7] There is no contradiction (or at least I have not found one) in imagining the existence of extraordinary persons satisfying conditions (a′)–(d′). Accordingly, there is no contradiction in supposing the existence of extraordinary persons who are justified in projecting the property grue with respect to emeralds after examining only emeralds before 2500. However, we are not such extraordinary people and there is no reason to believe that any such people exist, or physically speaking, could exist.

4. A Contrast with Goodman's Solution

In offering his own solution, Goodman, like me, allows the possibility (whether logical or physical) that persons exist who are justified in projecting the property grue with respect to emeralds after examining emeralds only before 2500.

Briefly, Goodman's solution is based on the idea that the term "green" is much better *entrenched* than "grue." The term "green" (as well as other terms true of the same class of things) has been used much more frequently than "grue" (or other co-extensive terms) in hypotheses of the form "All As are Bs" that have actually come to be adopted. Goodman's question is this: When is a hypothesis of the form "All As are Bs" projectible, that is, when is it confirmed by instances consisting of reports that particular As are Bs?

Suppose that two conflicting hypotheses "All As are Bs" and "All As are Cs" are such that all their examined instances are true. But suppose that the term B is much better entrenched than the term C. Then, according to Goodman, the hypothesis "All As are Cs" is not projectible. It receives no confirmation from its instances. Thus, although all the examined instances of the hypothesis "All emeralds are grue" are true, that hypothesis does not receive confirming support from those instances. The reason is that this hypothesis is "overridden" by the conflicting hypothesis "All emeralds are green" which (up to now) has equal numbers of examined instances but uses the better entrenched term "green" and conflicts with no hypotheses with still better entrenched terms. Under these circumstances examined instances of green emeralds confirm the hypothesis that all emeralds are green, whereas examined instances of grue emeralds fail to confirm the hypothesis that all emeralds are grue.

On this solution it is at least logically, if not physically, possible that persons exist for whom examined instances of grue emeralds confirm the hypothesis that all emeralds are grue, whereas examined instances of green emeralds fail to confirm the hypothesis that all emeralds are green. For such persons "grue" would be a better entrenched term than "green." It would be used more frequently than "green" by such persons in hypotheses of the form "All As are Bs" that have actually come to be adopted by such persons. So, for such persons, the hypothesis "All emeralds are green" would be overridden by the hypothesis "All emeralds are grue" which (until now) has equal numbers of examined instances but uses what is for them the better entrenched term "grue" and conflicts with no hypotheses with still better entrenched terms.

Goodman's solution appeals to entrenchment. Although all the emeralds examined so far are both green and grue, "green" is a much better entrenched term. It appears much more frequently than "grue" in hypotheses of the form "All As are Bs" that we have come to accept. This claim I do not want to deny. My question, however, is why this is so. Why have we accepted generalizations of the form "All As are green" or "All green things are B" much more frequently than "All As are grue" and "All grue things are B"? My solution offers an answer. (Goodman simply accepts that this is so.)

The answer is that for us grue is a disjunctive property of types (1) and (2) of section 2, whereas green is not. (For us, conditions (a)–(d) of section 3 hold.) Accordingly, for us, to generalize from examined instances of grue to hypotheses of the form "All As are grue" and "All grue things are B" (where A and B are not for us disjunctive properties of types (1) and (2)), we need to examine As (for "All As are grue") and grue things (for "All grue things are B") both before and after 2500. Since for us grue is a disjunctive property of type (2), in order to generalize we need to vary the instances and examine both things that satisfy the condition C of a property of type (2) and things that fail to satisfy C. Since green is not for us a disjunctive property of types (1) and (2), in order to generalize from examined instances of green to hypotheses of the form "All As are green" and "All green things are B" (where A and B are not for us disjunctive properties of types (1) and (2)) we do not need to examine As (for "all As are green") and green things (for "All green things are B") both before and after 2500, or ones that satisfy some corresponding condition C and others that fail to.

Accordingly, my solution is not based on the idea of entrenchment, which is really an idea about terms used in generalizations we have come to accept. It is based on the idea that for us, because grue is a disjunctive property of a certain sort, whereas green is not, in order to generalize from examined cases of emeralds that are grue we need to examine emeralds that satisfy one side of the disjunction and emeralds that satisfy the other.

5. Evidence

So far I have talked about generalizing from examined cases of green or grue to all cases (projecting these properties). Goodman, Carnap, and others who write about the grue paradox are concerned with the question of what counts as confirming evidence for a hypothesis. So, finally, on my solution, is the fact that

(e) All emeralds examined so far are green

evidence that

(h) All emeralds are always green?

To answer, we need to distinguish several concepts of evidence used in the sciences. In BE, I distinguish four such concepts, which I call (1) subjective, (2) ES (epistemic situation), (3) potential, and (4) veridical. Very briefly, (1) e is some person's (or group's) subjective evidence that h if the person (or group) believes that e is evidence that h, and if that person's reason for believing h true or probable is that e is true. That 24 hours ago Ann ate a pound of arsenic is my subjective evidence that she is now dead. It is what I take to be evidence; and my reason for believing she is dead is that she ate the arsenic. (2) e is ES-evidence that h, relative to a type of epistemic situation (a situation in which one knows or believes certain things), if anyone in that epistemic situation would be justified in believing that e is evidence that h. Relative to an epistemic situation containing the knowledge that arsenic is lethal, the fact that Ann ate the arsenic is ES-evidence that she is dead. (3) e is potential evidence that h if e provides a good reason to believe h, irrespective of epistemic situations. No matter what is assumed known or believed, the fact that Ann ate the arsenic is a good reason to believe she is dead, since arsenic is lethal. (4) e is veridical evidence that h if e is potential evidence that h, and h is true.[8]

Only the first concept is subjective: whether e is (subjective) evidence that h depends upon what some person or group in fact believes about e, h, and their relationship. The other types are objective. ES-evidence is relativized to a *type* of epistemic situation, not to the specific one of some individual or group; no one need be in an epistemic situation of that type. Finally, like the concepts of "sign" and "symptom," at least on one standard use of these, potential and veridical evidence are not relativized to any actual or potential epistemic situation.

In BE, I provide definitions for each of these concepts (which need not be given here). I argue that although potential evidence is the most basic concept (the others can be defined by reference to it), veridical evidence is what scientists seek. In the characterization of subjective evidence as what one believes to be evidence, the evidence one believes it to be is veridical; an analogous claim can be made for ES-evidence.

Our question, then, is whether information e above (concerning examined emeralds) is evidence that h (all emeralds are always green). Consider the simplest case first, subjective evidence. We green speakers (the normal folks who satisfy (a)–(d) of section 3 with respect to green and grue) believe that e is (veridical) evidence that h and that h is true; our reason for believing that h is true is that e is. In short, the fact that e is true is our subjective evidence that h is true. Similarly, if grue speakers existed (extraordinary but imaginary beings who satisfy (a′)–(d′)) the fact that all the emeralds examined so far are green (and hence grue) would be their subjective evidence that all emeralds are always grue.

In section 3 I claimed that a grue speaker is justified in projecting the property grue, while a green speaker is justified in projecting the property green. This is a case of ES-evidence, where the epistemic situation is understood as including a speaker's knowledge of definitions and of how to ascertain whether something is grue or green. A grue speaker's ES-evidence that all emeralds are always grue would be that all emeralds so far examined are grue. A green speaker's ES-evidence that all emeralds are always green is that all emeralds so far examined are green.

Is e potential evidence that h? Whether it is depends on the selection procedure used. I have characterized a selection procedure as a rule for determining how to test, or obtain evidence for or against, an hypothesis. In the case of our hypothesis h a selection procedure might include a rule for selecting emeralds to observe. If, for example, such a rule called for selecting emeralds only from a box containing green objects, then e would not be potential evidence that h. But a selection procedure for our hypothesis h may also include a rule for how to determine whether an emerald is green at a given time t.[9] Many such rules are possible, but let me concentrate on two.

SP(green)$_1$: Determine whether an emerald is green at a time t simply by looking at it at t, in good light, at a distance at which it can be seen clearly (etc.), and ascertaining whether it looks green.

SP(green)$_2$: Determine whether an emerald is green at t by looking at it at t and ascertaining whether it looks grue at t (the way our imagined grue speaker does) and t is prior to 2500 or whether it looks bleen at t and t is 2500 or later.

Suppose that all the emeralds selected for observation so far (before 2500) have been determined to be green. Is that fact potential evidence that all emeralds are always green? That depends not only on which selection procedure was used to select emeralds for observation but on which one was used to determine whether an emerald is green. Suppose that $SP(green)_2$ was used (e.g., by genuine grue speakers, who could not use $SP(green)_1$). Someone following $SP(green)_2$ and examining only emeralds before the year 2500 to determine whether they look grue and hence are green would need to wait until 2500 to examine emeralds to determine whether they look bleen after 2500 and hence are green. Such a person would need to do this in order to "vary the instances" to obtain genuine potential evidence that all emeralds are always green. If $SP(green)_2$ is really the selection procedure that was used for determining whether an emerald is green, then e is not potential evidence that h. By contrast, someone following $SP(green)_1$ and examining only emeralds before 2500 would not need to wait until 2500 to examine emeralds to determine whether they look bleen after 2500 and hence are green. The date 2500 plays no role in following $SP(green)_1$ the way it does in following $SP(green)_2$. If e were obtained by following $SP(green)_1$, e would be potential evidence that h.

Now, as Goodman loves to do, let us compare the situation with respect to the grue hypothesis. The question is whether

e′: All emeralds examined so far are determined to be grue

is potential evidence that

h′: All emeralds are always grue.

We need to say what selection procedure is being used. By analogy with the previous ones for green emeralds we have

$SP(grue)_1$: Determine whether an emerald is grue at t simply by looking at it at t in good light (etc.) and ascertaining whether it looks grue.

$SP(grue)_2$: Determine whether an emerald is grue at t by looking at it at t and ascertaining whether it looks green at t, where t is prior to 2500, or whether it looks blue at t and t is 2500 or later.

If $SP(grue)_2$ is used in obtaining the result e′, then e′ is not potential evidence that h′. Someone (such as us) following this selection procedure and examining only emeralds before 2500 to determine whether they look green and hence are grue would need to wait until 2500 to examine emeralds to determine whether they look blue and hence are grue. Such a person would need to do this in order to "vary the instances" to obtain genuine potential evidence that all emeralds are always grue. Using SP(grue)$_2$, examining emeralds only before 2500 would not suffice.

By contrast, a genuine grue speaker following SP(grue)$_1$ and examining emeralds only before 2500 would not need to wait until 2500 to examine emeralds to determine whether they look blue and hence are grue. Using SP(grue)$_1$, examining emeralds only before 2500 and determining that all of them are grue would allow e′ to be potential evidence that h′.

In short, (e) the fact that all emeralds observed so far are green is potential evidence that (h) all emeralds are always green, if selection procedure SP(green)$_1$ is used, but not SP(green)$_2$. And (e′) the fact that all emeralds observed so far are grue is potential evidence that (h′) all emeralds are always grue, if selection procedure SP(grue)$_1$ is used, but not SP(grue)$_2$. It should be emphasized that this is not to relativize the concept of potential evidence to a particular person or to a type of epistemic situation. If SP(green)$_1$ is used e is potential evidence that h, and if SP(grue)$_1$ is used e′ is potential evidence that h′, independently of who believes what. Nor are e and e′ just potential evidence for persons in epistemic situations of certain types.

Now, as a matter of fact, there are no grue speakers, that is, extraordinary persons who satisfy conditions (a′)–(d′) of section 3 for defining and identifying grue and green properties. There are just ordinary, everyday people like us, who satisfy conditions (a)–(d). So, even if it is logically possible that a selection procedure such as SP(grue)$_1$ is followed in determining whether an emerald is grue, and that a selection procedure such as SP(green)$_2$ is followed in determining whether an emerald is green, this will never happen (we confidently believe). In any real-life situation in which selection procedures involve actual observations of emeralds, SP(green)$_1$ and SP(grue)$_2$ will be followed, in which case the fact that all observed emeralds are determined to be green will be potential evidence that all emeralds are always green, and the fact that all observed emeralds are determined to be grue will not be potential evidence that all emeralds are always grue. If (as we also confidently believe) the green hypothesis is true, then the fact that all observed emeralds are determined to be green is veridical evidence for this hypothesis. If the hypothesis turns out to be false, then the fact about the observed emeralds is not veridical evidence that all emeralds are green.

Notes

1 For this essay I have used some of the material, with modifications, from Chapter 9 of my *The Book of Evidence* (New York: Oxford University Press, 2001). Hereafter this work will be referred to as *BE*.

2 Nelson Goodman, *Fact, Fiction, and Forecast* (Cambridge, MA: Harvard University Press, 4th edn, 1983).

3 Goodman's original definition is that "grue" applies to all things examined before some specific time T just in case they are green, and to other things just in case they are blue. In our response to Goodman, Steven Barker and I ("On the New Riddle of Induction," *Philosophical Review*, 69 (1960), 511–22) used the definition in the text (not Goodman's), except that (in 1960) we chose T to be the year 2000, which seems to have created a precedent for other writers on the subject. Since January 1, 2000 is now a memory, I have taken the liberty of pushing the date far into the future.

4 See note 3.

5 A perceptive discussion of the disjunctive character of grue is found in David H. Sanford, "A Grue Thought in a Bleen Shade: 'Grue' as a Disjunctive Predicate," in Douglas Stalker, ed., *GRUE! The New Riddle of Induction* (Chicago: Open Court, 1994), 173–92.

6 See James Hullett and Robert Schwartz, "Grue: Some Remarks," *Journal of Philosophy*, 64 (1967), 259–71.

7 Here I disagree with Judith Thomson, who claims that it is not even logically possible. See Judith Jarvis Thomson, "Grue," *Journal of Philosophy*, 63 (1966), 289–309.

8 A stronger concept of veridical evidence requires in addition that there be an explanatory connection between e and h, that is, that h correctly explain e, or that e correctly explain h, or that something correctly explain both h and e. See *BE*, Chapter 8.

9 The statement e can be understood in two ways: all the emeralds examined so far have been determined to be green; all the emeralds examined so far are in fact green (whether or not this has been determined). In what follows I confine my attention to the first. For a discussion of the second, see *BE*, Chapter 9.

21

N. Russell Hanson, "Seeing and Seeing As"

Norwood Russell Hanson (1924–1967) belonged to an influential group of post-positivist philosophers of science (which also included Thomas Kuhn, Paul Feyerabend, Stephen Toulmin, and others) who revolutionized the field in the late 1950s and early 1960s, by subjecting the Positivist dogmas to a devastating critique. Hanson's focus in this selection is the problem of *theory-ladenness of observation*.

In the last chapter we encountered four figures—a cube, a rhomboid, a staircase, and a tunnel—all of which displayed the phenomenon of reversible perspective. We also considered two drawings which, besides showing some variability in perspective, were marked by shifts in organization, or in aspect. These were called the "duck-rabbit" and the "wife-mother-in-law" respectively. In each case the question was asked, "Do we all see the same thing?" For there was no question here of a differing retinal reaction. The stimulus pattern is roughly the same for all onlookers. Nor is it easy to see how we could defensibly speak of our different reactions to these figures as being accompanied by different visual sensations, i.e., different sense-data. And yet, undeniably, different reports are forthcoming when we ask of people viewing these figures, "What do you see?"

We concluded the last chapter with a reiteration of our key question "Do the 13th Century and 20th Century astronomers see the same thing?" It is for the purpose of getting a better insight into the complications of this question that we will press our inquiry still further.

Let us begin with a few more variable figures: These vary, not in the perspective in which they may be perceived, but in the aspects they may present to a percipient.

N. Russell Hanson, *Perception and Discovery*, Chapter 6, 1969, pp. 91–110. San Francisco: Freeman. Figures included: From *Philosophical Investigations* by Wittgenstein, c. 1953.

Initially:

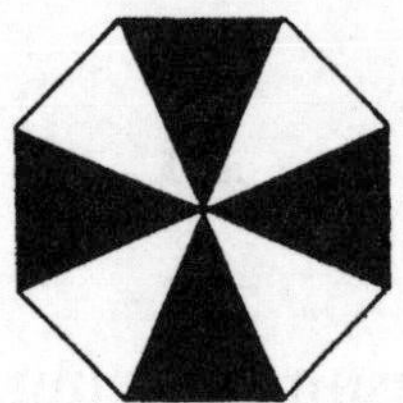

Figure 1

Some will see this as a white cross on a black ground.[1] Others will see this as a black cross on a white ground. But the difference cannot be accounted for by reference to different retinal reactions, for there need be no difference. Nor can it be accounted for by the suggestion that those mysterious entities, visual sense-data, are changing. For while I stare at such a figure, shifting from the seeing of a black cross to the seeing of a white cross, I am aware of no changes either in my retinal reaction or in my visual sense-data (whatever they are). Or if there *is* a shift in these latter I know of nothing in phenomenalism or in sense-datum theory to account for it. Indeed if I drew for you exactly what I saw when I reported "white cross on black ground," how would it differ from your drawing of what you see when you report "black cross on white ground"?

So too with Koehler's goblet:[2]

Figure 2

Again, our retinas may react normally to this. But while I see a Venetian goblet, you may see two men staring at each other. Have we seen different things? Of course we have. And yet if I draw my cup for you, you may say, "By Jove, that is exactly what I saw, two men in a staring contest." Or I may myself shift my attention from the cup to the faces. Does my retinal reaction shift? Do my sense-data change? There is nothing in sense-datum theory to suggest that my sense-datum, i.e., the "look(s)" of *Figure 2*, does change.

For clearly my private visual field is taken up with the same configuration of lines when I say I see a cup as it is when I say I see two faces. And yet it would be absurd to say that I saw the same thing in both cases.

In this respect *seeing* differs from *feeling*, as you would expect. For if I have had my right hand on a stove and my left hand in the refrigerator, when I plunge both hands into a basin of tepid water I will get a familiar variable reaction. Do my two hands feel the same thing? In an unimportant sense, yes; they are reacting normally to the tepid water. But it is much more natural to say that my hands feel different things: One feels the water as *hot*, the other feels it as *cold*. Different feelings, different sensations, different "sense-data" would be associated with each hand. These differences could be clearly and accurately described. But how to describe the difference between seeing a duck and seeing a rabbit in *Figure 3*? Or between any two aspects of the figures we have so far brought forward? To describe or draw such a figure in one of its aspects, say a duck or a cup, just *is* to describe or draw it in all its aspects (e.g., a rabbit or faces). Nonetheless we see different things here no less than we feel different things in the hot-cold experiment, even though the difference is not necessarily to be accounted for in terms of differing retinal reaction to stimuli, or in terms of having different pictures in the mind's eye.

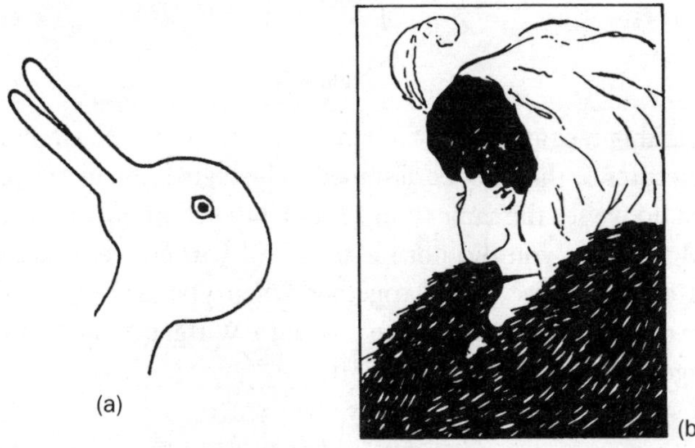

Figure 3

We have so far considered figures with reversible perspective and figures with variable aspects. I have dwelt on these because they seem to be clear cases in which we should wish to say that we saw different things, but where we might deny that this was due either to a difference in retinal reaction or to a difference in the features of the pictures registered in our private visual fields. It is in these cases too that we should probably deny that the differences in what we see are due to differences in how we interpret what we see. For, as Wittgenstein said, "To interpret is to think, to do something; seeing is a state."[3] Even Professor Price puts it that the perceptual act is not an *activity*. And the shifts in perspective and aspect that we have been considering might have occurred quite without thinking. Indeed, thinking hard will seldom enable one to see an aspect of a figure which he has been previously unable to notice.

I should like now to call up another group of figures that are variable in a rather less dramatic way. They are important, however, in the way that they continue to stress the *seeing as* component that has figured in all the examples so far. It is this largely overlooked component of our ordinary observations which will help us to see something more of the complexity of *observing*, *witnessing*, and *seeing* in scientific inquiry, and which will lead to a fuller appreciation of all that is involved in the situation wherein our two astronomers are witnessing the sun at dawn.

You may remember this one:

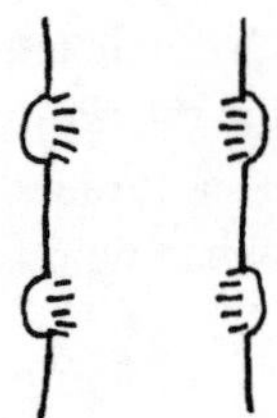

Figure 4

What is this meant to be? Your retinas and mine are similarly affected. Similar pictures of this may be assumed to be registering in our private visual fields. But do we see the same thing? I see a bear climbing up the other side of a tree. Most likely you did not see this. Did you notice, however, how the elements of this figure pulled together when you were told what I knew when drawing it? You might even say with Wittgenstein, "I see that it has not changed, and yet I see it differently . . ."[4]

And a student once suggested this one to me:

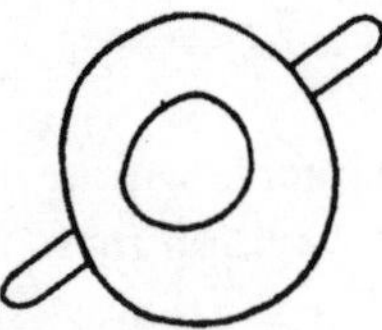

Figure 5

What do you see? A Mexican on a bicycle (seen from above)? Before I said that you might have seen just lines. But now, as Wittgenstein says, "[it] has a quite particular 'organization.'"[5]

What Wittgenstein calls here "organization" is really important, we will return to it repeatedly. We rarely see without such "organization" being operative, and yet this organization is nothing *seen* as are the lines and colors in a drawing. "Organization" is not out of the same concept-basket as are "lines" and "shapes" and "colors." This is the thin end of the wedge with which we may tumble the sense-datum account of seeing. For usually when we speak of seeing something we do so because our visual sense field is organized in certain ways. There is little in all this talk about private mental pictures that helps in any way our understanding of the organization requisite for seeing. This lacking, one might answer the question "What do you see?" with "What am I supposed to see?"—or even, "I see nothing," both of which might have been appropriate responses to the question following *Figure* 5.

Consider:

Figure 6(a)

in this context:

(b)

as follows:

(c)

The context clearly gives us the clue regarding which aspect of the duck-rabbit is appropriate: In such a context some people could not see the figure as a rabbit. Though in this context:

(d)

the figure may only come forward as a rabbit, e.g.:

(e)

It might even be argued, as Wittgenstein *does* argue, that the figure appearing in (e) has not the slightest similarity to the figure seen in (c), although they are congruent.[6] This flies in the face of sense-datum teaching.

Let us look further into the matter of context as it concerns aspect-vision or "seeing-as."

Of this square corner

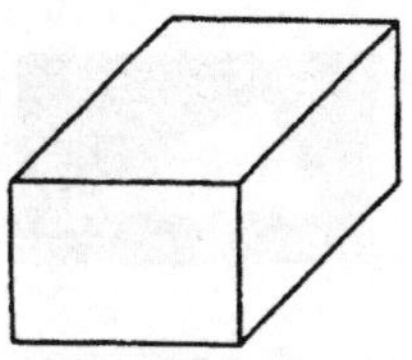

Figure 7

Wittgenstein wrote,

> You could imagine [this] appearing in several places in a book, a textbook for instance. In the relevant text something different is in

question every time: here a glass cube, there an inverted open box, there a wire frame of that shape, there three boards forming a solid angle. Each time the text supplies the interpretation of the illustration.

But we can also *see* the illustration now as one thing now as another. So we interpret it, and *see* it as we *interpret* it.[7]

In other words the appropriate aspect of *Figure 7* is brought out by the verbal context in which it appears, very much as one would have to talk and gesture around *Figure 3* to get an observer to see the rabbit when he had only been able to see the duck. The verbal context is, as it were, part of the illustration itself—a remark which, though it ought not to be taken literally, at least helps to show the sort of thing that brings out for a person one aspect of a visual object rather than another.

Wittgenstein also considers this triangle,

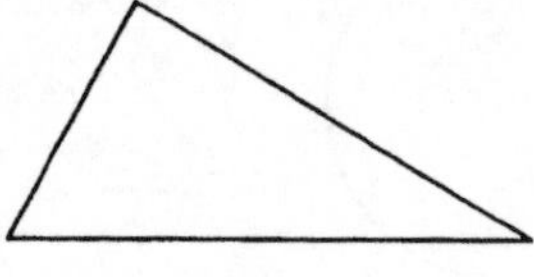

Figure 8

which he considers "... can be seen as a triangular hole, as a solid, as a geometrical drawing, as standing on its apex; as a mountain, as a wedge, as an arrow or a pointer, as an overturned object which is meant to stand on the shorter side of the right triangle, as a half parallelogram, and as various other things ... You can think now of *this*, now of *this* as you look at it, can regard it now as this, now as this, *and then you will see it now this way, now this* ..."[8]

Of course the context here is given in Wittgenstein's designations. For example:

"... triangular hole ..." does this to *Figure 8*

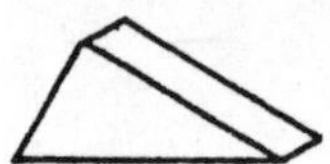

"... solid ..." does this

". . . geometrical drawing . . ." this

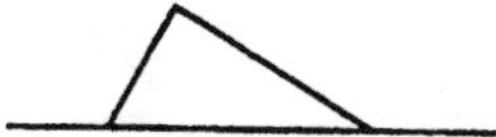

". . . standing on its base . . ."

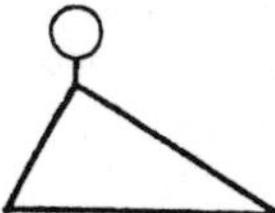

". . . hanging from its apex . . ."

". . . a mountain . . ."

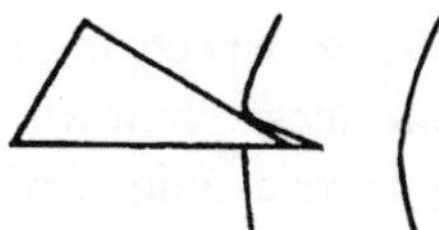

". . . a wedge . . ."

and so forth.

The context that brings an appropriate aspect of a figure or an object into focus, however, need not be set out explicitly in a paragraph or in a word. Such "contexts" are very often carried around with us in our heads, having been put there by intuition, experience, and reasoning. For example, the sequence in *Figure 9* could mean but one thing to the aeronautical engineer. We have the same retinal reactions and the same visual sensations as he does, do we not? But we would probably not see what he does, namely the sequence of airfoil types from the earliest days of heavier-than-air flight to the wing section of the present-day airplane. To see what the aeronaut sees, we would have to know what he knows. A novice sees what the specialist sees only in the way that a person who has never seen a duck or a rabbit, nor a picture of either, sees what we see when we look at the duck-rabbit. There are aspects of *Figure 9* to which the uninformed person will remain blind.

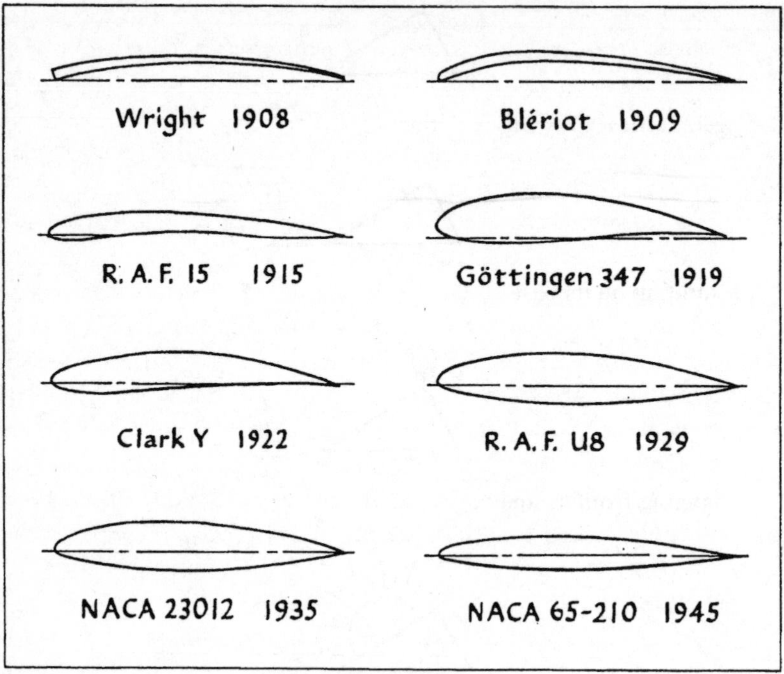

Figure 9

And he will remain blind to these aspects in just the way that he might be blind to the rabbit aspect of the duck-rabbit when the latter is surrounded by ducks. In both these cases he lacks a context within which he may see (in a significant way) what is before his eyes.

Try this one:

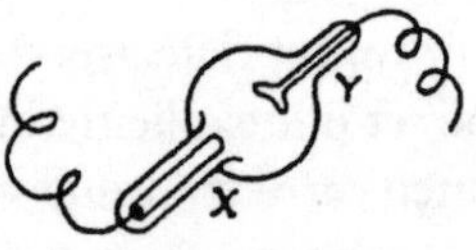

Figure 10

A trained natural scientist could only see this as one thing: an x-ray tube viewed from the cathode. Would a physicist and a non-scientist see the same thing when looking at *Figure 10*? The traditional, respectable answer to this runs: "Yes, they see the same thing, only the physicist interprets it in a way that the layman cannot." It is this "respectable" answer to the question, of course, that I have been at pains to unsettle. The answer is no more suitable here than in any of the other cases we have considered—indeed, it is positively harmful. We can agree again that the scientist and the layman have a normal retinal reaction to *Figure 10*. And we can assume that the pictures registered in their private visual fields are similar. But do they see the same thing?

Or consider just the physicist. On his first day at school years ago he had gazed in wonder at the glass and metal instruments on display in the lab. Now, after a long training in science at school and at university and in research, he returns to his old school and sees again that same x-ray tube that had so fired his imagination when he was a boy. Does he see the same thing now as he saw then? His eyes are still normal, his mental picture of the instrument is no different. But now he sees it in a very different context; he sees it in terms of electrical circuit theory, thermodynamic theory, information about the structure of metals and of glass, research into the nature of thermionic emission, optical transmission, refraction and diffraction, atomic theory, quantum theory, and relativity theory. This is a phenomenon we all know quite well. Compare the freshman's first view of his college with the senior's last view, or our first look under the hood of a newly purchased car with the same view ten exasperating years later.

"Oh yes, the physicist has learned all these things, doubtless," comes the "respectable" reply. "And they all figure in the interpretation the physicist puts upon what he sees—it is this interpretation that the layman is unable to make even though he sees exactly what the scientist sees."

But is the physicist doing any more than just seeing? As Wittgenstein says, interpreting is thinking, it is doing something. What is the physicist doing over and above what the layman does when he is seeing the x-ray tube? What do you do besides just looking and seeing when you notice the microscopes on the benches of the lab or when you see a galvanometer, or an automobile, or a close friend?

"Oh, it is just that in these familiar cases the interpretation takes up but a very short interval of time; it is all but instantaneous." So comes back the reply, and a typical philosopher's reply it is, too. It is out of the same bag of tricks that made sense-data, those final links in the perceptual causal chain, unlike all the other links, *mental, private, publicly unobservable*. These are, I feel, just dodges that philosophers have invented for the purpose of saving ideas for which they have formed a sentimental attachment. We all know very well what it is like to put an interpretation on what one sees. Artists do it, historians and journalists do it, Lysenko does it, and indeed at the frontiers of scientific research where the facts are thin and the problems thick everyone interprets what he sees. But the word *interpret* gets its bite and its use in these contexts precisely because they contrast with cases like the one where the physicist sees the x-ray tube, or where we see a bicycle. Insisting that even these last are situations involving interpretation is just another way of saying that only the apprehension of sense-data can count as *seeing*, everything in addition to that being interpretation, a saying that has been under attack since we began our inquiry.

Before a non-physicist could see *Figure 10* as a physicist sees it, before the elements of that picture will pull together, cohere, and "organize," he would have to learn a good deal of physics. It is not just that the physicist and the layman see the same thing but do not make the same thing out of it. The layman can make nothing out of it. And that is not just a figure of speech. I can make nothing out of the Arab word for *cat*, though my purely visual reaction to that word may not differ from that of an Arab child.

In the sense that I have been so far elaborating, the two do not see the same thing. To the question "What do you see?" the physicist will reply, "An x-ray tube with its cathode forward." The non-scientist may reply, "What am I supposed to see?" Both are quite appropriate answers. (In this connection it is interesting to note that very often the words "What do you see?" are used in posing the question "Can you identify the object before you?" To this question the two answers just given are comprehensive. It is not for nothing that "What do you see?" can be used in putting the "Can you identify . . . ?" question, a question which presupposes normal vision, but is calculated to test one's *knowledge*.)

Pierre Duhem puts the matter thus:

> Enter a laboratory; approach the table crowded with an assortment of apparatus, an electric cell, silk-covered copper wire, small cups of mercury, spools of wire, a mirror mounted on an iron bar; the experimenter is inserting into small openings the metal ends of ebony-headed pins; the iron oscillates, and the mirror attached to it throws a luminous band upon a celluloid scale; the forward-backward motion of this luminous spot enables the physicist to observe the minute oscillations of the iron bar. But ask him what he is doing. Will he answer "I am studying the oscillations of an iron bar which carries a mirror"? No, he will answer that he is measuring the electric resistance of the spools. If you are astonished, if you ask him what his words mean, *what relation they have with the phenomena he has been observing and which you have noted at the same time as he*, he will answer that your question requires a long explanation and that you should take a course in electricity.[9]

The physicist, in other words, must teach his visitor everything he knows before he can show him what he sees. Not until then will his visitor be supplied with an intellectual context sufficient for throwing into relief those aspects of the cluster of objects before his eyes that the physicist sees as an indication of the electrical resistance of the spools. There is nothing wrong with his eyes. He can see in the sense that he has normal vision, i.e., he is not blind. (This is the sense in which we *can* hear, even when we do not hear the ticking of the clock behind us, or when we do not hear the street noises during sleep, even though our ears are open and our auditory organs reacting normally to every acoustical vibration. This is only to say that we are not deaf. Still, we may not hear that the oboe is out of tune, something that will strike the trained musician at our side as painfully obvious.) The visitor cannot see what the physicist sees, even though the physicist's eyes are no better than his. He cannot see what the physicist sees in much the way that he may not be able to see a rabbit but only a duck in *Figure 3*, or only a wife but not a mother-in-law. He is, in a word, blind to what the physicist sees. The elements in his visual field, though perhaps similar or identical to the elements of the physicist's visual field in color, shape, arrangement, etc., are not organized conceptually for him as they are for the physicist. And this is much the same situation as we find when both you and I gaze at *Figure 3* but I see a rabbit and you see a duck. The conceptual organization of one's visual field is the all-important factor here. It is not something visually apprehended in the way that lines and shapes and colors are visually apprehended. It is rather the *way* in which lines, shapes, and colors are visually apprehended. And in all the cases we have been examining I have been inviting you to consider a given constellation of lines and shapes (what psychologists call "a stimulus pattern") and to consider further the different ways in which this given constellation or pattern can be apprehended visually, the different sorts of conceptual organization that can be accorded to that constellation. In short, the different ways it may be seen.

Of course, the reasons why these things are seen differently are not the same for every case we have examined. A thorough examination of that, however, is a task for the experimental psychologist, a title to which I can lay no claim whatever. We have been concerned here with a conceptual inquiry, and that is the province of philosophy. We have been asking, "What *is* our concept of *seeing*; might it not be more subtle, complex, and variable than 'classical' philosophers of science would have us believe?" We have not been directly concerned with the psychological questions "How do we *arrive* at the concepts of seeing we have got, and what causes this variability in what we see?"—though of course answers to these questions would mark more clearly the boundaries of our own inquiry. As Wittgenstein would have put it, "Here the psychological is a symbol of the logical."

What all this has been leading up to is the centrality of the notion of *seeing as* within our concept of *seeing*. You see it as a duck, I see it as a rabbit; the physicist sees it as an x-ray tube, the child sees it as a kind of complicated incandescent lamp bulb; the microscopist sees it as coelenterate mesoglea, the engineer as a kind of gooey, formless stuff. And how very relevant to every case of seeing is the knowledge of him who does the looking.

Goethe said that we see only what we know. In my opinion Goethe was right in a way that "classical" philosophers of science, with all their talk of *normal perceptions, sense-data, interpretations, logical constructions*, etc., were hopelessly wrong. The point of Goethe's remark should be within our reach now. I will try to secure it by means of discussion of what it is to *see as* . . . ; my argument will be that almost everything we usually call *seeing* involves as fundamental to it what I, following Wittgenstein, have called "*seeing as*."

I have just said that the reasons why people see things differently are not the same for all the cases we have considered. There is, however, one respect in which they do not differ, or so I shall argue. No case of seeing that we have considered is wholly independent of the knowledge of the percipient. I had to tell you what I knew about *Figure 4* before you could see it as I saw it, a bear climbing up the other side of a tree. And this bit of knowledge is intelligible only against the knowledge of what a bear is, what a tree is, and what climbing is. Almost everyone, of course, will see

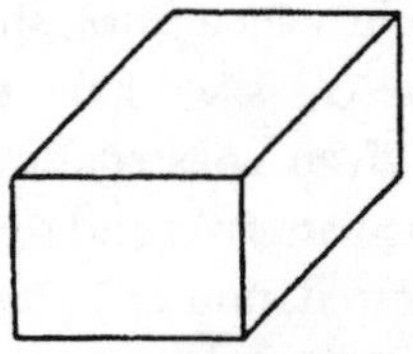

as a transparent box, or a wire-framed cube, viewed as from above, or as from below. But this need not mean that our observation of this figure is without a trace of any knowledge of the construction and properties of box-like objects and the functions of the lines used in representing such objects. On the contrary, this only goes to show that most people who are capable of experiencing the reversible perspective phenomenon of this figure, and this would of course exclude babies and dimwits as well as blind people, know enough, have learned enough, *to be able to see* this figure as a three-dimensional box, from above or from below. It is interesting to speculate as to whether a person ignorant of the existence and appearance of rabbits could see the duck-rabbit as anything but a duck. This speculation is no more inadmissible than Locke's conjecture that a man whose blindness had been cured by operation would be unable to identify and distinguish a cube from a sphere by sight alone, a conjecture admirably substantiated by Sherrington.[10] As Wittgenstein puts it, "You only see the duck and rabbit aspects if you are already conversant with the shapes of those two animals . . ."[11] Could a person who had never experienced a cup of any kind, much less one of the ornate Venetian variety, see anything but two faces in Koehler's drawing? And is nothing whatever required of us in order that the black and white crosses should alternately claim our attention? In every such case the traces of previous knowledge are to be found, and those traces figure in all the situations we have called *seeing*.

It is well known that babies, even those older than six months—the time when the retina has completely formed and a minimum of ocular coordination has been achieved—are capable of experiencing but very few of what we take to be the most ordinary visual experiences, like seeing a cloud in the sky. For all their delicate optical equipment, babies are not even in a position to be taken in by reversible perspectives or shifting aspects, much less galvanometers and x-ray tubes. They are in a "big, blooming, buzzing confusion," as William James once put it. The ophthalmologist Ida Mann likens this state to what we experience, or fail to experience perhaps, at the moment of waking when we "recapture our primitive amazement at the world for a few seconds." The usual bedroom things are before our eyes "but they look . . . bizarre and meaningless. Our brains are not as awake as our eyes . . ."[12] In such a state we could not be said to *see* anything, cubes or tubes, stairs or bears.

At this point it may be worth while to remark that some scientists and philosophers think the eye to be a kind of window in our skins. When the window is shut, as when the eyelids are closed or the cornea clouded, we cannot see. When the window is open, we can see. Normal retinal reaction on the one hand and sense-data on the other are given the title of seeing because of their apparently intimate relationship to the light coming in the window.

But the eye is not merely a transparent section of our skins. Part of it does develop embryologically from the skin, it is true; the lens and the cornea are epithelial. The retina and the optic nerve, however, are outgrowths of the brain. It could not alarm anyone, except a person with a theory to the contrary, to hear that alterations in the general state of the brain, alterations like learning what was not before known, or experiencing the heretofore inexperienced, could affect the whole character of seeing, particularly in its conceptual organization and in the singling out of "significant" aspects.

As Wittgenstein says, "'Now he is seeing it like *this*,' 'now like *that*' would only be said of someone *capable* of making certain applications of the figure quite freely . . . It is only if someone *can do*, has learnt, is master of, such-and-such, that it makes sense to say he has had *this* experience."[13]

Seeing a thing, therefore, is *seeing* it *as* this sort of thing, or as that sort of thing; we do not just *see* indeterminately or in general, as do infants and lunatics. And seeing a thing as *this* or *that* sort of thing presupposes a *knowledge* of *this* or *that* sort of thing. Our two astronomers would not say merely that they saw a brilliant yellow-white disc and leave it at that. What they see they see *as* the sun. And this presupposes a knowledge of what sort of thing the sun is, which digs up a nugget I buried three chapters ago. The knowledge of what the sun was in the 13th Century was very, very different from the knowledge of what the sun is now in the 20th Century. I will say no more about this now other than to suggest that the two astronomers are to the sun as you and I might be to the duck-rabbit when you see only a duck and I only a rabbit. The difference is in our conceptual organization of the elements of our visual experience. So too the sun, hills, and trees may be seen as in quite a different relation to the medieval scientist from the relation in which they appear to the modern astronomer. This is a point I will press further in the next chapter.

Here I wish to make it quite clear that I am not denying that there are a good many cases in scientific inquiry where the data before us are wonderfully confused, and about the nature of which we may not have an inkling. It occurs to me, however, that the importance of observation in such cases is overrated and its character is not enough understood. The model for such *seeing* is what we undergo in the oculist's office where we report on the apparent distance between the point of white light and the reference line seen with the left eye: "They are coming together now. There, the point is right on the line." Or the oculist will request, "Say when you can see a green light with your left eye." How similar to situations in microscopy where, when we are confronted with a totally new and unfamiliar phenomenon, we report our visual sensations in as lustreless and phenomenal a way as possible: "It has a green tint to it in this light, and those erratic jerky movements it makes along its longitudinal axis are noteworthy. Ah, there's another one, slightly longer and thicker; there are cilia-like appendages near the narrow end, and two darkened areas at its middle." So too the physicist who expresses a given experimental situation thus: "The needle is oscillating most erratically, I wonder what's up; and see that faint streak near the neon parabola, it looks almost like a reflection of the main parabola, and there are scintillations at the periphery of the cathode scope that have never before been dominant."

I certainly do not wish to say that these are not genuine cases of seeing. If I did I would be just as far off course as those who insist that these are the *only* genuine cases of seeing. What I would urge is that these observational situations have a point to them just because they contrast with our more usual cases of seeing. The language of shapes, color patches, oscillations, and pointer-readings is the language appropriate to the unsettled experimental situation, where confusion and perhaps even conceptual muddlement dominate. And the *seeing* that figures in such situations is of the sort where the observer *does not know what he is seeing*. He will not be satisfied until he does know, until his observations cohere and are intelligible as against the general background of his already accepted and established knowledge. And it is this latter kind of seeing that is the goal of observation. For it is largely in terms of it, and seldom in terms of merely phenomenal seeing, that new inquiry will proceed.

This is part of Goethe's meaning when he says that we see only what we know. New visual phenomena are noteworthy only against our accepted knowledge of the observable world. In psychologists' language, we are *set* to see, observe, notice, or attend to certain sorts of things, but not others. The ancient Greeks failed to notice thousands of things about the world that children now regard as commonplace, but this was not due to faulty vision or lack of curiosity. Galen's followers did not see that the middle wall of the heart was usually solid and not perforated. Physicists up until 1900 failed to detect the flaw in Galileo's proof that the acceleration of a freely falling body was proportional to the time and not the distance fallen. And Darwin himself remarked of an early expedition with a colleague, "Neither of us saw a trace of the wonderful glacial phenomena all around us; we did not notice plainly scored rocks, the perched boulders, the lateral and terminal moraines . . ."[14]

So of course it is often an essential step in the advancement of science to account for ourselves as observers in a strictly phenomenal way. Every great scientist has had to subject himself to the severities of a strict reporting of what lies in his visual field, of the shapes, lines, colors, and movements he sees.

But that is far from the end of the matter. Everyone who is forced by experimental difficulties and conceptual perplexities to observe his data as if he were in an oculist's office aims at coming to see his data in this other sense: where he knows what he is seeing, where he sees his data as it is (and not merely as it appears), where he can see that if a certain operation were performed on his data a certain other action would be observed, just as we see that if the first story of a tall building were demolished the upper stories would come crashing down.

The point is that coming to see one's data in the completely lustreless and noncommittal way that we see the objects of the oculist's test requires a highly specialized and rigorous training in science.[15] Learning to restrict and control one's vision in this way is a scientific accomplishment of the first magnitude, and it is far from being the birthright of every man who decides to study natural science. All of which is to say that *phenomenal* seeing is something acquired, something unusual, something different from our ordinary ways of seeing. Using *phenomenal seeing* as the typical, paradigm case of *seeing* is unjustified and misleading. Rather than our ordinary cases of seeing being logical constructions out of the research scientists' phenomenal variety of seeing, it is the latter which is a logical destruction of our ordinary kinds of seeing. It is something done in a calculated, systematic, premeditated way. But of course if *all* our seeing were carried on in this way we would collapse from exhaustion in a fortnight.

Hence I am not denying that "phenomenal" seeing is genuine seeing. I am urging that it is not the *only* genuine type of seeing, not the paradigm case of seeing, and indeed, it is only a case of seeing at all when considered against the more usual sort of seeing I have been discussing. The more usual sort of seeing is, as Goethe suggested, a seeing of what we know. It is, hence, a theory-laden operation, about which more will be said later, and hence relative in most respects to the observer's knowledge. It is this knowledge which in large measure affects what the observer will see things *as*. Wittgenstein put it this way:

> The concept of "seeing" makes a tangled impression . . . There is not one genuine proper case of [what is seen]—the rest being just vague, something which awaits clarification. . . . What we have . . . to do is to accept the everyday language-game, and to note false accounts of the matter *as* false. . . .[16]

I will try later to explore further the theory-loaded character of *seeing*, moving from the *seeing as* component we have been discussing to what I call *seeing that*. This will bring us to the large questions having to do with the interrelations between knowledge, language, and our ordinary observation.

Here may I commend to your reflections the story that Freud tells of the visitor to the fur shop who remarked on how wonderful it was that all the pelts had two holes in them just where the animal's eyes were situated.

Notes

1 L. Wittgenstein, *Philosophical Investigations* (New York: Macmillan, 1953), p. 207.
2 See *The Mind*, eds. John Rowan Wilson and the editors of *Life* (New York: Life Science Library series, Time, Inc., 1964), p. 15.
3 Wittgenstein, *Philosophical Investigations*, p. 212.
4 Ibid., p. 193.
5 Ibid., p. 196.
6 Ibid., p. 195.
7 Ibid., p. 193, Hanson's italics.
8 Ibid., p. 200.
9 Pierre Duhem, *The Aim and Structure of Physical Theory*, tr. P. P. Wiener (Princeton: Princeton University Press, 1954), p. 218.

10 See John Locke, *An Essay Concerning Human Understanding*, ed. A. S. Pringle-Pattison (Oxford: Clarendon Press, 1924), Book II, pp. 75–6.
11 Wittgenstein, *Philosophical Investigations*, p. 207.
12 I. Mann and A. Pirie, *The Science of Seeing* (New York: Penguin, 1946), p. 18.
13 Wittgenstein, *Philosophical Investigations*, pp. 208–9.
14 Charles Robert Darwin, *His Life Told in an Autobiographical Chapter*, ed. Francis Darwin (London: John Murray, 1902), p. 25.
15 See M. L. Johnson, "Seeing's Believing," *New Biology*, Vol. 15 (Oct., 1953), pp. 66–79ff.
16 Wittgenstein, *Philosophical Investigations*, p. 200.

22

W.V. Quine, "Two Dogmas of Empiricism"

Willard Van Orman Quine (1908–2001) is probably the most important twentieth-century American philosopher. His contributions to the philosophy of science include a seminal paper written in the early 1950s and reprinted below. In this paper, Quine attacks two fundamental assumptions of Logical Empiricism. His work influenced later post-positivist philosophers of science, most famously, Kuhn.

Modern empiricism has been conditioned in large part by two dogmas. One is a belief in some fundamental cleavage between truths which are *analytic*, or grounded in meanings independently of matters of fact, and truths which are *synthetic*, or grounded in fact. The other dogma is *reductionism*: the belief that each meaningful statement is equivalent to some logical construct upon terms which refer to immediate experience. Both dogmas, I shall argue, are ill-founded. One effect of abandoning them is, as we shall see, a blurring of the supposed boundary between speculative metaphysics and natural science. Another effect is a shift toward pragmatism.

1. Background for Analyticity

Kant's cleavage between analytic and synthetic truths was foreshadowed in Hume's distinction between relations of ideas and matters of fact, and in Leibniz's distinction between truths of reason and truths of fact. Leibniz spoke of the truths of reason as true in all possible worlds. Picturesqueness aside, this is to say that the truths of reason are those which could not possibly be false. In the same vein we hear analytic statements defined as statements whose denials are self-contradictory. But this definition has small

W.V. Quine, *From a Logical Point of View*, 1953, pp. 20–46. Cambridge, MA: Harvard University Press. Originally published in *Philosophical Review*, 1951, 60: 20–43.

explanatory value; for the notion of self-contradictoriness, in the quite broad sense needed for this definition of analyticity, stands in exactly the same need of clarification as does the notion of analyticity itself. The two notions are the two sides of a single dubious coin.

Kant conceived of an analytic statement as one that attributes to its subject no more than is already conceptually contained in the subject. This formulation has two shortcomings: it limits itself to statements of subject-predicate form, and it appeals to a notion of containment which is left at a metaphorical level. But Kant's intent, evident more from the use he makes of the notion of analyticity than from his definition of it, can be restated thus: a statement is analytic when it is true by virtue of meanings and independently of fact. Pursuing this line, let us examine the concept of *meaning* which is presupposed.

Meaning, let us remember, is not to be identified with naming.[1] Frege's example of "Evening Star" and "Morning Star," and Russell's of "Scott" and "the author of *Waverley*," illustrate that terms can name the same thing but differ in meaning. The distinction between meaning and naming is no less important at the level of abstract terms. The terms "9" and "the number of the planets" name one and the same abstract entity but presumably must be regarded as unlike in meaning; for astronomical observation was needed, and not mere reflection on meanings, to determine the sameness of the entity in question.

The above examples consist of singular terms, concrete and abstract. With general terms, or predicates, the situation is somewhat different but parallel. Whereas a singular term purports to name an entity, abstract or concrete, a general term does not; but a general term is *true of* an entity, or of each of many, or of none.[2] The class of all entities of which a general term is true is called the *extension* of the term. Now paralleling the contrast between the meaning of a singular term and the entity named, we must distinguish equally between the meaning of a general term and its extension. The general terms "creature with a heart" and "creature with kidneys," for example, are perhaps alike in extension but unlike in meaning.

Confusion of meaning with extension, in the case of general terms, is less common than confusion of meaning with naming in the case of singular terms. It is indeed a commonplace in philosophy to oppose intension (or meaning) to extension, or, in a variant vocabulary, connotation to denotation.

The Aristotelian notion of essence was the forerunner, no doubt, of the modern notion of intension or meaning. For Aristotle it was essential in men to be rational, accidental to be two-legged. But there is an important difference between this attitude and the doctrine of meaning. From the latter point of view it may indeed be conceded (if only for the sake of argument) that rationality is involved in the meaning of the word "man" while two-leggedness is not; but two-leggedness may at the same time be viewed as involved in the meaning of "biped" while rationality is not. Thus from the point of view of the doctrine of meaning it makes no sense to say of the actual individual, who is at once a man and a biped, that his rationality is essential and his two-leggedness accidental or vice versa. Things had essences, for Aristotle, but only linguistic forms have meanings. Meaning is what essence becomes when it is divorced from the object of reference and wedded to the word.

For the theory of meaning a conspicuous question is the nature of its objects: what sort of things are meanings? A felt need for meant entities may derive from an earlier failure to appreciate that meaning and reference are distinct. Once the theory of meaning is sharply separated from the theory of reference, it is a short step to recognizing as the primary business of the theory of meaning simply the synonymy of linguistic forms and the analyticity of statements; meanings themselves, as obscure intermediary entities, may well be abandoned.[3]

The problem of analyticity then confronts us anew. Statements which are analytic by general philosophical acclaim are not, indeed, far to seek. They fall into two classes. Those of the first class, which may be called *logically true*, are typified by:

(1) No unmarried man is married.

The relevant feature of this example is that it not merely is true as it stands, but remains true under any and all reinterpretations of "man" and "married." If we suppose a prior inventory of *logical* particles, comprising "no," "un-," "not," "if," "then," "and," etc., then in general a logical truth is a statement which is true and remains true under all reinterpretations of its components other than the logical particles.

But there is also a second class of analytic statements, typified by:

(2) No bachelor is married.

The characteristic of such a statement is that it can be turned into a logical truth by putting synonyms for synonyms; thus (2) can be turned into (1) by putting "unmarried man" for its synonym "bachelor." We still lack a proper characterization of this second class of analytic statements, and therewith of analyticity generally, inasmuch as we have had in the above description to lean on a notion of "synonymy" which is no less in need of clarification than analyticity itself.

In recent years Carnap has tended to explain analyticity by appeal to what he calls state-descriptions.[4] A state-description is any exhaustive assignment of truth values to the atomic, or noncompound, statements of the language. All other statements of the language are, Carnap assumes, built up of their component clauses by means of the familiar logical devices, in such a way that the truth value of any complex statement is fixed for each state-description by specifiable logical laws. A statement is then explained as analytic when it comes out true under every state description. This account is an adaptation of Leibniz's "true in all possible worlds." But note that this version of analyticity serves its purpose only if the atomic statements of the language are, unlike "John is a bachelor" and "John is married," mutually independent. Otherwise there would be a state-description which assigned truth to "John is a bachelor" and to "John is married," and consequently "No bachelors are married" would turn out synthetic rather than analytic under the proposed criterion. Thus the criterion of analyticity in terms of state-descriptions serves only for languages devoid of extralogical synonym-pairs, such as "bachelor" and "unmarried man"—synonym-pairs of the type which give rise to the "second class" of analytic statements. The criterion in terms of state-descriptions is a reconstruction at best of logical truth, not of analyticity.

I do not mean to suggest that Carnap is under any illusions on this point. His simplified model language with its state-descriptions is aimed primarily not at the general problem of analyticity but at another purpose, the clarification of probability and induction. Our problem, however, is analyticity; and here the major difficulty lies not in the first class of analytic statements, the logical truths, but rather in the second class, which depends on the notion of synonymy.

2. Definition

There are those who find it soothing to say that the analytic statements of the second class reduce to those of the first class, the logical truths, by *definition*; "bachelor," for example, is *defined* as "unmarried man." But how do we find that "bachelor" is defined as "unmarried man"? Who defined it thus, and when? Are we to appeal to the nearest dictionary, and accept the lexicographer's formulation as law? Clearly this would be to put the cart before the horse. The lexicographer is an empirical scientist, whose business is the recording of antecedent facts; and if he glosses "bachelor" as "unmarried man" it is because of his belief that there is a relation of synonymy between those forms, implicit in general or preferred usage prior to his own work. The notion of synonymy presupposed here has still to be clarified, presumably in terms relating to linguistic behavior. Certainly the "definition" which is the lexicographer's report of an observed synonymy cannot be taken as the ground of the synonymy.

Definition is not, indeed, an activity exclusively of philologists. Philosophers and scientists frequently have occasion to "define" a recondite term by paraphrasing it into terms of a more familiar vocabulary. But ordinarily such a definition, like the philologist's, is pure lexicography, affirming a relation of synonymy antecedent to the exposition in hand.

Just what it means to affirm synonymy, just what the interconnections may be which are necessary and sufficient in order that two linguistic forms be properly describable as synonymous, is far from clear; but, whatever these interconnections may be, ordinarily they are grounded in usage. Definitions reporting selected instances of synonymy come then as reports upon usage.

There is also, however, a variant type of definitional activity which does not limit itself to the reporting of preëxisting synonymies. I have in mind what Carnap calls *explication*—an activity to which philosophers are given, and scientists also in their more philosophical moments. In explication the purpose is not merely to paraphrase the definiendum into an outright synonym, but actually to improve upon the definiendum by refining or supplementing its meaning. But even explication, though not merely reporting a preëxisting synonymy between definiendum and definiens, does rest nevertheless on *other* pre-existing synonymies. The matter may be viewed as follows. Any word worth explicating has some contexts which, as wholes, are clear and precise enough to be useful; and the purpose of explication is to preserve the usage of these favored contexts while sharpening the usage of other contexts. In order that a given definition be suitable for purposes of explication, therefore, what is required is not that the definiendum in its antecedent usage be synonymous with the definiens, but just that each of these favored contexts of the definiendum, taken as a whole in its antecedent usage, be synonymous with the corresponding context of the definiens.

Two alternative definientia may be equally appropriate for the purposes of a given task of explication and yet not be synonymous with each other; for they may serve interchangeably within the favored contexts but diverge elsewhere. By cleaving to one of these definientia rather than the other, a definition of explicative kind generates, by fiat, a relation of synonymy between definiendum and definiens which did not hold before. But such a definition still owes its explicative function, as seen, to pre-existing synonymies.

There does, however, remain still an extreme sort of definition which does not hark back to prior synonymies at all: namely, the explicitly conventional introduction of novel notations for purposes of sheer abbreviation. Here the definiendum becomes synonymous with the definiens simply because it has been created expressly for the purpose of being synonymous with the definiens. Here we have a really transparent case of synonymy created by definition; would that all species of synonymy were as intelligible. For the rest, definition rests on synonymy rather than explaining it.

The word "definition" has come to have a dangerously reassuring sound, owing no doubt to its frequent occurrence in logical and mathematical writings. We shall do well to digress now into a brief appraisal of the role of definition in formal work.

In logical and mathematical systems either of two mutually antagonistic types of economy may be striven for, and each has its peculiar practical utility. On the one hand we may seek economy of practical expression—ease and brevity in the statement of multifarious relations. This sort of economy calls usually for distinctive concise notations for a wealth of concepts. Second, however, and oppositely, we may seek economy in grammar and vocabulary; we may try to find a minimum of basic concepts such that, once a distinctive notation has been appropriated to each of them, it becomes possible to express any desired further concept by mere combination and iteration of our basic notations. This second sort of economy is impractical in one way, since a poverty in basic idioms tends to a necessary lengthening of discourse. But it is practical in another way: it greatly simplifies theoretical discourse *about* the language, through minimizing the terms and the forms of construction wherein the language consists.

Both sorts of economy, though prima facie incompatible, are valuable in their separate ways. The custom has consequently arisen of combining both sorts of economy by forging in effect two languages, the one a part of the other. The inclusive language, though redundant in grammar and vocabulary, is economical in message lengths, while the part, called primitive notation, is economical in grammar and vocabulary. Whole and part are correlated by rules of translation whereby each idiom not in primitive notation is equated to some complex built up of primitive notation. These rules of translation are the so-called *definitions* which appear in formalized systems. They are best viewed not as adjuncts to one language but as correlations between two languages, the one a part of the other.

But these correlations are not arbitrary. They are supposed to show how the primitive notations can accomplish all purposes, save brevity and convenience, of the redundant language. Hence the definiendum and its definiens may be expected, in each case, to be related in one or another of the three ways lately noted. The definiens may be a faithful paraphrase of the definiendum into the narrower notation, preserving a direct synonymy[5] as of antecedent usage; or the definiens may, in the spirit of explication, improve upon the antecedent usage of the definiendum; or finally, the definiendum may be a newly created notation, newly endowed with meaning here and now.

In formal and informal work alike, thus, we find that definition—except in the extreme case of the explicitly conventional introduction of new notations—hinges on prior relations of synonymy. Recognizing then that the notion of definition does not hold the key to synonymy and analyticity, let us look further into synonymy and say no more of definition.

3. Interchangeability

A natural suggestion, deserving close examination, is that the synonymy of two linguistic forms consists simply in their interchangeability in all contexts without change of truth value—interchangeability, in Leibniz's phrase, *salva veritate*.[6] Note that synonyms so conceived need not even be free from vagueness, as long as the vaguenesses match.

But it is not quite true that the synonyms "bachelor" and "unmarried man" are everywhere interchangeable *salva veritate*. Truths which become false under substitution of "unmarried man" for "bachelor" are easily constructed with the help of "bachelor of arts" or "bachelor's buttons"; also with the help of quotation, thus:

"Bachelor" has less than ten letters.

Such counterinstances can, however, perhaps be set aside by treating the phrases "bachelor of arts" and "bachelor's buttons" and the quotation "bachelor" each as a single indivisible word and then stipulating that the interchangeability *salva veritate* which is to be the touchstone of synonymy is not supposed to apply to fragmentary occurrences inside of a word. This account of synonymy, supposing it acceptable on other counts, has indeed the drawback of appealing to a prior conception of "word" which can be counted on to present difficulties of formulation in its turn. Nevertheless some progress might be claimed in having reduced the problem of synonymy to a problem of wordhood. Let us pursue this line a bit, taking "word" for granted.

The question remains whether interchangeability *salva veritate* (apart from occurrences within words) is a strong enough condition for synonymy, or whether, on the contrary, some heteronymous expressions might be thus interchangeable. Now let us be clear that we are not concerned here with synonymy in the sense of complete identity in psychological associations or poetic quality; indeed no two expressions are synonymous in such a sense. We are concerned only with what may be called *cognitive* synonymy. Just what this is cannot be said without successfully finishing the present study; but we know something about it from the need which arose for it in connection with analyticity in §1. The sort of synonymy needed there was merely such that any analytic statement could be turned into a logical truth by putting synonyms for synonyms. Turning the tables and assuming analyticity, indeed, we could explain cognitive synonymy of terms as follows (keeping to the familiar example): to say that "bachelor" and "unmarried man" are cognitively synonymous is to say no more nor less than that the statement:

(3) All and only bachelors are unmarried men

is analytic.[7]

What we need is an account of cognitive synonymy not presupposing analyticity—if we are to explain analyticity conversely with help of cognitive synonymy as undertaken in §1. And indeed such an independent account of cognitive synonymy is at present up for consideration, namely, interchangeability *salva veritate* everywhere except within words. The question before us, to resume the thread at last, is whether such interchangeability is a sufficient condition for cognitive synonymy. We can quickly assure ourselves that it is, by examples of the following sort. The statement:

(4) Necessarily all and only bachelors are bachelors

is evidently true, even supposing "necessarily" so narrowly construed as to be truly applicable only to analytic statements. Then, if "bachelor" and "unmarried man" are interchangeable *salva veritate*, the result:

(5) Necessarily all and only bachelors are unmarried men

of putting "unmarried man" for an occurrence of "bachelor" in (4) must, like (4), be true. But to say that (5) is true is to say that (3) is analytic, and hence that "bachelor" and "unmarried man" are cognitively synonymous.

Let us see what there is about the above argument that gives it its air of hocus-pocus. The condition of interchangeability *salva veritate* varies in its force with variations in the richness of the language at hand. The above argument supposes we are working with a language rich enough to contain the adverb "necessarily," this adverb being so construed as to yield truth when and only when applied to an analytic statement. But can we condone a language which contains such an adverb? Does the adverb really make sense? To suppose that it does is to suppose that we have already made satisfactory sense of "analytic." Then what are we so hard at work on right now?

Our argument is not flatly circular, but something like it. It has the form, figuratively speaking, of a closed curve in space.

Interchangeability *salva veritate* is meaningless until relativized to a language whose extent is specified in relevant respects. Suppose now we consider a language containing just the following materials. There is an indefinitely large stock of one-place predicates (for example, "*F*" where "*Fx*" means that *x* is a man) and many-place predicates (for example, "*G*" where "*Gxy*" means that *x* loves *y*), mostly having to do with extralogical subject matter. The rest of the language is logical. The atomic sentences consist each of a predicate followed by one or more variables "*x*," "*y*," etc.; and the complex sentences are built up of the atomic ones by truth functions ("not," "and," "or," etc.) and quantification.[8] In effect such a language enjoys the benefits also of descriptions and indeed singular terms generally, these being contextually definable in known ways.[9] Even abstract singular terms naming classes, classes of classes, etc., are contextually definable in case the assumed stock of predicates includes the two-place predicate of class membership.[10] Such a language can be adequate to classical mathematics and indeed to scientific discourse generally, except in so far as the latter involves debatable devices such as contrary-to-fact conditionals or modal adverbs like "necessarily."[11] Now a language of this type is extensional, in this sense: any two predicates which agree extensionally (that is, are true of the same objects) are interchangeable *salva veritate*.[12]

In an extensional language, therefore, interchangeability *salva veritate* is no assurance of cognitive synonymy of the desired type. That "bachelor" and "unmarried man" are interchangeable *salva veritate* in an extensional language assures us of no more than that (3) is true. There is no assurance here that the extensional agreement of "bachelor" and "unmarried man" rests on meaning rather than merely on accidental matters of fact, as does the extensional agreement of "creature with a heart" and "creature with kidneys."

For most purposes extensional agreement is the nearest approximation to synonymy we need care about. But the fact remains that extensional agreement falls far short of cognitive synonymy of the type required for explaining analyticity in the manner of §1. The type of cognitive synonymy required there is such as to equate the synonymy of "bachelor" and "unmarried man" with the analyticity of (3), not merely with the truth of (3).

So we must recognize that interchangeability *salva veritate*, if construed in relation to an extensional language, is not a sufficient condition of cognitive synonymy in the sense needed for deriving analyticity in the manner of §1. If a language contains an intensional adverb "necessarily" in the sense lately noted, or other particles to the same effect, then interchangeability *salva veritate* in such a language does afford a sufficient condition of cognitive synonymy; but such a language is intelligible only in so far as the notion of analyticity is already understood in advance.

The effort to explain cognitive synonymy first, for the sake of deriving analyticity from it afterward as in §1, is perhaps the wrong approach. Instead we might try explaining analyticity somehow without appeal to cognitive synonymy. Afterward we could doubtless derive cognitive synonymy from analyticity satisfactorily enough if desired. We have seen that cognitive synonymy of "bachelor" and "unmarried man" can be explained as analyticity of (3). The same explanation works for any pair of one-place predicates, of course, and it can be extended in obvious fashion to many-place predicates. Other syntactical categories can also be accommodated in fairly parallel fashion. Singular terms may be said to be cognitively synonymous when the statement of identity formed by putting " = " between them is analytic. Statements may be said simply to be cognitively synonymous when their biconditional (the result of joining them by "if and only if") is analytic.[13] If we care to lump all categories into a single formulation, at the expense of assuming again the notion of "word" which was appealed to early in this section, we can describe any two linguistic forms as cognitively synonymous when the two forms are interchangeable (apart from occurrences within "words") *salva* (no longer *veritate* but) *analyticitate*. Certain technical questions arise, indeed, over cases of ambiguity or homonymy; let us not pause for them, however, for we are already digressing. Let us rather turn our backs on the problem of synonymy and address ourselves anew to that of analyticity.

4. Semantical Rules

Analyticity at first seemed most naturally definable by appeal to a realm of meanings. On refinement, the appeal to meanings gave way to an appeal to synonymy or definition. But definition turned out to be a will-o'-the-wisp, and synonymy turned out to be best understood only by dint of a prior appeal to analyticity itself. So we are back at the problem of analyticity.

I do not know whether the statement "Everything green is extended" is analytic. Now does my indecision over this example really betray an incomplete understanding, an incomplete grasp of the "meanings," of "green" and "extended"? I think not. The trouble is not with "green" or "extended," but with "analytic."

It is often hinted that the difficulty in separating analytic statements from synthetic ones in ordinary language is due to the vagueness of ordinary language and that the distinction is clear when we have a precise artificial language with explicit "semantical rules." This, however, as I shall now attempt to show, is a confusion.

The notion of analyticity about which we are worrying is a purported relation between statements and languages: a statement S is said to be *analytic for* a language L, and the problem is to make sense of this relation generally, that is, for variable "S" and "L". The gravity of this problem is not perceptibly less for artificial languages than for natural ones. The problem of making sense of the idiom "S is analytic for L," with variable "S" and "L," retains its stubbornness even if we limit the range of the variable "L" to artificial languages. Let me now try to make this point evident.

For artificial languages and semantical rules we look naturally to the writings of Carnap. His semantical rules take various forms, and to make my point I shall have to distinguish certain of the forms. Let us suppose, to begin with, an artificial language L_0 whose semantical rules have the form explicitly of a specification, by recursion or otherwise, of all the analytic statements of L_0. The rules tell us that such and such statements, and only those, are the analytic statements of L_0. Now here the difficulty is simply that the rules contain the word "analytic," which we do not understand! We understand what expressions the rules attribute analyticity to, but we do not understand what the rules attribute to those expressions. In short, before we can understand a rule which begins "A statement S is analytic for language L_0 if and only if . . . ," we must understand the general relative term "analytic for"; we must understand 'S is analytic for L" where "S" and "L" are variables.

Alternatively we may, indeed, view the so-called rule as a conventional definition of a new simple symbol "analytic-for-L_0," which might better be written untendentiously as "K" so as not to seem to throw light on the interesting word "analytic." Obviously any number of classes K, M, N, etc. of statements of L_0 can be specified for various purposes or for no purpose; what does it mean to say that K, as against M, N, etc., is the class of the "analytic" statements of L_0?

By saying what statements are analytic for L_0 we explain "analytic-for-L_0" but not "analytic," not "analytic for." We do not begin to explain the idiom "S is analytic for L" with variable "S" and "L," even if we are content to limit the range of "L" to the realm of artificial languages.

Actually we do know enough about the intended significance of "analytic" to know that analytic statements are supposed to be true. Let us then turn to a second form of semantical rule, which says not that such and such statements are analytic but simply that such and such statements are included among the truths. Such a rule is not subject to the criticism of containing the un-understood word "analytic"; and we may grant for the sake of argument that there is no difficulty over the broader term "true." A semantical rule of this second type, a rule of truth, is not supposed to specify all the truths of the language; it merely stipulates, recursively or otherwise, a certain multitude of statements which, along with others unspecified, are to count as true. Such a rule may be conceded to be quite clear. Derivatively, afterward, analyticity can be demarcated thus: a statement is analytic if it is (not merely true but) true according to the semantical rule.

Still there is really no progress. Instead of appealing to an unexplained word "analytic," we are now appealing to an unexplained phrase "semantical rule." Not every true statement which says that the statements of some class are true can count as a semantical rule—otherwise *all* truths would be "analytic" in the sense of being true according to semantical rules. Semantical rules are distinguishable, apparently, only by the fact of appearing on a page under the heading "Semantical Rules"; and this heading is itself then meaningless.

We can say indeed that a statement is *analytic-for-*L_0 if and only if it is true according to such and such specifically appended "semantical rules," but then we find ourselves back at essentially the same case which was originally discussed: "*S* is analytic-for-L_0 if and only if . . . " Once we seek to explain "*S* is analytic-for *L*" generally for variable "*L*" (even allowing limitation of "*L*" to artificial languages), the explanation "true according to the semantical rules of *L*" is unavailing; for the relative term "semantical rule of" is as much in need of clarification, at least, as "analytic for."

It may be instructive to compare the notion of semantical rule with that of postulate. Relative to a given set of postulates, it is easy to say what a postulate is: it is a member of the set. Relative to a given set of semantical rules, it is equally easy to say what a semantical rule is. But given simply a notation, mathematical or otherwise, and indeed as thoroughly understood a notation as you please in point of the translations or truth conditions of its statements, who can say which of its true statements rank as postulates? Obviously the question is meaningless—as meaningless as asking which points in Ohio are starting points. Any finite (or effectively specifiable infinite) selection of statements (preferably true ones, perhaps) is as much *a* set of postulates as any other. The word "postulate" is significant only relative to an act of inquiry; we apply the word to a set of statements just in so far as we happen, for the year or the moment, to be thinking of those statements in relation to the statements which can be reached from them by some set of transformations to which we have seen fit to direct our attention. Now the notion of semantical rule is as sensible and meaningful as that of postulate, if conceived in a similarly relative spirit—relative, this time, to one or another particular enterprise of schooling unconversant persons in sufficient conditions for truth of statements of some natural or artificial language *L*. But from this point of view no one signalization of a subclass of the truths of *L* is intrinsically more a semantical rule than another; and, if "analytic" means "true by semantical rules," no one truth of *L* is analytic to the exclusion of another.[14]

It might conceivably be protested that an artificial language *L* (unlike a natural one) is a language in the ordinary sense *plus* a set of explicit semantical rules—the whole constituting, let us say, an ordered pair; and that the semantical rules of *L* then are specifiable simply as the second component of the pair *L*. But, by the same token and more simply, we might construe an artificial language *L* outright as an ordered pair whose second component is the class of its analytic statements; and then the analytic statements of *L* become specifiable simply as the statements in the second component of *L*. Or better still, we might just stop tugging at our bootstraps altogether.

Not all the explanations of analyticity known to Carnap and his readers have been covered explicitly in the above considerations, but the extension to other forms is not hard to see. Just one additional factor should be mentioned which sometimes enters: sometimes the semantical rules are in effect rules of translation into ordinary language, in which case the analytic statements of the artificial language are in effect recognized as such from the analyticity of their specified translations in ordinary language. Here certainly there can be no thought of an illumination of the problem of analyticity from the side of the artificial language.

From the point of view of the problem of analyticity the notion of an artificial language with semantical rules is a *feu follet par excellence*. Semantical rules determining the analytic statements of an artificial language are of interest only in so far as we already understand the notion of analyticity; they are of no help in gaining this understanding.

Appeal to hypothetical languages of an artificially simple kind could conceivably be useful in clarifying analyticity, if the mental or behavioral or cultural factors relevant to analyticity—whatever they may be—were somehow sketched into the simplified model. But a model which takes analyticity merely as an irreducible character is unlikely to throw light on the problem of explicating analyticity.

It is obvious that truth in general depends on both language and extralinguistic fact. The statement "Brutus killed Caesar" would be false if the world had been different in certain ways, but it would also be false if the word "killed" happened rather to have the sense of "begat." Thus one is tempted to suppose in general that the truth of a statement is somehow analyzable into a linguistic component and a factual component. Given this supposition, it next seems reasonable that in some statements the factual component should be null; and these are the analytic statements. But, for all its a priori reasonableness, a boundary between analytic and synthetic statements simply has not been drawn. That there is such a distinction to be drawn at all is an unempirical dogma of empiricists, a metaphysical article of faith.

5. The Verification Theory and Reductionism

In the course of these somber reflections we have taken a dim view first of the notion of meaning, then of the notion of cognitive synonymy, and finally of the notion of analyticity. But what, it may be asked, of the verification theory of meaning? This phrase has established itself so firmly as a catch-word of empiricism that we should be very unscientific indeed not to look beneath it for a possible key to the problem of meaning and the associated problems.

The verification theory of meaning, which has been conspicuous in the literature from Peirce onward, is that the meaning of a statement is the method of empirically confirming or infirming it. An analytic statement is that limiting case which is confirmed no matter what.

As urged in §1, we can as well pass over the question of meanings as entities and move straight to sameness of meaning, or synonymy. Then what the verification theory says is that statements are synonymous if and only if they are alike in point of method of empirical confirmation or infirmation.

This is an account of cognitive synonymy not of linguistic forms generally, but of statements.[15] However, from the concept of synonymy of statements we could derive the concept of synonymy for other linguistic forms, by considerations somewhat similar to those at the end of §3. Assuming the notion of "word," indeed, we could explain any two forms as synonymous when the putting of the one form for an occurrence of the other in any statement (apart from occurrences within "words") yields a synonymous statement. Finally, given the concept of synonymy thus for linguistic forms generally, we could define analyticity in terms of synonymy and logical truth as in §1. For that matter, we could define analyticity more simply in terms of just synonymy of statements together with logical truth; it is not necessary to appeal to synonymy of linguistic forms other than statements. For a statement may be described as analytic simply when it is synonymous with a logically true statement.

So, if the verification theory can be accepted as an adequate account of statement synonymy, the notion of analyticity is saved after all. However, let us reflect. Statement synonymy is said to be likeness of method of empirical confirmation or infirmation. Just what are these methods which are to be compared for likeness? What, in other words, is the nature of the relation between a statement and the experiences which contribute to or detract from its confirmation?

The most naïve view of the relation is that it is one of direct report. This is *radical reductionism*. Every meaningful statement is held to be translatable into a statement (true or false) about immediate experience. Radical reductionism, in one form or another, well antedates the verification theory of meaning explicitly so called. Thus Locke and Hume held that every idea must either originate directly in sense experience or else be compounded of ideas thus originating; and taking a hint from Tooke we might rephrase this doctrine in semantical jargon by saying that a term, to be significant at all, must be either a name of a sense datum or a compound of such names or an abbreviation of such a compound. So stated, the doctrine remains ambiguous as between sense data as sensory events and sense data as sensory qualities; and it remains vague as to the admissible ways of compounding. Moreover, the doctrine is unnecessarily and intolerably restrictive in the term-by-term critique which it imposes. More reasonably, and without yet exceeding the limits of what I have called radical reductionism, we may take full statements as our significant units—thus demanding that our statements as wholes be translatable into sense-datum language, but not that they be translatable term by term.

This emendation would unquestionably have been welcome to Locke and Hume and Tooke, but historically it had to await an important reorientation in semantics—the reorientation whereby the primary vehicle of meaning came to be seen no longer in the term but in the statement. This reorientation, seen in Bentham and Frege, underlies Russell's concept of incomplete symbols defined in use;[16] also it is implicit in the verification theory of meaning, since the objects of verification are statements.

Radical reductionism, conceived now with statements as units, set itself the task of specifying a sense-datum language and showing how to translate the rest of significant discourse, statement by statement, into it. Carnap embarked on this project in the *Aufbau*.

The language which Carnap adopted as his starting point was not a sense-datum language in the narrowest conceivable sense, for it included also the notations of logic, up through higher set theory. In effect it included the whole language of pure mathematics. The ontology implicit in it (that is, the range of values of its variables) embraced not only sensory events but classes, classes of classes, and so on. Empiricists there are who would boggle at such prodigality. Carnap's starting point is very parsimonious, however, in its extralogical or sensory part. In a series of constructions in which he exploits the resources of modern logic with much ingenuity, Carnap succeeds in defining a wide array of important additional sensory concepts which, but for his constructions, one would not have dreamed were definable on so slender a basis. He was the first empiricist who, not content with asserting the reducibility of science to terms of immediate experience, took serious steps toward carrying out the reduction.

If Carnap's starting point is satisfactory, still his constructions were, as he himself stressed, only a fragment of the full program. The construction of even the simplest statements about the physical world was left in a sketchy state. Carnap's suggestions on this subject were, despite their sketchiness, very suggestive. He explained spatio-temporal point-instants as quadruples of real numbers and envisaged assignment of sense qualities to point-instants according to certain canons. Roughly summarized, the plan was that qualities should be assigned to point-instants in such a way as to achieve the laziest world compatible with our experience. The principle of least action was to be our guide in constructing a world from experience.

Carnap did not seem to recognize, however, that his treatment of physical objects fell short of reduction not merely through sketchiness, but in principle. Statements of the form "Quality *q* is at point-instant *x;y;z;t*" were, according to his canons, to be apportioned truth values in such a way as to maximize and minimize certain over-all features, and with growth of experience the truth values were to be progressively revised in the same spirit. I think this is a good schematization (deliberately oversimplified, to be sure) of what science really does; but it provides no indication, not even the sketchiest, of how a statement of the form "Quality *q* is at *x;y;z;t*" could ever be translated into Carnap's initial language of sense data and logic. The connective "is at" remains an added undefined connective; the canons counsel us in its use but not in its elimination.

Carnap seems to have appreciated this point afterward; for in his later writings he abandoned all notion of the translatability of statements about the physical world into statements about immediate experience. Reductionism in its radical form has long since ceased to figure in Carnap's philosophy.

But the dogma of reductionism has, in a subtler and more tenuous form, continued to influence the thought of empiricists. The notion lingers that to each statement, or each synthetic statement, there is associated a unique range of possible sensory events such that the occurrence of any of them would add to the likelihood of truth of the statement, and that there is associated also another unique range of possible sensory events whose occurrence would detract from that likelihood. This notion is of course implicit in the verification theory of meaning.

The dogma of reductionism survives in the supposition that each statement, taken in isolation from its fellows, can admit of confirmation or infirmation at all. My countersuggestion, issuing essentially from Carnap's doctrine of the physical world in the *Aufbau*, is that our statements about the external world face the tribunal of sense experience not individually but only as a corporate body.[17]

The dogma of reductionism, even in its attenuated form, is intimately connected with the other dogma—that there is a cleavage between the analytic and the synthetic. We have found ourselves led, indeed, from the latter problem to the former through the verification theory of meaning. More directly, the one dogma clearly supports the other in this way: as long as it is taken to be significant in general to speak of the confirmation and infirmation of a statement, it seems significant to speak also of a limiting kind of statement which is vacuously confirmed, *ipso facto*, come what may; and such a statement is analytic.

The two dogmas are, indeed, at root identical. We lately reflected that in general the truth of statements does obviously depend both upon language and upon extralinguistic fact; and we noted that this obvious circumstance carries in its train, not logically but all too naturally, a feeling that the truth of a statement is somehow analyzable into a linguistic component and a factual component. The factual component must, if we are empiricists, boil down to a range of confirmatory experiences. In the extreme case where the linguistic component is all that matters, a true statement is analytic. But I hope we are now impressed with how stubbornly the distinction between analytic and synthetic has resisted any straightforward drawing. I am impressed also, apart from prefabricated examples of black and white balls in an urn, with how baffling the problem has always been of arriving at any explicit theory of the empirical confirmation of a synthetic statement. My present suggestion is that it is nonsense, and the root of much nonsense, to speak of a linguistic component and a factual component in the truth of any individual statement. Taken collectively, science has its double dependence upon language and experience; but this duality is not significantly traceable into the statements of science taken one by one.

The idea of defining a symbol in use was, as remarked, an advance over the impossible term-by-term empiricism of Locke and Hume. The statement, rather than the term, came with Bentham to be recognized as the unit accountable to an empiricist critique. But what I am now urging is that even in taking the statement as unit we have drawn our grid too finely. The unit of empirical significance is the whole of science.

what good does this do?

6. Empiricism without the Dogmas

The totality of our so-called knowledge or beliefs, from the most casual matters of geography and history to the profoundest laws of atomic physics or even of pure mathematics and logic, is a man-made fabric which impinges on experience only along the edges. Or, to change the figure, total science is like a field of force whose boundary conditions are experience. A conflict with experience at the periphery occasions readjustments in the interior of the field. Truth values have to be redistributed over some of our statements. Re-evaluation of some statements entails re-evaluation of others, because of their logical interconnections—the logical laws being in turn simply certain further statements of the system, certain further elements of the field. Having reevaluated one statement we must reevaluate some others, which may be statements logically connected with the first or may be the statements of logical connections themselves. But the total field is so underdetermined by its boundary conditions, experience, that there is much latitude of choice as to what statements to reevaluate in the light of any single contrary experience. No particular experiences are linked with any particular statements in the interior of the field, except indirectly through considerations of equilibrium affecting the field as a whole.

If this view is right, it is misleading to speak of the empirical content of an individual statement—especially if it is a statement at all remote from the experiential periphery of the field. Furthermore it becomes folly to seek a boundary between synthetic statements, which hold contingently on experience, and analytic statements, which hold come what may. Any statement can be held true come what may, if we make drastic enough adjustments elsewhere in the system. Even a statement very close to the periphery can be held true in the face of recalcitrant experience by pleading hallucination or by amending certain statements of the kind called logical laws. Conversely, by the same token, no statement is immune to revision. Revision even of the logical law of the excluded middle has been proposed as a means of simplifying quantum mechanics; and what difference is there in principle between such a shift and the shift whereby Kepler superseded Ptolemy, or Einstein Newton, or Darwin Aristotle?

For vividness I have been speaking in terms of varying distances from a sensory periphery. Let me try now to clarify this notion without metaphor. Certain statements, though *about* physical objects and not sense experience, seem peculiarly germane to sense experience—and in a selective way: some statements to some experiences, others to others. Such statements, especially germane to particular experiences, I picture as near the periphery. But in this relation of "germaneness" I envisage nothing more than a loose association reflecting the relative likelihood, in practice, of our choosing one statement rather than another for revision in the event of recalcitrant experience. For example, we can imagine recalcitrant experiences to which we would surely be inclined to accommodate our system by reevaluating just the statement that there are brick houses on Elm Street, together with related statements on the same topic. We can imagine other recalcitrant experiences to which we would be inclined to accommodate our system by reevaluating just the statement that there are no centaurs, along with kindred statements. A recalcitrant experience can, I have urged, be accommodated by any of various alternative reevaluations in various alternative quarters of the total system; but, in the cases which we are now imagining, our natural tendency to disturb the total system as little as possible would lead us to focus our revisions upon these specific statements concerning brick houses or centaurs.

These statements are felt, therefore, to have a sharper empirical reference than highly theoretical statements of physics or logic or ontology. The latter statements may be thought of as relatively centrally located within the total network, meaning merely that little preferential connection with any particular sense data obtrudes itself.

As an empiricist I continue to think of the conceptual scheme of science as a tool, ultimately, for predicting future experience in the light of past experience. Physical objects are conceptually imported into the situation as convenient intermediaries—not by definition in terms of experience, but simply as irreducible posits[18] comparable, epistemologically, to the gods of Homer. For my part I do, qua lay physicist, believe in physical objects and not in Homer's gods; and I consider it a scientific error to believe otherwise. But in point of epistemological footing the physical objects and the gods differ only in degree and not in kind. Both sorts of entities enter our conception only as cultural posits. The myth of physical objects is epistemologically superior to most in that it has proved more efficacious than other myths as a device for working a manageable structure into the flux of experience.

Positing does not stop with macroscopic physical objects. Objects at the atomic level are posited to make the laws of macroscopic objects, and ultimately the laws of experience, simpler and more manageable; and we need not expect or demand full definition of atomic and subatomic entities in terms of macroscopic ones, any more than definition of macroscopic things in terms of sense data. Science is a continuation of common sense, and it continues the common-sense expedient of swelling ontology to simplify theory.

Physical objects, small and large, are not the only posits. Forces are another example; and indeed we are told nowadays that the boundary between energy and matter is obsolete. Moreover, the abstract entities which are the substance of mathematics—ultimately classes and classes of classes and so on up—are another posit in the same spirit. Epistemologically these are myths on the same footing with physical objects and gods, neither better nor worse except for differences in the degree to which they expedite our dealings with sense experiences.

The over-all algebra of rational and irrational numbers is underdetermined by the algebra of rational numbers, but is smoother and more convenient; and it includes the algebra of rational numbers as a jagged or gerrymandered part.[19] Total science, mathematical and natural and human, is similarly but more extremely underdetermined by experience. The edge of the system must be kept squared with experience; the rest, with all its elaborate myths or fictions, has as its objective the simplicity of laws.

Ontological questions, under this view, are on a par with questions of natural science.[20] Consider the question whether to countenance classes as entities. This, as I have argued elsewhere,[21] is the question whether to quantify with respect to variables which take classes as values. Now Carnap [3] has maintained that this is a question not of matters of fact but of choosing a convenient language form, a convenient conceptual scheme or framework for science. With this I agree, but only on the proviso that the same be conceded regarding scientific hypotheses generally. Carnap ([3], p. 32n) has recognized that he is able to preserve a double standard for ontological questions and scientific hypotheses only by assuming an absolute distinction between the analytic and the synthetic; and I need not say again that this is a distinction which I reject.[22]

The issue over there being classes seems more a question of convenient conceptual scheme; the issue over there being centaurs, or brick houses on Elm Street, seems more a question of fact. But I have been urging that this difference is only one of degree, and that it turns upon our vaguely pragmatic inclination to adjust one strand of the fabric of science rather than another in accommodating some particular recalcitrant experience. Conservatism figures in such choices, and so does the quest for simplicity.

Carnap, Lewis, and others take a pragmatic stand on the question of choosing between language forms, scientific frameworks; but their pragmatism leaves off at the imagined boundary between the analytic and the synthetic. In repudiating such a boundary I espouse a more thorough pragmatism. Each man is given a scientific heritage plus a continuing barrage of sensory stimulation; and the considerations which guide him in warping his scientific heritage to fit his continuing sensory promptings are, where rational, pragmatic.

Notes

1 See p. 9 of "On What There Is," in Quine 1953.

2 See p. 10 of "On What There Is."

3 See pp. 11f of "On What There Is" and pp. 107–15 of "Logic and the Reification of Universals," both in Quine 1953.

4 Carnap [1], pp. 9ff; [2], pp. 70ff.

5 According to an important variant sense of "definition," the relation preserved may be the weaker relation of mere agreement in reference; see p. 132 of "Notes on the Theory of Reference," in Quine 1953. But definition in this sense is better ignored in the present connection, being irrelevant to the question of synonymy.

6 Cf. Lewis [1], p. 373.

7 This is cognitive synonymy in a primary, broad sense. Carnap ([1], pp. 56ff) and Lewis ([2], pp. 83ff) have suggested how, once this notion is at hand, a narrower sense of cognitive synonymy which is preferable for some purposes can in turn be derived. But this special ramification of

concept-building lies aside from the present purposes and must not be confused with the broad sort of cognitive synonymy here concerned.

8 Pp. 81ff of "New Foundations for Mathematical Logic," in Quine 1953, contain a description of just such a language, except that there happens there to be just one predicate, the two-place predicate "ε".

9 See pp. 5–8 of "On What There Is"; also pp. 85f of "New Foundations for Mathematical Logic" and 166f of "Meaning and Existential Inference," all three in Quine 1953.

10 See p. 87 of "New Foundations . . ."

11 On such devices see also "Reference and Modality," in Quine 1953.

12 This is the substance of Quine [1], 121.

13 The "if and only if" itself is intended in the truth functional sense. See Carnap [3], p. 14.

14 The foregoing paragraph was not part of the present essay as originally published. It was prompted by Martin [see References] . . . .

15 The doctrine can indeed be formulated with terms rather than statements as the units. Thus Lewis describes the meaning of a term as "*a criterion in mind*, by reference to which one is able to apply or refuse to apply the expression in question in the case of presented, or imagined, things or situations" ([2], p. 133).—For an instructive account of the vicissitudes of the verification theory of meaning, centered however on the question of meaning*fulness* rather than synonymy and analyticity, see Hempel.

16 See p. 6 of "On What There Is."

17 This doctrine was well argued by Duhem, pp. 303–328. Or see Lowinger, pp. 132–140.

18 Cf. pp. 17f of "On What There Is."

19 Cf. p. 18 of "On What There Is."

20 "L'ontologie fait corps avec la science elle-même et ne peut en être separée." Meyerson, p. 439.

21 See pp. 12f of "On What There Is" and pp. 102 ff of "Logic and the Reification of Universals."

22 For an effective expression of further misgivings over this distinction, see White.

References

Carnap, Rudolf [1], *Meaning and Necessity* (Chicago: University of Chicago Press, 1947).

—— [2], *Logical Foundations of Probability* (Chicago: University of Chicago Press, 1950).

—— [3], "Empiricism, semantics, and ontology," *Revue internationale de philosophie* 4 (1950), 20–40. Reprinted in Linsky.

Duhem, Pierre, *La Théorie physique: son object et sa structure* (Paris, 1906).

Hempel, C. G. "Problems and changes in the empiricist criterion of meaning," *Revue internationale de philosophie 4* (1950), 41–63. Reprinted in Linsky.

Lewis, C. I. [1], *A Survey of Symbolic Logic* (Berkeley, 1918).

—— [2], *An Analysis of Knowledge and Valuation* (LaSalle, Ill.: Open Court, 1946).

Linsky, Leonard (ed.), *Semantics and the Philosophy of Language* (Urbana: University of Illinois Press, 1952).

Lowinger, Armand, *The Methodology of Pierre Duhem* (New York: Columbia University Press, 1941).

Martin, R. M., "On 'analytic,'" *Philosophical Studies 3* (1952), 42–47.

Meyerson, Émile, *Identité et réalité.* (Paris, 1908; 4th ed., 1932).

Quine, W. V. [1], *Mathematical Logic* (New York: Norton, 1940; Cambridge: Harvard University Press, 1947; rev. ed., Cambridge: Harvard University Press, 1951).

White, Morton, "The analytic and the synthetic: an untenable dualism," in Sidney Hook (ed.), *John Dewey: Philosopher of Science and Freedom* (New York: Dial Press, 1950), pp. 316–330. Reprinted in Linsky.

23

Larry Laudan and Jarrett Leplin, "Empirical Equivalence and Underdetermination"*

The authors' argument against the thesis of underdetermination is based on a careful consideration of how evidence bears on theory.

During this century, there emerged from the philosophical analysis of scientific theories two results invested with broad epistemological significance. By the 1920s, it was widely supposed that a perfectly general proof was available for the thesis that there are always empirically equivalent rivals to any successful theory. Secondly, by the 1940s and 1950s, it was thought that—in large part because of empirical equivalence—theory choice was radically underdetermined by any conceivable evidence. Whole theories of knowledge (e.g., W. V. Quine's[1]) have been constructed on the presumption that these results were sound; at the same time, fashionable recent repudiations of the epistemic project (e.g., Richard Rorty's) have been based on the assumption that these results are not only legitimate, but laden with broad implications for the theory of knowledge.

In this paper, we reject both the supposition of empirical equivalence and the inference from it to underdetermination. Not only is there no general guarantee of the possibility of empirically equivalent rivals to a given theory, but empirical equivalence itself is a problematic notion without safe application. Moreover, the empirical equivalence of a group of rival theories, should it obtain, would not by itself establish that they are underdetermined by the evidence. One of a number of empirically equivalent theories may be uniquely preferable on evidentially probative grounds. Having argued for these conclusions in the first two sections, respectively, we shall propose, in section III, a diagnosis of the difficulty that has impeded their recognition, and extract an attendant, positive moral for the prospects of epistemology.

L. Laudan and J. Leplin, "Empirical Equivalence and Underdetermination," *Journal of Philosophy*, 1991, 88: 449–72.

I. Problems with Empirical Equivalence

A. Inducements to Skepticism

The idea that theories can be empirically equivalent, that in fact there are indefinitely many equivalent alternatives to any theory, has wreaked havoc throughout twentieth-century philosophy. It motivates many forms of relativism, both ontological and epistemological, by supplying apparently irremediable pluralisms of belief and practice. It animates epistemic skepticism by apparently underwriting the thesis of underdetermination. In general, the supposed ability to supply an empirically equivalent rival to any theory, however well supported or tested, has been assumed sufficient to undermine our confidence in that theory and to reduce our preference for it to a status epistemically weaker than warranted assent.

Specifically, this supposed ability is the cornerstone of arguments for the inscrutability of reference and the indeterminacy of translation, which together insulate the epistemic agent by challenging the objectivity of criticism on which an entire philosophical culture has depended. It has spawned prominent, contemporary versions of empiricism, including those of Quine, Bas van Fraassen, and J. D. Sneed, which belie the promise of science to deliver theoretical knowledge. It encourages conventionalism in geometry through Hans Reichenbach's invocation of universal forces. It questions the possibility of ordinary knowledge of other minds through the contrivance of the inverted spectrum. It blocks inductive generalization through the stratagem of fashioning artificial universals to vie with natural kinds, as in Nelson Goodman's "grue" paradox, reducing the status of apparent laws to mere entrenchment.

The linkage between empirical equivalence and epistemic skepticism has roots that go back well beyond such contemporary manifestations. Hume reduced causal judgments to psychological projections of habit by offering coincidental concomitance as the empirically equivalent alternative to natural necessity. Descartes worried that, were our impressions illusory or demonically fabricated, what we take to be evidence would come out the same, inferring that empirical beliefs are unwarranted while those possibilities are open. Berkeley similarly exploited empirical equivalence to justify doubts about an external world. The implicit assumption in classical skepticism, operative still, is that no experience epistemically grounds a belief if that experience is strictly compatible with an alternative belief. Whence the method of undermining belief by constructing alternatives.

B. *An Argument against Empirical Equivalence*

We find the pervasiveness of this influence out of proportion to the conceptual credentials of the basic idea of empirical equivalence. By connecting three familiar and relatively uncontroversial theses, we can construct a simple argument to cast doubt on empirical equivalence in general, as a relation among scientific theories (and, by parity of reasoning, between any rival perspectives).

On the traditional view, theories are empirically equivalent just in case they have the same class of empirical, viz., observational, consequences.[2] A determination of empirical equivalence among theories therefore requires identifying their respective empirical consequence classes. As the empirical consequences of any statement are those of its logical consequences formulable in an observation language, these classes are (presumably proper) subsets of the logical consequence classes of theories. Central, therefore, to the standard notion of empirical equivalence are the notions of observational properties, the empirical consequences of a theory, and the logical consequences of a theory. We shall show that, when these concepts are properly understood, the doctrine of empirical equivalence loses all significance for epistemology.

Our three familiar theses are these:

Familiar thesis 1, the variability of the range of the observable (VRO):

> Any circumscription of the range of observable phenomena is relative to the state of scientific knowledge and the technological resources available for observation and detection.

In particular, entities or processes originally introduced by theory frequently achieve observable or "empirical" status as experimental methods and instruments of detection improve. Such variability applies to any viable distinction between observational and theoretical language.[3]

Familiar thesis 2, the need for auxiliaries in prediction (NAP):

> Theoretical hypotheses typically require supplementation by auxiliary or collateral information for the derivation of observable consequences.

While direct derivability of statements bearing evidentially on theory is not in principle precluded, auxiliaries are generally required for the derivation of epistemically significant results.[4]

Familiar thesis 3, the instability of auxiliary assumptions (IAA):

> Auxiliary information providing premises for the derivation of observational consequences from theory is unstable in two respects: it is defeasible and it is augmentable.

Auxiliary assumptions once sufficiently secure to be used as premises frequently come subsequently to be rejected, and new auxiliaries permitting the derivation of additional observational consequences frequently become available.

Our argument against empirical equivalence now proceeds as follows. As VRO makes clear, the decision to locate a logical consequence of a theory outside its empirical consequence class (on the grounds of the former's non-observational status) is subject to change. That class may increase, coming to incorporate an ever greater proportion of the theory's total consequence class.[5] This result already shows that findings of empirical equivalence are not reliably projectable, since we cannot reliably anticipate which of a theory's now unobservable consequences may become observable. But the problems with empirical equivalence run deeper than the inconstancy of the boundary of the observable. For even if it were possible to circumscribe the range of the observable relative to a state of science, we shall see that it would still be impossible so to circumscribe the range of auxiliary information available for use in deriving observational consequences.

By NAP, a theory's empirical consequence class must be allowed to include statements deducible from the theory only with the help of auxiliaries. One can distinguish the broad from the narrow class of a theory's empirical consequences, where the narrow class contains only observational statements implied by the theory in isolation from other theories and hypotheses. But NAP shows that it is the broad class, containing as well statements deducible only if the theory is conjoined with such auxiliaries, that matters epistemologically. Regardless of whether holists are right in contending that the narrow class is empty, it is a class of little epistemic moment. It is by the complement of the narrow with respect to the broad that theories are primarily tested, and a characterization of empirical equivalence limited to the narrow would have no such epistemological consequences as we are concerned to contest.

It follows by IAA that, apart from shifts in observational status, a theory's empirical consequence class may increase through augmentations to the theory's total consequence class. As new auxiliary information becomes available, new empirical consequences derived with its help are added. Of course, conditionals connecting the auxiliary statements newly used to the empirical statements newly derived were already present among the theory's logical consequences. But the detached empirical statements are not present until the auxiliaries on which their deducibility depends become available. So long as we include within a theory's empirical consequence class statements derivable from the theory only via auxiliaries, so long as we construe that class *broadly*—and we have argued that it must be so construed to reflect the realities of theory testing—the theory's logical consequence class will be augmentable in virtue of containing the empirical consequence class as a subset. The empirical consequence class can also diminish, again by IAA, as the rejection of needed auxiliaries discontinues the derivability of some of its members. Therefore, any determination of the empirical consequence class of a theory must be relativized to a particular state of science. We infer that empirical equivalence itself must be so relativized, and, accordingly, that any finding of empirical equivalence is both contextual and defeasible.

This contextuality shows that determinations of empirical equivalence are not a purely formal, a priori matter, but must defer, in part, to scientific practice. It undercuts any formalistic program to delimit the scope of scientific knowledge by reason of empirical equivalence, thereby defeating the epistemically otiose morals that empirical equivalence has been made to serve. The limitations on theoretical understanding that a defeasible empirical equivalence imposes need not be grievous. Nevertheless, we think there is still less to the notion of empirical equivalence than survives these concessions.

It has been widely supposed that one can, utilizing the resources of logic and semantics alone, "read off" the observable consequences of a theory. The mobility of the boundary of the observable has been regarded as an inconvenience, not as a fundamental challenge to the idea that, at least in principle, the consequences or content of a theory are unambiguously identifiable. On this view, enhancements to our observational repertoire do nothing to alter a theory's semantics; rather, they merely shift the line, within the class of a theory's logical consequences, between observational and nonobservable consequences. But NAP shows that there is an epistemic question here quite distinct from the logico-semantic one. Specifically, the availability of auxiliaries—auxiliaries crucial for determining what a theory's empirical consequences are—is neither a matter of logic nor semantics; it is inescapably epistemic. The determination that a given empirical statement, *e*, is an empirical consequence of a particular theory, T, depends on whether there are epistemically well-grounded collateral hypotheses that establish a suitable inferential link between T and *e*. The availability of such hypotheses is clearly a matter of evidential warrant. Once statements whose derivability requires auxiliaries are allowed to count as consequences, as they must by NAP, no statement can be disallowed as a (broad) consequence of a theory unless some statements are disallowed as auxiliaries. So, before deciding whether a derivation of an observation statement from a theory plus auxiliaries qualifies the statement for inclusion in the theory's empirical consequence class, we must assess the epistemic standing of the auxiliaries.

This makes clear that the epistemic bearing of evidence on theory is not a purely logical relation, but is subject to reinterpretation as science grows and may be indeterminate at a particular point in the process of growth. How well supported an auxiliary is by evidence available *now* may depend on findings made *later*—a problem exacerbated by the fact that standards of evidential support themselves are transformable by the fortunes of empirical beliefs.[6] This fact darkens the prospects even for time-indexed delineations of empirical consequence classes.

C. *Response to Anticipated Objections*

The response we anticipate to our argument is a challenge to its assumption that empirical consequence classes must be identified for their equivalence to be established. Can there not be a general argument to show that classes must be the same independently of determining their membership? An obvious suggestion is that logically or conceptually equivalent theories must have the same consequence class, whatever that class is. As we do not question the empirical equivalence of logically equivalent theories, we ignore this suggestion and assume henceforth that theories whose empirical equivalence is at issue are logically and conceptually distinct.

One approach to constructing a general argument is to invoke the Lowenheim–Skolem Theorem. This theorem asserts that any first-order, formal theory that has a model at all has a denumerable model. A standard proof uses terms involving individual constants indexed by the natural numbers as the domain of a model. But if the domain need only be a set of terms, it could just as well be any denumerable set whose members are proposed as the referents of those terms. So, in principle, such a theory has an infinite number of models.

The qualification "in principle" must be emphasized here, because there is no guarantee that the denumerable models of a consistent, formal theory are effectively constructable. The proof of the theorem relies on the completability of a consistent first-order theory, and the complete theory need not be axiomatic even if the theory it completes is axiomatic. Thus, any appeal to the Lowenheim–Skolem Theorem on behalf of empirical equivalence shows at most that equivalent theories exist in principle; it does not show that they are entertainable as alternatives.

But there is a more fundamental objection to the relevance of the theorem. Having multiple models of a formal theory does not mean having multiple theories of common empirical content. A physical theory, by virtue of being a physical theory, includes a semantic interpretation of its formal structure;

it is not simply a formal structure variously interpretable. A physical theory is inherently at least purportedly referential. Its referents may turn out not to exist, if nature fails to cooperate. But what its referents are if it has them is fixed by the theory itself; it is not a matter of optional interpretation. The reference-fixing devices of physical theory are rich and various. It does not require a philosophical theory of reference fixing to make the point that the semantic resources of physical theories transcend syntactic requirements on first-order formal theories. If, given the Lowenheim–Skolem Theorem, formal statements in first-order logic are referentially indeterminate, then a physical theory is not simply a set of formal statements in first-order logic.

Another approach is to construct an algorithm for generating empirical equivalents to a given physical theory, such as the Lowenheim–Skolem Theorem fails to do for formal theories. For example, there exist instrumentalist algorithms for excising the theoretical terms of a theory without empirical loss. Whether such algorithms are in fact successful is rendered highly dubious by the premises of our argument. It is by no means clear that a theory's instrumentalized version can match its capacity for empirical commitment, once the role of auxiliaries in fixing such commitment and the variability of the range of the observable are acknowledged. At most a theory's instrumentalized version can be held empirically equivalent to it relative to a circumscription of the observable and a presumed or intended domain of application. But while theories fix their own intended interpretations, they do not fix their own domains of application, nor the resources for detection of entities they posit. Algorithmically excised references may pick out entities that become detectable. New applications may arise with changes in collateral knowledge. Indeed, it is a measure of a theory's success when posited entities acquire a technological role, and applications for which the theory was not designed become possible.[7]

Be that as it may, what application of an instrumentalist algorithm to a theory produces is manifestly not an alternative *theory*. That is, the algorithm does not produce a rival representation of the world from which the same empirical phenomena may be explained and predicted. On the contrary, a theory's instrumentalized version posits nothing not posited by the theory, and its explanations, if any, of empirical phenomena deducible from it are wholly parasitic on the theory's own explanations. A theory's instrumentalized version cannot be a rival to it, because it is a logical consequence of the theory and is bound to be endorsed by anyone endorsing the theory. The challenge the instrumentalist poses is to justify endorsing more than the instrumentalized version, not to justify endorsing something instead of it. The point of the instrumentalist move is not to challenge the epistemic status of theory by demonstrating underdetermination among alternatives, but to argue that empirical evidence underdetermines theories individually by failing to discriminate between them and their instrumentalized versions. We know of no algorithm for generating genuine theoretical competitors to a given theory.[8]

The only other approach we know of to establishing empirical equivalence without identifying empirical content is to argue from cases. We propose an example, inspired by van Fraassen's in *The Scientific Image*, as representative. Let TN be Newtonian theory. Let R be the hypothesis that the center of mass of the expanding universe is at rest in absolute space. Let V be the hypothesis that the center of mass of the universe has constant absolute velocity *v*. Consider the claim that TN + R is empirically equivalent to TN + V.

This claim is based on the common TN component of the theories. It is Newtonian theory itself that assures us that unaccelerated absolute motion has no empirical consequences of a kind encompassed by the theory; that is, no consequences within mechanics. We can therefore bring two lines of criticism against the claim of empirical equivalence: either there is some other kind of consequence not envisioned within mechanics, or the underlying Newtonian assurance is wrong. The question is whether conceivable developments in scientific knowledge enable us to distinguish the theories empirically on one of these bases.

Van Fraassen's defense of his claim of empirical equivalence does not take account of these possibilities. He considers only extensions of the theories to further, late-breaking mechanical phenomena, arguing that whatever these may be, further equivalent extensions are constructable (op. cit., p. 47ff). Can a defense be provided to cover the possibilities that van Fraassen neglects?

Imagine that TN + V has nonmechanical consequences absent from TN + R. A new particle, "the bason," is independently hypothesized to arise with absolute motion. The positive absolute velocity of the universe represents energy available for bason creation, and basons will appear under certain conditions in principle realizable in the laboratory if TN + V is correct, but not if TN + R is correct. *v* can be measured by counting basons, so that variants of TN + V for different *v* are empirically distinguishable.

We can construct an extension of TN + V which agrees with TN + R in not predicting basons. Let TN + W be TN + V plus the hypothesis that there is a velocity *w* such that basons appear if and only if and to the extent that $v > w$. Then the absence of basons establishes only that *v* does not exceed *w*; it does not require R. The presence of basons still refutes TN + R, but TN + R can be supplemented to allow basons; perhaps they arise spontaneously. TN + R then lacks an explanation of bason production, such as TN + V provides. Something in the way of explanatory parity is achievable by adding to TN + R the hypothesis that what absolute motion produces is antibasons, which immediately annihilate basons. So the presence of basons is explained by the lack of absolute velocity. Still, TN + R does not explain the frequency of bason detection, as TN + V does. The observed frequency must simply be posited, as a constant determined by experiment, and this procedure is an admitted disadvantage relative to TN + V. But this comparison does not affect empirical equivalence.

The appeal to nonmechanical, differentiating phenomena can be defeated, because, if empirical equivalence holds within mechanics, it continues to hold for any extensions of mechanics in which the presence or absence of additional, nonmechanical phenomena is made to depend on the value of a mechanical property. This seems to be a general result. If theories T_1 and T_2 are equivalent with respect to properties $p_1, \ldots, p_n$, they have equivalent extensions for any enlarged class of properties $p_1, \ldots, p_n, q_1, \ldots, q_m$; where properties $q_1, \ldots, q_m$ are functions of $p_1, \ldots, p_n$. On the other hand, if $q_1, \ldots, q_m$ are not functions of $p_1, \ldots, q_n$, they cannot be used to discriminate between T_1 and T_2.

It remains to consider the common core TN of the allegedly equivalent theories of the example. Here our argument finds its proper vindication. The point to make is that this core, and with it the original rationale for regarding TN + R and TN + V as equivalent in the first place (that is, unextended to nonmechanical properties) is defeasible. We may have good or sufficient reason to regard theories as empirically equivalent, but there is no guarantee. That concession is all our argument requires. We do not deny the possibility that the world is such that equally viable, incompatible theories of it are possible. We do not deny the possibility of the world's being unamenable to epistemic investigation and adjudication, beyond a certain level. But whether or not the world is like that is itself an empirical question open to investigation. The answer cannot be preordained by a transcendent, epistemic skepticism.

It is noteworthy that contrived examples alleging empirical equivalence always invoke the relativity of motion; it is the impossibility of distinguishing apparent from absolute motion to which they owe their plausibility. This is also the problem in the pre-eminent historical examples, the competition between Ptolemy and Copernicus, which created the idea of empirical equivalence in the first place, and that between Einstein and H. A. Lorentz.[9] Either the relativity of motion is a physical discovery, founded on such evidence as the empirical success of Newtonian or relativity theory, or it is guaranteed conceptually. Both views have been advanced historically. Our reply to alleged examples of empirical equivalence takes the former view. We maintain not that there are no cases of empirical equivalence, but that the claim that there are is defeasible.[10] If the latter view is to be assumed, then we return to the caveat that we are not concerned to contest the empirical equivalence of conceptually equivalent theories.

II. Underdetermination

We have argued that the thesis that every empirically successful theory has empirically equivalent counterparts is precarious, at best. But for now let us suspend our incredulity about empirical equivalence and suppose that the thesis is sound. We wish to explore in this section what, if anything, then follows from the existence of empirically equivalent theories for general epistemology.

A number of deep epistemic implications, roughly collectable under the notion of "underdetermination," have been alleged for empirical equivalence. For instance, it is typical of recent empiricism to hold that evidence bearing on a theory, however broad and supportive, is impotent to single out that theory for acceptance, because of the availability or possibility of equally supported rivals. Instrumentalists argue that the existence of theoretically noncommittal equivalents for theories positing unobservable entities establishes the epistemic impropriety of deep-structure theorizing, and with it the failure of scientific realism. Some pragmatists infer that only non-epistemic dimensions of appraisal are applicable to theories, and that, accordingly, theory endorsement is not exclusive nor, necessarily, even preferential. One may pick and choose freely among theories whatever works for the problems at hand, so that the distinction between theories and models is lost. In a phrase, the thesis of underdetermination, denying the possibility of adequate evidential warrant for any theory, has become the epistemic corollary to the presumptively semantic thesis of empirical equivalence.

Against these positions, we shall argue that underdetermination does not in general obtain, not even under conditions of empirical equivalence. As we have seen, empirical equivalence is chiefly seen as a thesis about the *semantics* of theories; underdetermination, by contrast, is a thesis about the *epistemology* of theories. It has been supposed that, if theories possess the same empirical consequences, then they will inevitably be equally well (or ill) supported by those instances. We shall contest this supposition and, with it, the reduction of evidential relations to semantic relations, on which it rests. We dispute the ability of semantic considerations to resolve epistemic issues. But even allowing the epistemic dimension we have discerned in empirical equivalence, we shall find that the relative degree of evidential support for theories is not fixed by their empirical equivalence.

Specifically, we shall show, first, that significant evidential support may be provided for a theory by results that are not empirical consequences of the theory;[11] secondly, we shall show that (even) true empirical consequences need lend no evidential support to a theory. By this dual strategy, we propose to sever the presumed link between supporting instances and empirical consequences.[12] We shall show that being an empirical consequence of a hypothesis is neither necessary nor sufficient for being evidentially relevant to a hypothesis. These conclusions will establish that theories identical as to empirical consequences may be differentially supported, such that one is epistemically preferable to the other.

A. *Evidential Results that are not Consequences*

We begin by noting that instances of a generalization may evidentially support one another, although they are not consequences of one another. Previous sightings of black crows support the hypothesis that the next crow to be sighted will be black, although that hypothesis implies nothing about other crows. Supposing this evidential connection to be uncontroversial, we ask why, then, in the case of universal statements it should be supposed that evidential support is limited to logical consequences.

Is it that the evidential connection admitted to hold among singular statements is at best indirect, that it connects those statements only via a general statement that they instantiate? The thesis would then be that *direct* evidential support for a statement is limited to its logical consequences, and singular statements instantiating the same generalization support one another only in virtue of directly supporting that generalization. In short, where there appears to be evidential support for a statement, *s*, outside the range of *s*'s logical consequences, such support is parasitic on support of a general statement, *m*, which entails *s*, from *m*'s logical consequences.

We believe this to be an unperspicacious way of accounting for what goes on in singular inference. Often the evidential link between singular statements is stronger than the support available for a general intermediary, whose identification can, in any case, prove elusive. But even if this account worked, it should be noted straightaway that allowing a statement to accrue indirect empirical support in this fashion already undermines the claim that statements are confirmable only by their empirical consequences. This result alone suffices to establish that the class of empirical consequences of a statement and the class of its prospective confirming instances are distinct.

We began this discussion with the hackneyed case of black crows in order to show that the possibility of inferences of even the most mundane sort (from particular-to-particular) depend upon denying the thesis that evidential support accrues to a statement only via its positive instances. This claim becomes even clearer when one considers the manner in which real scientific theories garner empirical support. Consider, for instance, the theory of continental drift. It holds that every region of the earth's surface has occupied both latitudes and longitudes significantly different from those it now occupies. It is thereby committed to two general hypotheses:

H_1: There has been significant climatic variation throughout the earth, the current climate of all regions differing from their climates in former times.

H_2: The current alignment with the earth's magnetic pole of the magnetism of iron-bearing rock in any given region of the earth differs significantly from the alignment of the region's magnetic rocks from earlier periods.

During the 1950s and 1960s, impressive evidence from studies of remnant magnetism accumulated for H_2. Clearly, those data support H_1 as well, despite the fact that they are not consequences of H_1. Rather, by supporting H_2 they confirm the general drift theory, and thereby its consequence H_1.

Similar examples are readily adduced. Brownian motion supported the atomic theory—indeed, it was generally taken to demonstrate the existence of atoms—by being shown to support statistical mechanics. But, of course, Brownian motion is no consequence of atomic theory. The increase of mass with velocity, when achieved technologically in the 1920s, supported kinematic laws of relativity which to that point continued to be regarded with great suspicion. J. J. Thomson's cathode ray experiments in the 1890s were important evidence for a host of theoretical hypotheses about electricity that depended on Lorentz's electroatomism. Phenomena of heat radiation were used by J. C. Maxwell in the 1870s to support the kinetic molecular theory, which did not address the transmission of heat energy across the space intervening between bodies. The emergence in the 1920s of evidence showing heritable variation supported Darwin's hypothesis about the antiquity of the earth, although that hypothesis entailed nothing about biological variation. Contemporary observational astronomy is replete with indirect methods of calculating stellar distances, whereby general hypotheses in cosmology acquire support from facts they do not imply about internal compositions of stars.

A number of points are to be noted about such examples. First, by dating them we emphasize that they are not dismissable by invoking auxiliaries via which the evidence is derivable. One could not in the 1890s represent Thomson's results as consequences of electrical laws by making electroatomism an auxiliary. Despite Ludwig Boltzmann's pioneering work, statistical mechanics was too speculative in 1905 to qualify as an available auxiliary. Even taking an ahistorical view, it would be casuistical to represent evidence as a consequence of a hypothesis from which it is derivable via auxiliaries, if it is the auxiliaries rather than the hypothesis that really fuel the derivation. If a formal criterion is wanted, we may stipulate that a hypothesis be ineliminable from the derivation of what are to qualify as its consequences.[13]

Second, the more general theory via which the evidence supports a hypothesis of which it is not a consequence need not be very precise or specific. For example, the statistical mechanics that Brownian motion supported was more a program for interpreting phenomenological thermodynamics probabilistically than a developed theory. There can be good reason to believe that conceptually dissimilar hypotheses are related such that evidence for one supports the other, without possessing a well worked out or independently viable theory that connects them. Perhaps a theory that connected them has been discredited without the connection it effected being discredited. In this respect, nonconsequential evidence for general statements approximates the case of singular statements for which the inferential link proved elusive.

Third, we need not fear running afoul of familiar paradoxes of confirmation in taking evidence to confirm a hypothesis in virtue of supporting a more general statement that implies the hypothesis. The intuition that what increases our confidence in a statement thereby increases our confidence in what that statement entails is fundamentally sound. The difficulties that Carl Hempel, for example, extracted from his "special consequence condition" depend on a certain logical form for general laws and a simplistic criterion of confirmation—Nicod's criterion—which requires, in opposition to the position we have undertaken to defend (see section II.B), that all positive consequences be confirming. Much sophisticated reasoning in the natural sciences would be vitiated by restricting evidence relevant in assessing a theory to the entailments (via auxiliaries) of the theory. And any singular prediction would be so vitiated as well.

Finally, we need to acknowledge and take into account a subtlety of confirmation that might appear to challenge the force of nonconsequential evidence for our argument. There is an obvious way in which a statement not entailed by a theory can be evidence for the theory. The statement might imply another empirical statement that *is* entailed. Suppose, for example, that the theory entails a—perhaps indefinitely extendable—disjunction of which the statement is a disjunct. By implying a statement that is a consequence, the evidence, though not itself a consequence, fails to discriminate between the theory and any empirically equivalent theory. So showing that there can be evidence for a theory that is not a consequence of the theory does not suffice to show that empirically equivalent theories can be differentially supported.

Our examples, however, do not pose this problem. An example that does is to cite ten successive heads in tosses of a coin as evidence that the coin is biased. The hypothesis of bias does not entail the results of any given number of trials. But it does entail that something such as an improbably extended succession of heads will result over many trials. A hypothesis empirically equivalent to that of bias, were there one, would entail the same thing, and thereby be supported by any evidence thus exemplifying bias. On the other hand, the hypothesis of bias readily admits of evidential support from sources outside its consequence class that would not support purportedly equivalent hypotheses. An example is the information that the coin hypothesized to be biased was poured in a die cut by a chronically inebriated diemaker.

What, then, *is* the connection that we claim our examples to establish between nonconsequential evidence and differential support of empirically equivalent theories? We propose the following exemplar. Theoretical hypotheses H_1 and H_2 are empirically equivalent but conceptually distinct. H_1, but not H_2, is derivable from a more general theory T, which also entails another hypothesis H. An empirical consequence *e* of H is obtained. *e* supports H and thereby T. Thus, *e* provides indirect evidential warrant for H_1, of which it is not a consequence, without affecting the credentials of H_2. Thus, one of two empirically equivalent hypotheses or theories can be evidentially supported to the exclusion of the other by being incorporated into an independently supported, more general theory that does not support the other, although it does predict all the empirical consequences of the other. The view that assimilates evidence to consequences cannot, on pain of incoherence, accept the intuitive, uncontroversial principle that evidential support flows "downward" across the entailment relation.

We stress, however, that this is *only* an exemplar. There are modes of nonconsequential empirical support in science that do not invoke intermediate theories or generalizations. This has already been seen in the case of singular inference, and is present as well in the example from astronomy. Another type of example is the use of analogical reasoning. Analogical reasoning is often motivational, pertaining more to the heuristics of theory development than to confirmation. But sophisticated analogies can be evidentially probative.

Maxwell analogized a closed system of elastic particles to a contained gas, whereby the mathematical theory of collisions together with the observed properties of gases supports a molecular structure for gas. The power of the analogy to yield important, known properties of gases makes it reasonable to infer, Maxwell[14] thinks, that "minute parts" of gases are in rapid motion. Einstein supported his hypothesis of a quantum structure for radiation by analogizing the entropy of monochromatic radiation to the entropy of an ideal gas. He showed that a decrease in entropy corresponding to a reduction in volume for radiation of fixed energy has the same functional form as the decrease in entropy associated with a reduction in volume for an ideal gas. Evidence that warrants treating the gas statistically thereby supports a quantum structure for the gas. Yet such evidence sustains no logical relation to Einstein's hypothesis, nor does it support any theory that entails Einstein's hypothesis.

The many examples we have adduced may seem too obvious and frequent to dwell on. We welcome that reaction, but admonish many influential epistemological positions of the last half century for ignoring them. Karl Popper, for instance, presumes throughout his treatment of scientific methodology that evidence potentially relevant to the assessment of a hypothesis must come from its empirical consequence class. And many epistemologists follow Quine in supposing that the empirical equivalence of rival hypotheses renders inevitable their epistemic parity, reducing a choice among them to purely pragmatic considerations. In defiance of much celebrated epistemology, we claim that our examples establish that theories with exactly the same empirical consequences may admit of differing degrees of evidential support.

B. *Empirical Consequences that are not Evidential*

Establishing that evidential results need not be consequences is already enough to block the inference from empirical equivalence to underdetermination. But it is instructive to make the converse point as well. Suppose a televangelist recommends regular reading of scripture to induce puberty in young males. As evidence for his hypothesis (H) that such readings are efficacious, he cites a longitudinal study of 1000 males in Lynchburg, Virginia, who from the age of seven years were forced to read scripture for nine years. Medical examinations after nine years definitively established that all the subjects were pubescent by age sixteen. The putatively evidential statements supplied by the examinations are positive instances of H. But no one other than a resident of Lynchburg, or the like-minded, is likely to grant that the results support H.

This example has a self-serving aspect. That the televangelist has a pro-attitude toward H on grounds independent of the purported evidence he cites is already enough to make one wary; one need not recognize the flaws in the experimental design of the longitudinal study. In a case without this feature, a person hypothesizes that coffee is effective as a remedy for the common cold, having been convinced by finding that colds dissipate after several days of drinking coffee. The point here is that the very idea of experimental controls arises only because we recognize independently that empirical consequences need not be evidential; we recognize independently the need for additional conditions on evidence.

No philosopher of science is willing to grant evidential status to a result *e* with respect to a hypothesis H just because *e* is a consequence of H. That is the point of two centuries of debate over such issues as the independence of *e*, the purpose for which H was introduced, the additional uses to which H may be put, the relation of H to other theories, and so forth.

Results that test a theory and results that are obtainable as empirical consequences of the theory constitute partially nonoverlapping sets. Being an empirical consequence of a theory is neither necessary nor sufficient to qualify a statement as providing evidential support for the theory. Because of this, it is illegitimate to infer from the empirical equivalence of theories that they will fare equally in the face of any possible or conceivable evidence. The thesis of underdetermination, at least in so far as it is founded on presumptions about the possibility of empirical equivalence for theories—or "systems of the world"—stands refuted.

III. Formal Constraints on Epistemology

If the identification of empirical consequences with evidential support is so implausible, how has it managed to gain such a foothold? We suggest that a more persuasive, less readily dispelled confusion is ultimately responsible. That confusion, as we have intimated, is to misunderstand the relationship between semantics and epistemology, bringing the largely technical and formal machinery of semantics improperly to bear on epistemic issues.

Specifically, we wish to reveal and challenge the widespread—if usually implicit—conviction that epistemic relations are reducible to semantic relations. It is commonly supposed either that truth and meaning conditions just *are* justification conditions, or, at least, that they can be made to double as justification conditions.[15] Either way, epistemology is made the poor relation of a family of interconnections among semantic, syntactic, and epistemic concepts, and is left to make do with tools handed down from semantics. It seems to us that distinctively epistemic issues are left unresolved by such a presumed reduction, and that epistemic theses depending on it—such as the underdetermination thesis—are wrongheaded. We will first explain and illustrate the confusion we have diagnosed, then trace the mistaken assimilation of support to empirical consequences to it.

The problem originates in foundationalist epistemology—especially in Descartes's image of a mathematically rigorous, deductive structure for knowledge—and thus is not confined to empiricism. If the evidential relation is deductive, the evidence on which a knowledge claim is based must bear semantic relations to the claim sufficient to permit the deduction.

Perhaps the best modern illustration is the attempt by the logical empiricists to demarcate science by semantic means. Both "verifiability" and "falsifiability," the prime concepts in terms of which scientific status has been delimited, are tools of semantics. Demarcation criteria proposed by the Vienna Circle or its positivist disciples, and by the Popperians, are alike in depending on semantic analysis and syntactic form of statements. What is required for classification as verifiable or falsifiable is basically that a statement satisfy constraints as to logical form and be couchable in observation language. It was assumed on all sides that such conditions suffice to identify the class of statements that are properly the objects of scientific inquiry.

The problem that demarcation criteria were intended and offered to solve, however, was the epistemic one of judging the reasonableness of belief. What was called for was a distinction, impossible within semantics alone, between statements that are well-founded and those that are not. Whether one's target was the metaphysics, religion, or ethics that exercised the positivist, or the psychoanalysis, Marxism, or astrology that exercised the Popperian, it was the irrationality of credence that the target was to be convicted of, not impropriety of logical form or visual inaccessibility of subject matter.

One might think to defend the adequacy of semantic tools for the intended distinction by arguing that the relevant notion of "science" to be demarcated is not that of what passes muster by scientific standards, but merely that of what is up for grabs in scientific inquiry. After all, statements falsified by scientific inquiry are yet to be classed as scientific. But not only is a distinction between what *qualifies* as scientific and what does not basically epistemic; so too is a distinction between what is *worthy* of investigation or *entertainable* by scientific means and what is not. It is basically what we have already found it reasonable to believe that decides these things.

The demarcation problem of the logical empiricists arose as a variant on the logical positivists' program for distinguishing cognitive significance from emotive uses of language misleadingly given propositional form. Already at this level one may discern the assimilation of evidential to semantic relations. For the evaluative force of, e.g., ethical pronouncements that led positivists to disqualify them as genuine propositions is also present in epistemic pronouncements, and, derivatively, in science. Epistemology, it is now commonly recognized, is value-laden.[16] But science was the logical positivists' paradigm of cognitive significance; its propositional status, the ideal to which ethics, religion, and metaphysics futilely aspired. If epistemology, and science in particular, was to be salvaged, then epistemic evaluation would have to rest on semantic relations as the only factual alternative to value-free empirical relations.

Accounts of scientific explanation and confirmation proposed in the 1940s and 1950s exhibit the same priority of the semantic over the epistemic. Characteristically, these accounts dealt incidentally if at all with epistemic and pragmatic dimensions of confirmation and explanation, in favor of their syntax, logical structure, and semantics. Significantly, truth enters as a precondition of explanatory status, rather than as an attribute that it is rational to adduce (partly) in measure that explanations are achieved.

What made such approaches seem plausible, we suggest, was a linguistic view of conceptual analysis. To analyze a concept, one examined its use in language. To understand instances of its use in language, one identified the truth conditions for sentences. In particular, to analyze knowledge one identified the truth conditions for attributions of knowledge; for sentences of the form " . . . knows . . . " It turns out, however, that among the truth conditions for such a sentence is the truth of another sentence, that disquoted in the second position of the schema. It therefore looks like the semantic concept of truth is logically prior to the epistemic concept of knowledge.

Of course, this does not make the semantic concept sufficient for the epistemic one, but further developments tended to elevate semantics and syntax over the notions of evidential warrant and rationality of belief used in other truth conditions for knowledge attributions. The incompleteness of the list of truth conditions was manifested in a curious asymmetry between the truth and evidence conditions. If the truth condition is not met, no bolstering of the evidence is sufficient for knowing. But inconclusive evidence that leaves open the possibility of error can be sufficient for knowing, if only, as a matter of fact (or happenstance), the world cooperates. In many celebrated paradigms of knowing, the evidence needed does not seem all that strong.[17] Thus, attention focused more on the truth condition than the evidence condition—more on semantic than epistemic issues. Ironically, the recent emergence of reliability theory, which re-emphasizes the justificatory component of knowledge in the tradition of Gettier's challenge, underscores the paucity and defeasibility of the evidence on which ordinary knowledge relies. Add to this asymmetry the success of Tarski's theory of truth in contrast to the sorry state of theories of evidential warrant, and one has the makings of a semantic and syntactic orientation for epistemology.

Given this orientation, it was natural to approach the problem of warranting a hypothesis—the problem of testing—by attending to statements that bear syntactic and semantic relations to the hypothesis—to its instantiations. At least this was natural for empirical generalizations, whose instantiations are empirical statements. This approach then created so many internal problems and tasks—Hempel's paradoxes of confirmation across logical relations; Goodman's problem of projectability—that the possibility of warrant provided by statements syntactically and semantically independent of the hypothesis was lost sight of. Instantiations of theoretical hypotheses are not empirical, but an assimilation of support to consequences was somehow extrapolated for them, by supposing them in principle recastable in observational terms or, perhaps, by supposing their testability reducible to the testability of empirical generalizations. Such was the hold of the resulting picture, that the assimilation of support to consequences exceeded the confines of logical empiricism to capture the format of textbook characterizations of scientific method itself. Although written by a philosopher, Hempel, the following passage will strike every reader as stereotypical of standard accounts of empirical inquiry:

> First, from the hypothesis under test, suitable other statements are inferred which describe certain directly observable phenomena that should be found to occur under specifiable circumstances if the hypothesis is true; then those inferred statements are tested directly, i.e., by checking whether the specified phenomena do in fact occur; finally, the proposed hypothesis is accepted or rejected in the light of the outcomes of those tests.[18]

This is a pure and simple statement of a view that deserves to be called "consequentialism," viz., the thesis that hypotheses are to be tested exclusively by an exploration of the truth status of those empirically decidable statements which they entail. Consequentialism is closely connected historically with the ancient idea of "saving the phenomena," that is, of reconciling theory with aberrant phenomena by introducing auxiliaries permitting those phenomena to be derived. Although not the invention of empiricists, it held a special appeal for them in appearing to solve problems of meaning, truth, and justification in one go. Meaning conditions, truth conditions, and justification conditions became substantially the same thing. The pre-Tarskian slogan, "To know what *x* means is to know how to test *x*," became, after Tarski, "The truth conditions for *x* specify simultaneously the meaning of *x* and the test conditions for *x*."[19] And this cast of mind, we suggest, is behind the supposition that statements with the same empirical consequences must be on an equivalent epistemological footing.

It is remarkable, upon reflection, how much freight the entailment relation has been made to carry in recent philosophy. The early logical positivists identified the meaning of a statement with what that statement entailed in the observation language. The logical empiricists and Popper sought to demarcate scientific from nonscientific statements solely by whether or not they entailed observational statements (Rudolf Carnap) or negations of observation statements (Popper). The Tarskian account of truth held that a theory was true just in case every statement constituting its logical content was true, and in practice a statement's logical content was to be fixed by its entailments.

In twentieth-century epistemology, we find the notorious Nicod criterion, discussed at length by Hempel and other confirmation theorists of the 1950s and 1960s, presuming to define the evidential relation exclusively in terms of positive and negative instances of the hypothesis, where positive instances instantiate the hypothesis and negative instances instantiate its negation. Popper's influential account of theory testing held that genuine tests of a theory had to be drawn from statements in its empirical content class, a class defined as the set of empirical statements whose negations were entailed by the theory. Currently, van Fraassen's constructive empiricism defines its key evaluative concept—empirical adequacy—in terms of the observational structures that model a theory:

> . . . a theory is empirically adequate exactly if what it says about the observable things and events in this world, is true—exactly, if it "saves the phenomena" (op. cit., p. 12).

And that a structure models a theory is determined not, as we have remarked, definitionally, but by reference to the theory's entailments. Clark Glymour, an avowed scientific realist, develops a "boot-strapping" story about theory testing, which insists that the statements capable of testing a hypothesis imply its empirical consequences. And Quine repeatedly avers that the only central rule of scientific method is hypothetico-deduction.

This ubiquitous assimilation of a theory's test cases to its logical consequences in an observation language, as we have argued above, wrongly ignores some of the more salient ways of testing theories. Worse, it generously greases the slide from empirical equivalence to underdetermination and epistemic parity. Ironically, the limitation of a statement's justification conditions to its truth conditions represents a striking break with the traditional empiricist project. Prior to the emergence of neopositivism in the 1920s, the general idea about theory testing and evaluation was that there was a range of "phenomena" for which any theory in a particular field was epistemically accountable. (In planetary astronomy, for example, these phenomena would be observations of positions of the planets, sun, and moon.) A theory's success or failure was measured against these phenomena, and decided by the theory's ability to give an account of them. A theory was, of course, responsible for its entailments, but it was held equally accountable for all the relevant, established phenomena, and could not evade this responsibility by failing to address them. For a Newton, a Ptolemy, or a Mach, "saving the phenomena" meant being able to explain all the salient facts in the relevant domain.[20]

With the rise of neopositivism, the epistemic responsibilities of theories were radically reinterpreted. Theories became liable only for what they entailed. Failure to address relevant phenomena, or at least to be indirectly applicable to them, now emerges as a cheap way of protecting such success as a theory does achieve, rather than as a liability. Where empirical adequacy formerly meant the ability to explain and predict all the salient phenomena, it now requires only possession of none but true empirical consequences. Recall the passage lately quoted from van Fraassen. The radical character of the shift we are describing becomes immediately clear there when one notes his identification of "empirical adequacy," saying only true things about observable features of the world, and "saving the phenomena." Prior to our time, no one would have supposed, as does van Fraassen, that saving the phenomena amounts only to possessing an observable model. No one would have supposed, as does van Fraassen, that a theory is to be judged only against the correctness of its own observational commitments (be those commitments expressed in model-theoretic or propositional form), irrespective of the comprehensiveness of the class of such commitments, irrespective of the theory's applicability to problems independently raised. It is testimony to the pervasiveness of the thesis that epistemic assessment is reducible to semantics that van Fraassen's conflation of the hitherto quite disparate notions of empirical adequacy and saving the phenomena has gone unnoted.

Much epistemology in our day is arbitrarily and unreasonably constrained by these developments. Our concluding, positive moral is that epistemic warrant unfettered by semantics has rich and varied sources yet to be exploited.

Notes

* The authors wish to thank Bas van Fraassen, Alan Goldman, William Lycan, Jim Maffie, David Stump, and Mary Tiles for criticism.

1 The holistic theses of Pierre Duhem and Quine are probably the major instigators of this line of thought, as it appears in contemporary epistemology and philosophy of science. Quine's doctrines of the inscrutability of reference and the indeterminacy of translation are forms of underdetermination that depend on holism. See, e.g., "Epistemology Naturalized," *Ontological Relativity and Other Essays* (New York: Columbia, 1969), pp. 80ff.

The antirealist arguments of Arthur Fine and van Fraassen depend, in turn, on the thesis of underdetermination. See van Fraassen, *The Scientific Image* (New York: Oxford, 1980), ch. 3; and Fine, "Unnatural Attitudes: Realist and Instrumentalist Attachments to Science," *Mind*, XCV (April 1986): 149–79.

In the nineteenth century, both J. S. Mill and W. Whewell treated the possibility of empirical equivalence as an obstacle to scientific knowledge, differing as to whether and how it could be overcome.

2 Empirical equivalence can also be formulated in semantic terms: empirically equivalent theories have the same class of empirical models.

Although the point of the semantic approach is to achieve independence of theory from language, one still needs a criterion of empirical status or observability to formulate empirical equivalence. One also needs to circumscribe the class of models with which a theory is to be identified, and this will require some reference to the theory's axioms or basic assumptions. For the question whether or not some particular set of structures is a model for a theory is not answered definitionally in science. It is answered by attempting to *apply* the theory, by working out consequences. Collateral information is crucial to such applications. For these reasons, we believe that questions about empirical equivalence must involve all the elements we shall bring to bear, whether the notion is formulated in ours or in semantic terms.

3 In labeling VRO "relatively uncontroversial," we acknowledge that van Fraassen's empiricism rejects it. Van Fraassen claims that what is observable is determined by facts about human beings as organisms, not by the transitory state of knowledge of those facts—not by science or technology. This view does not affect the use we shall make of VRO, however, because *judgments* of empirical equivalence must depend on *judgments* of what is observable; they cannot invoke transcendent facts.

But further, we reject the implicit assumption that conditions of observability are fixed by physiology. Once it is decided what is to count as observing, physiology may determine what is observable. But physiology does not impose or delimit our concept of observation. We could possess the relevant physiological apparatus without possessing a concept of observation at all. The concept we do possess could perfectly well incorporate technological means of detection. In fact, the concept of observation has changed with science, and even to state that the (theory-independent) facts determine what is observable, van Fraassen must use a concept of observation that implicitly appeals to a state of science and technology.

4 An appeal to NAP in criticism of empirical equivalence may appear ironic in view of the role of holistic theses in fostering belief in empirical equivalence and underdetermination. It is indeed ironic that the problems NAP, in concert with other theses, poses for empirical equivalence have escaped the notice of holists.

5 Conceivably, it could also decrease through a shift in the status of a consequence from observational to nonobservational, although that is not the usual pattern in the sciences.

6 See our contributions to the symposium, "Normative Versions of Naturalized Epistemology," *Philosophy of Science*, LVII, 1 (1990): 20–34; 44–60.

7 Consider the stature gained by Newtonian mechanics through its unforeseen applicability to the motions of fluids, and unforeseeable applicability to electric current.

8 We discount here the trivial algorithm that, applied to any theory T committed to theoretical entities of type *r*, generates the "rival" T* that asserts the world to be observationally exactly as if T were true but denies the existence of *r*s. Its logical incompatibility with T is insufficient to qualify T* as a genuine rival, as T* offers no competing explanations and is totally parasitic on T for whatever virtues it does offer. For extended criticism of such devices, see Leplin, "Surrealism," *Mind*, XCVI (October 1987): 519–24.

9 Contemporary examples are also limited to space-time theories, raising the possibility of underdetermination for certain topological features of space-time. For example, it might be possible to obtain the same consequences from a dense but discreet space-time as are obtained by adding dimensions to continuous space-time. A more general treatment of relative motion would subsume it under topological considerations.

10 And we deny the omnibus a priori claim that every theory has empirically equivalent rivals.

11 Here we allow, as always, a role for auxiliaries in generating consequences.

12 In "Realism, Underdetermination, and a Causal Theory of Evidence," *Nous*, VII (March 1973) 1–12, Richard Boyd sought to drive a wedge between empirical consequences and supporting instances. But Boyd's case rests on the (in our view dubious) principle that "new theories should, prima facie, resemble current theories with respect to their accounts of causal relations among theoretical entities" (p. 8). Our argument will require no such restriction.

13 Logico-semantic trickery will not render it ineliminable so long as epistemic conditions are imposed on auxiliaries.

14 "Illustrations of the Dynamical Theory of Gases," in *The Scientific Papers of James Clerk Maxwell*, W. D. Niven, ed. (New York: Cambridge, 1890).

15 We do not mean to suggest that all epistemologists are guilty of conflating epistemic and semantic relations. But there are some influential theorists of knowledge, e.g., Quine, who do precisely this.

16 See Lycan, "Epistemic Value," *Synthese*, LXIV (1985): 137–64.

17 This is a striking feature of Gilbert Harman's examples in *Thought* (Princeton: University Press, 1973).

18 *Aspects of Scientific Explanation* (New York: Free Press, 1965), p. 83.

19 Compare Quine: ". . . epistemology now becomes semantics. For epistemology remains centered as always on evidence; and meaning remains centered as always on verification; and evidence is verification"—"Epistemology Naturalized," p. 89.

20 As Duhem, summarizing the state of instrumentalism in 1908, put it: "we now require that . . . [scientific theories] save all the phenomena of the inanimate universe together"—*To Save the Phenomena* (Chicago: University Press, 1969), p. 117.

24

Wesley Salmon, "Bayes's Theorem and the History of Science"

Although Bayesianism is a relatively new approach to the notion of confirmation of theory by data, it has already become a flourishing sub-industry in the philosophy of science. Salmon's article provides a critical introduction to it.

0. Introduction

In his splendid introduction to this volume, Herbert Feigl rightly stresses the central importance of the distinction between the *context of discovery* and the *context of justification*. These terms were introduced by Hans Reichenbach to distinguish the social and psychological facts surrounding the discovery of a scientific hypothesis from the evidential considerations relevant to its justification.[1] The folklore of science is full of dramatic examples of the distinction; e.g., Kepler's mystical sense of celestial harmony[2] versus the confrontation of the postulated orbits with the observations of Tycho; Kekulé's drowsing vision of dancing snakes[3] versus the laboratory confirmation of the hexagonal structure of the benzine ring; Ramanujan's visitations in sleep by the Goddess of Namakkal[4] versus his waking demonstrations of the mathematical theorems. Each of these examples offers a fascinating insight into the personality of a working scientist, and each provides a vivid contrast between those psychological factors and the questions of evidence that must be taken into account in order to assess the truth or probability of the result. Moreover, as we all learned in our freshman logic courses, to confuse the source of a proposition with the evidence for it is to commit the genetic fallacy.

If one accepts the distinction between discovery and justification as

W. Salmon, "Bayes's Theorem and The History of Science," in *Historical and Philosophical Perspectives of Science*, ed. R. Stuewer, 1970, pp. 68–86. Minneapolis: University of Minnesota Press.

viable, there is a strong temptation to maintain that this distinction marks the boundaries between history of science and philosophy of science. History is concerned with the *facts* surrounding the growth and development of science; philosophy is concerned with the *logical structure* of science, especially with the evidential relations between data and hypotheses or theories. As a matter of fact, Reichenbach described the transition from the context of discovery to the context of justification in terms of a *rational reconstruction*. On the one hand, the scientific innovator engages in thought processes that may be quite irrational or nonrational, manifesting no apparent logical structure: this is the road to discovery. On the other hand, when he wants to present his results to the community for judgment, he provides a reformulation in which the hypotheses and theories are shown in logical relation to the evidence that is offered in support of them: this is his rational reconstruction. The items in the context of discovery are *psychologically relevant* to the scientific conclusion; those in the context of justification are *logically relevant* to it. Since the philosopher of science is concerned with logical relations, not psychological ones, he is concerned with the rationally reconstructed theory, not with the actual process by which it came into being.

Views of the foregoing sort regarding the relations between the context of discovery and the context of justification have led to a conception of philosophy of science which might aptly be characterized as a "rational reconstructionist" or "logical reconstructionist" approach; this approach has been closely associated with the school of logical positivism, though by no means confined to it.[5] Critics of the reconstructionist view have suggested that it leaves the study of vital, living, growing science to the historian, while relegating philosophy of science to the dissection of scientific corpses—not the bodies of scientists, but of theories that have grown to the point of stagnation and ossification. According to such critics, the study of completed science is not the study of science at all. One cannot understand science unless he sees how it grows; to comprehend the logical structure of science, it is necessary to take account of scientific change and scientific revolution. Certain philosophers have claimed, consequently, that philosophy of science must deal with the logic of discovery as well as the logic of justification.[6] Philosophy of science, it has been said, cannot proceed apart from study of the history of science. Such arguments have led to a challenge of the very distinction between discovery and justification.[7] Application of this distinction, it is claimed, has led to the reconstructionist approach, which separates philosophy of science from real science, and makes philosophy of science into an unrealistic and uninteresting form of empty symbol manipulation.

The foregoing remarks make it clear, I hope, that the distinction between the context of discovery and the context of justification is a major focal point for any fundamental discussion of the relations between history of science and philosophy of science. As the dispute seems to shape up, the reconstructionists rely heavily upon the viability of a sharp distinction, and they apparently conclude that there is no very significant relation between the two disciplines. Such marriages as occur between them—e.g., the International Union of History and Philosophy of Science, the National Science Foundation Panel for History and Philosophy of Science, and the Departments of History and Philosophy of Science at Melbourne and Indiana—are all marriages of convenience. The anti-reconstructionists, who find a basic organic unity between the two disciplines, seem to regard a rejection of the distinction between discovery and justification as a cornerstone of their view. Whatever approach one takes, it appears that the distinction between the context of discovery and the context of justification is the first order of business.

I must confess at this point, if it is not already apparent, that I am an unreconstructed reconstructionist, and I believe that the distinction between the context of discovery and the context of justification is viable, significant, and fundamental to the philosophy of science. I do not believe, however, that this view commits me to an intellectual divorce from my historical colleagues; in the balance of this essay I should like to explain why. I shall not be concerned to argue in favor of the distinction, but shall instead try (1) to clarify the distinction, and repudiate certain common misconceptions of it, (2) to show that a clear analysis of the nature of scientific confirmation is essential to an understanding of the distinction, and that a failure to deal adequately with the logic of confirmation can lead to serious historical misinterpretations, and (3) to argue that an adequate conception of the logic of confirmation leads to basic, and largely unnoticed, logical functions of historical information. In other words, I shall be attempting to show how certain aspects of the relations between history and philosophy of science can be explicated within the reconstructionist framework. Some of my conclusions may appear idiosyncratic, but I shall take some pains along the way to argue that many of these views are widely shared.

1. The Distinction between Discovery and Justification

When one presents a distinction, it is natural to emphasize the differences between the two sorts of things, and to make the distinction appear more dichotomous than is actually intended. In the present instance, some commentators have apparently construed the distinction to imply that, first of all, a creative scientist goes through a great succession of irrational (or nonrational) processes, e.g., dreaming, being hit on the head, pacing the floor, or having dyspepsia, until a full-blown hypothesis is born. Only after these processes have terminated does the scientist go through the logical process of mustering and presenting his evidence so as to justify his hypothesis. Such a conception would, of course, be factually absurd; discovery and justification simply do not occur in that way. A more realistic account might go somewhat as follows. A scientist, searching for a hypothesis to explain some phenomenon, hits upon an idea, but soon casts it aside because he sees that it is inconsistent with some theory he accepts, or because it does not really explain the phenomenon in question. This phase undoubtedly involves considerable logical inference; it might, for instance, involve a mathematical calculation which shows that the explanation in question would not account for a result of the correct order of magnitude. After more searching around—in the meantime perhaps he attends a cocktail party and spends a restless night—he hits upon another idea, which also proves to be inadequate, but he sees that it can be improved by some modification or other. Again, by logical inference, he determines that his new hypothesis bears certain relations to the available evidence. He further realizes, however, that although his present hypothesis squares with the known facts, further modification would make it simpler and give it a wider explanatory range. Perhaps he devises and executes additional tests to check the applicability of his latest revision in new domains. And so it goes. What I am trying to suggest, by such science fiction, is that the processes of discovery and justification are intimately intertwined, with steps of one type alternating with steps of the other. There is no reason to conclude, from a distinction between the context of discovery and the context of justification, that the entire process of discovery must be completed before the process of justification can begin, and that the rational reconstruction can be undertaken only after the creative work has ended. Such conclusions are by no means warranted by the reconstructionist approach.

There is, moreover, no reason to suppose that the two contexts must be mutually exclusive. Not only may elements of the context of justification be temporally intercalated between elements of the context of discovery, but the two contexts may have items in common. The supposition that this cannot happen is perhaps the most widespread misunderstanding of the distinction between the two contexts. The most obvious example is the case in which a person or a machine discovers the answer to a problem by applying an algorithm, e.g., doing a sum, differentiating a polynomial, or finding a greatest common divisor. Empirical science also contains routine methods for finding answers to problems—which is to say, for discovering correct hypotheses. These are often the kinds of procedures that can be delegated to a technician or a machine, e.g., chemical analyses, ballistic testing, or determination of physical constants of a new compound. In such cases, the process of discovery and the process of justification may be nearly identical, though the fact that the machine blew a fuse, or the technician took a coffee break, could hardly qualify for inclusion in the latter context. Even though the two contexts are not mutually exclusive, the distinction does not vanish. The context of discovery consists of a number of items related to one another by psychological relevance, while the context of justification contains a number of items related to one another by (inductive and deductive) logical relevance. There is no reason at all why one and the same item cannot be both psychologically and logically relevant to some given hypothesis. Each context is a complex of entities all of which are interrelated in particular ways. The contexts are contrasted with one another, not on the ground that they can have no members in common, but rather on the basis of differences in the types of relations they incorporate. The fact that the two contexts can have items in common does not mean that the distinction is useless or perverse, for there *are* differences between logical and psychological relevance relations which are important for the understanding of science.

The problem of scientific discovery does not end with the thinking up of a hypothesis. One has also to discover evidence and the logical connections between the evidence and the hypothesis. The process of discovery is, therefore, involved in the very construction of the rational reconstruction. When the scientist publishes his hypothesis as acceptable, confirmed, or corroborated, along with the evidence and arguments upon which the claim is based, he is offering *his* rational reconstruction (the one he has *discovered*), and is presumably claiming that it is logically sound. This is a fact about the scientist; his evidence and arguments satisfy him. A critic—scientist or philosopher—might, of course, show that he has committed a logical or methodological error, and consequently, that his rational reconstruction is unsound. Such an occurrence would belong to the context of justification. However, even if the argument seems compelling to an entire scientific community, it may still be logically faulty. The convincing character of an argument is quite distinct from its validity; the former is a *psychological* characteristic, the latter is *logical*. Once more, even though there may be extensive overlap between the contexts of discovery and justification, it is important not to confuse them.

Considerations of the foregoing sort have led to serious controversy over the appropriate role of philosophy of science. On the other hand, it is sometimes claimed that philosophy of science must necessarily be a historically oriented empirical study of the methods scientists of the past and present have actually used, and of the canons they have accepted. On the other hand, it is sometimes maintained that such factual studies of the methods of science belong to the domain of the historian, and that the philosopher of science is concerned exclusively with logical and epistemological questions. Proponents of the latter view—which is essentially the reconstructionist approach—may appear quite open to the accusation that they are engaged in some sort of scholastic symbol mongering which has no connection whatever with actual science. To avoid this undesirable state of affairs, it may be suggested, we ought to break down the distinctions between history and philosophy, between psychology and logic, and, ultimately, between discovery and justification.

There is, I believe, a better alternative. While the philosopher of science may be basically concerned with abstract logical relations, he can hardly afford to ignore the actual methods that scientists have found acceptable. If a philosopher expounds a theory of the logical structure of science according to which almost all of modern physical science is methodologically unsound, it would be far more reasonable to conclude that the philosophical reasoning had gone astray than to suppose that modern science is logically misconceived. Just as certain empirical facts, such as geometrical diagrams or soap film experiments, may have great heuristic value for mathematics, so too may the historical facts of scientific development provide indispensable guidance for the formal studies of the logician. In spite of this, the philosopher of science is properly concerned with issues of logical correctness which cannot finally be answered by appeal to the history of science. One of the problems with which the philosopher of science might grapple is the question of what grounds we have for supposing scientific knowledge to be superior to other alleged types of knowledge, e.g., alchemy, astrology, or divination. The historian may be quick to reply that *he* has the means to answer that question, in terms of the relative success of physics, chemistry, and astronomy. It required the philosophical subtlety of David Hume to realize that such an answer involves a circular argument.[8] The philosopher of science, consequently, finds himself attempting to cope with problems on which the historical data may provide enormously useful guidance, but the solutions, if they are possible at all, must be logical, not historical, in character. The reason, ultimately, is that justification is a normative concept, while history provides only the facts.

I have been attempting to explain and defend the distinction between discovery and justification largely by answering objections to it, rather than by offering positive arguments. My attitude, roughly, is that it is such a plausible distinction to begin with, and its application yields such rich rewards in understanding, that it can well stand without any further justification. Like any useful tool, however, it must be wielded with some finesse; otherwise the damage it does may far outweigh its utility.

2. Bayes's Theorem and the Context of Justification

It would be a travesty to maintain, in any simpleminded way, that the historian of science is concerned only with matters of discovery, and not with matters of justification. In dealing with any significant case, say the replacement of an old theory by a new hypothesis, the historian will be deeply interested in such questions as whether, to what extent, and in what manner the old theory has been disconfirmed; and similarly, what evidence is offered in support of the new hypothesis, and how adequate it is. How strongly, he may ask, are factors such as national rivalry among scientists, esthetic disgust with certain types of theories, personal idiosyncrasies of influential figures, and other nonevidential factors operative? Since science aspires to provide objective knowledge of the world, it cannot be understood historically without taking very seriously the role of evidence in scientific development and change. Such historical judgments—whether a particular historical development was or was not rationally justified on the basis of the evidence available at the time—depend crucially upon the historian's understanding of the logic of confirmation and disconfirmation. If the historian seriously misunderstands the logic of confirmation, he runs the risk of serious historical misevaluation. And to the possible rejoinder that any historian worth his salt has a sufficiently clear intuitive sense of what constitutes relevant scientific evidence and what does not, I must simply reply that I am not convinced.

Perhaps the most widely held picture of scientific confirmation is one that had great currency in the nineteenth century; it is known as the hypothetico-deductive (H-D) method. According to this view, a scientific hypothesis is tested by deducing observational consequences from it, and seeing whether these consequences actually do transpire. If a given consequence does occur, it constitutes a confirming instance for the hypothesis; if it does not occur, it is a disconfirming instance. There are two rather immediate difficulties with this characterization, and they are easily repaired. First, a scientific hypothesis, by itself, ordinarily does not have any observational consequences; it is usually necessary to supply some empirically determined initial conditions to make it possible validly to deduce any observational consequences at all. For example, from Kepler's law of planetary motion alone, it is impossible to deduce the position of Mars at some future time, but with initial conditions on the motion of Mars at some earlier time, a prediction of the position is possible. Similarly, from Hooke's law alone it is impossible to predict the elongation of a spring under a given weight, but with an empirically determined coefficient of elasticity, the prediction can be deduced. Second, it is frequently, if not always, necessary to make use of auxiliary hypotheses in order to connect the observations with the hypothesis that is being tested.

For example, if a medical experimenter predicts that a certain bacillus will be found in the blood of a certain organism, he must conjoin to his medical hypothesis auxiliary hypotheses of optics which pertain to the operation of his microscope, for only in that way can he establish a deductive connection between what he observes under the microscope and the actual presence of the microorganism. With these additions, the H-D method can be schematized as follows:

H (hypothesis being tested)
A (auxiliary hypotheses)
I (initial conditions)

O (observational consequence)

Since we are not primarily interested in epistemological problems about the reliability of the senses, let us assume for the purposes of the present discussion that the initial conditions *I* have been established as true by observation and, in addition, that we can ascertain by observation whether the observational consequence O is true or false. Let us assume, moreover, that for purposes of the present test of our hypothesis *H*, the auxiliary hypotheses *A* are accepted as unproblematic.[9] With these simplifying idealizations, we can say that *H* implies *O*; consequently if O turns out to be false, it follows that *H* must be false—this is the deductively valid *modus tollens*. Given the truth of O, however, nothing follows deductively about the truth of *H*. To infer the truth of *H* from the truth of *O* in these circumstances is obviously the elementary deductive fallacy of *affirming the consequent*. According to the H-D view the truth of O does, nevertheless, tend to confirm or lend probability to *H*. Presumably, if enough of the right kinds of observational consequences are deduced and found by observation to be true—i.e., if enough observational predictions are borne out by experience—the hypothesis can become quite highly confirmed. Scientific hypotheses can never be completely and irrefutably verified in this manner, but they can become sufficiently confirmed to be scientifically acceptable. According to this H-D conception, induction—the logical relation involved in the confirmation of scientific hypotheses—is a kind of inverse of deduction. The fact that a true observational prediction follows deductively from a given hypothesis (in conjunction with initial conditions and auxiliary hypotheses) means, according to the H-D view, that a relation of inductive support runs in the reverse direction from O to *H*.

The H-D account of scientific confirmation is, it seems to me, woefully inadequate. The situation is nicely expressed in a quip attributed to Morris R. Cohen: A logic text is a book that contains two parts; in the first (on deduction) the fallacies are explained, and in the second (on induction) they are committed. Quite clearly, we need a more satisfactory account of scientific confirmation. Automatically transforming a deductive fallacy into a correct inductive schema may offer an appealing way to account for scientific inference, but certainly our forms of inductive inference ought to have better credentials than that. The main shortcomings of the H-D method are strongly suggested by the fact that, given *any* finite body of observational evidence, there are infinitely many hypotheses which are confirmed by it in exactly the same manner; that is, there are infinitely many alternative hypotheses that could replace our hypothesis *H* in the schema above and still yield a valid deduction. This point is obvious if one considers the number of curves that can be drawn through a finite set of points on a graph. Hence, Hooke's law, which says that a certain function is a straight line, and Kepler's first law, which says that a planetary orbit is an ellipse, could each be replaced by infinitely many alternatives that would give rise to precisely the same observational consequences as Hooke's and Kepler's laws respectively. As it stands, the H-D method gives us no basis whatever for claiming that either of these laws is any better confirmed by the available evidence than is any one of the infinitude of alternatives. Clearly it stands in dire need of supplementation.

When we look around for a more adequate account of scientific confirmation, it is natural to see whether the mathematical calculus of probability can offer any resources. If we claim that the process of confirmation is one of lending probability to a hypothesis in the light of evidence, it is reasonable to see whether there are any theorems on probability that characterize confirmation. If so, such a theorem would provide some sort of valid schema for formal confirmation relations. Theorems do not, of course, come labeled for their specific applications, but Bayes's theorem does seem well suited for this role.[10]

In order to illuminate the use of Bayes's theorem, let us introduce a simple game. This game is played with two decks of cards made up as follows: deck I contains eight red and four black cards; deck II contains four red and eight black cards. A turn begins with the toss of an ordinary die; if the side six appears the player draws from deck I, and if any other side comes up he draws from deck II. The draw of a red card constitutes a win. There is a simple way to calculate the probability of a win in this game. Letting P(A,B) stand for the probability *from* A *to* B (i.e., the probability of B, given A), and letting A stand for tosses of the die, B for draws from deck I, and C for draws resulting in red, the following formula yields the desired probability:

(1) $P(A,C) = P(A,B)P(A \,\&\, B,C) + P(A, \sim B)\, P(A \,\&\, \sim B,C)$

where the ampersand stands for "and" and the tilde preceding a symbol negates it. Probability expressions appearing in the formula are P(A,C), probability of a red card on a play of this game; P(A,B), probability of drawing from deck I (= 1/6); P(A, ~ B), probability of drawing from deck II (= 5/6); P(A & B,C), probability of getting red on a draw from deck I (= 2/3); P(A & ~ B,C), probability of getting red on a draw from deck II (= 1/3). The probability of a win on any given play is 7/18.

Suppose, now, that a player has just drawn a red card, but you failed to notice from which deck he drew. We ask, what is the probability that it was drawn from deck I? The probability we wish to ascertain is P(A & C,B), the probability that a play which resulted in a red card was one on which the die turned up six, and the draw was made from deck I. Bayes's theorem

$$(2) \quad P(A \,\&\, C,B) = \frac{P(A,B)P(A \,\&\, B,C)}{P(A,B)P(A \,\&\, B,C) + P(A, \sim B)P(A \,\&\, \sim B,C)}$$

supplies the answer. Substituting the available values on the right-hand side of the equation yields the value 2/7 for the desired probability. Note that although the probability of getting a red card if you draw from deck I is much greater than the probability of getting a red card if you draw from deck II, the probability that a given red draw came from deck I is much less than the probability that it came from deck II. This is because the vast majority of draws are made from deck II.

There is nothing controversial about either of the foregoing formulas, or about their application to games of the kind just described. The only difficulty concerns the legitimacy of extending the application of Bayes's theorem, formula (2), to the problem of confirmation of hypotheses. In order to see how that might go, let me redescribe the game, with some admitted stretching of usage. We can take the draw of a red card as an effect that can be produced in either of two ways, by throwing a six and drawing from deck I, or by tossing some other number and drawing from the other deck. There are, correspondingly, two causal hypotheses. When we ask for the probability that the red draw came from deck I, we are asking for the probability of the first of these hypotheses, given the evidence that a red card had been drawn. Looking now at the probability expressions that appear in Bayes's theorem, we have: P(A,B), the *prior probability* of the first hypothesis; P(A, ~ B), the *prior probability* that the first hypothesis does not hold; P(A & B,C), the probability of the effect (red card drawn) if the first hypothesis is correct; P(A & ~ B,C), the probability of the effect if the first hypothesis is incorrect; P(A & C,B), the *posterior probability* of the first hypothesis on the evidence that the effect has occurred. The probabilities P(A & B,C) and P(A & ~B,C) are called *likelihoods* of the two hypotheses, but it is important to note clearly that they are not probabilities of *hypotheses* but, rather, probabilities *of the effect*. It is the posterior probability that we seek when we wish to determine the probability of the hypothesis in terms of the given evidence.

In order to apply Bayes's theorem, we must have three probabilities to plug into the right-hand side of (2). Since the two prior probabilities must add up to one, it is sufficient to know one of them, but the likelihoods are independent, so we just have both of them. Thus, in order to compute the posterior probability of our hypothesis, we need its prior probability, the probability that we would get the evidence we have if it is true, and the probability that we would get the evidence we have if it were false. None of these three is dispensable, except in a few obvious special cases.[11]

When the H-D schema was presented, we stipulated that the hypothesis being tested implied the evidence, so in that case P(A & B,C) = 1. This value of one of the likelihoods does not determine a value for the posterior probability, and, indeed, the posterior probability can be arbitrarily small even in the case supplied by the H-D method. This fact shows the inadequacy of the H-D schema quite dramatically: even though the data confirm the hypothesis according to the H-D view, the posterior probability of the hypothesis in the light of the available evidence may be as small as you like—even zero in the limiting special case in which the prior probability of the hypothesis is zero.

If Bayes's theorem provides a correct formal schema for the logic of confirmation and disconfirmation of scientific hypotheses, it tells us that we need to take account of three factors in attempting to assess the degree to which a hypothesis is rendered probable by the evidence. Roughly, it says, we must consider how well our hypothesis explains the evidence we have (this is what the H-D schema requires), how well an alternative hypothesis might explain the same evidence, and the prior probability of the hypothesis. The philosophical obstacle that has always stood in the way of using Bayes's theorem to account for confirmation is the severe difficulty in understanding what a prior probability could be. I have argued elsewhere that it is essentially an assessment of what one might call the *plausibility* of the hypothesis, prior to, or apart from, the results of directly testing that hypothesis.[12] Without attempting to analyze what is meant by plausibility, I shall offer a few plausibility judgments of my own, just to illustrate the sort of thing I am talking about. For instance, I regard as quite implausible Velikovsky's hypotheses about the origin of Venus, any ESP theory that postulates transfer of information at a speed greater than that of light, and any teleological biological theory. Hypotheses of these kinds strike me as implausible because, in one way or another, they do not fit well with currently accepted scientific theory. I regard it as quite plausible that life originated on the face of the earth in accordance with straightforward physicochemical principles governing the formation of large "organic" molecules out of simpler inorganic ones. This does seem to fit well with what we know. You need not accept my plausibility judgments; you can supply your own. The only crucial issue is the existence of such prior probabilities for use in connection with Bayes's theorem.

Let us now return to the problems of the historian. I claim above that the analysis of the logic of confirmation could have a crucial bearing upon historical judgments. Having compared the H-D account of confirmation with the Bayesian analysis, we can see an obvious way in which this problem could arise. If a historian accepts the H-D analysis of confirmation, then there is no place for plausibility judgments in the logic of science—at least not in the context of justification. If such a historian finds plausibility considerations playing an important role historically in the judgments scientists render upon hypotheses, he will be forced to exclude them from the context of justification, and he may conclude that the course of scientific development is massively influenced by nonrational or nonevidential considerations. Such an "H-D historian" might well decide, along with the editors of *Harper's Magazine*, that it was scientific prejudice, not objective evaluation, that made the scientific community largely ignore Velikovsky's views.[13] He might similarly conclude that Einstein's commitment to the "principle of relativity" on the basis of plausibility arguments shows his views to have been based more upon preconceptions than upon objective evidence.[14] A "Bayesian historian," in contrast, will see these plausibility considerations as essential parts of the logic of confirmation, and he will place them squarely within the context of justification. The consequence is, I would say, that the historian of science who regards the H-D schema as a fully adequate characterization of the logical structure of scientific inference is in serious danger of erroneously excluding from the context of justification items that genuinely belong within it. The moral for the historian should be clear. There are considerations relating to the acceptance or rejection of scientific hypotheses which, on the H-D account, must be judged *evidentially* irrelevant to the truth or falsity of the hypothesis, but which are, nevertheless, used by scientists in making decisions about such acceptance or rejection. These same items, on the Bayesian account, become evidentially relevant. Hence, the judgment of whether scientists are making decisions on the basis of evidence, or on the basis of various psychological or social factors that are evidentially irrelevant, hinges crucially upon the question of whether the H-D or the Bayesian account of scientific inference is more nearly correct. It is entirely conceivable that one historian might attribute acceptance of a given hypothesis to nonrational considerations, while another might judge the same decision to have an entirely adequate rational basis. Which account is historically more satisfactory will depend mainly upon which account of scientific inference is more adequate. The historian can hardly be taken to be unconcerned with the context of justification, and with its differences from the context of discovery; indeed, if he is to do his job properly he must understand them very well.

3. The Status of Prior Probabilities

It would be rather easy, I imagine, for the historian, and others who are not intimately familiar with the technicalities of inductive logic and confirmation theory, to suppose that the H-D account of scientific inference is the correct one. This view is frequently expressed in the opening pages of introductory science texts, and in elementary logic books.[15] At the same time, it is important to raise the question of whether scientists in general—including the authors of the aforementioned introductory texts—actually comply with the H-D method in practice, or whether in fact they use something similar to the Bayesian approach sketched in the preceding section. I am strongly inclined to believe that the Bayesian schema comes closer than the H-D schema to capturing actual scientific practice, for it seems to me that scientists do make substantial use of plausibility considerations, even though they may feel somewhat embarrassed to admit it. I believe also that practicing scientists have excellent intuitions regarding what constitutes sound scientific methodology, but that they may not always be especially adept at fully articulating them. If we want the soundest guidance on the nature of scientific inference, we should look carefully at scientific practice, rather than the methodological pronouncements of scientists.

It is, moreover, the almost universal judgment of contemporary inductive logicians—the experts who concern themselves explicitly with the problems of confirmation of scientific hypotheses—that the simple H-D schema presented above is incomplete and inadequate. Acknowledging the well-known fact that there is very little agreement on which particular formulation among many available ones is most nearly a correct inductive logic, we can still see that among a wide variety of influential current approaches to the problems of confirmation, there is at least agreement in rejecting the H-D method. This is not the place to go into detailed discussions of the alternative theories, but I should like to mention five leading candidates, indicating how each demands something beyond what is contained in the H-D schema. In each case, I think, what needs to be added is closely akin to the plausibility considerations mentioned in the preceding section.

1. The most fully developed explicit confirmation theory available is Rudolf Carnap's theory of logical probability (degree of confirmation) contained in his monumental *Logical Foundations of Probability*.[16] In the systems of inductive logic he elaborated in that book, one begins with a formalized language and assigns a priori weights to all statements in that

language, including, of course, all hypotheses. It is very easy to show that Carnap's theory of confirmation is thoroughly Bayesian, with the a priori weights functioning precisely as the prior probabilities in Bayes's theorem. Although these systems had the awkward feature that general hypotheses all have prior probabilities, and consequently, posterior probabilities on any finite amount of evidence, equal to zero, Jaakko Hintikka has shown how this difficulty can be circumvented without fundamentally altering Carnap's conception of confirmation as a logical probability.[17]

2. Although not many exponents of the frequency theory of probability will agree that it even makes sense to talk about the probability of scientific hypotheses, those who do explicitly invoke Bayes's theorem for that purpose. Reichenbach is the leading figure in this school, although his treatment of the probability of hypotheses is unfortunately quite obscure in many important respects.[18] I have tried to clarify some of the basic points of misunderstanding.[19]

3. The important "Bayesian" approach to the foundations of statistics has become increasingly influential since the publication in 1954 of L. J. Savage's *The Foundations of Statistics*.[20] It has gained many adherents among philosophers as well as statisticians. This view is based upon a subjective interpretation of probability ("personal probability," as Savage prefers to say, in order to avoid confusion with earlier subjective interpretations), and it makes extensive use of Bayes's theorem. The prior probabilities are simply degrees of prior belief in the hypothesis, before the concrete evidence is available. The fact that the prior probabilities are so easily interpreted on this view means that Bayes's theorem is always available for use. Savage, himself, is not especially concerned with probabilities of general hypotheses, but those who are interested in such matters have a readymade Bayesian theory of confirmation.[21] On this view, the prior probabilities are subjective plausibility judgments.

4. Nelson Goodman, whose influential *Fact, Fiction, and Forecast* poses and attempts to resolve "the new riddle of induction," clearly recognizes that there is more to confirmation than mere confirming instances.[22] He attempts to circumvent the difficulties, which are essentially those connected with the H-D schema, by introducing the notion of "entrenchment" of terms that occur in hypotheses. Recognizing that a good deal of the experience of the human race becomes embedded in the languages we use, he brings this information to bear upon hypotheses that are candidates for confirmation. Although he never mentions Bayes's theorem or prior probabilities, the chapter in which he presents his solution can be read as a tract on the Bayesian approach to confirmation.

5. Sir Karl Popper rejects entirely the notions of confirmation and inductive logic.[23] His concept of *corroboration*, however, plays a central role in his

theory of scientific methodology. Although corroboration is explicitly regarded as nonprobabilistic, it does offer a measure of how well a scientific hypothesis has stood up to tests. The measure of corroboration involves such factors as simplicity, content, and testability of hypotheses, as well as the seriousness of the attempts made to falsify them by experiment. Although Popper denies that a highly corroborated hypothesis is highly probable, the highly corroborated hypothesis does enjoy a certain status: it may be chosen over its less corroborated fellows for further testing, and if practical needs arise, it may be used for purposes of prediction. The important point, for the present discussion, is that Popper rejects the H-D schema, and introduces additional factors into his methodology that play a role somewhat analogous to our plausibility considerations.

The foregoing survey of major contemporary schools of thought on the logic of scientific confirmation strongly suggests not only that the naive H-D schema is *not universally accepted* nowadays by inductive logicians as an adequate characterization of the logic of scientific inference, but also that it is *not even a serious candidate* for that role. Given the wide popular acceptance of the H-D method, it seems entirely possible that significant numbers of historians of science may be accepting a view of confirmation that is known to be inadequate, and one which differs from the current serious contending views in ways that can have a profound influence upon historical judgments. It seems, therefore, that the branch of contemporary philosophy of science that deals with inductive logic and confirmation theory may have some substantive material that is highly relevant to the professional activities of the historian of science.

It is fair to say, I believe, that one of the most basic points on which the leading contemporary theories of confirmation differ from one another is with regard to the nature of the prior probabilities. As already indicated, the logical theorist takes the prior probability as an a priori assessment of the hypothesis, the personalist takes the prior probability as a measure of subjective plausibility, the frequentist must look at the prior probability as some sort of success frequency for a certain type of hypothesis, Goodman would regard the prior probability as somehow based upon linguistic usage, and Popper (though he violently objects to regarding it as a prior probability) needs something like the potential explanatory value of the hypothesis. In addition, I should remark, N. R. Hanson held plausibility arguments to belong to the logic of discovery, but I have argued that, on his own analysis, they have an indispensable role in the logic of justification.[24]

This is not the place to go into a lengthy analysis of the virtues and shortcomings of the various views on the nature of prior probabilities.[25] Rather, I should like merely to point out a consequence of my view that is quite germane to the topic of the conference. If one adopts a frequency view of probability, and attempts to deal with the logic of confirmation by way of Bayes's theorem (as I do), then he is committed to regarding the prior probability as some sort of frequency—e.g., the frequency with which hypotheses relevantly similar to the one under consideration have enjoyed significant scientific success. Surely no one would claim that we have reliable statistics on such matters, or that we can come anywhere near assigning precise numerical values in a meaningful way. Fortunately, that turns out to be unnecessary; it is enough to have very, very rough estimates. But this approach does suggest that the question of the plausibility of a scientific hypothesis has something to do with our experience in dealing with scientific hypotheses of similar types. Thus, I should say, the reason I would place a rather low plausibility value on teleological hypotheses is closely related to our experience in the transitions from teleological to mechanical explanations in the physical and biological sciences and, to some extent, in the social sciences. To turn back toward telelogical hypotheses would be to go against a great deal of scientific experience about what kinds of hypotheses work well scientifically. Similarly, when Watson and Crick were enraptured with the beauty of the double helix hypothesis for the structure of the DNA molecule, I believe their reaction was more than purely esthetic.[26] Experience indicated that hypotheses of that degree of simplicity tend to be successful, and they were inferring that it had not only beauty, but a good chance of being correct. Additional examples could easily be exhibited.

If I am right in claiming not only that prior probabilities constitute an indispensable ingredient in the confirmation of hypotheses and the context of justification, but also that our estimates of them are based upon empirical experience with scientific hypothesizing, then it is evident that the history of science plays a crucial, but largely unheralded, role in the current scientific enterprise. The history of science is, after all, a chronicle of our past experience with scientific hypothesizing and theorizing—with learning what sorts of hypotheses work and what sorts do not. Without the Bayesian analysis, one could say that the study of the history of science might have some (at least marginal) heuristic value for the scientist and philosopher of science, but on the Bayesian analysis, the data provided by the history of science constitute, *in addition*, an essential segment of the evidence relevant to the confirmation or disconfirmation of hypotheses. Philosophers of science and creative scientists ignore this fact at their peril.

Notes

1 *Experience and Prediction* (Chicago: University of Chicago Press, 1938), section 1. I have offered an elementary discussion of the distinction in *Logic* (Englewood Cliffs, N.J.: Prentice-Hall, 1963), sections 1–3.

2 A. Pannekoek, *A History of Astronomy* (New York: Interscience, 1961), p. 235.

3 J. R. Partington, *A History of Chemistry* (London: Macmillan, 1964), IV, 553ff.

4 G. H. Hardy *et al.*, *Collected Papers of Srinivasa Ramanujan* (Cambridge: Cambridge University Press, 1927), p. xii.

5 Reichenbach, for example, was not a logical positivist; indeed, he was one of the earliest and most influential critics of that school.

6 See, for example, N. R. Hanson, "Is There a Logic of Discovery?" in *Current Issues in the Philosophy of Science*, ed. Herbert Feigl and Grover Maxwell (New York: Holt, Rinehart and Winston, 1961), pp. 20–35.

7 E.g., Thomas S. Kuhn, *The Structure of Scientific Revolutions* (Chicago: University of Chicago Press, 1962), p. 9.

8 For further elaboration of this point see my *Foundations of Scientific Inference* (Pittsburgh: University of Pittsburgh Press, 1967), pp. 5–17.

9 This is, of course, an unrealistic assumption, for as Pierre Duhem pointed out, in many cases the appropriate move upon encountering a disconfirming case is rejection or modification of an auxiliary hypothesis, rather than rejection of the principal hypothesis. This point does not affect the present discussion.

10 I have discussed the Bayesian conception of confirmation at some length in *Foundations of Scientific Inference*, pp. 108–131 (Bayes's theorem is deduced within the formal calculus on pp. 58–62), and in "Inquiries into the Foundations of Science," in *Vistas in Science*, ed. David L. Arm (Albuquerque: University of New Mexico Press, 1968), pp. 1–24. The latter article is the less technical of the two.

11 Viz., if $P(A,B) = 0$ or $P(A \& B,C) = 0$, then $P(A \& C,B) = 0$; if $P(A,\sim B) = 0$ or $P(A \& \sim B,C) = 0$, then $P(A \& C,B) = 1$. Also, if $P(A,C) = 0$, the fraction becomes indeterminate, for, by (1), that is the denominator in (2).

12 *Foundations of Scientific Inference* and "Inquiries into the Foundations of Science."

13 This case is discussed in "Inquiries into the Foundations of Science."

14 See Albert Einstein, "Autobiographical Notes," in *Albert Einstein: Philosopher-Scientist*, ed. Paul Arthur Schilpp (New York: Tudor, 1949), pp. 2–95.

15 In my *Logic*, section 23, I have tried, without introducing any technicalities of the probability calculus, to offer an introductory Bayesian account of the confirmation of hypotheses.

16 Chicago: University of Chicago Press, 1950.

17 Jaakko Hintikka, "A Two-Dimensional Continuum of Inductive Methods," in *Aspects of Inductive Logic*, ed. Jaakko Hintikka and Patrick Suppes (Amsterdam: North-Holland, 1966), pp. 113–132.

18 Hans Reichenbach, *The Theory of Probability* (Berkeley and Los Angeles: University of California Press, 1949), section 85.

19 *Foundations of Scientific Inference*, pp. 115ff.

20 New York: Wiley, 1954. An excellent exposition is found in Ward Edwards, Harold Lindman, and Leonard J. Savage, "Bayesian Statistical Inference for Psychological Research," *Psychological Review*, 70 (1963), 193–242.

21 Sir Harold Jeffreys illustrates an explicitly Bayesian approach to the probability of hypotheses; see his *Scientific Inference* (Cambridge: Cambridge University Press, 1957), and *Theory of Probability* (Oxford: Clarendon Press, 1939).

22 First edition (Cambridge, Mass.: Harvard University Press, 1955); second edition (Indianapolis: Bobbs-Merrill, 1965).

23 *The Logic of Scientific Discovery* (New York: Basic Books, 1959). It is to be noted that, in spite of the title of his book, Popper accepts the distinction between discovery and justification, and explicitly declares that he is concerned with the latter but not the former.

24 *Foundations of Scientific Inference*, pp. 111–114, 118.

25 This is done in the items mentioned in note 10.

26 James D. Watson, *The Double Helix* (New York: New American Library, 1969). This book provides a fascinating account of the *discovery* of an important scientific hypothesis, and it illustrates many of the points I have been making. Perhaps if literary reviewers had had a clearer grasp of the distinction between the context of discovery and the context of justification they would have been less shocked at the emotions reported in the narrative.

QUESTIONS

1 What are the old and new riddles of induction? Are you convinced by Achinstein's response to the "grue paradox"?

2 Popper thinks that defending an empirically falsified theory by making suitable modifications in it is never permissible as it violates the "code" of scientific rationality. But the history of science shows that scientists have done this frequently. Did they behave "irrationally"? If you think not, discuss some conditions under which it may be rational to defend an empirically "refuted" theory by making modifications in it.

3 Popper seems to have two different complaints about Darwin's evolutionary theory: (a) that the theory's prediction of the gradual character of species modification is not borne out by the fossil record; and (b) that the theory's central notion of *adaptation* or *fitness* is circular. The passage from Darwin suggests that Darwin himself was aware of (a) and had suggested some

ways of explaining the non-gradual character of the fossil record. Where does the matter stand today? Reflect on (b) and try to respond to Popper's charge of circularity.

4 Many philosophers of science take it almost for granted nowadays that all observations are theory-laden. Others think that is true only in a trivial and uninteresting sense, but in a non-trivial and interesting sense there is a level of scientific experience at which it can be described in a theoretically neutral language operating with such notions as pointer readings, etc. Take a stance in this debate and support your view.

5 Despite their insistence on the "holistic" nature of empirical tests (i.e., that theories confront observations as wholes), neither Quine nor his precursor Duhem thought that scientific progress could actually be paralyzed by holistic considerations. Duhem, for example, noted that scientists usually succeed in deciding which element of a theoretical system to abandon in light of adverse data. Although such decisions "do not impose themselves with the same implacable rigor that the prescriptions of logic do," he wrote, "we may find it childish and unreasonable for the . . . physicist to maintain obstinately at any cost, at the price of continual repairs and many tangled-up stays, the worm-eaten columns of a building tottering in every part, when by razing these columns it would be possible to constrict a simple, elegant, and solid system" (Duhem 1954, p. 217). Comment on this passage while taking into account other "anti-holistic" arguments due to Laudan and Leplin.

FURTHER READING

Hume's classic treatment of the problem of induction is found in his famous *Inquiry Concerning Human Understanding* (Hume 1974). The locus classicus for the "new riddle of induction" and the "grue paradox" is Goodman (1983). Papers discussing various aspects of the paradox are collected in Stalker (1994). Achinstein's new arguments against it are based on a general theory of evidence developed in his work, *The Book of Evidence* (2001).

Popper first introduced his falsificationist methodology in 1934 in *Logik der Forschung* (English trans. Popper 1959). As an example of a favorable reception of his theory by working scientists, see Bondi and Kilmister (1959). Popper's student Imre Lakatos sought to incorporate some of the former's insights (without going to the extremes) in his "Methodology of Scientific Research Programs" (see Lakatos and Musgrave 1970; this volume also contains valuable contributions by Kuhn, Feyerabend, and Popper himself). For a discussion of Popper's views on Darwinism, see Ruse (1988b).

The Duhem–Quine thesis and the notion of underdetermination have generated extensive literature. For some earlier articles see Harding (1976). More recent contributions include Greenwood (1990), Balashov (1994) and Hoefer and Rosenberg (1994).

For useful introductions and (sometimes fairly technical) discussions of Bayesian confirmation theory and other probabilistic approaches to evidence, see Horwich (1982), Earman (1992), Howson and Urbach (1993), Maher (1993), and Mayo (1996).

PART VI

SCIENCE IN CONTEXT: THE CHALLENGE OF HISTORY AND SOCIOLOGY

INTRODUCTION

If observational evidence underdetermines theories, we need an explanation of what has determined the succession of theories that characterizes science's history. Even more, for philosophy's purposes, we need a justification for the claim that these observationally unsupported theories are epistemically rational and reasonable ones to adopt. Clearly, Empiricism cannot by itself do this, as its resources in justification are limited to observation.

An important historian of science, Thomas Kuhn, was among the first to explore the history of science for these non-observational factors that explain theory choice, and to consider how they might justify it as well. His book, *The Structure of Scientific Revolutions*, sought to explore the character of scientific change—how theories succeed one another—with a view to considering what explains and what justifies the replacement of one theory by another. The Logical Empiricists and their post-positivist successors held that theories succeed one another by reduction, which preserves what is correct in an earlier theory, and so illuminates the history of science as progress. The reader may remember that these claims were discussed and elaborated in Nagel's and Feyerabend's papers in Part III and Laudan's paper in Part IV above.

Kuhn's research challenges this idea. By introducing considerations from psychology and sociology, as well as history, Kuhn reshaped the landscape in the philosophy of science and made it take seriously the idea that science is not a disinterested pursuit of the truth, successively cumulating in the direction of greater approximation to the truth, as guided by unambiguous observational test. Kuhn's shocking conclusion suggests that science is as creative an undertaking as painting or music, and not to be viewed as more objectively progressive, correct, or approximating to some truth about the world, than these other human activities. The history of science is the history of change, but not progress; in a sense that Kuhn defends, we are no nearer the truth about the nature of things nowadays than we were in Aristotle's time. These radical conclusions represent a great challenge to contemporary philosophy of science.

Dudley Shapere's paper, a review of *The Structure of Scientific Revolutions*, both summarizes and exemplifies the strong reaction within traditional philosophy of science to these views. Kuhn's "Objectivity, Value Judgment, and Theory Choice" was written some ten years after his original book. It clarifies and emphasizes some of the themes of that work. But it also qualifies some of the more radical claims propounded in his earlier writings.

Kuhn's doctrines have generally been interpreted so as to give rise to Relativism, the idea that there are no truths, or at least nothing can be asserted to be true independent of some point of view, and that disagreements among points of view are irreconcilable. The result of course is to deprive science of a position of strength from which it can defend its findings as more well justified than those of pseudoscience; it also undermines the claims of the so-called "hard sciences"—physics and chemistry—to greater authority for their findings, methods, standards of argument and explanation, and strictures on theory construction, than can be claimed by the "soft sciences" and the humanities. Post-modernists and deconstructionists took much support from a radical interpretation of Kuhn's doctrines for the Relativism they embraced.

Among sociologists of science especially, a movement emerged to argue that the same factors which explain scientific successes must also explain scientific failures, and this deprives facts about the world—as reported in the results of observations and experiments—of their decisive role in explaining the success of science. This approach to understanding scientific change is exemplified in David Bloor's "The Strong Program in the Sociology of Knowledge."

The doctrines of Bloor and other sociologists of scientific knowledge had a liberating effect on the social and behavioral sciences and other disciplines which had hitherto sought acceptance by aping "scientific methods" as described by empiricist philosophers of science, but no longer felt the need to do so. The sociological and even more the political focus on science revealed its traditional associations with the middle classes, with capitalism, its blindness towards the interests of women, and indifference to minorities. Elisabeth Anderson's paper "Feminist Epistemology" provides a particularly effective example of the effects of a broadened vision in philosophy of science, without, however, succumbing to the epistemic Relativism which Kuhn did so much to encourage.

As Ernan McMullin argues in "The Social Dimensions of Science," the perspective on science from which the philosophy of science operates can accommodate important insights about the social forces that underlie much scientific change, without undermining the objectivity of scientific methods and the rationality of scientific change. His approach should reassure us that in the end the doctrine that science is not a distinctive body of knowledge, one which attains higher standards of objectivity and reliability than other methods, is not sustainable. This conclusion, however, requires that we return to the fundamental problems in epistemology, the philosophy of language, and metaphysics in order to see where philosophy went wrong and led some of the more radically inclined followers of Kuhn to such extreme conclusions.

25

Dudley Shapere, "The Structure of Scientific Revolutions"

Dudley Shapere was among the first to react to Kuhn's groundbreaking book, *The Structure of Scientific Revolutions*, in a detailed review reprinted below. The review combines full appreciation of the significance of Kuhn's work with incisive criticism and, along with other early critical responses, prompted Kuhn to reflect more on the issues raised in the book in its second edition and in subsequent work. (See "Postscript" to the second edition of *The Structure* and the next selection.)

This important book[1] is a sustained attack on the prevailing image of scientific change as a linear process of ever-increasing knowledge, and an attempt to make us see that process of change in a different and, Kuhn suggests, more enlightening way. In attacking the "concept of development-by-accumulation," Kuhn presents numerous penetrating criticisms not only of histories of science written from that point of view, but also of certain philosophical doctrines (mainly Baconian and positivistic philosophies of science, particularly verification, falsification, and probabilistic views of the acceptance or rejection of scientific theories) which he convincingly argues are associated with that view of history. In this review, I will not deal with those criticisms or with the details of the valuable case studies with which Kuhn tries to support his views; rather, I will concentrate on certain concepts and doctrines which are fundamental to his own interpretation of the development and structure of science. His view, while original and richly suggestive, has much in common with some recent antipositivistic reactions among philosophers of science—most notably, Feyerabend, Hanson, and Toulmin—and inasmuch as it makes explicit, according to Kuhn, "some of the new historiography's implications" (p. 3), it is bound to exert a very

D. Shapere, "The Structure of Scientific Revolutions," *Philosophical Review*, 1964, 73: 383–94.

wide influence among philosophers and historians of science alike. It is therefore a view which merits close examination.

Basic to Kuhn's interpretation of the history of science is his notion of a paradigm. Paradigms are "universally recognized scientific achievements that for a time provide model problems and solutions to a community of practitioners" (p. x). Because a paradigm is "at the start largely a promise of success discoverable in selected and still incomplete examples" (pp. 23–24), it is "an object for further articulation and specification under new or more stringent conditions" (p. 23); hence from paradigms "spring particular coherent traditions of scientific research" (p. 10) which Kuhn calls "normal science." Normal science thus consists largely of "mopping-up operations" (p. 24) devoted to actualizing the initial promise of the paradigm "by extending the knowledge of those facts that the paradigm displays as particularly revealing, by increasing the extent of the match between those facts and the paradigm's predictions, and by further articulation of the paradigm itself" (p. 24). In this process of paradigm development lie both the strength and weakness of normal science: for though the paradigm provides "a criterion for choosing problems that, while the paradigm is taken for granted, can be assumed to have solutions" (p. 37), on the other hand those phenomena "that will not fit the box are often not seen at all" (p. 24). Normal science even "often suppresses fundamental novelties because they are necessarily subversive of its basic commitments. Nevertheless, so long as those commitments retain an element of the arbitrary, the very nature of normal research ensures that novelty shall not be suppressed for very long" (p. 5). Repeated failures of a normal-science tradition to solve a problem or other anomalies that develop in the course of paradigm articulation produce "the tradition-shattering complements to the tradition-bound activity of normal science" (p. 6).

The most pervasive of such tradition-shattering activities Kuhn calls "scientific revolutions."

> Confronted with anomaly or with crisis, scientists take a different attitude toward existing paradigms, and the nature of their research changes accordingly. The proliferation of competing articulations, the willingness to try anything, the expression of explicit discontent, the recourse to philosophy and to debate over fundamentals, all these are symptoms of a transition from normal to extraordinary research [p. 90].
>
> Scientific revolutions are inaugurated by a growing sense . . . that an existing paradigm has ceased to function adequately in the exploration of an aspect of nature to which that paradigm itself had previously led the way [p. 91].

The upshot of such crises is often the acceptance of a new paradigm:

> Scientific revolutions are here taken to be those non-cumulative developmental episodes in which an older paradigm is replaced in whole or in part by an incompatible new one [p. 91].

This interpretation of scientific development places a heavy burden indeed on the notion of a paradigm. Although in some passages we are led to believe that a community's paradigm is simply "a set of recurrent and quasi-standard illustrations of various theories," and that these are "revealed in its textbooks, lectures, and laboratory exercises" (p. 43), elsewhere we find that there is far more to the paradigm than is contained, at least explicitly, in such illustrations. These "accepted examples of actual scientific practice . . . include law, theory, application, and instrumentation together" (p. 10). A paradigm consists of a "strong network of commitments—conceptual, theoretical, instrumental, and methodological" (p. 42); among these commitments are "quasi-metaphysical" ones (p. 41). A paradigm is, or at least includes, "some implicit body of intertwined theoretical and methodological belief that permits selection, evaluation, and criticism" (pp. 16–17). If such a body of beliefs is not implied by the collection of facts (and, according to Kuhn, it never is), "it must be externally supplied, perhaps by a current metaphysic, by another science, or by personal and historical accident" (p. 17). Sometimes paradigms seem to be patterns (sometimes in the sense of archetypes and sometimes in the sense of criteria or standards) upon which we model our theories or other work ("from them as models spring particular coherent traditions"); at other times they seem to be themselves vague theories which are to be refined and articulated. Most fundamentally, though, Kuhn considers them as not being rules, theories, or the like, or a mere sum thereof, but something more "global" (p. 43), from which rules, theories, and so forth are abstracted, but to which no mere statement of rules or theories or the like can do justice. The term "paradigm" thus covers a range of factors in scientific development including or somehow involving laws and theories, models, standards, and methods (both theoretical and instrumental), vague intuitions, explicit or implicit metaphysical beliefs (or prejudices). In short, anything that allows science to accomplish anything can be a part of (or somehow involved in) a paradigm.

Now, historical study does bear out the existence of guiding factors which are held in more or less similar form, to greater or less extent, by a multitude of scientists working in an area over a number of years. What must be asked is whether anything is gained by referring to such common factors as "paradigms," and whether such gains, if any, are offset by confusions that ensue because of such a way of speaking. At the very outset, the explanatory value of the notion of a paradigm is suspect: for the truth of the thesis that shared paradigms are (or are behind) the common factors guiding scientific research appears to be guaranteed, not so much by a close examination of actual historical cases, however scholarly, as by the breadth of definition of the term "paradigm." The suspicion that this notion plays a determinative role in shaping Kuhn's interpretation of history is strengthened by his frequent remarks about what *must* be the case with regard to science and its development: for example, "No natural history can be interpreted in the absence of at least some . . . belief" (pp. 16–17); "Once a first paradigm through which to view nature has been found, there is no such thing as research in the absence of any paradigm" (p. 79); "no experiment can be conceived without some sort of theory" (p. 87); "if, as I have already urged, there can be no scientifically or empirically neutral system of language or concepts, then the proposed construction of alternate tests and theories must proceed from within one or another paradigm-based tradition" (p. 145). Such views appear too strongly and confidently held to have been extracted from a mere investigation of how things *have* happened.

Still greater perplexities are generated by Kuhn's view that paradigms cannot, in general, be formulated adequately. According to him, when the historian tries to state the rules which scientists follow, he finds that "phrased in just that way, or in any other way he can imagine, they would almost certainly have been rejected by some members of the group he studies" (p. 44). Similarly, there may be many versions of the same theory. It would appear that, in Kuhn's eyes, the concepts, laws, theories, rules, and so forth that are common to a group are just not common enough to guarantee the coherence of the tradition; therefore he concludes that the paradigm, "the concrete scientific achievement" that is the source of that coherence, must not be identified with, but must be seen as "prior to the various concepts, laws, theories, and points of view that may be abstracted from it" (p. 11). (It is partly on the basis of this argument that Kuhn rejects the attempt by philosophers of science to formulate a "logic" of science in terms of precise rules.) Yet if it is true that all that can be said about paradigms and scientific development can and must be said only in terms of what are mere "abstractions" from paradigms, then it is difficult to see what is gained by appealing to the notion of a paradigm.

In Kuhn's view, however, the fact that paradigms cannot be described adequately in words does not hinder us from recognizing them: they are open to "direct inspection" (p. 44), and historians can "agree in their *identification* of a paradigm without agreeing on, or even attempting to produce, a full *interpretation* or *rationalization* of it" (p. 44). Yet the feasibility of a historical inquiry concerning paradigms is exactly what is brought into question by the scope of the term "paradigm" and the inaccessibility of particular paradigms to verbal formulation. For on the one hand, as we have seen, it is *too* easy to identify a paradigm; and on the other hand, it is not easy to determine, in particular cases treated by Kuhn, what the paradigm is supposed to have been in that case. In most of the cases he discusses, it is the theory that is doing the job of posing problems, providing criteria for selection of data, being articulated, and so forth. But of course the theory is not the paradigm, and we might assume that Kuhn discusses the theory because it is as near as he can get in words to the inexpressible paradigm. This, however, only creates difficulties. In the case of "what is perhaps our fullest example of a scientific revolution" (p. 132), for instance, what was "assimilated" when Dalton's theory (paradigm) became accepted? Not merely the laws of combining proportions, presumably, but something "prior to" them. Was it, then, the picture of matter as constituted of atoms? But contrary to the impression Kuhn gives, that picture was never even nearly universally accepted: from Davy to Ostwald and beyond there was always a very strong faction which "regarded it with misgiving, or with positive dislike, or with a constant hope for an effective substitute" (J. C. Gregory, *A Short History of Atomism* [London, 1931], p. 93), some viewing atoms as convenient fictions, others eschewing the vocabulary of atoms entirely, preferring to talk in terms of "proportions" or "equivalents." (It is noteworthy that Dalton was presented with a Royal Medal, not unequivocally for his development of the atomic theory, but rather "for his development of the Theory of Definite Proportions, usually called the Atomic Theory of Chemistry"; award citation, quoted in Gregory, p. 84.) No, it was certainly not atoms to which the most creative chemists of the century were "committed"—unless (contrary to his general mode of expression) Kuhn means that they were "committed" to the atomic theory because they—most of them—used it even though they did not believe in its truth. Further, what else was "intertwined" in this behind-the-scenes paradigm? Did it include, for instance, some inexpressible Principle of Uniformity of Nature or Law of Causality? Is this question so easy to answer—a matter of "direct inspection"—after all these years of philosophical dispute? One begins to doubt that paradigms are open to "direct inspection," or else to be amazed at Professor Kuhn's eyesight. (And why is it that such historical facts should be open to direct inspection, whereas scientific facts must always be seen "through" a paradigm?) But if there are such difficulties, how can historians know that they agree in their identification of the paradigms present in historical episodes, and so determine that "the same" paradigm persists through a long sequence of such episodes? They cannot, by hypothesis, compare their formulations. Suppose they disagree: how is their dispute to be resolved?

On the other hand, where do we draw the line between different paradigms and different articulations of the same paradigm? It is natural and common to say that Newton, d'Alembert, Lagrange, Hertz, Hamilton, Mach, and others formulated different versions of classical mechanics; yet certainly some of these formulations involved different "commitments"—for example, some to forces, others to energy, some to vectorial, others to variational principles. The distinction between paradigms and different articulations of a paradigm, and between scientific revolutions and normal science, is at best a matter of degree, as is commitment to a paradigm: expression of explicit discontent, proliferation of competing articulations, debate over fundamentals are all more or less present throughout the development of science; and there are always guiding elements which are more or less common, even among what are classified as different "traditions." This is one reason why, in particular cases, identification of "the paradigm" is so difficult: not just because it is hard to see, but because looking for the guiding elements in scientific activity is not like looking for a unitary entity that either is there or is not.

But furthermore, the very reasons for supposing that paradigms (nevertheless) exist are unconvincing. No doubt some theories are very similar—so similar that they can be considered to be "versions" or "different articulations" of one another (or of "the same subject"). But does this imply that there must be a common "paradigm" of which the similar theories are incomplete expressions and from which they are abstracted? No doubt, too, many expressions of methodological rules are not as accurate portrayals of scientific method as they are claimed to be; and it is possible that Kuhn is right in claiming that no such portrayal can be given in terms of any one set of precise rules. But such observations, even if true, do not compel us to adopt a *mystique* regarding a single paradigm which guides procedures, any more than our inability to give a single, simple definition of "game" means that we must have a unitary but inexpressible idea from which all our diverse uses of "game" are abstracted. It may be true that "The coherence displayed by the research tradition . . . may not imply even the existence of an underlying body of rules and assumptions" (p. 46); but neither does it imply the existence of an underlying "paradigm."

Finally, Kuhn's blanket use of the term "paradigm" to cover such a variety of activities and functions obscures important differences between those activities and functions. For example, Kuhn claims that "an apparently arbitrary element . . . is always a formative ingredient" (p. 4) of a paradigm; and, indeed, as we shall see shortly, this is a central aspect of his view of paradigms and scientific change. But is the acceptance or rejection of a scientific theory "arbitrary" in the same sense that acceptance or rejection of a standard (to say nothing of a metaphysical belief) is? Again, Newtonian and Hertzian formulations of classical mechanics are similar to one another, as are the Einstein, Whitehead, Birkhoff, and Milne versions of relativity, and as are wave mechanics and matrix mechanics. But there are significant differences in the ways in and degrees to which these theories are "similar"—differences which are masked by viewing them all equally as different articulations of the same paradigm.

There are, however, deeper ways in which Kuhn's notion of a paradigm affects adversely his analysis of science; and it is in these ways that his view reflects widespread and important tendencies in both the history and philosophy of science today.

Because a paradigm is

> the source of the methods, problem-field, and standards of solution accepted by any mature scientific community at any given time, . . . the reception of a new paradigm often necessitates a redefinition of the corresponding science. . . . And as the problems change, so, often, does the standard that distinguishes a real scientific solution from a mere metaphysical speculation, word game, or mathematical play. The normal-scientific tradition that emerges from a scientific revolution is not only incompatible but often actually incommensurable with that which has gone before [p. 102].

Thus the paradigm change entails "changes in the standards governing permissible problems, concepts, and explanations" (p. 105). In connection with his view that concepts or meanings change from one theory (paradigm) to another despite the retention of the same terms, Kuhn offers an argument whose conclusion is both intrinsically important and crucial to much of his book. This argument is directed against the "positivistic" view that scientific advance is cumulative, and that therefore earlier sciences are derivable from later; the case he considers is the supposed deducibility of Newtonian from Einsteinian dynamics, subject to limiting conditions. After summarizing the usual derivation, Kuhn objects that

> the derivation is spurious, at least to this point. Though the [derived statements] are a special case of the laws of relativistic mechanics, they are not Newton's Laws. Or at least they are not unless those laws are reinterpreted in a way that would have been impossible until after Einstein's work. . . . The physical referents of these Einsteinian concepts are by no means identical with those of the Newtonian concepts that bear the same name. (Newtonian mass is conserved; Einsteinian is convertible with energy. Only at low relative velocities may the two be measured in the same way, and even then they must not be conceived to be the same.) . . . The argument has still not done what it purported

> to do. It has not, that is, shown Newton's Laws to be a limiting case of Einstein's. For in the passage to the limit it is not only the forms of the laws that have changed. Simultaneously we have had to alter the fundamental structural elements of which the universe to which they apply is composed [pp. 100–101].

But Kuhn's argument amounts simply to an assertion that despite the derivability of expressions which are in every formal respect identical with Newton's Laws, there remain differences of "meaning." What saves this from begging the question at issue? His only attempt to support his contention comes in the parenthetical example of mass; but this point is far from decisive. For one might equally well be tempted to say that the "concept" of mass (the "meaning" of "mass") has remained the same (thus accounting for the deducibility) even though the *application* has changed. Similarly, rather than agree with Kuhn that "the Copernicans who denied its traditional title 'planet' to the sun . . . were changing the meaning of 'planet'" (p. 127), one might prefer to say that they changed only the application of the term. The real trouble with such arguments arises with regard to the cash difference between saying, in such cases, that the "meaning" has changed, as opposed to saying that the "meaning" has remained the same though the "application" has changed. Kuhn has offered us no clear analysis of "meaning" or, more specifically, no criterion of change of meaning; consequently it is not clear why he classifies such changes as changes of meaning rather than, for example, as changes of application. This is not to say that no such criterion could be formulated, or that a distinction between change of meaning and change of application could not be made, or that it might not be very profitable to do so for certain purposes. One might, for example, note that there are statements that can be made, questions that can be raised, views that may be suggested as possibly correct, within the context of Einsteinian physics that would not even have made sense—would have been self-contradictory—in the context of Newtonian physics. And such differences might (for certain purposes) be referred to with profit as changes of meaning, indicating, among other things, that there are differences between Einsteinian and Newtonian terms that are not brought out by the deduction of Newtonian-like statements from Einsteinian ones. But attributing such differences to alterations of "meaning" must not blind one to any resemblances there might be between the two sets of terms. Thus it is not so much Kuhn's conclusion that is objectionable as, first, the fact that it is based, not on any solid argument, but on the feature of meaning dependence which Kuhn has built into the term "paradigm" (scientists see the world from different points of view, through different paradigms, and therefore see different things through different paradigms); and second, the fact that this feature leads him to a distorted portrayal of the relations between different scientific theories. For Kuhn's term "paradigm," incorporating as it does the view that statements of fact are (to use Hanson's expression) theory-laden, and as a consequence the notion of (in Feyerabend's words) meaning variance from one theory or paradigm to another, calls attention excessively to the differences between theories or paradigms, so that relations that evidently do exist between them are in fact passed over or denied.

The significance of this point emerges fully when we ask about the grounds for accepting one paradigm as better than another. For if "the differences between successive paradigms are both necessary and irreconcilable" (p. 102), and if those differences consist in the paradigms' being "incommensurable"—if they disagree as to what the facts are, and even as to the real problems to be faced and the standards which a successful theory must meet—then what are the two paradigms disagreeing about? And why does one win? There is little problem for Kuhn in analyzing the notion of progress within a paradigm tradition (and, indeed, he notes, such evolution is the source of the prevailing view of scientific advance as "linear"); but how can we say that "progress" is made when one paradigm replaces another? The logical tendency of Kuhn's position is clearly toward the conclusion that the replacement is not cumulative, but is mere change: being "incommensurable," two paradigms cannot be judged according to their ability to solve the same problems, or deal with the same facts, or meet the same standards. "If there were but one set of scientific problems, one world within which to work on them, and one set of standards for their solution, paradigm competition might be settled more or less routinely. . . . But . . . The proponents of competing paradigms are always at least slightly at cross-purposes" (pp. 146–147). Hence "the competition between paradigms is not the sort of battle that can be resolved by proofs" (p. 147), but is more like a "conversion experience" (p. 150). In fact, in so far as one can compare the weights of evidence of two competing paradigms—and, on Kuhn's view that after a scientific revolution "the whole network of fact and theory . . . has shifted" (p. 140), one must wonder how this can be done at all—the weight of evidence is more often in favor of the older paradigm than the new (pp. 155–156). "What occurred was neither a decline nor a raising of standards, but simply a change demanded by the adoption of a new paradigm" (p. 107). "In these matters neither truth nor error is at issue" (p. 150); indeed, Kuhn's view of the history of science implies that "We may . . . have to relinquish the notion, explicit or implicit, that changes of paradigm carry scientists and those who learn from them closer and closer to the truth" (p. 169).

Kuhn is well aware of the relativism implied by his view, and his common sense and feeling for history make him struggle mightily to soften the dismal conclusion. It is, for instance, only "often" that the reception of a new paradigm necessitates a redefinition of the corresponding science. Proponents of different paradigms are only "at least partially" at cross-purposes. Though they "see different things when they look from the same point in the same direction," this is "not to say that they can see anything they please. Both are looking at the world" (p. 149). It is only "in some areas" that "they see different things" (p. 149). But these qualifications are more the statement of the problems readers will find with Kuhn's views than the solutions of those problems. And it is small comfort to be told, in the closing pages of the book, that "a sort of progress will inevitably characterize the scientific enterprise" (p. 169), especially if that "progress," whether or not it is aimed toward final truth, is not at least an advance over past error. Nor will careful readers feel reassured when they are asked, rhetorically, "What better criterion [of scientific progress] than the decision of the scientific group could there be?" (p. 169). For Kuhn has already told us that the decision of a scientific group to adopt a new paradigm is not based on good reasons; on the contrary, what counts as a good reason is determined by the decision.

A view such as Kuhn's had, after all, to be expected sooner or later from someone versed in the contemporary treatment of the history of science. For the great advances in that subject since Duhem have shown how much more there was to theories that were supposedly overthrown and superseded than had been thought. Historians now find that "the more carefully they study, say, Aristotelian dynamics, phlogistic chemistry, or caloric thermodynamics, the more certain they feel that those once current views of nature were, as a whole, neither less scientific nor more the product of human idiosyncrasy than those current today" (p. 2). Yet perhaps that deep impression has effected too great a reaction; for that there is more to those theories than was once thought does not mean that they are immune to criticism—that there are not *good* reasons for their abandonment and replacement by others. And while Kuhn's book calls attention to many mistakes that have been made regarding the (good) reasons for scientific change, it fails itself to illuminate those reasons, and even obscures the existence of such reasons. We must, as philosophers of science, shape our views of the development and structure of scientific thought in the light of what we learn from science and its history. But until historians of science achieve a more balanced approach to their subject—neither too positivistic nor too relativistic—philosophers must receive such presentations of evidence with extremely critical eyes.

Certainly there is a vast amount of positive value in Kuhn's book. Besides making many valid critical remarks, it does bring out, through a wealth of case studies, many common features of scientific thought and activities which make it possible and, for many purposes, revealing to speak of "traditions" in science; and it points out many significant differences between such traditions. But Kuhn, carried away by the logic of his notion of a paradigm, glosses over many important differences between scientific activities classified as being of the same tradition, as well as important continuities between successive traditions. He is thus led to deny, for example, that Einsteinian dynamics is an advance over Newtonian or Aristotelian dynamics in a sense more fundamental than can consistently be extracted from his conceptual apparatus. If one holds, without careful qualification, that the world is seen and interpreted "through" a paradigm, or that theories are "incommensurable," or that there is "meaning variance" between theories, or that all statements of fact are "theory-laden," then one may be led all too readily into relativism with regard to the development of science. Such a view is no more implied by historical facts than is the opposing view that scientific development consists solely of the removal of superstition, prejudice, and other obstacles to scientific progress in the form of purely incremental advances toward final truth. Rather, I have tried to show, such relativism, while it may seem to be suggested by a half-century of deeper study of discarded theories, is a *logical* outgrowth of conceptual confusions, in Kuhn's case owing primarily to the use of a blanket term. For his view is made to appear convincing only by inflating the definition of "paradigm" until that term becomes so vague and ambiguous that it cannot easily be withheld, so general that it cannot easily be applied, so mysterious that it cannot help explain, and so misleading that it is a positive hindrance to the understanding of some central aspects of science; and then, finally, these excesses must be counterbalanced by qualifications that simply contradict them. There are many other facets of Kuhn's book that deserve attention—especially his view that a paradigm "need not, and in fact never does, explain all the facts with which it can be confronted" (p. 18), and his suggestion that no paradigm ever could be found which would do so. But the difficulties that have been discussed here indicate clearly that the expanded version of this book which Kuhn contemplates will require not so much further historical evidence (p. xi) as—at the very least—more careful scrutiny of his tools of analysis.

Note

1 *The Structure of Scientific Revolutions*. By Thomas S. Kuhn. (Chicago, University of Chicago Press, 1962. Pp. xiv, 172.) All page references, unless otherwise noted, are to this work.

26

Thomas Kuhn, "Objectivity, Value Judgment, and Theory Choice"

Thomas Kuhn's (1922–1996) emphasis in this paper is on value judgments employed by scientists in situations of theory choice. He identifies some characteristics—theoretical values—that he thinks any good theory ought to possess: accuracy, consistency, scope, simplicity, fruitfulness, and shows how their various combinations were taken into account by scientists at some crucial moments, when they had to discriminate among competing theoretical alternatives. Does it make theory choice an *objective* process, contrary to what Kuhn claimed earlier? You may guess that the answer is no.

In the penultimate chapter of a controversial book first published fifteen years ago, I considered the ways scientists are brought to abandon one time-honored theory or paradigm in favor of another. Such decision problems, I wrote, "cannot be resolved by proof." To discuss their mechanism is, therefore, to talk "about techniques of persuasion, or about argument and counterargument in a situation in which there can be no proof." Under these circumstances, I continued, "lifelong resistance [to a new theory] . . . is not a violation of scientific standards. . . . Though the historian can always find men—Priestley, for instance—who were unreasonable to resist for as long as they did, he will not find a point at which resistance becomes illogical or unscientific."[1] Statements of that sort obviously raise the question of why, in the absence of binding criteria for scientific choice, both the number of solved scientific problems and the precision of individual problem solutions should increase so markedly with the passage of time. Confronting that issue, I sketched in my closing chapter a number of characteristics that scientists share by virtue of the training which licenses their membership in one or another community of specialists. In the absence of criteria able to dictate

T. Kuhn, *The Essential Tension: Selected Studies in Scientific Tradition and Change*, 1977, pp. 320–39. Chicago: The University of Chicago Press.

the choice of each individual, I argued, we do well to trust the collective judgment of scientists trained in this way. "What better criterion could there be," I asked rhetorically, "than the decision of the scientific group?"[2]

A number of philosophers have greeted remarks like these in a way that continues to surprise me. My views, it is said, make of theory choice "a matter for mob psychology."[3] Kuhn believes, I am told, that "the decision of a scientific group to adopt a new paradigm cannot be based on good reasons of any kind, factual or otherwise."[4] The debates surrounding such choices must, my critics claim, be for me "mere persuasive displays without deliberative substance."[5] Reports of this sort manifest total misunderstanding, and I have occasionally said as much in papers directed primarily to other ends. But those passing protestations have had negligible effect, and the misunderstandings continue to be important. I conclude that it is past time for me to describe, at greater length and with greater precision, what has been on my mind when I have uttered statements like the ones with which I just began. If I have been reluctant to do so in the past, that is largely because I have preferred to devote attention to areas in which my views diverge more sharply from those currently received than they do with respect to theory choice.

What, I ask to begin with, are the characteristics of a good scientific theory? Among a number of quite usual answers I select five, not because they are exhaustive, but because they are individually important and collectively sufficiently varied to indicate what is at stake. First, a theory should be accurate: within its domain, that is, consequences deducible from a theory should be in demonstrated agreement with the results of existing experiments and observations. Second, a theory should be consistent, not only internally or with itself, but also with other currently accepted theories applicable to related aspects of nature. Third, it should have broad scope: in particular, a theory's consequences should extend far beyond the particular observations, laws, or subtheories it was initially designed to explain. Fourth, and closely related, it should be simple, bringing order to phenomena that in its absence would be individually isolated and, as a set, confused. Fifth—a somewhat less standard item, but one of special importance to actual scientific decisions—a theory should be fruitful of new research findings: it should, that is, disclose new phenomena or previously unnoted relationships among those already known.[6] These five characteristics—accuracy, consistency, scope, simplicity, and fruitfulness—are all standard criteria for evaluating the adequacy of a theory. If they had not been, I would have devoted far more space to them in my book, for I agree entirely with the traditional view that they play a vital role when scientists must choose between an established theory and an upstart competitor. Together with others of much the same sort, they provide *the* shared basis for theory choice.

Nevertheless, two sorts of difficulties are regularly encountered by the men who must use these criteria in choosing, say, between Ptolemy's astronomical theory and Copernicus's, between the oxygen and phlogiston theories of combustion, or between Newtonian mechanics and the quantum theory. Individually the criteria are imprecise: individuals may legitimately differ about their application to concrete cases. In addition, when deployed together, they repeatedly prove to conflict with one another; accuracy may, for example, dictate the choice of one theory, scope the choice of its competitor. Since these difficulties, especially the first, are also relatively familiar, I shall devote little time to their elaboration. Though my argument does demand that I illustrate them briefly, my views will begin to depart from those long current only after I have done so.

Begin with accuracy, which for present purposes I take to include not only quantitative agreement but qualitative as well. Ultimately it proves the most nearly decisive of all the criteria, partly because it is less equivocal than the others but especially because predictive and explanatory powers, which depend on it, are characteristics that scientists are particularly unwilling to give up. Unfortunately, however, theories cannot always be discriminated in terms of accuracy. Copernicus's system, for example, was not more accurate than Ptolemy's until drastically revised by Kepler more than sixty years after Copernicus's death. If Kepler or someone else had not found other reasons to choose heliocentric astronomy, those improvements in accuracy would never have been made, and Copernicus's work might have been forgotten. More typically, of course, accuracy does permit discriminations, but not the sort that lead regularly to unequivocal choice. The oxygen theory, for example, was universally acknowledged to account for observed weight relations in chemical reactions, something the phlogiston theory had previously scarcely attempted to do. But the phlogiston theory, unlike its rival, could account for the metals' being much more alike than the ores from which they were formed. One theory thus matched experience better in one area, the other in another. To choose between them on the basis of accuracy, a scientist would need to decide the area in which accuracy was more significant. About that matter chemists could and did differ without violating any of the criteria outlined above, or any others yet to be suggested.

However important it may be, therefore, accuracy by itself is seldom or never a sufficient criterion for theory choice. Other criteria must function as well, but they do not eliminate problems. To illustrate I select just two—consistency and simplicity—asking how they functioned in the choice between the heliocentric and geocentric systems. As astronomical theories both Ptolemy's and Copernicus's were internally consistent, but their relation to related theories in other fields was very different. The stationary central earth was an essential ingredient of received physical theory, a tight-knit body of doctrine which explained, among other things, how stones fall, how water pumps function, and why the clouds move slowly across the skies. Heliocentric astronomy, which required the earth's motion, was inconsistent with the existing scientific explanation of these and other terrestrial phenomena. The consistency criterion, by itself, therefore, spoke unequivocally for the geocentric tradition.

Simplicity, however, favored Copernicus, but only when evaluated in a quite special way. If, on the one hand, the two systems were compared in terms of the actual computational labor required to predict the position of a planet at a particular time, then they proved substantially equivalent. Such computations were what astronomers did, and Copernicus's system offered them no labor-saving techniques; in that sense it was not simpler than Ptolemy's. If, on the other hand, one asked about the amount of mathematical apparatus required to explain, not the detailed quantitative motions of the planets, but merely their gross qualitative features—limited elongation, retrograde motion, and the like—then, as every schoolchild knows, Copernicus required only one circle per planet, Ptolemy two. In that sense the Copernican theory was the simpler, a fact vitally important to the choices made by both Kepler and Galileo and thus essential to the ultimate triumph of Copernicanism. But that sense of simplicity was not the only one available, nor even the one most natural to professional astronomers, men whose task was the actual computation of planetary position.

Because time is short and I have multiplied examples elsewhere, I shall here simply assert that these difficulties in applying standard criteria of choice are typical and that they arise no less forcefully in twentieth-century situations than in the earlier and better-known examples I have just sketched. When scientists must choose between competing theories, two men fully committed to the same list of criteria for choice may nevertheless reach different conclusions. Perhaps they interpret simplicity differently or have different convictions about the range of fields within which the consistency criterion must be met. Or perhaps they agree about these matters but differ about the relative weights to be accorded to these or to other criteria when several are deployed together. With respect to divergences of this sort, no set of choice criteria yet proposed is of any use. One can explain, as the historian characteristically does, why particular men made particular choices at particular times. But for that purpose one must go beyond the list of shared criteria to characteristics of the individuals who make the choice. One must, that is, deal with characteristics which vary from one scientist to another without thereby in the least jeopardizing their adherence to the canons that make science scientific. Though such canons do exist and should be discoverable (doubtless the criteria of choice with which I began are among them), they are not by themselves sufficient to determine the decisions of individual scientists. For that purpose the shared canons must be fleshed out in ways that differ from one individual to another.

Some of the differences I have in mind result from the individual's previous experience as a scientist. In what part of the field was he at work when confronted by the need to choose? How long had he worked there; how successful had he been; and how much of his work depended on concepts and techniques challenged by the new theory? Other factors relevant to choice lie outside the sciences. Kepler's early election of Copernicanism was due in part to his immersion in the Neoplatonic and Hermetic movements of his day; German Romanticism predisposed those it affected toward both recognition and acceptance of energy conservation; nineteenth-century British social thought had a similar influence on the availability and acceptability of Darwin's concept of the struggle for existence. Still other significant differences are functions of personality. Some scientists place more premium than others on originality and are correspondingly more willing to take risks; some scientists prefer comprehensive, unified theories to precise and detailed problem solutions of apparently narrower scope. Differentiating factors like these are described by my critics as subjective and are contrasted with the shared or objective criteria from which I began. Though I shall later question that use of terms, let me for the moment accept it. My point is, then, that every individual choice between competing theories depends on a mixture of objective and subjective factors, or of shared and individual criteria. Since the latter have not ordinarily figured in the philosophy of science, my emphasis upon them has made my belief in the former hard for my critics to see.

What I have said so far is primarily simply descriptive of what goes on in the sciences at times of theory choice. As description, furthermore, it has not been challenged by my critics, who reject instead my claim that these facts of scientific life have philosophic import. Taking up that issue, I shall begin to isolate some, though I think not vast, differences of opinion. Let me begin by asking how philosophers of science can for so long have neglected the subjective elements which, they freely grant, enter regularly into the actual theory choices made by individual scientists? Why have these elements seemed to them an index only of human weakness, not at all of the nature of scientific knowledge?

One answer to that question is, of course, that few philosophers, if any, have claimed to possess either a complete or an entirely well-articulated list of criteria. For some time, therefore, they could reasonably expect that further research would eliminate residual imperfections and produce an algorithm able to dictate rational, unanimous choice. Pending that achievement, scientists would have no alternative but to supply subjectively what the best current list of objective criteria still lacked. That some of them might still do so even with a perfected list at hand would then be an index only of the inevitable imperfection of human nature.

That sort of answer may still prove to be correct, but I think no philosopher still expects that it will. The search for algorithmic decision procedures has continued for some time and produced both powerful and illuminating results. But those results all presuppose that individual criteria of choice can be unambiguously stated and also that, if more than one proves relevant, an appropriate weight function is at hand for their joint application. Unfortunately, where the choice at issue is between scientific theories, little progress has been made toward the first of these desiderata and none toward the second. Most philosophers of science would, therefore, I think, now regard the sort of algorithm which has traditionally been sought as a not quite attainable ideal. I entirely agree and shall henceforth take that much for granted.

Even an ideal, however, if it is to remain credible, requires some demonstrated relevance to the situations in which it is supposed to apply. Claiming that such demonstration requires no recourse to subjective factors, my critics seem to appeal, implicitly or explicitly, to the well-known distinction between the contexts of discovery and of justification.[7] They concede, that is, that the subjective factors I invoke play a significant role in the discovery or invention of new theories, but they also insist that that inevitably intuitive process lies outside of the bounds of philosophy of science and is irrelevant to the question of scientific objectivity. Objectivity enters science, they continue, through the processes by which theories are tested, justified, or judged. Those processes do not, or at least need not, involve subjective factors at all. They can be governed by a set of (objective) criteria shared by the entire group competent to judge.

I have already argued that that position does not fit observations of scientific life and shall now assume that that much has been conceded. What is now at issue is a different point: whether or not this invocation of the distinction between contexts of discovery and of justification provides even a plausible and useful idealization. I think it does not and can best make my point by suggesting first a likely source of its apparent cogency. I suspect that my critics have been misled by science pedagogy or what I have elsewhere called textbook science. In science teaching, theories are presented together with exemplary applications, and those applications may be viewed as evidence. But that is not their primary pedagogic function (science students are distressingly willing to receive the word from professors and texts). Doubtless *some* of them were *part* of the evidence at the time actual decisions were being made, but they represent only a fraction of the considerations relevant to the decision process. The context of pedagogy differs almost as much from the context of justification as it does from that of discovery.

Full documentation of that point would require longer argument than is appropriate here, but two aspects of the way in which philosophers ordinarily demonstrate the relevance of choice criteria are worth noting. Like the science textbooks on which they are often modeled, books and articles on the philosophy of science refer again and again to the famous crucial experiments: Foucault's pendulum, which demonstrates the motion of the earth; Cavendish's demonstration of gravitational attraction; or Fizeau's measurement of the relative speed of sound in water and air. These experiments are paradigms of good reason for scientific choice; they illustrate the most effective of all the sorts of argument which could be available to a scientist uncertain which of two theories to follow; they are vehicles for the transmission of criteria of choice. But they also have another characteristic in common. By the time they were performed no scientist still needed to be convinced of the validity of the theory their outcome is now used to demonstrate. Those decisions had long since been made on the basis of significantly more equivocal evidence. The exemplary crucial experiments to which philosophers again and again refer would have been historically relevant to theory choice only if they had yielded unexpected results. Their use as illustrations provides needed economy to science pedagogy, but they scarcely illuminate the character of the choices that scientists are called upon to make.

Standard philosophical illustrations of scientific choice have another troublesome characteristic. The only arguments discussed are, as I have previously indicated, the ones favorable to the theory that, in fact, ultimately triumphed. Oxygen, we read, could explain weight relations, phlogiston could not; but nothing is said about the phlogiston theory's power or about the oxygen theory's limitations. Comparisons of Ptolemy's theory with Copernicus's proceed in the same way. Perhaps these examples should not be given since they contrast a developed theory with one still in its infancy. But philosophers regularly use them nonetheless. If the only result of their doing so were to simplify the decision situation, one could not object. Even historians do not claim to deal with the full factual complexity of the situations they describe. But these simplifications emasculate by making choice totally unproblematic. They eliminate, that is, one essential element of the decision situations that scientists must resolve if their field is to move ahead. In those situations there are always at least some good reasons for each possible choice. Considerations relevant to the context of discovery are then relevant to justification as well; scientists who share the concerns and sensibilities of the individual who discovers a new theory are ipso facto likely to appear disproportionately frequently among that theory's first supporters. That is why it has been difficult to construct algorithms for theory choice, and also why such difficulties have seemed so thoroughly worth resolving. Choices that present problems are the ones philosophers of science need to understand. Philosophically interesting decision procedures must function where, in their absence, the decision might still be in doubt.

That much I have said before, if only briefly. Recently, however, I have recognized another, subtler source for the apparent plausibility of my critics' position. To present it, I shall briefly describe a hypothetical dialogue with one of them. Both of us agree that each scientist chooses between competing theories by deploying some Bayesian algorithm which permits him to compute a value for $p(T,E)$, i.e., for the probability of a theory T on the evidence E available both to him and to the other members of his professional group at a particular period of time. "Evidence," furthermore, we both interpret broadly to include such considerations as simplicity and fruitfulness. My critic asserts, however, that there is only one such value of p, that corresponding to objective choice, and he believes that all rational members of the group must arrive at it. I assert, on the other hand, for reasons previously given, that the factors he calls objective are insufficient to determine in full any algorithm at all. For the sake of the discussion I have conceded that each individual has an algorithm and that all their algorithms have much in common. Nevertheless, I continue to hold that the algorithms of individuals are all ultimately different by virtue of the subjective considerations with which each must complete the objective criteria before any computations can be done. If my hypothetical critic is liberal, he may now grant that these subjective differences do play a role in determining the hypothetical algorithm on which each individual relies during the early stages of the competition between rival theories. But he is also likely to claim that, as evidence increases with the passage of time, the algorithms of different individuals converge to the algorithm of objective choice with which his presentation began. For him the increasing unanimity of individual choices is evidence for their increasing objectivity and thus for the elimination of subjective elements from the decision process.

So much for the dialogue, which I have, of course, contrived to disclose the non sequitur underlying an apparently plausible position. What converges as the evidence changes over time need only be the values of p that individuals compute from their individual algorithms. Conceivably those algorithms themselves also become more alike with time, but the ultimate unanimity of theory choice provides no evidence whatsoever that they do so. If subjective factors are required to account for the decisions that initially divide the profession, they may still be present later when the profession agrees. Though I shall not here argue the point, consideration of the occasions on which a scientific community divides suggests that they actually do so.

My argument has so far been directed to two points. It first provided evidence that the choices scientists make between competing theories depend not only on shared criteria—those my critics call objective—but also on idiosyncratic factors dependent on individual biography and personality. The latter are, in my critics' vocabulary, subjective, and the second part of my argument has attempted to bar some likely ways of denying their philosophic import. Let me now shift to a more positive approach, returning briefly to the list of shared criteria—accuracy, simplicity, and the like—with which I began. The considerable effectiveness of such criteria does not, I now wish to suggest, depend on their being sufficiently articulated to dictate the choice of each individual who subscribes to them. Indeed, if they were articulated to that extent, a behavior mechanism fundamental to scientific advance would cease to function. What the tradition sees as eliminable imperfections in its rules of choice I take to be in part responses to the essential nature of science.

As so often, I begin with the obvious. Criteria that influence decisions without specifying what those decisions must be are familiar in many aspects of human life. Ordinarily, however, they are called, not criteria or rules, but maxims, norms, or values. Consider maxims first. The individual who invokes them when choice is urgent usually finds them frustratingly vague and often also in conflict one with another. Contrast "He who hesitates is lost" with "Look before you leap," or compare "Many hands make light work" with "Too many cooks spoil the broth." Individually maxims dictate different choices, collectively none at all. Yet no one suggests that supplying children with contradictory tags like these is irrelevant to their education. Opposing maxims alter the nature of the decision to be made, highlight the essential issues it presents, and point to those remaining aspects of the decision for which each individual must take responsibility himself. Once invoked, maxims like these alter the nature of the decision process and can thus change its outcome.

Values and norms provide even clearer examples of effective guidance in the presence of conflict and equivocation. Improving the quality of life is a value, and a car in every garage once followed from it as a norm. But quality of life has other aspects, and the old norm has become problematic. Or again, freedom of speech is a value, but so is preservation of life and property. In application, the two often conflict, so that judicial soul-searching, which still continues, has been required to prohibit such behavior as inciting to riot or shouting fire in a crowded theater. Difficulties like these are an appropriate source for frustration, but they rarely result in charges that values have no function or in calls for their abandonment. That response is barred to most of us by an acute consciousness that there are societies with other values and that these value differences result in other ways of life, other decisions about what may and what may not be done.

I am suggesting, of course, that the criteria of choice with which I began function not as rules, which determine choice, but as values, which influence it. Two men deeply committed to the same values may nevertheless, in particular situations, make different choices as, in fact, they do. But that difference in outcome ought not to suggest that the values scientists share are less than critically important either to their decisions or to the development of the enterprise in which they participate. Values like accuracy, consistency, and scope may prove ambiguous in application, both individually and collectively; they may, that is, be an insufficient basis for a *shared* algorithm of choice. But they do specify a great deal: what each scientist must consider in reaching a decision, what he may and may not consider relevant, and what he can legitimately be required to report as the basis for the choice he has made. Change the list, for example by adding social utility as a criterion, and some particular choices will be different, more like those one expects from an engineer. Subtract accuracy of fit to nature from the list, and the enterprise that results may not resemble science at all, but perhaps philosophy instead. Different creative disciplines are characterized, among other things, by different sets of shared values. If philosophy and engineering lie too close to the sciences, think of literature or the plastic arts. Milton's failure to set *Paradise Lost* in a Copernican universe does not indicate that he agreed with Ptolemy but that he had things other than science to do.

Recognizing that criteria of choice can function as values when incomplete as rules has, I think, a number of striking advantages. First, as I have already argued at length, it accounts in detail for aspects of scientific behavior which the tradition has seen as anomalous or even irrational. More important, it allows the standard criteria to function fully in the earliest stages of theory choice, the period when they are most needed but when, on the traditional view, they function badly or not at all. Copernicus was responding to them during the years required to convert heliocentric astronomy from a global conceptual scheme to mathematical machinery for predicting planetary position. Such predictions were what astronomers valued; in their absence, Copernicus would scarcely have been heard, something which had happened to the idea of a moving earth before. That his own version convinced very few is less important than his acknowledgement of the basis on which judgments would have to be reached if heliocentricism were to survive. Though idiosyncrasy must be invoked to explain why Kepler and Galileo were early converts to Copernicus's system, the gaps filled by their efforts to perfect it were specified by shared values alone.

That point has a corollary which may be more important still. Most newly suggested theories do not survive. Usually the difficulties that evoked them are accounted for by more traditional means. Even when this does not occur, much work, both theoretical and experimental, is ordinarily required before the new theory can display sufficient accuracy and scope to generate widespread conviction. In short, before the group accepts it, a new theory has been tested over time by the research of a number of men, some working within it, others within its traditional rival. Such a mode of development, however, *requires* a decision process which permits rational men to disagree, and such disagreement would be barred by the shared algorithm which philosophers have generally sought. If it were at hand, all conforming scientists would make the same decision at the same time. With standards for acceptance set too low, they would move from one attractive global viewpoint to another, never giving traditional theory an opportunity to supply equivalent attractions. With standards set higher, no one satisfying the criterion of rationality would be inclined to try out the new theory, to articulate it in ways which showed its fruitfulness or displayed its accuracy and scope. I doubt that science would survive the change. What from one viewpoint may seem the looseness and imperfection of choice criteria conceived as rules may, when the same criteria are seen as values, appear an indispensable means of spreading the risk which the introduction or support of novelty always entails.

Even those who have followed me this far will want to know how a value-based enterprise of the sort I have described can develop as a science does, repeatedly producing powerful new techniques for prediction and control. To that question, unfortunately, I have no answer at all, but that is only another way of saying that I make no claim to have solved the problem of induction. If science did progress by virtue of some shared and binding algorithm of choice, I would be equally at a loss to explain its success. The lacuna is one I feel acutely, but its presence does not differentiate my position from the tradition.

It is, after all, no accident that my list of the values guiding scientific choice is, as nearly as makes any difference, identical with the tradition's list of rules dictating choice. Given any concrete situation to which the philosopher's rules could be applied, my values would function like his rules, producing the same choice. Any justification of induction, any explanation of why the rules worked, would apply equally to my values. Now consider a situation in which choice by shared rules proves impossible, not because the rules are wrong but because they are, as rules, intrinsically incomplete. Individuals must then still choose and be guided by the rules (now values) when they do so. For that purpose, however, each must first flesh out the rules, and each will do so in a somewhat different way even though the decision dictated by the variously completed rules may prove unanimous. If I now assume, in addition, that the group is large enough so that individual differences distribute on some normal curve, then any argument that justifies the philosopher's choice by rule should be immediately adaptable to my choice by value. A group too small, or a distribution excessively skewed by external historical pressures, would, of course, prevent the argument's transfer.[8] But those are just the circumstances under which scientific progress is itself problematic. The transfer is not then to be expected.

I shall be glad if these references to a normal distribution of individual differences and to the problem of induction make my position appear very close to more traditional views. With respect to theory choice, I have never thought my departures large and have been correspondingly startled by such charges as "mob psychology," quoted at the start. It is worth noting, however, that the positions are not quite identical, and for that purpose an analogy may be helpful. Many properties of liquids and gases can be accounted for on the kinetic theory by supposing that all molecules travel at the same speed. Among such properties are the regularities known as Boyle's and Charles's law. Other characteristics, most obviously evaporation, cannot be explained in so simple a way. To deal with them one must assume that molecular speeds differ, that they are distributed at random, governed by the laws of chance. What I have been suggesting here is that theory choice, too, can be explained only in part by a theory which attributes the same properties to all the scientists who must do the choosing. Essential aspects of the process generally known as verification will be understood only by recourse to the features with respect to which men may differ while still remaining scientists. The tradition takes it for granted that such features are vital to the process of discovery, which it at once and for that reason rules out of philosophical bounds. That they may have significant functions also in the philosophically central problem of justifying theory choice is what philosophers of science have to date categorically denied.

What remains to be said can be grouped in a somewhat miscellaneous epilogue. For the sake of clarity and to avoid writing a book, I have throughout this paper utilized some traditional concepts and locutions about the viability of which I have elsewhere expressed serious doubts. For those who know the work in which I have done so, I close by indicating three aspects of what I have said which would better represent my views if cast in other terms, simultaneously indicating the main directions in which such recasting should proceed. The areas I have in mind are: value invariance, subjectivity, and partial communication. If my views of scientific development are novel—a matter about which there is legitimate room for doubt—it is in areas such as these, rather than theory choice, that my main departures from tradition should be sought.

Throughout this paper I have implicitly assumed that, whatever their initial source, the criteria or values deployed in theory choice are fixed once and for all, unaffected by their participation in transitions from one theory to another. Roughly speaking, but only very roughly, I take that to be the case. If the list of relevant values is kept short (I have mentioned five, not all independent) and if their specification is left vague, then such values as accuracy, scope, and fruitfulness are permanent attributes of science. But little knowledge of history is required to suggest that both the application of these values and, more obviously, the relative weights attached to them have varied markedly with time and also with the field of application. Furthermore, many of these variations in value have been associated with particular changes in scientific theory. Though the experience of scientists provides no philosophical justification for the values they deploy (such justification would solve the problem of induction), those values are in part learned from that experience, and they evolve with it.

The whole subject needs more study (historians have usually taken scientific values, though not scientific methods, for granted), but a few remarks will illustrate the sort of variations I have in mind. Accuracy, as a value, has with time increasingly denoted quantitative or numerical agreement, sometimes at the expense of qualitative. Before early modern times, however, accuracy in that sense was a criterion only for astronomy, the science of the celestial region. Elsewhere it was neither expected nor sought. During the seventeenth century, however, the criterion of numerical agreement was extended to mechanics, during the late eighteenth and early nineteenth centuries to chemistry and such other subjects as electricity and heat, and in this century to many parts of biology. Or think of utility, an item of value not on my initial list. It too has figured significantly in scientific development, but far more strongly and steadily for chemists than for, say, mathematicians and physicists. Or consider scope. It is still an important scientific value, but important scientific advances have repeatedly been achieved at its expense, and the weight attributed to it at times of choice has diminished correspondingly.

What may seem particularly troublesome about changes like these is, of course, that they ordinarily occur in the aftermath of a theory change. One of the objections to Lavoisier's new chemistry was the roadblocks with which it confronted the achievement of what had previously been one of chemistry's traditional goals: the explanation of qualities, such as color and texture, as well as of their changes. With the acceptance of Lavoisier's theory such explanations ceased for some time to be a value for chemists; the ability to explain qualitative variation was no longer a criterion relevant to the evaluation of chemical theory. Clearly, if such value changes had occurred as rapidly or been as complete as the theory changes to which they related, then theory choice would be value choice, and neither could provide justification for the other. But, historically, value change is ordinarily a belated and largely unconscious concomitant of theory choice, and the former's magnitude is regularly smaller than the latter's. For the functions I have here ascribed to values, such relative stability provides a sufficient basis. The existence of a feedback loop through which theory change affects the values which led to that change does not make the decision process circular in any damaging sense.

About a second respect in which my resort to tradition may be misleading, I must be far more tentative. It demands the skills of an ordinary language philosopher, which I do not possess. Still, no very acute ear for language is required to generate discomfort with the ways in which the terms "objectivity" and, more especially, "subjectivity" have functioned in this paper. Let me briefly suggest the respects in which I believe language has gone astray. "Subjective" is a term with several established uses: in one of these it is opposed to "objective," in another to "judgmental." When my critics describe the idiosyncratic features to which I appeal as subjective, they resort, erroneously I think, to the second of these senses. When they complain that I deprive science of objectivity, they conflate that second sense of subjective with the first.

A standard application of the term "subjective" is to matters of taste, and my critics appear to suppose that that is what I have made of theory choice. But they are missing a distinction standard since Kant when they do so. Like sensation reports, which are also subjective in the sense now at issue, matters of taste are undiscussable. Suppose that, leaving a movie theater with a friend after seeing a western, I exclaim: "How I liked that terrible potboiler!" My friend, if he disliked the film, may tell me I have low tastes, a matter about which, in these circumstances, I would readily agree. But, short of saying that I lied, he cannot disagree with my report that I liked the film or try to persuade me that what I said about my reaction was wrong. What is discussable in my remark is not my characterization of my internal state, my exemplification of taste, but rather my *judgment* that the film was a potboiler. Should my friend disagree on that point, we may argue most of the night, each comparing the film with good or great ones we have seen, each revealing, implicitly or explicitly, something about how he *judges* cinematic merit, about his aesthetic. Though one of us may, before retiring, have persuaded the other, he need not have done so to demonstrate that our difference is one of judgment, not taste.

Evaluations or choices of theory have, I think, exactly this character. Not that scientists never say merely, I like such and such a theory, or I do not.

subj: judgment, not taste

After 1926 Einstein said little more than that about his opposition to the quantum theory. But scientists may always be asked to explain their choices, to exhibit the bases for their judgments. Such judgments are eminently discussable, and the man who refuses to discuss his own cannot expect to be taken seriously. Though there are, very occasionally, leaders of scientific taste, their existence tends to prove the rule. Einstein was one of the few, and his increasing isolation from the scientific community in later life shows how very limited a role taste alone can play in theory choice. Bohr, unlike Einstein, did discuss the bases for his judgment, and he carried the day. If my critics introduce the term "subjective" in a sense that opposes it to judgmental—thus suggesting that I make theory choice undiscussable, a matter of taste—they have seriously mistaken my position.

Turn now to the sense in which "subjectivity" is opposed to "objectivity," and note first that it raises issues quite separate from those just discussed. Whether my taste is low or refined, my report that I liked the film is objective unless I have lied. To my judgment that the film was a potboiler, however, the objective–subjective distinction does not apply at all, at least not obviously and directly. When my critics say I deprive theory choice of objectivity, they must, therefore, have recourse to some very different sense of subjective, presumably the one in which bias and personal likes or dislikes function instead of, or in the face of, the actual facts. But that sense of subjective does not fit the process I have been describing any better than the first. Where factors dependent on individual biography or personality must be introduced to make values applicable, no standards of factuality or actuality are being set aside. Conceivably my discussion of theory choice indicates some limitations of objectivity, but not by isolating elements properly called subjective. Nor am I even quite content with the notion that what I have been displaying are limitations. Objectivity ought to be analyzable in terms of criteria like accuracy and consistency. If these criteria do not supply all the guidance that we have customarily expected of them, then it may be the meaning rather than the limits of objectivity that my argument shows.

Turn, in conclusion, to a third respect, or set of respects, in which this paper needs to be recast. I have assumed throughout that the discussions surrounding theory choice are unproblematic, that the facts appealed to in such discussions are independent of theory, and that the discussions' outcome is appropriately called a choice. Elsewhere I have challenged all three of these assumptions, arguing that communication between proponents of different theories is inevitably partial, that what each takes to be facts depends in part on the theory he espouses, and that an individual's transfer of allegiance from theory to theory is often better described as conversion than as choice. Though all these theses are problematic as well as controversial, my commitment to them is undiminished. I shall not now defend them, but must at least attempt to indicate how what I have said here can be adjusted to conform with these more central aspects of my view of scientific development.

For that purpose I resort to an analogy I have developed in other places. Proponents of different theories are, I have claimed, like native speakers of different languages. Communication between them goes on by translation, and it raises all translation's familiar difficulties. That analogy is, of course, incomplete, for the vocabulary of the two theories may be identical, and most words function in the same ways in both. But some words in the basic as well as in the theoretical vocabularies of the two theories—words like "star" and "planet," "mixture" and "compound," or "force" and "matter"—do function differently. Those differences are unexpected and will be discovered and localized, if at all, only by repeated experience of communication breakdown. Without pursuing the matter further, I simply assert the existence of significant limits to what the proponents of different theories can communicate to one another. The same limits make it difficult or, more likely, impossible for an individual to hold both theories in mind together and compare them point by point with each other and with nature. That sort of comparison is, however, the process on which the appropriateness of any word like "choice" depends.

Nevertheless, despite the incompleteness of their communication, proponents of different theories can exhibit to each other, not always easily, the concrete technical results achievable by those who practice within each theory. Little or no translation is required to apply at least some value criteria to those results. (Accuracy and fruitfulness are most immediately applicable, perhaps followed by scope. Consistency and simplicity are far more problematic.) However incomprehensible the new theory may be to the proponents of tradition, the exhibit of impressive concrete results will persuade at least a few of them that they must discover how such results are achieved. For that purpose they must learn to translate, perhaps by treating already published papers as a Rosetta stone or, often more effective, by visiting the innovator, talking with him, watching him and his students at work. Those exposures may not result in the adoption of the theory; some advocates of the tradition may return home and attempt to adjust the old theory to produce equivalent results. But others, if the new theory is to survive, will find that at some point in the language-learning process they have ceased to translate and begun instead to speak the language like a native. No process quite like choice has occurred, but they are practicing the new theory nonetheless. Furthermore, the factors that have led them to risk the conversion they have undergone are just the ones this paper has underscored in discussing a somewhat different process, one which, following the philosophical tradition, it has labelled theory choice.

Notes

1 *The Structure of Scientific Revolutions*, 2nd ed. (Chicago, 1970), pp. 148, 151–52, 159. All the passages from which these fragments are taken appeared in the same form in the first edition, published in 1962.

2 Ibid., p. 170.

3 Imre Lakatos, "Falsification and the Methodology of Scientific Research Programmes," in I. Lakatos and A. Musgrave, eds., *Criticism and the Growth of Knowledge* (Cambridge, 1970), pp. 91–195. The quoted phrase, which appears on p. 178, is italicized in the original.

4 Dudley Shapere, "Meaning and Scientific Change," in R. G. Colodny, ed., *Mind and Cosmos: Essays in Contemporary Science and Philosophy*, University of Pittsburgh Series in the Philosophy of Science, vol. 3 (Pittsburgh, 1966), pp. 41–85. The quotation will be found on p. 67.

5 Israel Scheffler, *Science and Subjectivity* (Indianapolis, 1967), p. 81.

6 The last criterion, fruitfulness, deserves more emphasis than it has yet received. A scientist choosing between two theories ordinarily knows that his decision will have a bearing on his subsequent research career. Of course he is especially attracted by a theory that promises the concrete successes for which scientists are ordinarily rewarded.

7 The least equivocal example of this position is probably the one developed in Scheffler, *Science and Subjectivity*, chap. 4.

8 If the group is small, it is more likely that random fluctuations will result in its members sharing an atypical set of values and therefore making choices different from those that would be made by a larger and more representative group. External environment—intellectual, ideological, or economic—must systematically affect the value system of much larger groups, and the consequences can include difficulties in introducing the scientific enterprise to societies with inimical values or perhaps even the end of that enterprise within societies where it had once flourished. In this area, however, great caution is required. Changes in the environment where science is practiced can also have fruitful effects on research. Historians often resort, for example, to differences between national environments to explain why particular innovations were initiated and at first disproportionately pursued in particular countries, e.g., Darwinism in Britain, energy conservation in Germany. At present we know substantially nothing about the minimum requisites of the social milieux within which a sciencelike enterprise might flourish.

27

David Bloor, "The Strong Programme in the Sociology of Knowledge"

David Bloor is one of the founders of the "strong program in the sociology of knowledge," a movement originated at the University of Edinburgh in the 1970s. It has since grown into a powerful school of thought variously referred to as "sociology of scientific knowledge" and "social constructionism," and exercising considerable intellectual influence in many quarters, both within and outside philosophy and sociology departments. Bloor's book, from which this selection is taken, is programmatic.

Can the sociology of knowledge investigate and explain the very content and nature of scientific knowledge? Many sociologists believe that it cannot. They say that knowledge as such, as distinct from the circumstances surrounding its production, is beyond their grasp. They voluntarily limit the scope of their own enquiries. I shall argue that this is a betrayal of their disciplinary standpoint. All knowledge, whether it be in the empirical sciences or even in mathematics, should be treated, through and through, as material for investigation. Such limitations as do exist for the sociologist consist in handing over material to allied sciences like psychology or in depending on the researches of specialists in other disciplines. There are no limitations which lie in the absolute or transcendent character of scientific knowledge itself, or in the special nature of rationality, validity, truth or objectivity.

It might be expected that the natural tendency of a discipline such as the sociology of knowledge would be to expand and generalize itself: moving from studies of primitive cosmologies to that of our own culture. This is precisely the step that sociologists have been reluctant to take. Again, the sociology of knowledge might well have pressed more strongly into the area currently occupied by philosophers, who have been allowed to take upon

D. Bloor, *Knowledge and Social Imagery*, 2nd edn, 1991, pp. 3–23. Chicago: The University of Chicago Press.

themselves the task of defining the nature of knowledge. In fact sociologists have been only too eager to limit their concern with science to its institutional framework and external factors relating to its rate of growth or direction. This leaves untouched the nature of the knowledge thus created (cf. Ben-David 1971; DeGré 1967; Merton 1964; Stark 1958).

What is the cause for this hesitation and pessimism? Is it the enormous intellectual and practical difficulties which would attend such a programme? Certainly these must not be underestimated. A measure of their extent can be gained from the effort that has been expended on the more limited aims. But these are not the reasons that are in fact advanced. Is the sociologist at a loss for theories and methods with which to handle scientific knowledge? Surely not. His own discipline provides him with exemplary studies of the knowledge of other cultures which could be used as models and sources of inspiration. Durkheim's classic study *The Elementary Forms of the Religious Life* shows how a sociologist can penetrate to the very depths of a form of knowledge. What is more Durkheim dropped a number of hints as to how his findings might relate to the study of scientific knowledge. The hints have fallen on deaf ears.

The cause of the hesitation to bring science within the scope of a thorough-going sociological scrutiny is lack of nerve and will. It is believed to be a foredoomed enterprise. Of course, the failure of nerve has deeper roots than this purely psychological characterization suggests, and these will be investigated later. Whatever the cause of the malady, its symptoms take the form of a priori and philosophical argumentation. By these means sociologists express their conviction that science is a special case, and that contradictions and absurdities would befall them if they ignored this fact. Naturally philosophers are only too eager to encourage this act of self-abnegation (e.g. Lakatos 1971; Popper 1966).

It will be the purpose of this book to combat these arguments and inhibitions. For this reason the discussions which follow will sometimes, though not always, have to be methodological rather than substantive. But I hope they will be positive in their effect. Their aim is to put weapons in the hands of those engaged in constructive work to help them attack critics, doubters and sceptics.

I shall first spell out what I call the strong programme in the sociology of knowledge. This will provide the framework within which detailed objections will then be considered. Since a priori arguments are always embedded in background assumptions and attitudes it will be necessary to bring these to the surface for examination as well. This will be the second major topic and it is here that substantial sociological hypotheses about our conception of science will begin to emerge. The third major topic will concern what is perhaps the most difficult of all the obstacles to the sociology of knowledge, namely mathematics and logic. It will transpire that the problems of principle involved are not, in fact, unduly technical. I shall indicate how these subjects can be studied sociologically.

The Strong Programme

The sociologist is concerned with knowledge, including scientific knowledge, purely as a natural phenomenon. The appropriate definition of knowledge will therefore be rather different from that of either the layman or the philosopher. Instead of defining it as true belief—or perhaps, justified true belief—knowledge for the sociologist is whatever people take to be knowledge. It consists of those beliefs which people confidently hold to and live by. In particular the sociologist will be concerned with beliefs which are taken for granted or institutionalized, or invested with authority by groups of people. Of course knowledge must be distinguished from mere belief. This can be done by reserving the word "knowledge" for what is collectively endorsed, leaving the individual and idiosyncratic to count as mere belief.

Our ideas about the workings of the world have varied greatly. This has been true within science just as much as in other areas of culture. Such variation forms the starting point for the sociology of knowledge and constitutes its main problem. What are the causes of this variation, and how and why does it change? The sociology of knowledge focuses on the distribution of belief and the various factors which influence it. For example: how is knowledge transmitted; how stable is it; what processes go into its creation and maintenance; how is it organized and categorized into different disciplines or spheres?

For sociologists these topics call for investigation and explanation and they will try to characterize knowledge in a way which accords with this perspective. Their ideas therefore will be in the same causal idiom as those of any other scientist. Their concern will be to locate the regularities and general principles or processes which appear to be at work within the field of their data. The aim will be to build theories to explain these regularities. If these theories are to satisfy the requirement of maximum generality they will have to apply to both true and false beliefs, and as far as possible the same type of explanation will have to apply in both cases. The aim of physiology is to explain the organism in health and disease; the aim of mechanics is to understand machines which work and machines which fail; bridges which stand as well as those which fall. Similarly the sociologist seeks theories which explain the beliefs which are in fact found, regardless of how the investigator evaluates them.

Some typical problems in this area which have already yielded interesting findings may serve to illustrate this approach. First, there have been studies of the connections between the gross social structure of groups and the general form of the cosmologies to which they have subscribed. Anthropologists have found the social correlates, and the possible causes of our having anthropomorphic and magical world-views as distinct from impersonal and naturalistic ones (Douglas 1966, 1970). Second, there have been studies which have traced the connections between economic, technical and industrial developments and the content of scientific theories. For example, the impact of practical developments in water and steam technology on the content of theories in thermodynamics has been studied in great detail. The causal link is beyond dispute (Kuhn 1959; Cardwell 1971). Third, there is much evidence that features of culture which usually count as non-scientific greatly influence both the creation and the evaluation of scientific theories and findings. Thus Eugenic concerns have been shown to underlie and explain Francis Galton's creation of the concept of the coefficient of correlation in statistics. Again the general political, social and ideological standpoint of the geneticist Bateson has been used to explain his role of sceptic in the controversy over the gene theory of inheritance (Coleman 1970; Cowan 1976; MacKenzie 1981). Fourth, the importance that processes of training and socialization have in the conduct of science is becoming increasingly documented. Patterns of continuity and discontinuity, of reception and rejection, appear to be explicable by appeal to these processes. An interesting example of the way in which a background in the requirements of a scientific discipline influences the assessment of a piece of work is afforded by Lord Kelvin's criticisms of the theory of evolution. Kelvin calculated the age of the sun by treating it as an incandescent body cooling down. He found that it would have burnt itself out before evolution could have reached its currently observable state. The world is not old enough to have allowed evolution to have run its course, so the theory of evolution must be wrong. The assumption of geological uniformity, with its promise of vast stretches of time, had been rudely pulled from beneath the biologist's feet. Kelvin's arguments caused dismay. Their authority was immense and in the 1860s they were unanswerable, they followed with convincing rigour from convincing physical premises. By the last decade of the century the geologists had plucked up courage to tell Kelvin that he must have made a mistake. This newfound courage was not because of any dramatic new discoveries; indeed, there had been no real change in the evidence available. What had happened in the interim was a general consolidation in geology as a discipline with a mounting quantity of detailed observation of the fossil record. It was this growth which caused a variation in the assessments of probability and plausibility: Kelvin simply must have left some vital but unknown factor out of consideration. It was only with the understanding of the sun's nuclear sources of energy that his physical argument could be faulted. Geologists and biologists had no foreknowledge of this, they simply had not waited for an answer (Rudwick 1972; Burchfield 1975). This example also serves to make another point. It deals with social processes internal to science, so there is no question of sociological considerations being confined to the operation of external influences.

Finally, mention must be made of a fascinating and controversial study of the physicists of Weimar Germany. Forman (1971) uses their academic addresses to show them taking up the dominant, antiscientific "Lebensphilosophie" surrounding them. He argues "that the movement to dispense with causality in physics which sprang up so suddenly and blossomed so luxuriantly in Germany after 1918, was primarily an effort by German physicists to adapt the content of their science to the values of their intellectual environment" (p. 7). The boldness and interest of this claim derives from the central place of acausality in modern quantum theory.

The approaches that have just been sketched suggest that the sociology of scientific knowledge should adhere to the following four tenets. In this way it will embody the same values which are taken for granted in other scientific disciplines. These are:

(1) It would be causal, that is, concerned with the conditions which bring about belief or states of knowledge. Naturally there will be other types of causes apart from social ones which will cooperate in bringing about belief.
(2) It would be impartial with respect to truth and falsity, rationality or irrationality, success or failure. Both sides of these dichotomies will require explanation.
(3) It would be symmetrical in its style of explanation. The same types of cause would explain, say, true and false beliefs.
(4) It would be reflexive. In principle its patterns of explanation would have to be applicable to sociology itself. Like the requirement of symmetry this is a response to the need to seek for general explanations. It is an obvious requirement of principle because otherwise sociology would be a standing refutation of its own theories.

These four tenets, of causality, impartiality, symmetry and reflexivity, define what will be called the strong programme in the sociology of knowledge. They are by no means new, but represent an amalgam of the more optimistic and scientistic strains to be found in Durkheim (1938), Mannheim (1936) and Znaniecki (1965).

In what follows I shall try to maintain the viability of these tenets against criticism and misunderstanding. What is at stake is whether the strong programme can be pursued in a consistent and plausible way. Let us therefore turn to the main objections to the sociology of knowledge to draw out the full significance of the tenets and to see how the strong programme stands up to criticism.

The Autonomy of Knowledge

One important set of objections to the sociology of knowledge derives from the conviction that some beliefs do not stand in need of any explanation, or do not stand in need of a causal explanation. This feeling is particularly strong when the beliefs in question are taken to be true, rational, scientific or objective.

When we behave rationally or logically it is tempting to say that our actions are governed by the requirements of reasonableness or logic. The explanation of why we draw the conclusion we do from a set of premises may appear to reside in the principles of logical inference themselves. Logic, it may seem, constitutes a set of connections between premises and conclusions and our minds can trace out these connections. As long as someone is being reasonable then the connections themselves would seem to provide the best explanation for the beliefs of the reasoner. Like an engine on rails, the rails themselves dictate where it will go. It is as if we can transcend the directionless push and pull of physical causality and harness it, or subordinate it, to quite other principles and let these determine our thoughts. If this is so then it is not the sociologist or the psychologist but the logician who will provide the most important part of the explanation of belief.

Of course, when someone makes mistakes in their reasoning then logic itself is no explanation. A lapse or deviation may be due to the interference of a whole variety of factors. Perhaps the reasoning is too difficult for the limited intelligence of the reasoner, perhaps he or she is inattentive, or too emotionally involved in the subject under discussion. As when a train goes off the rails, a cause for the accident can surely be found. But we neither have, nor need, commissions of enquiry into why accidents do not happen.

Arguments such as these have become a commonplace in contemporary analytical philosophy. Thus in *The Concept of Mind* (1949) Ryle says: "Let the psychologist tell us why we are deceived, but we can tell ourselves and him why we are not deceived" (p. 308). This approach may be summed up by the claim that nothing makes people do things that are correct but something does make, or cause, them to go wrong (cf. Hamlyn 1969; Peters 1958).

The general structure of these explanations stands out clearly. They all divide behaviour or belief into two types: right and wrong, true or false, rational or irrational. They then invoke sociological or psychological causes to explain the negative side of the division. Such causes explain error, limitation and deviation. The positive side of the evaluative divide is quite different. Here logic, rationality and truth appear to be their own explanation. Here psycho-social causes do not need to be invoked.

Applied to the field of intellectual activity these views have the effect of making a body of knowledge an autonomous realm. Behaviour is to be explained by appeal to the procedures, results, methods and maxims of the activity itself. It makes successful and conventional intellectual activity appear self-explanatory and self-propelling. It becomes its own explanation. No expertise in sociology or psychology is required: only expertise in the intellectual activity itself.

A currently fashionable version of this position is to be found in Lakatos's (1971) theory about how the history of science ought to be written. This theory was explicitly meant to have implications for the sociology of science as well. The first prerequisite, says Lakatos, is that a philosophy or methodology of science be chosen. These are accounts of what science ought to be, and of what steps in it are rational. The chosen philosophy of science becomes the framework on which hangs all the subsequent work of explanation. Guided by this philosophy it ought to be possible to display science as a process which exemplifies its principles and develops in accord with its teachings. In as far as this can be done then science has been shown to be rational in the light of that philosophy. This task, of showing that science embodies certain methodological principles, Lakatos calls either "rational reconstruction" or "internal history." For example, an inductivist methodology would perhaps stress the emergence of theories out of an accumulation of observations. It would therefore focus on events like Kepler's use of Tycho Brahe's observations when formulating the laws of planetary motion.

It will never be possible, however, to capture all of the diversity of actual scientific practice by this means. Lakatos therefore insists that internal history will always need to be supplemented by an "external history." This looks after the irrational residue. It is a matter which the philosophical historian will hand over to the "external historian" or the sociologist. Thus, from an inductivist standpoint the role of Kepler's mystical beliefs about the majesty of the sun would require a nonrational or external explanation.

The points to notice about this approach are first that internal history is self-sufficient and autonomous. To exhibit the rational character of a scientific development is sufficient explanation in itself of why the events took place. Second, not only are rational reconstructions autonomous; they also have an important priority over external history or sociology. The latter merely close the gap between rationality and actuality. This task is not even defined until internal history has had its say. Thus:

> internal history is primary, external history only secondary, since the

> most important problems of external history are defined by internal history. External history either provides non-rational explanation of the speed, locality, selectiveness, etc. of historical events as interpreted in terms of internal history; or when history differs from its rational reconstruction, it provides an empirical explanation of why it differs. But the rational aspect of scientific growth is fully accounted for by one's logic of scientific discovery. (1971, p. 9)

Lakatos then answers the question of how to decide which philosophy should dictate the problems of external history or sociology. Alas for externalists the answer represents yet a further humiliation. Not only is their function derivative, it now transpires that the best philosophy of science, according to Lakatos, is one which minimizes this role. Progress in philosophy of science is to be measured by the amount of actual history which can be exhibited as rational. The better the guiding methodology the more of actual science is rendered safe from the indignity of empirical explanation. The sociologist is allowed a crumb of comfort from the fact that Lakatos is only too pleased to grant that there will always be some irrational events in science that no philosophy will ever be able or willing to rescue. He instances here unsavoury episodes of Stalinist intervention in science, like the Lysenko affair in biology.

These refinements, however, are less important than the general structure of the position. It does not matter how the central principles of rationality are chosen, or how they might change. The central point is that, once chosen, the rational aspects of science are held to be self-moving and self-explanatory. Empirical or sociological explanations are confined to the irrational.

What can it mean to say that nothing makes people do or believe things which are rational or correct? Why in that case does the behaviour take place at all? What prompts the internal and correct functioning of an intellectual activity if the search for psychological and sociological causes is only deemed appropriate in the case of irrationality or error? The theory that must tacitly underlie these ideas is a goal-directed or teleological vision of knowledge and rationality.

Suppose that it is assumed that truth, rationality and validity are our natural goals and the direction of certain natural tendencies with which we are endowed. We are rational animals and we naturally reason justly and cleave to the truth when it comes within our view. Beliefs that are true then clearly require no special comment. For them, their truth is all the explanation that is needed of why they are believed. On the other hand, this self-propelling progress towards truth may be impeded or deflected and here natural causes must be located. These will account for ignorance, error, confused reasoning and any impediment to scientific progress.

Such a theory makes a great deal of sense of what is written in this area even if it seems implausible at first sight to impute it to contemporary thinkers. It even appears to have intruded itself into the thinking of Karl Mannheim. Despite his determination to set up causal and symmetrical canons of explanation, his nerve failed him when it came to such apparently autonomous subjects as mathematics and natural science. This failure expressed itself in passages such as the following, from *Ideology and Utopia*:

> The existential determination of thought may be regarded as a demonstrated fact in those realms of thought in which we can show . . . that the process of knowing does not actually develop historically in accordance with immanent laws, that it does not follow only for the "nature of things" or from "pure logical possibilities," and that it is not driven by an "inner dialectic." On the contrary, the emergence and the crystallization of actual thought is influenced in many decisive points by extra-theoretical factors of the most diverse sort (1936, p. 239).

Here social causes are being equated with "extra-theoretical" factors. But where does this leave behaviour conducted in accord with the inner logic of a theory or governed by theoretical factors? Clearly it is in danger of being excluded from sociological explanation because it functions as the base-line for locating those things which do require explanation. It is as if Mannheim slipped into sharing the sentiments expressed in the quotations from Ryle and Lakatos and said to himself, "When we do what is logical and proceed correctly, nothing more needs to be said." But to see certain sorts of behaviour as unproblematic is to see them as natural. In this case what is natural is proceeding correctly, that is via or towards the truth. So here too the teleological model is probably at work.

How does this model of knowledge relate to the tenets of the strong programme? Clearly it violates them in a number of serious ways. It relinquishes a thorough-going causal orientation. Causes can only be located for error. Thus the sociology of knowledge is confined to the sociology of error. In addition it violates the requirements of symmetry and impartiality. A prior evaluation of the truth or rationality of a belief is called for before it can be decided whether it is to be counted as self-explanatory or whether a causal theory is needed. There is no doubt that if the teleological model is true then the strong programme is false.

The teleological and causal models, then, represent programmatic alternatives which quite exclude one another. Indeed, they are two opposed metaphysical standpoints. This may make it appear that it is necessary to decide at the outset which is true. Doesn't the sociology of knowledge depend on the teleological view being false? So doesn't this have to be established before the strong programme dare proceed? The answer is "no." It is more sensible to look at matters the other way round. It is unlikely that any decisive, independent grounds could be adduced "a priori" to prove the truth or falsity of such major metaphysical alternatives. Where objections and arguments are proposed against one of the two theories it will be found that they depend on and presuppose the other, and so beg the question at issue. All that can be done is to check the internal consistency of the different theories and then see what happens when practical research and theorizing is based upon them. If their truth can be decided at all it will only be after they have been adopted and used, not before. So the sociology of knowledge is not bound to eliminate the rival standpoint. It only has to separate itself from it, reject it, and make sure that its own house is in logical order.

These objections to the strong programme are thus not based on the intrinsic nature of knowledge but only on knowledge viewed from the standpoint of the teleological model. Reject that model and all its associated distinctions, evaluations and asymmetries go with it. It is only if that model has a unique claim to attention that its corresponding patterns of explanation are binding upon us. Its mere existence, and the fact that some thinkers find it natural to use it, do not endow it with probative force.

In its own terms the teleological model is no doubt perfectly consistent and there are perhaps no logical reasons why anyone should prefer the causal approach to the goal-directed view. There are, however, methodological considerations which may influence the choice in favour of the strong programme.

If explanation is allowed to hinge on prior evaluations, then the causal processes that are thought to operate in the world will come to reflect the pattern of these evaluations. Causal processes will be made to etch out the pattern of perceived error, throwing into relief the shape of truth and rationality. Nature will take on a moral significance, endorsing and embodying truth and right. Those who indulge their tendencies to offer asymmetrical explanations will thus have every opportunity to represent as natural what they take for granted. It is an ideal recipe for turning one's gaze away from one's own society, values and beliefs and attending only to deviations from them.

Care is needed not to overstate this point, for the strong programme does exactly the same thing in certain respects. It is also based on values, for example: the desire for generality of a specific kind and for a conception of the natural world as morally empty and neutral. So it too insists on giving nature a certain role with respect to morality, albeit of a negative kind. This means that it too represents as natural what it takes for granted.

What may be said, however, is that the strong programme possesses a certain kind of moral neutrality, namely the same kind as we have learned to associate with all the other sciences. It also imposes on itself the need for the same kind of generality as other sciences. It would be a betrayal of these values, of the approach of empirical science, to choose to adopt the teleological view. Obviously these are not reasons which could compel anyone to adopt the causal view. For some they may be precisely the reasons that would incline them to reject causality and adopt asymmetrical, teleological conceptions. But these points do make clear the ramifications of the choice and expose those values that are going to inform the approach to knowledge. From this type of confrontation, then, the sociology of knowledge can proceed, if it so chooses, without let or hindrance.

The Argument from Empiricism

The premise underlying the teleological model was that causality is associated with error or limitation. This represents an extreme form of asymmetry and so stands as the most radical alternative to the strong programme with its insistence on symmetrical styles of explanation. It may be, however, that the strong programme can be criticized from a less extreme standpoint. Is it not plausible to say that some causes bring about erroneous belief whilst others bring about true belief? If it further transpires that certain types of causes are systematically correlated with true and false belief, respectively, then here is another basis for rejecting the symmetrical standpoint of the strong programme.

Consider the following theory: social influences produce distortions in our beliefs whilst the uninhibited use of our faculties of perception and our sensory-motor apparatus produce true beliefs. This praise for experience as a source of knowledge can be seen as encouraging individuals to rely on their own physical and psychological resources for getting to know the world. It is a statement of faith in the power of our animal capacities for knowledge. Give these full play and their natural, but causal, operation will yield knowledge tested and tried in practical interaction with the world. Depart from this path, rely on one's fellows, and one will be prey to superstitious stories, myth and speculation. At best these stories will be second-hand belief rather than first-hand knowledge. At worst the motives behind them will be corrupt, the product of liars and tyrants.

It is not difficult to recognize this picture. It is a version of Bacon's warning to avoid the Idols of the Market Place and the Theatre. Much of standard empiricism represents a refined and rarefied statement of this

Misuse of term 'causality'

approach to knowledge. Although the current fashion amongst empiricist philosophers is to avoid the psychological rendering of their theory the basic vision is not too dissimilar to that sketched above. I shall therefore refer to the above theory without more ado as empiricism.

If empiricism is correct then once again the sociology of knowledge is really the sociology of error, belief or opinion, but not knowledge as such. This conclusion is not as extreme as that derived from the teleological model of knowledge. It amounts to a division of labour between the psychologist and sociologist where the former would deal with real knowledge, the latter with error or something less than knowledge. The total enterprise would nevertheless be naturalistic and causal. There is therefore no question, as there was with the teleological model, of being confronted with a choice between a scientific perspective and a standpoint which embodies quite different values. Here the battle has to be fought entirely within science's own territory. Is the boundary between truth and error correctly drawn by this empiricist conception of knowledge? There are two shortcomings in empiricism which suggest that it is not.

First, it would be wrong to assume that the natural working of our animal resources always produces knowledge. They produce a mixture of knowledge and error with equal naturalness, and through the operation of one and the same type of cause. For example, a medium level of anxiety will often increase the learning and successful performance of a task compared with a very low level, but the performance will then drop again if the anxiety level gets too high. As a laboratory phenomenon the point is fairly general. A certain level of hunger will facilitate an animal's retention of information about its environment, as in a rat's learning of a laboratory maze for food. A very high level of hunger may well produce urgent and successful learning of the whereabouts of food, but it will lower the natural ability to pick up cues which are irrelevant to the current, overriding concern. These examples suggest that different causal conditions may indeed be associated with different patterns of true and false belief. However, they do not show that different types of cause correlate simply with true and false belief. In particular they show that it is incorrect to put psychological causes all on one side of this divide, as naturally leading to truth.

No doubt this shortcoming could be corrected. Perhaps all that the counter-examples show is that psychological learning mechanisms have an optimum working arrangement and that they produce error when they are thrown out of focus. It may be insisted that when our perceptual apparatus is operating under normal conditions, and performing its functions properly, then it brings about true belief. This revision of the doctrine may be granted because there is a far more important objection to it to be considered.

The crucial point about empiricism is its individualistic character. Those aspects of knowledge which each of us can and has to furnish for himself may be adequately explained by this type of model. But how much of man's knowledge, and how much of his science is built up by the individual relying simply on the interaction of the world with his animal capacities? Probably very little. The important question is: what analysis is to be given to the remainder? It is plausible to say that the psychological approach leaves out of account the social component of knowledge.

Does not individual experience, as a matter of fact, take place within a framework of assumptions, standards, purposes and meanings which are shared? Society furnishes the mind of the individual with these things and also provides the conditions whereby they can be sustained and reinforced. If the individual's grasp of them wavers, there are agencies ready to remind him; if his view of the world begins to deviate there are mechanisms which encourage realignment. The necessities of communication help to sustain collective patterns of thought in the individual psyche. As well as the individual's sensory experience of the natural world, there is, then, something that points beyond that experience, that provides a framework for it and gives it a wider significance. It fills out the individual's sense of what that overall Reality is, that his experience is experience of.

The knowledge of a society designates not so much the sensory experience of its individual members, or the sum of what may be called their animal knowledge. It is, rather, their collective vision or visions of Reality. Thus the knowledge of our culture, as it is represented in our science, is not knowledge of a reality that any individual can experience or learn about for himself. It is what our best attested theories, and our most informed thoughts tell us is the case, despite what the appearances may say. It is a story woven out of the hints and glimpses that we believe our experiments offer us. Knowledge then, is better equated with Culture than Experience.

If this designation of the word "knowledge" is accepted then the distinction between truth and error is not the same as the distinction between (optimum) individual experience and social influence. Rather it becomes a distinction within the amalgam of experiences and socially mediated beliefs that make up the content of a culture. It is a discrimination between rival mixtures of experience and belief. The same two ingredients occur in true and false beliefs and so the way is open for symmetrical styles of explanation which invoke the same types of cause.

One way of putting this point which may assist its recognition and acceptance is to say that what we count as scientific knowledge is largely "theoretical." It is largely a theoretical vision of the world that, at any given time, scientists may be said to know. It is largely to their theories that scientists must repair when asked what they can tell us about the world. But theories and theoretical knowledge are not things which are given in our experience. They are what give meaning to experience by offering a story about what underlies, connects and accounts for it. This does not mean that theory is unresponsive to experience. It is, but it is not given along with the experience it explains, nor is it uniquely supported by it. Another agency apart from the physical world is required to guide and support this component of knowledge. The theoretical component of knowledge is a social component, and it is a necessary part of truth, not a sign of mere error.

Two major sources of opposition to the sociology of knowledge have now been discussed and both have been rejected. The teleological model was indeed a radical alternative to the strong programme but there is not the slightest compulsion to accept it. The empiricist theory is implausible as a description of what we in fact count as our knowledge. It provides some of the bricks but is silent on the designs of the varying edifices that we build with them. The next step will be to relate these two positions to what is perhaps the most typical of all objections to the sociology of knowledge. This is the claim that it is a self-refuting form of relativism.

The Argument from Self-Refutation

If someone's beliefs are totally caused and if there is necessarily within them a component provided by society then it has seemed to many critics that these beliefs are bound to be false or unjustified. Any thorough-going sociological theory of belief then appears to be caught in a trap. For are not sociologists bound to admit that their own thoughts are determined, and in part even socially determined? Must they not therefore admit that their own claims are false in proportion to the strength of this determination? The result appears to be that no sociological theory can be general in its scope otherwise it would reflexively enmesh itself in error and destroy its own credibility. The sociology of knowledge is thus itself unworthy of belief or it must make exceptions for scientific or objective investigations and hence confine itself to the sociology of error. There can be no self-consistent, causal and general sociology of knowledge, especially not scientific knowledge.

It can be seen at once that this argument depends on one or the other of the two conceptions of knowledge discussed above, namely the teleological model or a form of individualistic empiricism. The conclusion follows, and it only follows, if these theories are first granted. This is because the argument takes as its premise their central idea that causation implies error, deviation or limitation. This premise may be in the extreme form that any causation destroys credibility or in the weaker form that only social causation has this effect. One or the other is crucial for the argument.

These premises have been responsible for a plethora of feeble and badly argued attacks on the sociology of knowledge. Mostly the attacks have failed to make explicit the premises on which they rest. If they had, their weakness would have been more easily exposed. Their apparent strength has derived from the fact that their real basis was hidden or simply unknown. Here is an example of one of the much better forms of this argument which does make quite clear the standpoint from which it derives.

Grünwald, an early critic of Mannheim, is explicit in his statement of the assumption that social determination is bound to enmesh a thinker in error. In the introduction to Mannheim's *Essays on the Sociology of Knowledge* (1952) Grünwald is quoted as saying "it is impossible to make any meaningful statement about the existential determination of ideas without having any Archimedean point beyond all existential determination . . ." (p. 29). Grünwald goes on to draw the conclusion that any theory, such as Mannheim's, which suggests that all thought is subject to social determination must refute itself. Thus: "No long argument is needed to show beyond doubt that this version of sociologism, too, is a form of scepticism and therefore refutes itself. For the thesis that all thinking is existentially determined and cannot claim to be true claims itself to be true" (p. 29).

This would be a cogent objection against any theory that did indeed assert that existential determination implied falsity. But its premise should be challenged for what it is: a gratuitous assumption and an unrealistic demand. If knowledge does depend on a vantage point outside society and if truth does depend on stepping above the causal nexus of social relations, then we may give them up as lost.

There are a variety of other forms of this argument. One typical version is to observe that research into the causation of belief is itself offered to the world as being correct and objective. Therefore, the argument goes, the sociologist assumes that objective knowledge is possible, so not everybody's beliefs can be socially determined. As the historian Lovejoy (1940) put it: "Even they, then, necessarily presuppose possible limitations or exceptions to their generalisations in the act of defending them" (p. 18). The limitations the "sociological relativists" are said necessarily to presuppose are designed to make room for criteria of factual truth and valid inference. So this objection, too, depends on the premise that factual truth and valid inference would be violated by beliefs that are determined, or at least socially determined.

Because these arguments have become so taken for granted their formulation has become abbreviated and routine. They can now be given in such condensed versions as the following, provided by Bottomore (1956): "For if all propositions are existentially determined and no proposition is absolutely true, then this proposition itself, if true, is not absolutely true, but is existentially determined" (p. 52).

The premise, that causation implies error, on which all these arguments depend has been exposed and rejected. The arguments can therefore be disposed of along with them. Whether a belief is to be judged true or false has nothing to do with whether it has a cause.

The Argument from Future Knowledge

Social determinism and historical determinism are closely related ideas. Those who believe there are laws governing social processes and societies will wonder if there are also laws governing their historical succession and development. To believe that ideas are determined by social milieu is but one form of believing that they are, in some sense, relative to the actor's historical position. It is therefore not surprising that the sociology of knowledge has been criticized by those who believe that the very idea of historical laws is based on error and confusion. One such critic is Karl Popper (1960). It will be the purpose of this section to refute these criticisms as far as they may be applied to the sociology of knowledge.

The reason why the search for laws is held to be wrong is that if they could be found they would imply the possibility of prediction. A sociology which furnished laws could permit the prediction of future beliefs. In principle it would seem to be possible to know what the physics of the future would be like just as it is possible to predict future states of a mechanical system. If the laws of the mechanism are known along with a knowledge of its initial position, and the masses and forces on its parts, then all the future positions may be predicted.

Popper's objection to this ambition is partly informal and partly formal. He informally observes that human behaviour and society just do not furnish the same spectacle of repeated cycles of events as do some limited portions of the natural world. So long-term predictions are hardly realistic. This much may be certainly granted.

The nub of the argument, however, is a logical point about the nature of knowledge. It is impossible, says Popper, to predict future knowledge. The reason is that any such prediction would itself amount to the discovery of that knowledge. The way we behave depends on what we know so behaviour in the future will depend on this unpredictable knowledge and this too will be unpredictable. This argument appears to depend on a peculiar property of knowledge and to result in a gulf between the natural sciences and the social sciences in as far as they dare to touch humans as knowers. It suggests that the aspirations of the strong programme with its search for causes and laws is misguided and that something more modestly empirical is called for. Perhaps sociology should again restrict itself to no more than a chronicle of errors or a catalogue of external circumstances which help or hinder science.

In fact the point which Popper makes is a correct though trite one which, properly understood, merely serves to emphasize the similarities rather than the differences between the social and the natural sciences. Consider the following argument which moves along exactly the same steps as Popper's but would, if correct, prove that the physical world is unpredictable. This will jerk our critical faculties into action. The argument is this: It is impossible to make predictions in physics which utilize or refer to physical processes of which we have no knowledge. But the course of the physical world will depend in part on the operation of these unknown factors. Therefore the physical world is unpredictable.

Naturally the objection will be raised that all that this proves is that our predictions will often be wrong, not that nature is unpredictable. Our predictions will be falsified in as far as they fail to take into account relevant facts that we did not know were involved. Exactly the same rejoinder can be made to the argument against historical laws. Really Popper is offering an inductive argument based on our record of ignorance and failure. All that it points to is that our historical and sociological predictions will usually be false. The reason for this is correctly located by Popper. It is that people's future actions will often be contingent on things which they will know, but which we do not know now, and of which we therefore take no account when we make the prediction. The correct conclusion to be drawn for the social sciences is that we are unlikely to make much headway predicting the behaviour and beliefs of others unless we know at least as much as they do about their situation. There is nothing in the argument which need discourage the sociologist of knowledge from developing conjectural theories on the basis of empirical and historical case studies and testing them by further studies. Limited knowledge and the vast scope for error will ensure that these predictions will mostly be false. On the other hand the fact that social life depends on regularity and order gives grounds for hope that some progress will be possible. It is worth remembering that Popper himself sees science as an endless vista of refuted conjectures. Since this vision was not intended to intimidate natural scientists there is no reason why it should appear in this light when it is applied to the social sciences—despite the fact that this is how Popper has chosen to present it.

But still the objection must be met: doesn't the social world present us with mere trends and tendencies and not the genuine law-like regularity of the natural world? Trends, of course, are merely contingent and superficial drifts rather than reliable necessities within phenomena. The answer is that this distinction is spurious. Take the orbiting planets, which are the usual symbols of law rather than trend. In fact the solar system is a mere physical tendency. It endures because nothing disturbs it. There was a time when it did not exist and it is easy to imagine how it might be disrupted: a large gravitating body could pass close by it, or the sun could explode. Nor do the basic laws of nature even require the planets to move in ellipses. They only happen to orbit round the sun because of their conditions of origin and formation. Whilst obeying the same law of attraction their trajectories could be very different. No: the empirical surface of the natural world is dominated by tendencies. These tendencies wax and wane because of an underlying tustle of laws, conditions and contingencies. Our scientific understanding seeks to tease out those laws which, as we are prone to say, are "behind" observable states of affairs. The contrast between the natural and social worlds on which the objection depends fails to compare like with like. It compares the laws found to underlie physical tendencies with the purely empirical surface of social tendencies.

Interestingly, the word "planet" originally meant "wanderer." Planets attracted attention precisely because they did not conform to the general tendencies visible in the night sky. Kuhn's historical study of astronomy, *The Copernican Revolution* (1957), is a record of just how difficult it was to find regularities beneath the tendencies. Whether there are any underlying social laws is a matter for empirical enquiry, not philosophical debate. Who knows what wandering, aimless, social phenomena will turn into symbols of law-like regularity? The laws that do emerge may well not govern massive historical tendencies, for these are probably complex blends like the rest of nature. The law-like aspects of the social world will deal with the factors and processes which combine to produce empirically observable effects. Professor Mary Douglas's brilliant anthropological study *Natural Symbols* (1970) shows what such laws may look like. The data are incomplete, her theories are still evolving, like all scientific works it is provisional, but patterns can be glimpsed.

In order to bring the discussion of laws and predictions down to earth it may be useful to conclude with an example. This will show the sort of law the sociologist of science actually looks for. It will also help to clarify the abstract terminology of "law" and "theory," which has little practical currency in the conduct of either the sociology or history of science.

The search for laws and theories in the sociology of science is absolutely identical in its procedure with that of any other science. This means that the following steps are to be found. Empirical investigation will locate typical and recurrent events. Such investigation might itself have been prompted by some prior theory, the violation of a tacit expectation or practical needs. A theory must then be invented to explain the empirical regularity. This will formulate a general principle or invoke a model to account for the facts. In doing so it will provide a language with which to talk about them and may sharpen perception of the facts themselves. The scope of the regularity may be seen more clearly once an explanation of its first vague formulation has been attempted. The theory or model may, for example, explain not only why the empirical regularity occurs but also why, sometimes, it does not occur. It may act as a guide to the conditions on which the regularity depends and hence the causes for deviation and variation. The theory, therefore, may prompt more refined empirical researches which in turn may demand further theoretical work: the rejection of the earlier theory or its modification and elaboration.

All of these steps may be seen in the following case. It has often been noted that priority disputes about discoveries are a common feature of science. There was a famous dispute between Newton and Leibniz over the invention of the calculus; there was bitterness over the discovery of the conservation of energy; Cavendish, Watt and Lavoisier were involved in the dispute over the chemical composition of water; biologists like Pasteur, medical men like Lister, mathematicians like Gauss, physicists like Faraday and Davy all became embroiled in priority disputes. The approximately true generalization can thus be formulated: discoveries prompt priority disputes.

It is quite possible to sweep this empirical observation aside and declare it to be irrelevant to the true nature of science. Science as such, it may be said, develops according to the inner logic of scientific enquiry and these disputes are mere lapses, mere psychological intrusions into rational procedures. However, a more naturalistic approach would simply take the facts as they are and invent a theory to explain them. One theory which has been proposed to explain priority disputes sees science as working by an exchange system. "Contributions" are exchanged for "recognition" and status—hence all those eponymous laws like Boyle's Law and Ohm's Law. Because recognition is important and scarce there will be struggles for it, hence priority disputes (Merton 1957; Storer 1966). The question then arises of why it is not obvious who has made a certain contribution: why is it possible for the matter to become one of dispute at all? Part of the answer is that because science depends so much on published and shared knowledge, a number of scientists are often in a position to make similar steps. The race will be a close one between near equals. But second, and more important, is the fact that discoveries involve more than empirical findings. They involve questions of theoretical interpretation and reinterpretation. The changing meaning of empirical results provides rich opportunities for misunderstanding and misdescription.

The discovery of oxygen will illustrate these complexities (Toulmin 1957). Priestley is frequently credited with the discovery of oxygen, but this is not how he saw the matter. For him the new gas that he isolated was dephlogisticated air. It was a substance intimately connected with combustion processes as conceived in terms of the phlogiston theory. It required the rejection of that theory and its replacement by Lavoisier's account of combustion before scientists saw themselves as dealing with a gas called oxygen. It is the theoretical components of science which give scientists the terms in which they see their own and other's actions. Hence those descriptions of actions which are involved in the imputation of a discovery are precisely the ones which become problematic when important discoveries are taking place.

Now it should be possible to offer an account of why some discoveries are less prone to create priority disputes than others. The original empirical generalization can be refined. This refinement, however, will not be a simple or arbitrary limitation on the scope of the generalization. Rather, it will take the form of a discrimination between different types of discovery prompted by the above reflections on the exchange theory. This allows for an improved statement of the empirical law: discoveries at times of theoretical change prompt priority disputes; those at times of theoretical stability do not.

Naturally the matter does not rest here. First, the refined version of the law has to be checked to see if it is empirically plausible. This, of course, means checking a prediction about the beliefs and behaviour of scientists. Second, another theory needs to be developed to make sense of the new law. There is no need to go into more detail although the point may be made that a theory has been formulated which performs this task. It is provided by T. S. Kuhn in his paper "The Historical Structure of Scientific Discovery" (1962a) and his book *The Structure of Scientific Revolutions* (1962b). More will be said about this view of science in a subsequent chapter.

It does not matter for the present whether the exchange model, or Kuhn's account of science, is correct. What is at issue is the general way in which empirical findings and theoretical models relate, interact and develop. The point is that they work here in exactly the same way as they do in any other science.

References

Ben-David, J. *The Scientist's Role in Society*. Englewood Cliffs, N.J.: Prentice-Hall, 1971.

Bottomore, T. B. "Some Reflections on the Sociology of Knowledge." *British Journal of Sociology* 7, no. 1 (1956): 52–58.

Burchfield, J. D. *Lord Kelvin and the Age of the Earth*. London: Macmillan, 1975.

Cardwell, D. S. L. *From Watt to Clausius*. London: Heinemann, 1971.

Coleman, W. "Bateson and Chromosomes: Conservative Thought in Science." *Centaurus* 15, no. 3–4 (1970): 228–314.

Cowan, R. S. "Francis Galton's Statistical Ideas: The Influence of Eugenics." *Isis* 63 (1976): 509–28.

DeGré, G. *Science as a Social Institution*. New York: Random House, 1967.

Douglas, Mary. *Purity and Danger: An Analysis of Concepts of Pollution and Taboo*. London: Routledge & Kegan Paul, 1966.

—— *Natural Symbols*. London: Barrie & Jenkins, 1970.

Durkheim, E. *The Elementary Forms of the Religious Life*. Trans. J. W. Swain. London: Allen and Unwin, 1915.

—— *The Rules of Sociological Method*, 8th Edition. Trans. S. A. Soloway and J. H. Mueller. New York: The Free Press, 1938.

Forman, P. "Weimar Culture, Causality, and Quantum Theory, 1918–1927: Adaptation by German Physicists and Mathematicians to a Hostile Intellectual Environment." In *Historical Studies in the Physical Sciences*, vol. 3, edited by R. McCormmach, 1–115. Philadelphia: University of Pennsylvania Press, 1971.

Hamlyn, D. W. *The Psychology of Perception*. London: Routledge & Kegan Paul, 1969.

Kuhn, T. S. *The Copernican Revolution*. Cambridge, Mass.: Harvard University Press, 1957.

—— "Energy Conservation as an Example of Simultaneous Discovery." In *Critical Problems in the History of Science*, edited by M. Clagett. Madison: University of Wisconsin Press, 1959.

—— "The Historical Structure of Scientific Discovery." *Science* 136 (1962a): 760–64.

—— *The Structure of Scientific Revolutions*. Chicago: University of Chicago Press, 1962b.

Lakatos, I. "History of Science and Its Rational Reconstructions." In *Boston Studies*, v. 8, edited by R. C. Buck and R. S. Cohen. Dordrecht: Reidel, 1971.

Lovejoy, A. O. "Reflections on the History of Ideas." *Journal of the History of Ideas* 1, no. 1 (1940): 3–23.

MacKenzie, D. *Statistics in Britain, 1865–1930: The Social Construction of Scientific Knowledge*. Edinburgh: Edinburgh University Press, 1981.

Mannheim, K. *Ideology and Utopia*. Trans. with an introduction by L. Wirth and E. Shils. London: Routledge & Kegan Paul, 1936.

—— *Essays on the Sociology of Knowledge*. London: Routledge & Kegan Paul, 1952.

Merton, R. K. "Priorities in Scientific Discoveries." *American Sociological Review* 22, no. 6 (1957): 635–59.

—— *Social Theory and Social Structure*. London: Collier-Macmillan, 1964.

Peters, R. S. *The Concept of Motivation*. London: Routledge & Kegan Paul, 1958.

Popper, K. R. *The Poverty of Historicism*. London: Routledge & Kegan Paul, 1960.

—— *The Open Society and Its Enemies*, vol. 2. London: Routledge & Kegan Paul, 1966.

Rudwick, M. J. S. *The Meaning of Fossils*. London: Macdonald, 1972.

Ryle, G. *The Concept of Mind*. London: Hutchinson, 1949.

Stark, W. *The Sociology of Knowledge*. London: Routledge & Kegan Paul, 1958.

Storer, N. W. *The Social System of Science*. New York: Holt, Rinehart & Winston, 1966.

Toulmin, S. "Crucial Experiments: Priestley and Lavoisier." *Journal of the History of Ideas* 18 (1957): 205–20.

Znaniecki, F. *The Social Role of the Man of Knowledge*. New York: Octagon Books, 1965.

28

Elizabeth Anderson, "Feminist Epistemology: An Interpretation and a Defense"

Feminist epistemology and philosophy of science is an integral part of Feminism—an intellectual, social, and political movement too well known to need special introduction. Elizabeth Anderson's article describes and situates feminist epistemology in the spectrum of different approaches to the nature of human knowledge, including, of course, science.

Feminist epistemology has often been understood as the study of feminine "ways of knowing." But feminist epistemology is better understood as the branch of naturalized, social epistemology that studies the various influences of norms and conceptions of gender and gendered interests and experiences on the production of knowledge. This understanding avoids dubious claims about feminine cognitive differences and enables feminist research in various disciplines to pose deep internal critiques of mainstream research.

Feminist epistemology is about the ways gender influences what we take to be knowledge. Consider impersonal theoretical and scientific knowledge, the kind of knowledge privileged in the academy. Western societies have labeled this kind of knowledge "masculine" and prevented women from acquiring and producing it, often on the pretext that it would divert their vital energies from their "natural" reproductive labor (Hubbard 1990; Schiebinger 1989). Theoretical knowledge is also often tailored to the needs of mostly male managers, bureaucrats, and officials exercising power in their role-given capacities (H. Rose 1987; Smith 1974; Collins 1990). Feminist epistemologists claim that the ways gender categories have been used to understand the character and status of theoretical knowledge, whether men or women have produced and applied this knowledge, and whose interests it

E. Anderson, "Feminist Epistemology: An Interpretation and a Defense," *Hypatia*, 1995, 10: 50–84 (with some cuts). Indiana University Press.

has served have often had a detrimental impact on its content. For instance, feminist epistemologists suggest that various kinds of practical know-how and personal knowledge (knowledge that bears the marks of the knower's biography and identity), such as the kinds of untheoretical knowledge that mothers have of children, are undervalued when they are labeled "feminine." Given the androcentric need to represent the "masculine" as independent of the "feminine," this labeling has led to a failure to use untheoretical knowledge effectively in theoretical reasoning (Smith 1974; H. Rose 1987).

Traditional epistemology finds these claims of feminist epistemology to be highly disturbing, if not plainly absurd. Some feminist epistemologists in turn have rejected empiricism (Harding 1986) or even traditional epistemology as a whole (Flax 1983) for its seeming inability to comprehend these claims. I argue, contrary to these views, that a naturalized empiricist epistemology offers excellent prospects for advancing a feminist epistemology of theoretical knowledge.

The project of feminist epistemology with respect to theoretical knowledge has two primary aims (Longino 1993a). First, it endeavors to explain the achievements of feminist criticism of science, which is devoted to revealing sexism and androcentrism in theoretical inquiry. An adequate feminist epistemology must explain what it is for a scientific theory or practice to be sexist and androcentric, how these features are expressed in theoretical inquiry and in the application of theoretical knowledge, and what bearing these features have on evaluating research. Second, the project of feminist epistemology aims to defend feminist scientific practices, which incorporate a commitment to the liberation of women and the social and political equality of all persons. An adequate feminist epistemology must explain how research projects with such moral and political commitments can produce knowledge that meets such epistemic standards as empirical adequacy and fruitfulness. I will argue that these aims can be satisfied by a branch of naturalized, social epistemology that retains commitments to a modest empiricism and to rational inquiry. Feminist naturalized epistemologists therefore demand no radical break from the fundamental internal commitments of empirical science. They may propose changes in our conceptions of what these commitments amount to, or changes in our methods of inquiry. But these can be derived from the core concept of reason, conjoined with perhaps surprising yet empirically supported hypotheses about social or psychological obstacles to achieving them, and the social and material arrangements required for enabling better research to be done. . . .

The Gendered Division of Theoretical Labor

Feminist critics of science have carefully documented the history of women's exclusion from theoretical inquiry (Rossiter 1982; Schiebinger 1989). Although formal barriers to women's entry into various academic disciplines are now illegal in the United States, informal barriers at all levels remain. Girls are socialized by parents and peers to avoid studying or excelling in subjects considered "masculine," such as mathematics and the natural sciences. Teachers and school counselors actively discourage girls from pursuing these subjects (Curran 1980, 30–32). The classroom climate in mixed-gender schools favors boys. Teachers pay more attention and offer more encouragement to white boys than to girls, solicit their participation more, and expect them to achieve more, especially in mathematics courses (Becker 1981; AAUW 1992). Boys marginalize girls in class by interruption and sexual harassment (AAUW 1992). These behaviors in mixed-gender schools have a detrimental impact on girls' academic ambitions and performance. Girls in all-girl schools express a wider diversity of academic interests and perform better academically than girls in mixed-gender schools (Curran 1980, 34). The disadvantage to women's academic performance and interests from attending mixed-gender schools extends to college. The predominantly male faculty in mixed-gender colleges support women students' academic ambitions less than male and female faculty at women's colleges. Women's colleges produce 50 percent more high-achieving women relative to the number of their female graduates than coeducational institutions (Tidball 1980). Graduate schools present women with informal barriers or costs to advancement, including sexual harassment and exclusion from networks of male mentors and colleagues often vital to the advancement of aspiring academics (Reskin 1979; S. Rose 1989).

Women who overcome these obstacles and obtain advanced degrees are not treated as equals once they enter academic positions. Women whose qualifications are comparable to their male colleagues get lower pay, less research support, jobs in less prestigious institutions, lower-ranking positions, and positions that assign more and lower-level teaching (Astin and Beyer 1973; Fox 1981). The prestige of the graduate institution, publications, and having one's work cited aid men's career advancement much more than women's (Rosenfeld 1981). Women in scientific and engineering professions with publication rates equal to those of their male peers have higher unemployment rates, lower starting salaries, and lower academic rank than men. These differences cannot be explained by the greater impact on women of marriage and children (Vetter 1981). The National Science Foundation (1984) found that after adjusting for factors such as women interrupting their careers to take care of children, half the salary differential between male and female scientists could be explained only by sex discrimination.

The gendered division of theoretical labor does not simply prevent women from doing research or getting published. It fits into a broader gendered structure of epistemic authority which assigns greater credibility, respect, and importance to men's than women's claims. Laboratory, field, and natural experiments alike show that the perceived gender of the author influences people's judgments of the quality of research, independent of its content. Psychologists M. A. Paludi and W. D. Bauer (1983) found that a group told that a paper's author was "John T. McKay" assigned it a much higher average ranking than a group told that the same paper's author was "Joan T. McKay." A group told that its author was "J. T. McKay" rated the paper between the other groups' evaluations, reflecting the suspicion that the author was a woman trying to conceal her gender identity. Academics are no less disposed than others to judge the quality of work higher simply because they believe a man has done it. L. S. Fidell (1970) sent vitae identical in all but name to heads of psychology departments that advertised open rank positions. The jobs the psychologists said they would offer to the purportedly male applicant were higher-ranking than those they were willing to offer to the purportedly female applicant. When the Modern Language Association reviewed papers submitted for their meetings with authors' names attached, men's submissions were accepted at significantly higher rates than women's. After the MLA instituted blind reviewing of papers, women's acceptance rates rose to equality with men's (Lefkowitz 1979).

The concerns raised by the influence of sexist norms on the division of theoretical labor and epistemic authority are not simply matters of justice. Feminist epistemology asks what impact these injustices toward women students and researchers have had on the content, shape, and progress of theoretical knowledge. In some cases, sex discrimination in the academy has demonstrably retarded the growth of knowledge. It took more than three decades for biologists to understand and recognize the revolutionary importance of Barbara McClintock's discovery of genetic transposition. Her attempts to communicate this discovery to the larger scientific community met with incomprehension and disdain. This failure can be partly explained by the fact that no biology department was willing to hire her for a permanent position despite her distinguished record of discoveries and publications. Lacking the opportunities such a position would have provided to recruit graduate students to her research program, McClintock had no one else doing research like hers who could replicate her results or help communicate them to a wider scientific community (Keller 1983).

Cases such as McClintock's demonstrate that the gendered structure of theoretical labor and cognitive authority sometimes slows the progress of knowledge. But does it change the content or shape of knowledge or the direction of knowledge growth? If the gender of the knower is irrelevant to the content of what is investigated, discovered, or invented, then the impact of removing sex discrimination would be to add to the pace of knowledge growth by adding more inquirers and by raising the average level of talent and dedication in the research community. Feminist epistemology would then recommend strictly "gender-blind" changes in the processes by which research jobs get assigned and epistemic authority distributed. The MLA's adoption of blind reviewing of papers to reduce cognitive bias due to sexism in the evaluation of research represents an exemplary application of this side of feminist epistemology. It is logically on a par with the institution of double-blind testing in drug research to reduce cognitive bias due to wishful thinking.

But if the gender of the inquirer makes a difference to the content of what is accepted as knowledge, then the exclusion and undervaluation of women's participation in the theoretical inquiry does not merely set up randomly distributed roadblocks to the improvement of understanding. It imparts a systematic bias on what is taken to be knowledge. If the gender of the inquirer makes a difference to what is known, then feminist epistemology would not confine its recommendations to purely gender-blind reforms in our knowledge practices. It could recommend that these knowledge practices actively seek gender diversity and balance among inquirers and actively attend to the gender of the researchers in evaluating their products.

The gender of the researcher is known to make a difference to what is known in certain areas of social science. In survey research, subjects give different answers to questions depending on the perceived gender of the interviewer (Sherif 1987, 47–48). The perceived race of the interviewer also influences subjects' responses. It is a highly significant variable accounting for subjects' responses to questions about race relations (Schuman and Hatchett 1974). In anthropology, informants vary their responses depending on the gender of the anthropologist. In many societies, male anthropologists have less access to women's social worlds than female anthropologists do (Leacock 1982). The race of the researcher affects access to social worlds as well. Native Americans sometimes grant Asian anthropologists access to religious rituals from which they ban whites (Pai 1985).

Where the perceived gender and race of the researcher are variables influencing the phenomena being observed or influencing access to the phenomena, sound research design must pay attention to the gender and racial makeup of the researchers. In survey research, these effects can be analytically excised by ensuring a gender balanced and racially diverse research team and then statistically isolating the variations in responses due to factors other than subjects' responses to the characteristics of the interviewers. In anthropology, the method of reflexive sociology, instead of attempting to analyze away these effects, treats them as a subject of study in their own right. It advises researchers to interpret what informants tell them not as straightforward native observation reports on their own culture, but as reflections of a strategic interaction between informant and researcher and between the informant and other members of the community being studied (Bourdieu 1977). To obtain a complete representation of informants' report strategies with respect to gender, both male and female researchers must interact with both male and female informants and consider why informants varied their responses according to their own and the researcher's gender (see Bell, Caplan, and Karim [1993] for exemplary cases of feminist reflexive anthropology). Similar reasoning applies to factors such as race, class, nationality, and sexual orientation. So reflexive sociology, like survey research, requires a diversity of inquirers to obtain worthwhile results.

The phenomena just discussed concern the *causal* impact of the gender of the researcher on the *object* of knowledge. Many feminist epistemologists claim that the gender of the inquirer influences the *character* of knowledge itself by another route, which travels through the subjectivity of the researcher herself. The gender of the researcher influences what is known not just through her influence on the object of knowledge but by what are claimed to be gender-specific or gender-typical cognitive or affective dispositions, skills, knowledge, interests, or methods that she brings to the study of the object. The variety of claims of this type must be sorted through and investigated with great care. Some are local and modest. No one disputes that personal knowledge of what it is like to be pregnant, undergo childbirth, suffer menstrual cramps, and have other experiences of a female body is specific to women. Gynecology has certainly progressed since women have entered the field and have brought their personal knowledge to bear on misogynist medical practices. The claims get more controversial the more global they are in scope. Some people claim that women have gender-typical "ways of knowing," styles of thinking, methodologies, and ontologies that globally govern or characterize their cognitive activities across all subject matters. For instance, various feminist epistemologists have claimed that women think more intuitively and contextually, concern themselves more with particulars than abstractions, emotionally engage themselves more with individual subjects of study, and frame their thoughts in terms of a relational rather than an atomistic ontology (Belenky, Clinchy, Goldberger and Tarule 1986; Gilligan 1982; H. Rose 1987; Smith 1974; Collins 1990).

There is little persuasive evidence for such global claims (Tavris 1992, chap. 2). I believe the temptation to accept them is based partly on a confusion between gender symbolism—the fact that certain styles of thinking are *labeled* "feminine"—and the actual characteristics of women. It is also partly due to the lack of more complex and nuanced models of how women entering certain fields have changed the course of theorizing for reasons that seem connected to their gender or their feminist commitments. I will propose an alternative model toward the end of this essay, which does not suppose that women theorists bring some shared feminine difference to all subjects of knowledge. Controversies over supposed global differences in the ways men and women think have tended to overshadow other highly interesting work in feminist epistemology that does not depend on claims that men and women think in essentially different ways. The influence of gendered concepts and norms in our knowledge practices extends far beyond the ways male and female individuals are socialized and assigned to different roles in the division of labor. To see this, consider the role of *gender symbolism* in theoretical knowledge.

Gender Symbolism (I): The Hierarchy of Knowledge

It is a characteristic of human thought that our concepts do not stay put behind the neat logical fences philosophers like to erect for them. Like sly coyotes, they slip past these flimsy barriers to range far and wide, picking up consorts of all varieties, and, in astonishingly fecund acts of miscegenation shocking to conceptual purists, leave offspring who bear a disturbing resemblance to the wayward parent and inherit the impulse to roam the old territory. The philosophical guardians of these offspring, trying to shake off the taint of sexual scandal but feeling guilty about the effort, don't quite know whether to cover up a concept's pedigree or, by means of the discovery/justification distinction, deny that it matters. The latter strategy can work only if, like keepers of a zoo, the philosophers can keep their animals fenced in. Feminist epistemologists track these creatures sneaking past their fences while their keepers dream of tamed animals happy to remain confined.

The most cunning and promiscuous coyotes are our gender concepts. In a manner befitting their own links to sex, they will copulate with *anything*. Feminist epistemologists note that there is hardly any conceptual dichotomy that has not been modeled after and in turn used to model the masculine/feminine dichotomy: mind/body, culture/nature, reason/emotion, objective/subjective, tough-minded/soft-hearted, and so forth. These scandalous metaphorical unions generate conceptions of knowledge, science, and rational inquiry, as well as conceptions of the objects of these inquiries, that are shaped in part by sexist views about the proper relations between men and women. Feminist epistemologists investigate how these conceptions are informed and distorted by sexist imagery. They also consider how alternative conceptions are suppressed by the limits imposed by sexism on the imagination, or by the sexist or androcentric interests served by their present symbolic links to gender (Rooney 1991).

Gender symbolism appears on at least two levels of our knowledge practices: in the construction of a hierarchy of prestige and authority among kinds and fields of knowledge and in the content of theoretical inquiry itself. Consider first the ways different kinds and fields of knowledge are gendered. At the most general level, impersonal theoretical knowledge is coded "masculine." Personal knowledge—the kind of knowledge that is inseparable from the knower's identity, biography, and emotional experiences—is coded "feminine." Theoretical knowledge is thought to be masculine in part because it lays claims to objectivity, which is thought to be achieved through the rigorous exclusion from thought of feminine subjectivity—of emotions, particularity, interests, and values. These uses of gender symbolism have *epistemic* import because they structure a hierarchy of prestige and cognitive authority among kinds of knowledge, and hence of knowers, that is homologous with the gender hierarchy. As men in sexist society express contempt for women and enjoy higher prestige than women, so do theoretical knowers express contempt for those with "merely" personal knowledge of the same subject matters, and enjoy higher prestige than they. Echoing the sexist norms that women must obey men but men need not listen to women, the gender-coded hierarchy of knowledge embodies the norm that personal knowledge must submit to the judgments of impersonal theoretical knowledge, while theoretical knowledge has nothing to learn from personal knowledge and may ignore its claims.

These epistemic norms cannot withstand reflective scrutiny. Successful theorizing deeply depends on personal knowledge, particularly embodied skills, and often depends on emotional engagement with the subjects of study (Polanyi 1958; Keller 1983, 1985). Cora Diamond's (1991) insightful discussion of Vicki Hearne's personal knowledge as an animal trainer provides a particularly fine illustration of this point. Hearne's writings (1982) expose the failures of knowledge that occur when theorists ignore the experiences, skills, and language of animal trainers. In her animal training classes, Hearne observed that people's success in training their pets was inversely related to their training in the behavioral sciences. The anthropomorphic and value-laden language of animal trainers enables them to understand what animals are doing in ways not readily accessible to the impersonal, behavioristic language favored by most behavioral scientists. And their skills and personal knowledge of the animals they work with empower trainers to elicit from animals considerably more complex and interesting behaviors than scientists elicit. These powers are not irrelevant to theorizing about animals. Reflecting on Hearne's story about the philosopher Ray Frey, Diamond writes:

> [Frey] attempted to set up a test for his dog's capacity to rank rational desires. When, in order to see how the dog would rank desires, he threw a stick for his dog . . . and at the same time put food before the dog, the dog stood looking at him. Frey could not see that the dog wanted to know what Frey wanted him to do; Frey's conception of the dog as part of an experimental set-up (taken to include two possible desired activities but not taken to include queer behavior by the dog's master), with Frey as the observer, blocked his understanding. Frey's past experience with his dog did not feed an understanding of how the dog saw him; he could not grasp his own failure, as the dog's master, to make coherent sense, so could not see the dog as responding to that failure to make sense. (Diamond 1991, 1014n. 15)

Diamond diagnoses this epistemic failure as the product of Frey's attachment to a theory of knowledge that distrusts personal experience on the ground that it is distorted by the subject's emotional engagement with the object of knowledge. The theory supposes that we can't achieve objective knowledge of our object through such engagement because all it will offer is a reflection of the subject's own emotions. Subjectivity merely projects qualities onto the object and does not reveal qualities of the object. But the theory is mistaken. Love and respect for another being, animal or person, and trust in the personal experiences of engagement that are informed by such love and respect may be essential both for drawing out and for grasping that being's full potentialities. One of the reasons why behaviorists tend to elicit such boring behavior from animals and humans is that they don't give them the opportunities to exhibit a more impressive repertoire of behaviors that respect for them would require them to offer.

The gender-coded hierarchy of knowledge extends to specific subject matters and methods within theoretical knowledge. The natural sciences are "harder," more like the male body and hence more prestigious, than the social sciences or the "soft" humanities, supposed to be awash in feminine emotionality and subjectivity. Mathematics is coded masculine and is the language of physics, the most prestigious science. Through their closer association with physics, quantitative subfields of biology and the social sciences enjoy higher prestige than subfields of the same discipline or branch of science employing a qualitative, historical, or interpretive methodology. Experimentation asserts more control over subjects of study than observation does. So experimental subfields in biology and psychology are coded masculine and command more cognitive authority than observational subfields of the same disciplines. Values are designated feminine. So normative subfields in philosophy such as ethics and political philosophy enjoy less prestige than supposedly nonnormative fields such as philosophy of language and mind. Social interpretation is thought to be a feminine skill. So interpretive anthropology is designated less masculine, scientific, and rigorous than physical anthropology, which deals with "hard" facts like fossil bones. In each of these cases, the socially enforced norm for relations between fields of knowledge mirrors that of the relations between husband and wife in the ideal patriarchal family: the masculine science is autonomous from and exercises authority over the feminine science, which is supposedly dependent on the former's pronouncements to know what it should think next.

This gendered hierarchy of theoretical subfields produces serious cognitive distortions. Carolyn Sherif (1987) has investigated how the hierarchy of prestige generates cognitive biases in psychology. Forty years ago, experimental psychology dominated developmental and social psychology. The gendered character of this difference in cognitive authority is not difficult to read. Experimental psychologists, by imitating the methods of the "hard" sciences through manipulating quantified variables, claim some of the prestige of the natural sciences. Developmental and social psychologists engage in labor that looks more like the low-status labor conventionally assigned to women. Developmental psychologists work with children; social psychologists deal with human relationships, and forty years ago usually did so in settings not under the control of the researcher. Following the norm that "masculine" sciences need not pay attention to findings in "feminine" sciences, which it is assumed cannot possibly bear on their more "fundamental" research, experimental psychology has a history of constructing experiments that, like Ray Frey's, ignore the ways the social context of the experiment itself and the social relation between experimenter and subject influence outcomes. The result has been a history of findings that lack robustness because they are mere artifacts of the experimental situation. In experimental research on sex differences, this error has taken the form of ascribing observed differences in male and female behavior under experimental conditions to innate difference in male and female psychology rather than to the ways the experiment has socially structured the situation so as to elicit different responses from men and women.

The notorious claim in experimental psychology that women are more suggestible than men offers an instructive illustration of the perils of ignoring social psychology (Sherif 1987, 49–50). The original experiments that confirmed the hypothesis of greater suggestibility involved male researchers trying to persuade men and women to change their beliefs with respect to subject matters oriented to stereotypical male interests. Unaware of how their own gender-typical interests had imparted a bias in the selection of topics of persuasion, the predominantly male researchers confidently reported as a sex difference in suggestibility what was in fact a difference in suggestibility owing to the degree of interest the subjects had in the topics. Differences in the gender-typed cognitive authority of the researcher also affect subjects' responses. Men are more open to the suggestions of a female researcher when the topic is coded feminine, while women are more open to the suggestions of a male researcher when the topic is coded masculine.

Cognitive distortions due to the gender-coding of types and fields of knowledge are strictly separable from any claims about differences in the ways men and women think. Although it is true that the "feminine" sciences and subfields attract more women researchers than the "masculine" sciences do, the differences in cognitive authority between the various sciences and subfields were modeled on differences in social authority between men and women before women constituted a significant proportion of the researchers in any field. Men still predominate even in fields of study that are designated feminine. And scientists' neglect of personal knowledge deprives many men who engage in stereotypically male activities of cognitive authority. For example, animal behaviorists ignore the personal knowledge male policemen have about their police dogs (Diamond 1991). For these reasons, Diamond and Sherif have questioned how gender figures into the cognitive distortions instituted by the hierarchy of knowledge and by scientistic conceptions of objectivity.[1] By shifting our focus from gender structure and supposed gender differences in ways of knowing to gender symbolism, we can see how ideas about gender can distort the relations between forms of knowledge independently of the gender of the knower. In the light of the cognitive distortions caused by the gender-coding of types and domains of knowledge, feminist naturalized epistemologists should recommend that we no longer model the relations between different kinds of knowledge on a sexist view of the authority relations between men and women.

Gender Symbolism (II): The Content of Theories

Gender symbolism figures in the *content* of theories as well as in their relations of cognitive authority whenever conceptions of human gender relations or gendered characteristics are used to model phenomena that are not gendered. Biology is particularly rich with gender symbolism—in models of gamete fertilization, nucleus–cell interaction, primatology, and evolutionary theory (Biology and Gender Study Group 1988; Haraway 1989; Keller 1985, 1992). Evelyn Fox Keller, a mathematical biologist and feminist philosopher of science, has explored gender symbolism in evolutionary theory most subtly (Keller 1992). Consider the fact that evolutionary theory tends to delineate the unit of natural selection, the entity accorded the status of an "individual," at the point where the theorist is willing to use complex and cooperative rather than competitive models of interaction. Among individuals, antagonistic competition predominates and mutualistic interactions are downplayed. The individual is considered "selfish" in relation to other individuals. Thus, theories that take the gene to be the unit of selection characterize the gene as a ruthless egoist ready to sacrifice the interests of its host organism for the sake of reproducing itself (Dawkins 1976). Where the organism is taken to be the unit of selection, it is represented as selfishly competitive with respect to other individual organisms. But within the individual, cooperation among constitutive parts prevails. Cooperation is modeled after the family, often a patriarchal family. The cells of an individual organism cooperate because of the bonds of kinship: they share the same genes. The constitutive parts of an individual cell cooperate because they are ruled by a wise and benevolent patriarch, the "master molecule" DNA, which autonomously tells all the other parts of the cell what to do, solely on the basis of information it contains within itself. Thus, evolutionary theory models the biological world after a sexist and androcentric conception of liberal society, in which the public sphere is governed by competition among presumably masculine selfish individuals and the private sphere of the family is governed by male heads of households enforcing cooperation among its members (Keller 1992, chap. 8). This model is not rigidly or consistently applied in evolutionary theory, but it does mark theoretical tendencies that can be traced back to the fact that Darwin modeled his theory of natural selection after Malthus's dismal model of capitalist society.

Taken by itself, that evolutionary theory employs a sexist ideology of liberal society to model biological phenomena does not have any straightforward normative implications. Defenders of the theory can appeal to the discovery/justification distinction here: just because a theory had its origins in politically objectionable ideas or social contexts does not mean that it is false or useless. Evolutionary theory is extraordinarily fruitful and empirically well confirmed. The model-theoretic view of theories, widely used by feminist empiricists and feminist postmodernists to analyze the roles of gender in the construction of theoretical knowledge, affirms the epistemic legitimacy of any coherent models, hence of any coherent sexist models, in science (Longino 1993b; Haraway 1986).

In the model-theoretic view, scientific theories propose elaborate metaphors or models of phenomena. Their virtues are empirical adequacy, simplicity, clarity, and fruitfulness. Theories are empirically adequate to the extent that the relations among entities in the model are homologous with the observed relations among entities in the world. Empirically adequate models offer a satisfactory explanation of phenomena to the extent that they model unfamiliar phenomena in ways that are simple, perspicuous and analytically tractable. They are fruitful to the extent that they organize inquirers' conceptions of their subjects in ways that suggest lines of investigation that uncover novel phenomena that can be accommodated by further refinements of the model. Empiricists place no a priori constraints on the things that may constitute useful models for phenomena. Anything might be an illuminating model for anything else. So, empiricists can offer no a priori epistemic objections to modeling nongendered phenomena after gendered ones, even if the models are overtly sexist or patriarchal. Such models may well illuminate and effectively organize important aspects of the objects being studied.

So the trouble with using sexist gender symbolism in theoretical models is not that the models are sexist. The trouble lies rather in the extraordinary political salience and rhetorical power of sexist gender ideology, which generates numerous cognitive distortions. Keller has carefully delineated several such distortions in evolutionary theory, especially with respect to its privileging of models of competitive over cooperative or mutualist interactions among organisms. First, to the extent that political ideology incorporates false conceptual identities and dichotomies, a scientific model borrowing its vocabulary and structure is likely to overlook the alternatives suppressed by that ideology or to elide distinctions between empirically distinct phenomena. The ideology of possessive individualism falsely identifies autonomy with selfishness and falsely contrasts self-interest with cooperation. When used to model phenomena in evolutionary biology, it leads to a false identification of peaceful, passive consumption activity with violent, competitive behavior, and to a neglect of mutualist interactions among organisms. Thus, the mathematical tools of population biology and mathematical ecology are rarely used to model cooperation among organisms although they could do so; in contrast with sociobiology, these mathematical subfields of biology have even neglected the impact of sexual intercourse and parenting behavior on the fitness of organisms (Keller 1992, 119–21). Although the technical definition of competition avoids false identities and dichotomies, biologists constantly turn to its colloquial meanings to explain their findings and frame research questions. In this way, "the use of a term with established colloquial meaning in a technical context permits the simultaneous transfer and denial of its colloquial connotations" (Keller 1992, 121). When the language used in a model has particularly strong ideological connotations, the cognitive biases it invites are particularly resistant to exposure and criticism.

The symbolic identification of the scientific with a masculine outlook generates further cognitive distortions. The ideology of masculinity, in representing emotion as feminine and as cognitively distorting, falsely assimilates emotion-laden thoughts—and even thoughts about emotions—to sentimentality. In identifying the scientific outlook with that of a man who has outgrown his tutelage, cut his dependence on his mother, and is prepared to meet the competitive demands of the public sphere with a clear eye, the ideology of masculinity tends to confuse seeing the natural world as indifferent in the sense of devoid of teleological laws with seeing the social world as hostile in the sense of full of agents who pursue their interests at others' expense (Keller 1992, 116–18). This confusion tempts biologists into thinking that the selfishness their models ascribe to genes and the ruthless strategic rationality their models ascribe to individual organisms (mere metaphors, however theoretically powerful) are more "real" than the actual care a dog expresses toward her pups. Such thoughts also reflect the rhetoric of unmasking base motivations behind policies that seem to be benevolent, a common if overused tactic in liberal politics and political theory. The power of this rhetoric depends on an appearance/reality distinction that has no place where the stakes are competing social *models* of biological phenomena, whose merits depend on their metaphorical rather than their referential powers. Thus, to the extent that the theoretical preference for competitive models in biology is underwritten by rhetoric borrowed from androcentric political ideologies, the preference reflects a confusion between models and reality as well as an unjustified intrusion of androcentric political loyalties into the scientific enterprise.

These are not concerns that can be relieved by deploying the discovery/justification distinction. To the extent that motivations tied to acquiring a masculine-coded prestige as a theorist induce mathematical ecologists to overlook the epistemic defects of models of natural selection that fail to consider the actual impact of sexual selection, parenting, and cooperative interactions, they distort the context of justification itself. Some of the criteria of justification, such as simplicity, are also distorted in the light of the androcentric distinction between public and private values. For example, simplicity in mathematical biology has been characterized so as to prefer explanations of apparently favorable patterns of group survival in terms of chance to explanations in terms of interspecific feedback loops, if straightforward individualistic mechanisms are not available to explain them (Keller 1992, 153). Finally, to the extent that gender ideologies inform the context of discovery by influencing the direction of inquiry and development of mathematical tools, they prevent the growth of alternative models and the tools that could make them tractable, and hence they bias our views of what is "simple" (Keller 1992, 160). The discovery/justification distinction, while useful when considering the epistemic relation of a theory to its confirming or disconfirming evidence, breaks down once we consider the relative merits of alternative theories. In the latter context, any influence that biases the development of the field of alternatives will bias the evaluation of theories. A theoretical approach may appear best justified not because it offers an adequate model of the world but because androcentric ideologies have caused more thought and resources to be invested in it than in alternatives.

So feminist naturalized epistemologists should offer a complex verdict on gender symbolism in the content of theories. They should leave open the possibility that gendered models of ungendered phenomena may be highly illuminating and successful, and hence legitimately used in theoretical inquiry. The impressive explanatory successes of evolutionary theory demonstrate this. At the same time, the ideological power of gender symbolism sometimes gets the better of otherwise careful theorists. It can generate conceptual confusion in ways that are hard to detect, and obscure theoretical possibilities that may be worth pursuing. The most reliable way to tell when the use of gender symbolism is generating such cognitive distortions is to critically investigate the gender ideology it depends on and the role this ideology plays in society. In other words, theorists who use gendered models would do well to consider how feminist theory can help them avoid cognitive distortion. Feminist naturalized epistemologists therefore should recommend that theorists attracted to gendered models of ungendered phenomena proceed with caution, in consultation with feminist theorists. It recommends an important change in the cognitive authority of disciplines, through its demonstration that biologists have something to learn from feminist theory after all.

Androcentrism

A knowledge practice is androcentric if it reflects an orientation geared to specifically or typically male interests or male lives. Androcentrism can appear in a knowledge practice in at least two ways: in the content of theories or research programs and in the interests that lead inquirers to frame their research in certain terms or around certain problems. Feminists in the natural and social sciences have advanced feminist epistemology most fully and persuasively by exposing androcentrism in the content of social-scientific and biological theories.

The content of theories can be androcentric in several ways. A theory may reflect the view that males, male lives, or "masculinity" set the norm for humans or animals generally. From this point of view, females, their lives, or "feminine" characteristics are represented as problematic, deviations from the norm, and hence in need of a type of explanation not required for their male counterparts. Androcentrism of this sort often appears in the ways theoretical questions are framed. For decades, psychological and biological research about sex differences has been framed by the question, "Why are women different from men?" and the presumed sex difference has cast women in a deviant position. Researchers have been preoccupied with such

questions as why girls are more suggestible, less ambitious, less analytically minded, and have lower self-esteem than boys. Let us leave aside the fact that all these questions are based on unfounded beliefs about sex differences (Maccoby and Jacklin 1974). Why haven't researchers asked why boys are less responsive to others, more pushy, less synthetically minded, and more conceited than girls? The framing of the problem to be investigated reflects not just a commitment to asymmetrical explanation of men's and women's characteristics, but to an evaluation of women's differences as dimensions of inferiority (Tavris 1992, chap. 1). It is thus sexist as well as androcentric.

Another way in which the content of theories can be androcentric is in describing or defining phenomena from the perspective of men or typically male lives, without paying attention to how they would be described differently if examined from the point of view of women's lives. Economists and political scientists have traditionally defined class and socioeconomic status from the point of view of men's lives: a man's class or socioeconomic status is defined in terms of his own occupation or earnings, whereas a women's status is defined in terms of her father's or husband's occupation or earnings. Such definitions obscure the differences in power, prestige, and opportunities between male managers and their homemaker wives, and between homemaker wives and female managers (Stiehm 1983). They also prevent an analysis of the distinctive economic roles and status of full-time homemakers and of adult independent unmarried women. The distinction between labor and leisure, central to standard economic analyses of the supply of wage labor, also reflects the perspective of male heads of households (Waring 1990). Classically, the distinction demarcates the public from the private spheres by contrasting their characteristic activities as having negative versus positive utility, or instrumental versus intrinsic value, or as controlled by others versus freely self-directed. From the standpoint of the lives of women with husbands or children, these demarcations make no sense. These women are not at leisure whenever they are not engaged in paid labor. Professional women often find much of their unpaid work to constitute a drudgery from which paid labor represents an escape with positive intrinsic value. Middle-class and working-class women who engage in paid labor and who cannot afford to hire others to perform their household tasks and child care are better represented as engaged in (sometimes involuntary) dual-career or double-shift labor than in trading off labor for leisure. Full-time mothers and homemakers often view what some consider to be their leisure activities as highly important work in its own right, even if it is unpaid.

The androcentrism implicit in the standard economic definition of productive labor has profound implications for national income accounting, the fundamental conceptual framework for defining and measuring what counts as economically relevant data for macroeconomic theory. It effectively excludes women's gender-typical unpaid domestic labor from gross national product (GNP) calculations, making women's work largely invisible in the economy. In the advanced industrialized nations, economists explain this omission by arguing that GNP figures properly measure only the economic value of production for market exchange. In developing nations, where only a modest proportion of productive activity shows up in market exchanges, economists have long recognized the uselessness of measures of national production that look only at the market; so they impute a market value to various unmarketed domestic production activities associated with subsistence agriculture, home construction, and the like. But which of these household activities do economists choose to count as productive? In practice, they have defined the "production boundary" in such societies by imposing an obsolete Western androcentric conception of the household. They assume that households consist of a productive primary producer, the husband, who supports a wife engaged in "housework," which is assumed to be economically unimportant or unproductive. "Housework" has no clear definition in societies where most production takes place within the household. So economists apply the concept of "housework" to whatever productive activities a society conventionally assigns to women. Thus, women's unmarketed labor in these societies counts as productive only if men usually perform it too, whereas men's unmarketed labor is usually counted in the national income statistics regardless of its relation to women's labor (Waring 1990, 74–87). The result is that in Africa, where women do 70 percent of the hoeing and weeding of subsistence crops, 80 percent of crop transportation and storage, and 90 percent of water and fuel collecting and food processing, these vital activities rarely appear in the national income accounts (Waring 1990, 84). Here, androcentrism is built into the very data for economic theorizing, in a such a way that women's gender-typical activities become invisible.

Even when a theory does not go so far as to define the phenomena in a way that excludes female activities, it may still be androcentric in assuming that male activities or predicaments are the sole or primary sources of important changes or events. Until recently, primatologists focused almost exclusively on the behavior of male primates. They assumed that male sexual and dominance behaviors determined the basic structure of primate social order, and that the crucial social relationships among troop-dwelling primates that determined the reproductive fitness of individuals and maintained troop organization were between the dominant male and other males. The assumption followed from a sociobiological argument that claimed to show that females of any species will typically be the "limiting resource" for reproduction: most females will realize an equal and maximum reproductive potential, while males will vary enormously in their reproductive fitness. Natural selection, the driving force of evolutionary change, would therefore operate primarily on male characteristics and behavior (Hrdy 1986).

These assumptions were not seriously challenged until women, some inspired by the feminist movement, started entering the field of primatology in substantial numbers in the mid-1970s. Many studied female–female and female–infant interactions, female dominance and cooperative behavior, and female sexual activity. By turning their focus from male to female behaviors and relationships, they found that infant survival varied enormously, depending on the behavior and social status of the mothers, that troop survival itself sometimes depended on the eldest female (who would teach others the location of distant water holes that had survived droughts), and that female-directed social and sexual behaviors play key roles in maintaining and changing primate social organizations (Hrdy 1981; Haraway 1989). Today the importance of female primates is widely recognized and studied by both male and female primatologists.

What normative implications should be drawn about the epistemic status of androcentric theories? Some feminist epistemologists propose that theory can proceed better by viewing the world through the eyes of female agents. Gynocentric theory can be fun. What could be a more amusing retort to a study that purports to explain why women lack self-esteem than a study that explains why men are conceited? It can also be instructive. Richard Wrangham (1979) has proposed a gynocentric model of primate social organization that has achieved widespread recognition in primatology. The model assumes the centrality of female competition for food resources, and predicts how females will space themselves (singly or in kin-related groups) according to the distribution of the foods they eat. Males then space themselves so as to gain optimum access to females. The model is gynocentric both in defining the core of primate social groups around female kin-relations rather than around relations to a dominant male and in taking the situation of females to constitute the primary variable that accounts for variations in male and general primate social organization. According to the feminist primatologist Sarah Hrdy (1981, 126), Wrangham's model offers the best available explanation of primate social organization.

The three androcentric theoretical constructs mentioned correspond to three different ways in which a theory could be "gender-centric": in taking one sex or gender to set the norm for both, in defining central concepts with respect to the sex or gender-typical characteristics, behaviors, or perspectives of males or females alone, and in taking the behaviors, situation, or characteristics of one sex or gender to be causally central in determining particular outcomes. These logical differences in gender-centric theorizing have different epistemic implications. As Wrangham's theory shows, gynocentric causal models can sometimes be superior to androcentric models. Whether they are superior in any particular domain of interest is an empirical question. It can only be answered by comparing rival gender-centric models to one another and to models that do not privilege either male- or female-typical activities or situations in their causal accounts, but rather focus on activities and situations common to both males and females. An important contribution of feminist scholarship in the social sciences and biology has been to show that the activities and situations of females have been far more causally important in various domains than androcentric theories have recognized.

The other two types of gender-centrism are much more problematic than this causal type. A theory that takes one gender to set the norm for both must bear an explanatory burden not borne by theories that refuse to represent difference as deviance. It must explain why an asymmetrical explanation is required for male- and female-specific characteristics. Given the dominant background assumption of modern science that the cosmos does not have its own telos, it is hard to justify any claim that one gender naturally sets the norm for both. Claims about norms must be located in human value judgments, which is to say that the only justification for normative gender-centrism would have to lie in a substantive sexist moral or political theory. As we shall see below, empiricism does not rule out the use of value judgments as background assumptions in scientific theories. Nevertheless, this analysis of normative gender-centrism suggests why feminists should not be satisfied with a table-turning, "why men are so conceited" type of gynocentric theorizing. Posing such questions may expose the androcentrism of standard ways of framing research problems in sex-differences research to healthy ridicule. But because feminists are interested in upholding the equality of all persons, not the domination of women over men, they have no interest in claiming that women set the norm for humans generally.

Theories that tailor concepts to the activities or positions specific to or typical of one gender only and then apply them to everyone are straightforwardly empirically inadequate. As the case of androcentric definitions of class showed, they obscure actual empirical differences between men and women and between differently situated women. As the case of the labor/leisure distinction showed, they overgeneralize from the typical situation of one gender to that of both. When conceptually androcentric theories guide public policy, the resulting policies are usually sexist, since theories cannot respond to phenomena they make invisible. Thus, when GNP statistics fail to count women's labor as productive, and public policies aim to increase GNP, they may do so in ways that fail to improve the well-being of women and their families and may even reduce it. In Malawi and Lesotho, where women grow most of the food for domestic consumption, foreign aid projects have provided agricultural training to the men who have no use for it, and offered only home economics education to women (Waring 1990, 232, 234). In the Sahel, a USAID drought-relief project forced women into economic dependency on men by replacing only men's cattle herds, on the androcentric assumption that women did not engage in economically significant labor (Waring 1990, 176–77).

Feminist naturalized epistemologists therefore pass different judgments on different kinds of gender-centrism in theoretical inquiry. Conceptual gender-centrism is plainly inadequate in any society with overlapping gender roles, because it leads to overgeneralization and obscures the differences between empirically distinct phenomena. It could work only in societies where men and women inhabit completely and rigidly segregated spheres, and only for concepts that apply exclusively to one or the other gender in such a society. Normative gender-centrism either depends on a problematic cosmic teleology or on sexist values. This does not automatically make it epistemically inadequate, but it does require the assumption of an explanatory burden (why men's and women's traits do not receive symmetrical explanatory treatment) that non-gender-centric theories need not assume. In addition, its dependence on sexist values gives theorists who repudiate sexism sufficient reason to conduct inquiry that is not normatively gender-centric. Finally, causal gender-centrism may or may not be empirically justified. Some events do turn asymmetrically on what men or women do, or on how men or women are situated.

The chief trap in causal gender-centrism is the temptation to reify the domain of events that are said to turn asymmetrically on the actions or characteristics of one or the other gender. The selection of a domain of inquiry is always a function of the interests of the inquirer.[2] Failure to recognize this may lead androcentric theorists to construct their domain of study in ways that confine it to just those phenomena that turn asymmetrically on men's activities. They may therefore declare as an objective fact that, say, women have little causal impact on the "economy," when all that is going on is that they have not taken any interest in women's productive activities, and so have not categorized those activities as "economic." Feminist naturalized epistemologists caution against the view that domains of inquiry demarcate natural kinds. Following Quine, they question supposed conceptual barriers between natural and social science, analytic and synthetic knowledge, personal and impersonal knowledge, fact and value (Nelson 1990, chap. 3). Their empiricist commitments enable them to uncover surprising connections among apparently distant points in the web of belief. If naturalized epistemologists use space-age technology to explore the universe of knowledge, feminist naturalized epistemologists could be said to specialize in the discovery of wormholes in that universe. Gender and science are not light-years apart after all; subspace distortions in our cognitive apparatus permit surprisingly rapid transport from one to the other, but feminist navigators are needed to ensure that we know the route we are travelling and have reason to take it.

Sexism in Scientific Theories

One frequently traveled route between gender and science employs normative assumptions about the proper relations between men and women, or about the respective characteristics and interests of men and women, in the content or application of scientific theories. When a theory asserts that women are inferior to men, properly subordinated to men, or properly confined to gender-stereotyped roles, or when it judges or describes women according to sexist or double standards, the content of the theory is sexist. When people employ such assumptions in applying theories, the application of the theory is sexist. Naturalized feminist epistemology considers how our evaluations of theories should change once their sexism is brought to light.

The application of theories can be sexist in direct or indirect ways. Theories may be used to provide direct ideological justification for patriarchal structures. Steven Goldberg (1973) uses his theory of sex differences in aggression to justify a gendered division of labor that deliberately confines women to low-prestige occupations. More usually, the application of theories is indirectly sexist in taking certain sexist values for granted rather than trying to justify them. For example, research on oral contraceptives for men and women uses a double standard for evaluating the acceptability of side effects. Oral contraceptives for men are disqualified if they reduce libido, but oral contraceptives for women are not rejected for reducing women's sexual desire.

In a standard positivist analysis, neither form of sexism in the application of theories has any bearing on the epistemic value of the theories in question. That a theory is used to support unpopular political programs does not show that the theory is false. At most, it reflects a failure of the proponents of the program to respect the logical gap between fact and value. But opponents of the program fail to respect this gap in attacking a theory for the uses to which it is put. According to this view, theories supply facts that all persons must accept, regardless of their political commitments. That a theory is indirectly applied in sexist ways provides even less ground for attacking its content. The question of truth must be strictly separated from the uses to which such truths are put.

Naturalized epistemology does not support such a sanguine analysis of theories that are applied in sexist ways. "Successful" technological applications of theories are currently taken to provide evidence of their epistemic merits. If knowledge is power, then power is a criterion of adequate understanding. The prevailing interpretation of this criterion does not consider whose power is enhanced by the theory and whose interests are served by it. Feminists urge that these considerations be taken explicitly into account when one evaluates whether technological applications of theories supply evidence of an adequate understanding of the phenomena they control (Tiles 1987). It may be true that certain drugs would be effective in controlling the phenomena of women's hormonal cycles that are currently designated as pathologies constitutive of premenstrual syndrome. Such control may come at the expense of women's interests, not just because of undesirable side effects but also because the legitimation of drug treatment reinforces the medicalization of women's complaints, as if these complaints were symptoms to be medicated rather than as claims on others to change their behavior (Zita 1989). Doctors may be satisfied that such a "successful" drug treatment of PMS supplies evidence that the theory it applies provides them with an adequate understanding of women's menstrual cycles. But should women be satisfied with this understanding? Suppose the phenomena associated with PMS could also be eliminated, or revalued, by widespread acceptance of feminist conceptions of women's bodies or by egalitarian changes that would make social arrangements less frustrating to women. (This would be possible if women's symptoms of distress in PMS were partly caused by misogynist social expectations that represent women's menstrual cycles as pathological.) Such a successful "technological" application of feminist theory would provide women with an understanding of their own menstrual cycles that would empower them. Where the sexist medical technology would enable women to adapt their bodies to the demands of a sexist society, the feminist technology would empower women to change society so that their bodies were no longer considered "diseased." Thus, applications of theories may influence the content of theories whenever "success" in application is taken to justify the theory in question. Sexist or feminist values may inform criteria of success in application, which may in turn inform competing criteria of adequate understanding. The epistemic evaluation of theories therefore cannot be sharply separated from the interests their applications serve.

Feminist naturalized epistemology also rejects the positivist view that the epistemic merits of theories can be assessed independently of their direct ideological applications (Longino 1990; Antony 1993; Potter 1993). Although any acceptable ideology must make sure that it does not fly in the face of facts, theories do not merely state facts but organize them into systems that tell us what their significance is. Theories logically go beyond the facts; they are "underdetermined" by all the empirical evidence that is or ever could be adduced in their favor (Quine 1960, 22). The evidential link between an observed fact and a theoretical hypothesis can only be secured by background auxiliary hypotheses. This leaves open the logical possibility that ideological judgments may not be implications of an independently supported theory but figure in the justification of the theory itself, by supplying evidential links between empirical observations and hypotheses.

A particularly transparent example of this phenomenon may be found in theories about sex differences in intelligence. Girls scored significantly higher than boys on the first Stanford–Binet IQ tests developed by Lewis Terman. To correct for this "embarrassment," Terman eliminated portions of the test where girls scored higher than boys and inserted questions on which boys scored higher than girls. The substitution was considered necessary to ensure the validity of the test against school grades, the only available independent measures of children's intelligence, which did not differ by gender. But Terman did not adjust his test to eliminate sex differences on subtests of the IQ, such as those about quantitative reasoning. These differences seemed unproblematic because they conformed to prevailing ideological assumptions about appropriate gender roles (Mensh and Mensh 1991, 68–69). Today, that IQ scores are good predictors of a child's school grades is still taken to provide key evidence for the claim that differences in IQ scores measure differences in children's innate intelligence. But the evidential link tying school grades to this theoretical claim depends on the background value judgment that schools provide fair educational opportunities to all children with respect to all fields of study. Those schools that discourage girls from pursuing math and science assume that girls have inferior quantitative reasoning ability; they do not recognize that lack of encouragement can cause relatively lower performance on math tests.

From a positivist point of view, this reasoning is defective on two counts. First, it is circular to claim that IQ tests demonstrate innate sex differences in quantitative reasoning ability when the assumption of innate sex differences is built into the background hypotheses needed to validate the tests. Second, no reasoning is scientifically sound that incorporates value judgments into the background assumptions that link observations to theory. The salience of positivist views of science as well as their usefulness to feminists in criticizing research about sex differences has tempted some feminists to use the positivist requirement that science be value-free to discredit all scientific projects that incorporate sexist values in the explicit or implicit content of their theories. But this appropriation of positivism puts at odds the two aims of feminist epistemology—to criticize sexist science and to promote feminist science. If incorporating sexist values into scientific theories is illegitimate on positivist grounds, then so is incorporating feminist values into scientific theories (Longino 1993a, 259).

Feminist naturalized epistemologists offer a more nuanced response to the presence of value judgments in scientific inference. Even "good science" can incorporate such value judgments. The logical gap between theory and observation ensures that one cannot in principle rule out the possibility that value judgments are implicit in the background assumptions used to argue that a given observation constitutes evidence for a given hypothesis (Longino 1990). From the perspective of an individual scientist, it is not unreasonable to use any of one's firm beliefs, including beliefs about values, to reason from an observation to a theory. Nor does the prospect of circularity threaten the scientific validity of one's reasoning, as long as the circle of reasoning is big enough. In a coherent web of belief, every belief offers some support for every other belief, and no belief is perfectly self-supporting. Theories that incorporate value judgments can be scientifically sound as long as they are empirically adequate.

This reasoning underwrites the legitimacy of feminist scientific research, which incorporates feminist values into its theories. Such values may be detected in the commitment of feminist researchers to regard women as intelligent agents, capable of reflecting on and changing the conditions that presently constrain their actions. This commitment tends to support a theoretical preference for causal models of female behavior that highlight feedback loops between their intentional states and their social and physical environments, and that resist purely structuralist accounts of female "nature" that leave no room for females to resist their circumstances or maneuver among alternate possibilities (Longino 1989, 210–13; Haraway 1989, chap. 13). In contrast, most behaviorist and some sociobiological theories favor models that highlight linear causal chains from fixed physiological or physical conditions to determinate behaviors, and that emphasize the structural constraints on action. The epistemic values of simplicity, prediction, and control might seem to support linear, structural causal models. But we have seen that control at least is a contested value; the kinds of control taken to warrant claims of adequate understanding depend on substantive value judgments about the importance of particular human interests. Is adequate understanding achieved when a theory empowers scientists to control women's lives, or when it empowers women to control their own lives? Rival interpretations of the other epistemic values also depend on contested nonepistemic values. The kind of simplicity one favors depends on one's aesthetic values. In any event, other epistemic values, such as fruitfulness, appear to favor complex, nonlinear causal models of human behavior. Such models support experiments that generate novel behaviors disruptive of presumed structural constraints on action.

Naturalized feminist epistemology thus permits scientific projects that incorporate feminist values into the content and application of theories. It does not provide methodological arguments against the pursuit of sexist theories. It does claim, however, that it is irrational for theorists to pursue sexist research programs if they do not endorse sexist values. Moral and political arguments about the rationality of particular values may therefore have a bearing on the rationality of pursuing particular research programs. In addition, the objectivity of science demands that the background assumptions of research programs be exposed to criticism. A scientific community composed of inquirers who share the same background assumptions is unlikely to be aware of the roles these assumptions play in licensing inferences from observations to hypotheses, and even less likely to examine these assumptions critically. Naturalized epistemology therefore recommends that the scientific community include a diversity of inquirers who accept different background assumptions. A community of inquirers who largely accept sexist values and incorporate them into their background assumptions could enhance the objectivity of the community's practice by expanding its membership to include researchers with feminist commitments (Longino 1993a, 267–69).

The Local Character of Naturalized Feminist Epistemology

In reading the project of feminist epistemology along naturalized, empiricist lines, I have tried to show how its interest and critical power do not depend on the global, transcendental claims that all knowledge is gendered or that rationality as a regulatory epistemic ideal is masculine. Naturalized feminist epistemologists may travel to distant locations in the universe of belief, but they always remain inside that universe and travel from gender to science by way of discrete, empirically discovered paths. They have an interest in constructing new paths to empirically adequate, fruitful, and useful forms of feminist science and in breaking up other paths that lead to cognitively and socially unsatisfactory destinations. All the paths by which naturalized epistemologists find gender to influence theoretical knowledge are local, contingent, and empirically conditioned. All the paths by which they propose to change these influences accept rationality as a key epistemic ideal and empirical adequacy as a fundamental goal of acceptable theories. This ideal and this goal are in principle equally open to pursuit by male and female inquirers, but may be best realized by mixed-gender research communities. Naturalized epistemologists find no persuasive evidence that indicates that all women inquirers bring some shared global feminine difference in ways of thinking to all subjects of study nor that such a feminine difference gives us privileged access to the way the world is.

In rejecting global, transcendental claims about differences in the ways men and women think, naturalized feminist epistemologists do not imply that the entry and advancement of significant numbers of women into scientific communities makes no systematic difference to the knowledge these communities produce. But, following their view of inquiry as a social, not an individual, enterprise, they credit the improvements in knowledge such entry produces to the greater diversity and equality of membership in the scientific community rather than to any purportedly privileged subject position of women as knowers (Tuana 1992; Longino 1993a). Men and women do have *some* gender-specific experiences and personal knowledge due to their different socialization and social status. We have seen that such experiences and forms of knowledge can be fruitfully brought to bear upon theoretical inquiry. So it should not be surprising that women researchers have exposed and criticized androcentrism in theories much more than men have. The diversity and equality of inquirers help ensure that social models do not merely reflect or fit the circumstances of a narrow demographic segment of the population when they are meant to apply to everyone. They correct a cognitive bias commonly found among inquirers belonging to all demographic groups, located in the habit of assuming that the way the world appears to oneself is the way it appears to everyone.

This survey of some findings of naturalized feminist epistemology has also identified improvements in knowledge that have or would come about through the entry of *feminist* theorists into various fields, and through revisions in the systems of cognitive authority among fields that would bring the findings of feminist theorists to bear upon apparently distant subjects.[3] We have seen that the use of gender symbolism to model nonhuman phenomena is fraught with cognitive traps. So it should not be surprising that feminist researchers, who make it their business to study the contradictions and incoherences in our conceptions of gender, can improve theories by exposing and clearing up the confusions they inherit from the gender ideologies they use as models. By pursuing feminist research in the humanities, social sciences, and biology, feminist researchers also pose challenges to prevailing theories. Here again, the kinds of changes we should expect in theoretical knowledge from the entry of feminist researchers into various fields do not typically consist in the production of specifically feminist ontologies, methodologies, standpoints, paradigms, or doctrines. Feminist contributions to theorizing are more usefully conceived as altering the field of theoretical possibilities (Haraway 1986, 81, 96). Research informed by feminist commitments makes new explanatory models available, reframes old questions, exposes facts that undermine the plausibility of previously dominant theories, improves data-gathering techniques, and shifts the relations of cognitive authority among fields and theories. In these and many other ways, it reconfigures our assessments of the prospects and virtues of various research programs. Without claiming that women, or feminists, have a globally different or privileged way of knowing, naturalized feminist epistemology explains how feminist theory can productively transform the field of theoretical knowledge.

Notes

I wish to thank Ann Cudd, Sally Haslanger, Don Herzog, David Hills, Peter Railton, Justin Schwartz, Miriam Solomon, and the faculties at the Law Schools of Columbia University, the University of Chicago, and Northwestern University for helpful comments and criticisms.

1 Diamond (1991, 1009) writes that the exclusion of animal trainers' knowledge from the realm of authoritative knowledge "cannot in any very simple way be connected to gender." Pointing out that the terms "hard" and "soft" as applied to forms of knowledge are used by "men trying to put down other men," Sherif argues that for this reason it is "particularly misleading" to infer that these terms symbolize "masculine" and "feminine" (1987, 46–47). I would have thought that her observation supports the gendered reading, since a standard way for men to put down other men is to insinuate that they are feminine.

2 The interests at stake need not be self-interests or even ideological interests of a broader sort. One might just be curious about how rainbows form, without seeking this knowledge for the sake of finding out how to get the proverbial pot of gold at the end. Curiosity is one kind of interest we can express in a phenomenon.

3 The question of the impact of feminist theorists on knowledge is distinct from but related to the question of the impact of women theorists on knowledge. Not all women theorists are feminists, and some feminist theorists are men. At the same time, there could be no genuine feminist theory that was conducted by men alone. Feminist theory is theory committed to the liberation and equality of women. These goals can only be achieved through the exercise of women's own agency, especially in defining and coming to know themselves. Feminist theory is one of the vehicles of women's agency in pursuit of these goals, and therefore cannot realize its aims if it is not conducted by women. So it should not be surprising that most of the transformations of knowledge induced by feminist theory were brought about by women.

References

Alcoff, Linda, and Elizabeth Potter, eds. 1993. *Feminist epistemologies*. New York: Routledge.

American Association of University Women. 1992. *The AAUW report: How schools shortchange girls*. Prepared by Wellesley College Center for Research on Women.

Antony, Louise. 1993. Quine as feminist: The radical import of naturalized epistemology. In *A mind of one's own: Feminist essays on reason and objectivity*. See Antony and Witt 1993.

Antony, Louise, and Charlotte Witt. 1993. *A mind of one's own: Feminist essays on reason and objectivity* Boulder, CO. Westview.

Astin, Helen S., and Alan E. Beyer. 1973. Sex discrimination in academe. In *Academic women on the move*, ed. Alice. S. Rossi and Ann Calderwood. New York: Russell Sage Foundation.

Becker, Joanne Rossi. 1981. Differential treatment of females and males in mathematics classes. *Journal for Research in Mathematics Education* 12(1): 40–53.

Belenky, Mary, Blythe Clinchy, Nancy Goldberger, and Jill Tarule. 1986. *Women's ways of knowing*. New York: Basic Books.

Bell, Diane, Pat Caplan, and Wazir Karim, eds. 1993. *Gendered fields: Women, men, and ethnography*. New York: Routledge.

Biology and Gender Study Group. 1988. The importance of feminist critique for contemporary cell biology. *Hypatia* 3(1): 61–76.

Bourdieu, Pierre. 1977. *Outline of a theory of practice*. Cambridge: Cambridge University Press.

Collins, Patricia Hill. 1990. *Black feminist thought: Knowledge, consciousness, and the politics of empowerment*. Boston: Unwin Hyman.

Curran, Libby. 1980. Science education: Did she drop out or was she pushed? In *Alice through the microscope*, ed. Brighton Women and Science Group. London: Virago.

Dawkins, Richard. 1976. *The selfish gene*. New York: Oxford University Press.

Diamond, Cora. 1991. Knowing tornadoes and other things. *New Literary History* 22: 1001–15.

Fidell, L. S. 1970. Empirical verification of sex discrimination in hiring practices in psychology. *American Psychologist* 25(12): 1094–98.

Flax, Jane. 1983. Political philosophy and the patriarchal unconscious. In *Discovering reality*. See Harding and Hintikka 1983.

Fox, Mary Frank. 1981. Sex segregation and salary structure in academia. *Sociology of Work and Occupations* 8(1): 39–60.

Gilligan, Carol. 1982. *In a different voice*. Cambridge: Harvard University Press.

Goldberg, Steven. 1973. *The inevitability of patriarchy*. New York: Morrow.

Haraway, Donna. 1986. Primatology is politics by other means. In *Feminist approaches to science*, ed. Ruth Bleier. New York: Pergamon.

—— 1989. *Primate visions*. New York: Routledge.

Harding, Sandra. 1986. *The science question in feminism*. Ithaca: Cornell University Press.

Harding, Sandra, and Merrill B. Hintikka, eds. 1983. *Discovering Reality*. Dodrecht, Holland. D. Reidel:

Hearne, Vicki. 1982. *Adam's task*. New York: Vintage.

Hrdy, Sarah. 1981. *The woman that never evolved*. Cambridge: Harvard University Press.

—— 1986. Empathy, polyandry, and the myth of the coy female. In *Feminist approaches to science*, ed. Ruth Bleier. New York: Pergamon.

Hubbard, Ruth. 1990. *The politics of women's biology*. New Brunswick, NJ: Rutgers University Press.

Keller, Evelyn Fox. 1983. *A feeling for the organism*. New York: Freeman.

—— 1985. The force of the pacemaker concept in theories of aggregation in cellular slime mold. In *Reflections on gender and science*. New Haven: Yale University Press.

—— 1992. *Secrets of life, secrets of death*. New York: Routledge.

Leacock, Eleanor. 1982. *Myths of male dominance*. New York: Monthly Review Press.

Lefkowitz, M. R. 1979. Education for women in a man's world. *Chronicle of Higher Education*, 6 August, p. 56.

Longino, Helen. 1989. Can there be a feminist science? In *Women, knowledge, and reality*, ed. Ann Garry and Marilyn Pearsall. Boston: Unwin Hyman.

—— 1990. *Science as social knowledge*. Princeton, NJ: Princeton University Press.

—— 1993a. Essential tensions—Phase two: Feminist, philosophical, and social studies of science. In *A mind of one's own*. See Antony and Witt 1993.

—— 1993b. Subjects, power, and knowledge: Description and prescription in feminist philosophies of science. In *Feminist epistemologies*. See Alcoff and Potter 1993.

Maccoby, Eleanor, and Carol Jacklin. 1974. *The psychology of sex differences*. Stanford: Stanford University Press.

Mensh, Elaine, and Harry Mensh. 1991. *The IQ mythology*. Carbondale: Southern Illinois University Press.

National Science Foundation. 1984. *Women and minorities in science and engineering*.

Nelson, Lynn. 1990. *Who knows? From Quine to a feminist empiricism*. Philadelphia: Temple University Press.

Pai, Hyung Il. 1985. (Anthropologist, University of California, Santa Barbara). Personal communication.

Paludi, Michele Antoinette, and William D. Bauer. 1983. Goldberg revisited: What's in an author's name. *Sex Roles* 9(3): 287–390.

Polanyi, Michael. 1958. *Personal knowledge*. Chicago: University of Chicago Press.

Potter, Elizabeth. 1993. Gender and epistemic negotiation. In *Feminist epistemologies*. See Alcoff and Potter 1993.

Quine, W. V. O. 1960. *Word and object*. Cambridge: MIT Press.

Reskin, Barbara. 1979. Academic sponsorship and scientists' careers. *Sociology of Education* 52(3): 129–46.

Rooney, Phyllis. 1991. Gendered reason: Sex metaphor and conceptions of reason. *Hypatia* 6(2): 77–103.

Rose, Hilary. 1987. Hand, brain, and heart: A feminist epistemology for the natural sciences. In *Sex and scientific inquiry*, ed. Sandra Harding and Jean O'Barr. Chicago: University of Chicago Press.

Rose, Suzanna. 1989. Women biologists and the "old boy" network. *Women's Studies International Forum* 12(3): 349–54.

Rosenfeld, Rachel. 1981. Academic career mobility for psychologists. In *Women in scientific and engineering professions*, ed. Violet Haas and Carolyn Perrucci. Ann Arbor: University of Michigan Press.

Rossiter, Margaret. 1982. *Women scientists in America: Struggles and strategies to 1940*. Baltimore: Johns Hopkins University Press.

Schiebinger, Londa. 1989. *The mind has no sex?* Cambridge: Harvard University Press.

Schuman, Howard, and Shirley Hatchett. 1974. *Black racial attitudes: Trends and complexities*. Ann Arbor: University of Michigan Press.

Sherif, Carolyn. 1987. Bias in psychology. In *Feminism and methodology*, ed. Sandra Harding. Bloomington: Indiana University Press.

Smith, Dorothy. 1974. Women's perspective as a radical critique of sociology. *Sociological Inquiry* 44(1): 7–13.

Stiehm, Judith. 1983. Our Aristotelian hangover. In *Discovering reality*. See Harding and Hintikka 1983.

Tavris, Carol. 1992. *The mismeasure of woman*. New York: Simon and Schuster.

Tidball, M. Elizabeth. 1980. Women's colleges and women achievers revisited. *Signs* 5(3): 504–17.

Tiles, Mary. 1987. A science of Mars or of Venus? *Philosophy* 62(July): 293–306.

Tuana, Nancy. 1992. The radical future of feminist empiricism. *Hypatia* 7(1): 100–14.

Vetter, Betty. 1981. Changing patterns of recruitment and employment. In *Women in scientific and engineering professions*, ed. Violet Haas and Carolyn Perrucci. Ann Arbor: University of Michigan Press.

Waring, Marilyn. 1990. *If women counted*. San Francisco: HarperCollins.

Wrangham, Richard. 1979. On the evolution of ape social systems. *Biology and Social Life: Social Sciences Information* 18: 335–68.

Zita, Jacquelyn. 1989. The premenstrual syndrome: "Dis-easing" the female cycle. In *Feminism and science*, ed. Nancy Tuana. Bloomington: Indiana University Press.

29

Ernan McMullin, "The Social Dimensions of Science"

As Ernan McMullin notes, "the social dimensions of science must be taken seriously. But there [is] little agreement as to what this advice amounts to."

One of Bacon's most intriguing works was a short essay, left unfinished at his death in 1626, on how the work of natural philosophy might be organized. It is written in the form of a fable:

> This fable my Lord devised, to the end that he might exhibit therein a model or description of a college instituted for the interpreting of nature and the producing of great and marvellous works for the benefit of men, under the name of Salomon's House, or the College of the Six Days' Works. . . . Certainly the model is more vast and high than can possibly be imitated in all things; notwithstanding, most things therein are within men's power to effect.[1]

Bacon describes a large island called Bensalem in the Pacific where natural philosophy has progressed much more than it has in Europe. This progress has been in part due to the cohesive social organization given it by "Salomon's House" where all the activities of research have been concentrated. The visitor is told: "The end of our Foundation is the knowledge of causes and secret motions of things, and the enlarging of the bounds of human empire to the affecting of all things possible."[2] The "Father" of the House describes the caves where mining experiments are carried on, the high towers where meteorological phenomena are systematically observed, the mineral springs whose effects on health are studied, the gardens and farms where different strains of plants and animals are transformed by careful

E. McMullin (ed.), *The Social Dimensions of Science*, 1992, pp. 12–26. Notre Dame: University of Notre Dame Press.

cross-breeding, the furnaces that produce a "diversity of heats," "perspective-houses" where experiments on optical phenomena are carried on, "sound-houses" where "we practice and demonstrate all sounds and their generation," "perfume-houses" where the objects of smell and taste are catalogued, "engine-houses" for the construction of machines, including machines of war, "houses of deceits of the senses" where illusion is studied, and a "mathematical house" for making instruments for astronomical and geometrical use.

The emphasis is clearly on technology, on transformation for practical use. It is not clear (any more than it is in the *New Organon*) how the "secret motions of things" are to be discovered. Bacon describes induction as though it were simply a matter of generalizing relationships between observables. But this technique alone could never attain to the latent processes and configurations "which for the most part escape the sense," on which he takes observable phenomena to depend.[3] The role that theory would soon take on in this regard still lies below Bacon's horizon. There is more than a hint in his prospectus of alchemy, of techniques that "endow bodies with new natures." Research is organized, oddly, to our eyes, around the distinctions between the human senses of sight, sound, and smell. The role of mathematics is minor, and typically, is tied to the construction of instruments.

One can see how all this mirrors the prescriptions scattered throughout Bacon's works and the natural histories on which the new inductive science is to be based. Even more significant is the description of how the work is to be carried on. There is a sharp division of labor. Instead of single investigators (like a Descartes or a Galileo) who carry on research, if not single-handedly, at least with only the distant assistance of collaborators and predecessors, there are "Depredators" who collect experiments from books, "Merchants of Light" who travel in secret abroad to bring back accounts of experiments performed in other parts, "Mystery Men" who study the practices of the different arts and sciences, "Pioneers" who try new experiments, "Compilers" who draw together the findings of the previous groups, "Dowry Men" who investigate practical applications of the knowledge thus far gained, "Lamps" who on the basis of all the foregoing devise further experiments "of a higher light," "Inoculators" who execute the actual experiments, and finally "Interpreters of Nature" who "raise the former discoveries by experiments into greater observations, axioms, and aphorisms."[4] And all is to be carried on in secret: "All take an oath of secrecy for the concealing of those [inventions] which we think fit to keep secret though some of those we do reveal sometimes to the state, and some not."[5]

As a lawyer, Bacon had been much concerned with the new practice of allowing "patents," open letters (*litterae patentes*) under the Royal seal, supposedly for rewarding technological innovation, but in fact much misused by both Elizabeth and James I as patronage for their favorite courtiers to whom monopolies over such common substances as lead, iron, and saltpetre had been granted. In 1615, during Bacon's tenure as Lord Chancellor, a celebrated legal judgment was handed down ("The cloth-workers of Ipswich") in which it was held that the Crown could not grant a monopoly except where technological innovation was involved. In the following year, Bacon loyally defended the King's right to award profitable monopolies to his favorites; Parliament had to be convened to deal with the consequent political unrest. The Statute of Monopolies of 1623, the basis of modern English and U.S. patent law, was its answer; by then, Bacon had been disgraced.[6]

Those were years, then, when the matter of technological innovation was in the forefront of discussion, and certainly a preoccupation on Bacon's part. This is only one of the factors that go to explain the emphasis in the organization of Salomon's House, an organization so widely at variance with the one that someone like Descartes or Galileo, working in the more theoretical traditions of natural philosophy, would have given. Bacon's experience with the crafts and his tendency to view technological change in broadly alchemical categories, gave him a perspective altogether different from that of the "mechanical philosophers" of the next generation. This perspective was one which required large-scale and costly collaborative effort, needing therefore the assistance of the Crown. Despite Bacon's best efforts, this assistance was never forthcoming. The Royal Society chartered by a later monarch was on a far less ambitious scale than Salomon's House, and fostered notions of research importantly different from those of Bacon. Not until our own century, when for the first time natural science began in a significant way to shape and to be shaped by technological change, would there be "Foundations" with the ability to accomplish the goals of Salomon's House, though with types of organization very different from the one he recommended. The giant industrial firm of today is closer in that respect, perhaps, to the model in which Bacon embodied his hopes than was the Royal Society or the later scientific academies that claimed his mantle.[7]

Salomon's House was an object of hope, a utopia in Mannheim's sense of that term: an imagined state of affairs that seeks to legitimize new beliefs and practices by projecting an order different from the existing one. The *New Atlantis* was not a work of sociology. It was a speculative construction in social terms of an institution that did not yet exist, but one that Bacon's vision of natural philosophy appeared to require and thus to legitimate. The book was also a work of propaganda, designed to influence those in the best position to ensure that such an institution would come to pass. What links it with the modern discipline of the sociology of science is that Bacon (like

Robert Merton) saw natural science as an activity that could best be described in organizational terms.

Many have found utopian elements (in a different sense of "utopian") in Merton's work also, his stress on "the purity of science," in particular:

> One sentiment which is assimilated by the scientist from the very outset of his training pertains to the purity of science. Science must not suffer itself to become the handmaiden of theology or economy or state. The function of this sentiment is likewise to preserve the autonomy of science. For if such extra-scientific criteria of the value of science as presumable consonance with religious doctrines or economic utility or political appropriateness are adopted, science becomes acceptable only in so far as it meets these criteria. In other words, as the pure science element is eliminated, science becomes subject to the direct control of other institutional agencies and its place in society becomes increasingly uncertain. The persistent repudiation by scientists of the application of utilitarian norms to their work has as its chief function the avoidance of this danger, which is particularly marked at the present time.[8]

And he goes on to characterize "the ethos of science" by a set of values that include disinterestedness, universalism, organized skepticism, and humility.[9] Merton was perfectly well aware that these values are by no means always characteristic of the way in which scientists conduct their affairs, but he clearly regarded them as norms which on the whole are observed.[10] His concern with how science *should* be carried on has been challenged by a later generation of sociologists of scientific knowledge. And his assessment of the "purity" of the motives actually animating the efforts of scientists would not be shared by many of his more skeptical successors, who would quarrel with both the language and the implications of his claim that the social processes of science act to maintain the integrity of the cognitive processes through an impersonal moral ethos and a universalistic distribution of rewards. In the *New Atlantis*, Bacon conjured up an imaginary society on the authority of his own vision of how natural philosophy and technological change might fruitfully be conjoined. Merton had a simpler task. The society that he described already existed, indeed had existed for several centuries, even if it did not always live up to the ideals he projected for it.

. . .

Mention of Merton returns us from our excursion in the byways of the seventeenth century to the more travelled highways of the present. In these concluding pages we shall sketch in outline some of the developments that have in a short half century created a new discipline, or rather, clusters of disciplines, focused on the study of science as a complex form of social activity. Prior to that, there had been C. S. Peirce, of course, who had already set about redefining science in partially social terms. And there were the founders of modern sociology, Weber and Mannheim, in particular, who extended the concepts and techniques of the newly forming discipline of sociology to activities clustered under the loose label of "knowledge." They tended, however, to exempt the natural sciences from the scope of their analysis. So great at that time was the authority and the cumulative weight of sciences like physics and chemistry that inquiry into the particularities of knowledge-formation in such fields seemed unlikely to yield fruit. The ascendancy of logical positivism in the thirties and forties further reinforced the classical emphasis on science as basically atemporal and asocial.

The beginnings of a specifically sociological approach to the activity of science, as we have just seen, can be traced to the work of Merton in the U.S. From the thirties onwards, Merton and his students made use of the standard tools of sociological analysis to investigate the workings of the scientific community in functionalist terms. Merton shared with the earlier sociologists of knowledge the assumption that the natural sciences are objectively certified by means of methods that are presumptively universal in scope. His goal was to understand the social dynamics that enabled such a process to operate. The "Merton school" shaped the growth of sociology of science as a discipline in the U.S. during the sixties and seventies, and remains one of the main influences in the broader "social studies of science" enterprise today.

Kuhn's challenge to the older "asocial" image of science ran deeper. He argued that science is, first and foremost, the characteristic set of activities of a particular social group. Even though it is legitimate to abstract a set of propositions of which a physics text, say, might be composed, it must be understood that these propositions cannot be understood or evaluated in isolation from the particular community producing them. A paradigm is not simply a theory. It involves its own mechanisms for training students to see the world in a particular way; it requires its own methods of assessment of the claims it makes. It cannot be understood without detailed study of the social processes sustaining it. Historical analysis does not confirm the universalistic ethos presupposed by the earlier sociologists of knowledge.

In the light of later developments, Kuhn now seems relatively conservative in intention. For example, he insisted that the "mature" sciences are relatively insulated from the larger society of which the scientists are members. The methods of training and assessment characteristic of a mature science ensure this. Thus, the content of the typical theory in fields such as mechanics or chemistry is not influenced to any appreciable degree by the values and assumptions of the larger political and social groupings to which scientists belong.

Again, though there is no logic to coerce theory-choice in science, there are "values" like empirical accuracy, consistency, and fertility, which carry over from one paradigm to the next and provide a guide to the scientist in evaluating problem-solutions of a sort that can be called "objective," provided it be kept in mind that they leave wide room for disagreement.[11] Kuhn, like Merton before him, focused attention on what was *distinctive* about the scientific community, on the ways that scientists are trained, on the manner in which they communicate, on the ways in which they formulate and solve problems, and so forth. Though the notion of paradigm might be (and was) exported to fields distant from science, the mechanics of paradigm change he described were specific to a quite definitely marked-off group, practitioners of the natural sciences.

In Europe, the growth of "science studies" was to lead in more radical directions. Though originally inspired in part by Kuhn, European sociologists of science of the seventies, especially in Britain, were not as inclined as he had been to allow any special privilege to scientific knowledge or to scientific methods. The most influential group, proponents of the "strong program" in the sociology of knowledge, many of them associated with the University of Edinburgh, argued that knowledge is to be construed, in science as elsewhere, as accepted (and not as true) belief. The form of analysis used by the sociologist of science ought not depend, then, on the truth or falsity of the scientific claim that is being investigated. Barnes and Bloor stressed the theoretical character of scientific knowledge, and argued that theories are imposed upon reality rather than deriving from it.[12] Thus, even though the natural world exerts some constraint upon scientific theory, it is the various commitments that scientists bring to their theorizing that are decisive in explaining the outcome. The truth of the outcome is not a relevant factor in that explanation. The commitments themselves can best be understood in social terms. It is through the existence of a complex web of social relationships that the scientific community is enabled to carry through an enterprise that could never be completed in terms of logic and "pure" observation alone.

Philosophers of science have always supposed that their schemes could allow one to understand the historical activity of the scientist, that they can reveal what was *really* going on as science progressed. This is just what proponents of the strong program denied. They argued that philosophy of science had to rely on a false inductivist epistemology and an untenable essentialism in order to get under way. Once these be replaced, philosophy of science is seen to be incapable of carrying through its aims. It must itself be abandoned, or at least thoroughly reconstituted within a properly sociological structure.

Barnes and Bloor saw their analysis as *causal*: ideas and reasons were taken to be causes of human behavior and were to be understood in lawlike terms, like other causes. They took the goal of sociology of science to be the discovery of general regularities and systematic social relationships, just as Ben David had earlier done, but unlike him they supposed these regularities to be constitutive even of the *content* of scientific knowledge. The implicit causal determinism of this view appeared to clash with the notion of an "interest" as an active construction of the social world. In later work, Barnes has weakened the claim to lawlikeness, allowing room for the particularities of social context and the perceptions of individual interests.

One field that has held a particular challenge for sociologists of science has been mathematics. From Plato to the present day, mathematics has seemed the paradigm of an eternal and necessary structure, the very antithesis of a contingent social product. Must it, therefore, be held exempt from sociological analysis? To allow this would be a serious limitation of scope. Building on the earlier work of Wittgenstein and Lakatos, Bloor argued that the historical development of mathematics has been influenced in quite fundamental ways by social factors, and that the appearance of objectivity and inevitability this history possesses for us is an artifact. Critics of Bloor's thesis have suggested that mathematics is more (not less) open to constructivist interpretations than are the physical sciences, since it does not have the constraint on it that empirical anomaly imposes on the latter. So that even if Bloor's reading of the history of mathematics were to be conceded, this would by no means license an *a fortiori* inference in regard to the applicability of constructivist categories to the history of the physical sciences.

So far, the emphasis has been on sociology. But much of the force of the sociological case has derived from case histories directed to specific episodes in the past history of science, particularly episodes involving controversy.[13] The more "internalist" historiography of science of earlier decades has been, if not replaced, then at the least greatly augmented by a broader sort of historiography that emphasizes social influences, and tends to see in the science of a particular place and time a reflection of the prevailing institutional structures and the popular ideologies. A typical chapter on French science in the late eighteenth century, for example, is headed: "Aristocratic science, 1789–1793," "Democratic science, 1793–1795," "Bureaucratic science, 1795–1799," "Imperial science, 1799–1815."

Likewise, a book on Newtonian science claims to find in the new mechanics an image of the political structures of the ruling establishment of early eighteenth-century England. A history of statistical theory argues that in its nineteenth-century beginnings this theory was shaped in content as well as in the sequence of its development by the social interests of the rising British middle class. The earlier schematic work of Marxist historians of science like Bernal has been succeeded by detailed studies of the role played by politics and class-interest in the shaping of the "big science" of our own century.

In the last decade, sociology of science itself has continued to diversify. One lively new direction is toward studies of "laboratory life," using the methods of the anthropologists, analogous (in the words of Latour and Woolgar) to those of the "intrepid explorer of the Ivory Coast, who having studied the belief system or material production of 'savage minds' by living with tribesmen, sharing their hardships and almost becoming one of them, eventually returns with a body of observations which he can present as a preliminary research report."[14] This trend reflects a dissatisfaction with scientists' own reports of what goes on, which are idealized in a number of ways. There is little agreement as yet among the ethnographers themselves as to what the most profitable lines of approach to scientific practice might be. In particular, how can ethnographers claim to give an objective account of science "as it happens," if they deny a similar ability to natural scientists in *their* pursuit of natural knowledge? An ethnographic study, it would appear, cannot be taken as a *report*, a way of setting straight what is *really* going on in the laboratory. Woolgar speaks of a "reflexive ethnography," one whose aim is "to gain insight into general processes of reasoning practices," rather than giving the news about what is happening in the laboratory.

Despite the fact that sociologists of science have been applying their skills to the community of scientists for more than fifty years past, no serious effort seems as yet to have been made to apply these same skills to sociologists of science themselves. There is much work to be done here, as social historians and sociologists apply themselves to the question of why the field of sociology of science has exploded in the way it has, in the last two decades especially. To what extent is the marked "social turn" in history of science journals itself to be understood in social terms, by the premium set in matters of academic advancement on novel formulations, for instance? What sorts of interest prompt the strongly critical approach to science adopted by exponents of SSK [Sociology of Scientific Knowledge]? Does it, for example, in some way reflect the anti-science sentiments characteristic of large segments of British, German, and French society since World War II? These are important questions from the perspective of the new sociology of science itself, since they bear on the status of the knowledge-claims being advanced. When Shapin and Schaffer claim to have "shown" that Boyle's science reflected certain features of the society of which he was so active a member, does their "showing" reflect, in some significant way, particular features of *their* society? If not, why not? If so, what does this do to the interest of the claim they are making? How these issues are to be dealt with has not yet been clarified satisfactorily.[15]

A development of a different sort is that of scientometrics, or the "science of science," as it is sometimes called in Europe. Already in the 1960s Derek Price was studying the growth patterns of science, using as indicators the numbers of scientists and the production of scientific articles. A topic of special interest is the research network, or the "invisible college," and a research instrument of choice is the citation index. The dominance of "elites" is easily shown by this means. In any field, a small proportion of the workers produces the bulk of the literature and monopolizes the rewards. The more eminent the scientist, the more likely he or she is to get credit for work shared with other less well-known scientists (what Merton called the "Matthew effect"). The more eminent the scientist, the more likely it is that he or she studied with another eminent scientist. And so forth. Scientometrics is often linked with policy studies, on the assumption that the development of science and technology for political ends can best be planned if some kind of quantitative instruments are available for measuring and perhaps even predicting its directions and rates of growth.

In most of these sociological studies the stress is on negotiation and compromise. But some sociologists prefer to emphasize the role of conflict, and underline the alienation and exploitation that may accompany scientific development. Though broadly Marxist in inspiration, this sort of emphasis is in a certain tension with the commitment to science that was so central to Marx's own thinking. It tends rather to be anti-science; it sees science as potentially oppressive since it is always at the service of the reigning political orthodoxy. Radical sociology of science is usually part of a larger political program; it is unashamedly ideological but in its defense argues that the supposed ideological neutrality of science is in any case an illusion.

One of the liveliest areas of discussion at present is that of feminist theories of science. Uncontroversial is the claim that research in some specific parts of science (notably primatology and brain physiology) has in the past embodied a caricature of women or of the male/female relationship. Equally uncontroversial is the assertion that women have been, until recently, largely excluded from active participation in the doing of science, and that significant barriers still exist. Debate begins to arise, however, when the claim is made that the methods and goals of science, as it is currently practiced, betray a strong masculine bias: patriarchal, authoritarian, deterministic, purportedly neutral, detached from nature, lending itself to destructive and oppressive application. Feminists divide over the preferred alternative.[16] Is it a science free of bias, or is it one where the masculine is replaced by a feminine bias? Is bias acceptable provided it is the *right* bias? Many criticize the notion of a value-free science, and advocate a science which is guided by feminist values throughout. Political commitments to a certain view of human action, for example, would be allowed to determine the choice between alternative theories in such fields as neuro-anatomy.

Some feminist theorists argue that there is a distinctively feminine world-view which is characterized by wholism, interaction, and complexity, and which derives from the female sensibility or temperament; it would support a very different sort of science. Other feminists strongly disagree, but would still want to transform science in the direction of wholism and a more interactive view of relationships, not because of a uniquely feminine form of validating insight but because the resultant science would, in their view, be better as science. Finally, since the androcentric structures of science as it is now practiced are rooted (it is claimed) in the equally androcentric structures of the larger society, some feminists conclude that the only way in which a truly feminist science can ever be achieved is to work for the radical transformation of the larger society first. All of these views are premised on a particular sociological analysis of science, as it is currently carried on. They are, in effect, radical sociologies of science, more politically radical than most versions of SSK since their primary aim is to transform entirely the way in which science is carried on. Yet they are less epistemically radical than the constructivist versions of SSK since their defenders generally believe that a "truer" science is, in fact, attainable.

And so, finally, back once again to Steve Woolgar, whose constructivism is as far from the tradition of Bacon and Whewell as the string could stretch. He is emphatic about his ontology. The discovery of pulsars at Cambridge:

> undermines the standard presumption about the existence of the object prior to its discovery. The argument is not just that social networks mediate between the object and observational work done by the participants. Rather, the social network constitutes the object (or lack of it).[17]

The problem, then, in his eyes is to try to explain how discoverers manage to convince themselves they *have* discovered something "out there." As far as he is concerned: "realist ontology is a *post hoc* justification of existing institutional arrangements."[18] He is equally explicit about his epistemology:

> The representational practices [of the scientist] constitute the objects of the world, rather than being a reflection of (or arising from) them. . . . We think the objects precede and give rise to their representation precisely because this is the way we happen to organize our perceptions of the world.[19]

Scientists, in effect, "rewrite history so as to give the discovered object its ontological foundation."[20] But no matter how hard they try, "facts and objects in the world are inescapably textual constructions." And then a final thrust: "Relativism has not yet been pushed far enough. Proponents of relativism (both within and beyond SSK) are still wedded to an objectivist ontology, albeit one slightly displaced."[21]

Not yet pushed far enough? Only *slightly* displaced? One suspects a genial attempt to *épater les philosophes*. But here a serious issue is joined, and the arguments of Bacon and Whewell and their realist descendants have to face the constructivist counterarguments head-on, without any distracting talk of the social networks that sustain each side. May the (epistemically) best side win!

In a short fifty years, a set of interlocking (and often competing) disciplines have sprung up, each claiming to provide an insight into what scientists do, most of them challenging the older assumption that science, if properly executed, reflects the world of nature. In this rapid sketch, we have passed over developments in philosophy associated with Wittgenstein, Quine, Heidegger, and others, which have helped to accelerate the decline of the older asocial model of science. At this point, there would be little disagreement about the claim that the social dimensions of science must be taken seriously. But there would be just as little agreement as to what this advice amounts to.

Notes

1 William Rawley in his Introduction to the first English edition of the *New Atlantis, Francis Bacon: A Selection of His Works*, ed. S. Warhaft (New York: Odyssey, 1965), 418.

2 Ibid., 447.

3 *New Organon*, trans. James Spedding *et al.* (Boston: Taggard and Thompson, 1863), Book II, aph. 6.

4 *New Atlantis*, 456.

5 Ibid.

6 Ernan McMullin, "Openness and secrecy in science: Some notes on early history," *Science, Technology and Human Values 10* (1985): 14–23.

7 The academies *did*, of course, encourage systematic experimentation and the development of the technologies this experimentation required. The numerous affinities between them and the model sketched so imaginatively, if somewhat prematurely, by Bacon are dealt with in illuminating detail in *Salomon's House Revisited: The Organization and Institutionalization of Science*, ed. T. Frängsmyr (Canton, Mass.: Science History, 1990).

8 Robert K. Merton, *Social Theory and Social Structure* (New York: Free Press, 1957), 549.
9 Ibid., 552–561.
10 "Priorities in scientific discovery," *American Sociological Review 22* (1957): 635–659.
11 Thomas Kuhn, "Objectivity, value judgment, and theory choice," in *The Essential Tension* (Chicago: University of Chicago Press, 1977), 320–339.
12 Barry Barnes, *Scientific Knowledge and Sociological Theory* (London: Routledge, 1974); B. Barnes, *Interests and the Growth of Knowledge* (London: Routledge, 1977); David Bloor, *Knowledge and Social Imagery* (London: Routledge, 1976).
13 For an extensive review and bibliography, see Steven Shapin, "History of science and its social reconstructions," *History of Science 20* (1982): 157–211.
14 Bruno Latour and Steve Woolgar, *Laboratory Life: The Construction of Scientific Facts* (Princeton: Princeton University Press, 1986), 28.
15 See Barry Gruenberg, "The problem of reflexivity in the sociology of science," *Philosophy of the Social Sciences 8* (1978): 321–343, and the essays in *Knowledge and Reflexivity: New Frontiers in the Sociology of Knowledge*, ed. Steve Woolgar (London: Sage, 1988).
16 See Helen Longino, *Science as Social Knowledge* (Princeton: Princeton University Press, 1990).
17 *Science: The Very Idea* (London: Tavistock, 1988), 65.
18 Ibid., 67.
19 Ibid.
20 Ibid., 69.
21 Ibid., 98.

QUESTIONS

1 In his paper Kuhn says that he was misunderstood by his critics who accused him of irrationalism and of assimilating philosophy of science to "mob psychology." Such criticisms can be found, for example, in Shapere's review of Kuhn's book. Do you find Kuhn's response to these charges in "Objectivity, Value Judgment, and Theory Choice" adequate? Support your answer.

2 Bloor defines the "strong program in the sociology of knowledge" in terms of the four tenets: causality, impartiality, symmetry, and reflexivity. What does each of them involve? Take a critical stance against some them.

3 Feminist epistemology is a systematic attempt to examine and make explicit ways in which gender categories and ideologies of gender influence

the process of knowledge acquisition and its results. According to feminist theorists, such influences or "gender biases" are diverse and operate at various levels of inquiry. Some feminists think that to be objective, human knowledge in general, and scientific knowledge in particular, should be *liberated* of the influences and biases of that sort. Others think that is not possible, since knowledge, by its very nature, is *ineradicably* and *essentially* biased. Hence, rather than trying to liberate science from gender biases (which would be a futile task), we should develop a distinctly feminist science (incorporating specifically feminine ways of viewing the world), to make up for the masculine bias, which has dominated science throughout its history. The second project appears to be more radical than the first and may include some political agenda as well. Which of these two projects, if any, do you consider to be appropriate? Support your answer.

FURTHER READING

The third edition of Kuhn's book is now available (Kuhn 1996). Gutting (1980) and Horwich (1993) are collections of articles debating different aspects of Kuhn's conception of science.

The most influential works on social constructionism and the sociology of scientific knowledge include Barnes (1974), Pickering (1984, 1992), Shapin and Schaffer (1985), Latour and Woolgar (1986), Woolgar (1988), and Bloor (1991). For critical analyses, see Brown (1984), Koertge (1998), and Hacking (1999). The so-called "Sokal affair" may be of special interest in this connection (see *The Sokal Hoax: The Sham that Shook the Academy*, edited by the editors of Lingua Franca, Lincoln, University of Nebraska Press, 2000; and Sokal and Bricmont 1999).

The following items are must-reading for anyone interested in the feminist philosophy of science: Longino (1990), Harding (1991), Keller and Longino (1996), and Kourany (1998).

BIBLIOGRAPHY

Achinstein, Peter (2001) *The Book of Evidence*, New York: Oxford University Press.

Armstrong, David (1983) *What Is Law of Nature?* Cambridge: Cambridge University Press.

Ayer, A.J. (1946) *Language, Truth and Logic*, 2nd edn, New York: Dover.

Ayer, A.J. (ed.) (1959) *Logical Positivism*, Glencoe, IL: Free Press.

Balashov, Yuri (1994) "Duhem, Quine, and the Multiplicity of Scientific Tests," *Philosophy of Science* 61: 608–28.

Barnes, Barry (1974) *Scientific Knowledge and Sociological Theory*, London: Routledge & Kegan Paul.

Barnes, Barry, Bloor, David, and John Henry (1996) *Scientific Knowledge: A Sociological Analysis*, Chicago: The University of Chicago Press.

Beauchamp, Tom L. and Alex Rosenberg (1981) *Hume and the Problem of Causation*, Oxford: Oxford University Press.

Bloor, David (1991) *Knowledge and Social Imagery*, 2nd edn, Chicago: The University of Chicago Press.

Bondi, Hermann and C.W. Kilmister (1959) "The Impact of 'Logik der Forschung,'" *The British Journal for the Philosophy of Science* 10: 55–7.

Boyd, Richard, Gasper, Philip, and J.D. Trout (eds) (1991) *The Philosophy of Science*, Cambridge, MA: The MIT Press.

Brown, James Robert (1984) *Scientific Rationality: The Sociological Turn*, Dordrecht and Boston: Reidel.

Cartwright, Nancy (1983) *How the Laws of Physics Lie*, Oxford: Clarendon Press.

Churchland, Paul and Clifford Hooker (eds) (1985) *Images of Science: Essays on Realism and Empiricism*, Chicago: The University of Chicago Press.

Curd, Martin and J.A. Cover (1998) *Philosophy of Science: The Central Issues*, New York: Norton.

Cushing, James T. (1994) *Quantum Mechanics: Historical Contingency and the Copenhagen Hegemony*, Chicago: The University of Chicago Press.

Darwin, Charles (1979) *On the Origin of Species*, New York: Avenel (first published 1859).

Dawkins, Richard (1986) *The Blind Watchmaker*, New York: Norton.

Dowe, Phil (2000) *Physical Causation*, Cambridge and New York: Cambridge University Press.

Duhem, Pierre (1954) *The Aim and Structure of Physical Theory*, trans. Philip P. Wiener, Princeton: Princeton University Press.

Earman, John (1992) *Bayes or Bust? A Critical Examination of Bayesian Confirmation Theory*, Cambridge, MA: The MIT Press.

Feyerabend, Paul (1975) *Against Method*, London: Verso.

Fine, Arthur (1984) "The Natural Ontological Attitude," in Leplin 1984, pp. 83–107.

Fine, Arthur (1986) "Unnatural Attitudes: Realist and Instrumentalist Attachments to Science," *Mind* 95: 149–79.

Fine, Arthur (1991) "Piecemeal Realism," *Philosophical Studies* 61: 79–96.

Fine, Arthur (1996) *The Shaky Game: Einstein, Realism, and the Quantum Theory*, 2nd edn, Chicago: The University of Chicago Press.

Friedman, Michael (1974) "Explanation and Scientific Understanding," *The Journal of Philosophy* 71: 5–19.

Friedman, Michael (1999) *Reconsidering Logical Positivism*, Cambridge and New York: Cambridge University Press.

Giere, Ronald N. and Alan W. Richardson (eds) (1997) *Origins of Logical Positivism*, Minneapolis: University of Minnesota Press.

Glymour, Clark (1980) *Theory and Evidence*, Princeton, NJ: Princeton University Press.

Goodman, Nelson (1983) *Fact, Fiction and Forecast*, 4th edn, Cambridge, MA: Harvard University Press (first published 1948).

Greenwood, J.D. (1990) "Two Dogmas of Neo-Empiricism: the 'Theory-Informity' of Observation and the Duhem–Quine Thesis," *Philosophy of Science* 57: 553–74.

Gutting, Gary (1980) *Paradigms and Revolutions*, Notre Dame, IN: University of Notre Dame Press.

Hacking, Ian (1983) *Representing and Intervening*, Cambridge: Cambridge University Press.

Hacking, Ian (1999) *The Social Construction of What?* Cambridge, MA: Harvard University Press.

Hardin, Clyde L. and Alex Rosenberg (1982) "In Defense of Convergent Realism," *Philosophy of Science* 49: 604–15.

Harding, Sandra (ed.) (1976) *Can Theories Be Refuted? Essays on the Duhem–Quine Thesis*, Dordrecht: Reidel.

Harding, Sandra (1991) *Whose Science? Whose Knowledge? Thinking from Women's Lives*, Ithaca, NY: Cornell University Press.

Heisenberg, Werner (1958) *Physics and Philosophy*, New York: Harper & Row.

Hempel, Carl G. (1965) *Aspects of Scientific Explanation*, New York: Free Press.

Hempel, Carl G. (1966) *Philosophy of Natural Science*, Englewood Cliffs, NJ: Prentice-Hall.

Hempel, Carl G. and Paul Oppenheim (1948) "Studies in the Logic of Explanation," *Philosophy of Science* 15: 567–79.

Hoefer, Carl and Alex Rosenberg (1994) "Empirical Equivalence, Underdetermination and Systems of the World," *Philosophy of Science* 61: 592–607.

Horwich, Paul (1982) *Probability and Evidence*, Cambridge: Cambridge University Press.

Horwich, Paul (ed.) (1993) *World Changes: Thomas Kuhn and the Nature of Science*, Cambridge, MA: The MIT Press.

Howson, Colin and Peter Urbach (1993) *Scientific Reasoning: The Bayesian Approach*, 2nd edn, La Salle, IL: Open Court.

Hull, David L. (1974) *Philosophy of Biological Science*, Englewood Cliffs, NJ: Prentice-Hall.

Hull, David L. and Michael Ruse (eds) (1998) *The Philosophy of Biology*, Oxford and New York: Oxford University Press.

Hume, David (1974) *Inquiry Concerning Human Understanding*, Indianapolis: Hackett.

Keller, Evelyn Fox and Helen E. Longino (eds) (1996) *Feminism and Science*, Oxford and New York: Oxford University Press.

Kitcher, Philip (1981) "Explanatory Unification," *Philosophy of Science* 48: 507–31.

Kitcher, Philip (1993) *The Advancement of Science: Science without Legend, Objectivity without Illusions*, New York: Oxford University Press.

Koertge, Noretta (ed.) (1998) *A House Built on Sand: Exposing Postmodernist Myths About Science*, New York: Oxford University Press.

Kosso, Peter (1998) *Appearance and Reality: An Introduction to the Philosophy of Physics*, New York: Oxford University Press.

Kourany, Janet A. (ed.) (1998), *Philosophy in a Feminist Voice: Critiques and Reconstructions*, Princeton, NJ: Princeton University Press.

Kripke, Saul A. (1972) "Naming and Necessity," in D. Davidson and G. Harman (eds.), *Semantics of Natural Language*, Dordrecht: Reidel, pp. 253–355.

Kuhn, Thomas (1957) *The Copernican Revolution*, Cambridge, MA: Harvard University Press.

Kuhn, Thomas (1977) *The Essential Tension*, Chicago: The University of Chicago Press.

Kuhn, Thomas S. (1996) *The Structure of Scientific Revolutions*, 3rd edn, Chicago: The University of Chicago Press.

Lakatos, Imre and Alan Musgrave (eds) (1970) *Criticism and the Growth of Knowledge*, Cambridge: Cambridge University Press.

Latour, Bruno and Steve Woolgar (1986) *Laboratory Life: The Construction of Scientific Facts*, 2nd edn, Princeton, NJ: Princeton University Press.

Laudan, Larry (1977) *Progress and its Problems: Toward a Theory of Scientific Growth*, Berkeley: University of California Press.

Leplin, Jarrett (ed.) (1984) *Scientific Realism*, Berkeley: University of California Press.

Leplin, Jarrett (1997) *A Novel Defense of Scientific Realism*, New York: Oxford University Press.

Levins, Richard and Richard Lewontin (1985) *The Dialectical Biologist*, Cambridge, MA: Harvard University Press.

Lewis, David (1973) *Counterfactuals*, Oxford: Basil Blackwell.

Lewis, David (1983) "New Work for a Theory of Universals," *Australasian Journal of Philosophy* 59: 343–77.

Longino, Helen E. (1990) *Science as Social Knowledge: Values and Objectivity in Scientific Inquiry*, Princeton, NJ: Princeton University Press.

Maher, Patrick (1993) *Betting on Theories*, Cambridge: Cambridge University Press.

Maxwell, Grover (1962) "The Ontological Status of Theoretical Entities," in H. Feigl and G. Maxwell (eds), *Scientific Explanation, Space, and Time*, vol. 3, *Minnesota Studies in the Philosophy of Science*, Minneapolis: University of Minnesota Press, pp. 3–15.

Mayo, Deborah (1996) *Error and the Growth of Experimental Knowledge*, Chicago: The University of Chicago Press.

McMullin, Ernan (1970) "The History and Philosophy of Science: A Taxonomy," in Roger H. Stuewer (ed.), *Historical and Philosophical Perspectives of Science*, Minneapolis: University of Minnesota Press, pp. 12–67.

McMullin, Ernan (1991) "Comment: Selective Anti-Realism," *Philosophical Studies* 61: 97–108.

Nagel, Ernest (1961) *The Structure of Science: Problems in the Logic of Scientific Explanation*, New York: Harcourt, Brace & World, Inc. (reprinted by Hackett in 1979).

Newton-Smith, William (1981) *The Rationality of Science*, London: Routledge.

Pickering, Andrew (1984) *Constructing Quarks: A Sociological History of Particle Physics*, Chicago: The University of Chicago Press.

Pickering, Andrew (ed.) (1992) *Science as Practice and Culture*, Chicago: The University of Chicago Press.

Pitt, Joseph C. (ed.) (1988) *Theories of Explanation*, Oxford: Oxford University Press.

Popper, Karl R. (1959) *The Logic of Scientific Discovery*, New York: Basic Books (first published in German in 1934).

Putnam, Hilary (1975) "The Meaning of 'Meaning,'" in H. Putnam, *Mind, Language and Reality*, Cambridge: Cambridge University Press, pp. 215–71.

Putnam, Hilary (1981) *Reason, Truth and History*, Cambridge: Cambridge University Press.

Putnam, Hilary (1983) *Realism and Reason, Philosophical Papers*, vol. 3, Cambridge: Cambridge University Press.

Putnam, Hilary (1987) *The Many Faces of Realism*, La Salle, IL: Open Court.

Quine, Willard V.O. (1953) *From a Logical Point of View*, Cambridge, MA: Harvard University Press.

Quine, Willard V.O. (1961) *Word and Object*, Cambridge, MA: The MIT Press.

Reichenbach, Hans (1951) *The Rise of Scientific Philosophy*, Berkeley: University of California Press.

Rosenberg, Alex (1985) *The Structure of Biological Science*, Cambridge: Cambridge University Press.

Rosenberg, Alex (1992) *Philosophy of Social Science*, Boulder: Westview.

Ruben, David-Hillel (ed.) (1993) *Explanation*, Oxford: Oxford University Press.

Ruse, Michael (1988a) *The Philosophy of Biology Today*, Albany: State University of New York Press.

Ruse, Michael (ed.) (1988b) *But Is It Science? The Philosophical Question in the Creation/Evolution Controversy*, Buffalo, NY: Prometheus Books.

Salmon, Wesley (1984) *Scientific Explanation and the Causal Structure of the World*, Princeton, NJ: Princeton University Press.

Salmon, Wesley (1998) *Causality and Explanation*, New York: Oxford University Press.

Salmon, Wesley and Philip Kitcher (eds) (1989) *Scientific Explanation*, vol. 3, *Minnesota Studies in the Philosophy of Science*, Minneapolis: University of Minnesota Press.

Schilpp, Paul (ed.) (1949) *Albert Einstein: Philosopher-Scientist*, Evanston, IL: Library of Living Philosophers.

Shapin, Steven and Simon Schaffer (1985) *Leviathan and the Air-Pump: Hobbes, Boyle, and the Experimental Life*, Princeton, NJ: Princeton University Press.

Sokal, Alan D. and Jean Bricmont (1999) *Fashionable Nonsense: Postmodern Intellectuals' Abuse of Science*, New York: Picador USA.

Stalker, Douglas (1994) *GRUE! The New Riddle of Induction*, La Salle, IL: Open Court.

Suppe, Fredrick (1977) *The Structure of Scientific Theories*, 2nd edn, Urbana: University of Illinois Press.

van Fraassen, Bas (1980) *The Scientific Image*, Oxford: Clarendon Press.

van Fraassen, Bas (1989) *Laws and Symmetries*, Oxford: Clarendon Press.

Weinert, Friedel (ed.) (1995) *Laws of Nature: Essays on the Philosophical, Scientific and Historical Dimensions*, Berlin and New York: Walter de Gruyter.

Woolgar, Steve (1988) *Science, the Very Idea*, Chichester and London: Ellis Horwood and Tavistock Publications.

INDEX

a priori 27, 366, 384n, 397–9, 439, 447, 471
abduction 214, 227; *see also* retroduction
abductive inference 227
abstract: entity(-ies) 341, 358; term(s) 341
accidental generalizations: *see* law(s): and accidental generalizations
accuracy (as a theoretical value/virtue) 421n, 422–3, 429–31, 433, 435–6, 494; *see also* predictive: accuracy
Achinstein, Peter 75, 88n, 183n, 188n, 230n, 286, 307–20, 402, 403n
action at a distance 101, 251, 276n
ad hoc: assumption(s) 298; hypothesis(-es) 270
Adams, John 168
Adler, Alfred 295–7, 299
aether 216, 218; caloric 218; electromagnetic 224–5; gravitational 219; optical 218, 224–5; physiological 219; theory(-ies) of 217, 219; *see also* ether
alchemy 390, 490–1
analytic: knowledge 478; statement 341–2, 347, 349–53, 355–7, 359; truth: *see* truth: analytic; *see also* analyticity
analyticity 342–3, 346–53, 360n; *see also* analytic
Anderson, Alan R. 70n
Anderson, Elizabeth 408, 459–89
androcentrism 473–9
anthropology 100, 102, 441, 455, 463–4, 496; interpretive 468; reflexive 464
anti-realism: *see* realism and anti-realism
Antony, Louise 480
Arago, François 57
argument(s): deductive 51, 72, 81, 242, 285–6, *see also* deduction; inductive 51, 72, 246, 256, 285–6, 454, *see also* induction; retroductive 254, 267
Aristotelian/Aristotle's: dynamics 187n, 419–20; physics 166, 169; theory of motion 146–7, 149–50; theory of science 66
Aristotle 24, 39, 61–2, 65–6, 146–7, 149–50, 257–8, 269, 301n, 341–2, 357, 407, 419–20
Arm, David L. 401n
Armstrong, David 43, 118, 124n, 126n
Aspect, Alain 101
Asquith, Peter 247n
assumption(s): auxiliary 30, 278n, 287–8, 298, 364–8, 373, 380, 384n; background 107–8, 477, 481–3; basic 139, 383n; fundamental 129, 133
Astin, Helen S. 461
astrology 24, 295–6, 298–9, 301n, 378, 390
astronomy 12, 17, 245, 250, 288, 296, 373, 375, 381, 390, 430, 433, 455, 490; ancient 224, 257; Copernican 423–5, 427; heliocentric 424, 430; mathematical 258; medieval 224;

modern 336; Ptolemaic 164, 258, 270, 423–4, 427
astrophysics 252, 254, 269
atomic theory(-ies) of matter 198, 206, 209, 216–17, 224, 234–5, 237–8, 241, 331, 356, 373, 414; *see also* atomism
atomism 219, 414; *see also* atomic theory(-ies) of matter
auxiliaries: *see* auxiliary
auxiliary: assumption(s): *see* assumption(s): auxiliary; hypothesis(-es): *see* hypothesis(-es): auxiliary
Avogadro, Amedeo 138
axiom(s) 129–31, 383n, 490; *see also* postulate(s)
axiomatic: approach 131; system(s) 129–30, 132n; theory 367
Ayer, Alfred J. 35n, 117

Bacon, Francis 410, 448, 489–91, 498–9
Balashov, Yuri 404n
Balmer, John Jacob 72
Banks, Joseph 185n
Barker, Stephen 308, 320n
Barnes, Barry 494–5, 500n, 501n
Bateson, Gregory 441
Bauer, William D. 462
Bayes, Thomas 270, 288, 385–402, 404n, 428
Bayesianism 385n; *see also* confirmation: Bayesian theory of
Bayes's theorem 270, 288, 385–402
Becher, Johann 184n
Becker, Joanne Rossi 461
Beckner M. 59, 64, 69n
Belenky, Mary 464
Bell, Diane 464
Bell, John S. 101, 252, 277n
Bellarmine, Cardinal Robert 227
Belnap, Nuel 67, 70n
Ben-David, Joseph 429, 495
Benedetti, Giambattista 187n
Bentham, Jeremy 354, 356
Berkeley, George 194, 250, 363
Bernal, John Desmond 496
Berofsky, B. 125n
Beyer, Alan E. 461
biology 4, 6, 99–100, 189n, 267, 400, 433, 442, 445, 456, 462, 467, 469, 471–3, 477, 484; mathematical 472; molecular 269; philosophy of 22–34
Birkhoff, Garrett 416
Black, Joseph 218
bleen 311–13, 317–18
Bloor, David 408, 438–58, 494–5, 500, 501n
Boerhaave, Herman 218
Bohm, David 101
Bohr, Niels 72, 101, 134–7, 216, 252, 254, 256, 270, 280n, 435; the complementary principle of 252
Boltzmann, Ludwig 68, 70n, 254, 373
Bondi, Hermann 403n
Bottomore, T. B. 452
boundary and/or initial condition(s) 123
Bourdieu, Pierre 464
Boyd, Richard 211–12, 215, 219–21, 227, 229, 230n, 266, 279n, 384n
Boyle, Robert 49, 72, 156, 238, 432, 456, 496
Bradie, Michael 272, 280n
Brahe, Tycho 385, 444
Braithwaite, R. B. 92, 115, 125n
Bricmont, Jean 501n
bridge principle(s)/law(s): *see* correspondence: rule(s)
Brittan, G. G. 278n
Brodbeck, May 55n, 92
Brody, Baruch 62, 64, 70n
Bromberger, Sylvain 63, 66, 70n, 73, 93
Brown, James 501n
Brownian motion 373–4
Brush, Stephen 70n
Bunge, Mario 161n
Burchfield, J. D. 442
Burian, Richard 217, 230n
Buridan, Jean 186n, 187n
Burnet, John 161n
Butts, Robert 280n

Caplan, Pat 464
Cardwell, D. S. L. 441
Carnap, Rudolf 8n, 308, 316, 343–4, 350, 352, 354–5, 359, 360n, 380, 397–8, 401n

Carroll, Lewis 25
Cartesian(s) 250; *see also* Descartes
Cartwright, Nancy 70n, 277n, 281n
causal: account 266; category(-ies) 250; chains 482; concept(s) 240; explanation: *see* explanation: causal; interaction(s) 92n, 95, 100; mechanism(s) 96; power(s) 43–4, 96; process(es) 92n, 95–6, 100, 447; property(-ies) 240; relation(s) 96, 168, 220, 230n, 262, 384n, 441; structure of the world 71, 88
causality: *see* causation
causation 31–3, 41, 44, 62, 65, 79, 92–3, 95–6, 442–3, 448, 451–2, 500; Hume's analysis/critique of 80, 93, 96, 116, 119, 121, 363
cause(s) 66, 79, 93, 100, 116, 150, 258–9, 269, 289, 293, 443, 446, 448–50, 453, 456
Cavendish, Henry 95, 173–4, 181–2, 184n, 185n, 187n, 218, 427, 456
Charles, Jacques 72, 156, 432
chemistry 4, 12, 32, 72, 198, 209, 216, 218, 224, 226, 266, 268, 390, 408, 414, 419, 423, 433–4, 493
Chew, Geoffrey 255–6, 277n
Churchland, Paul 88n, 89n, 281n
Clagett, Marshall 161n, 187n, 188n
Clark, R. 65
Clinchy, Blythe 464
Coffa, J. Alberto 74
cognitive: significance 25–6, 30, 117, 300, 378; status of theories 194, 197–210
Cohen, Morris R. 392
Cohen, Robert S. 183n, 280n
coherence 248
Coleman, W. 441
Collins, Patricia Hill 459, 464
Colodny, Robert 69n, 183n
Compton, Arthur 96
Compton effect/scattering 96
Conant, James B. 184n
conceptual: analysis 6, 379; change 129–89, 268; *see also* scientific: change, theoretical: change; framework 359; relativism 163n, 164–6, 169, 171, 175, 182, 187n, 189, *see also* relativism; scheme 359, 430
concrete: entity(-ies) 341; term(s) 341
confirmation: Bayesian theory of 396, 398, 400, 401n, 404n, 428; bootstrap theory of 381; Carnap's theory of 398; degree of 51, 225, 397; the logic of 387, 391, 395–6, 399–400; of theory(-ies) and hypothesis(-es) by evidence 131, 143–4, 234, 279n, 285–404, paradoxes of 379; probabilistic theory(-ies) of 410, *see also* confirmation: Bayesian theory of; theory(-ies) of 215, 397, 399
conjecture(s) 286, 294–301, 454; *see also* hypothesis(-es)
consistency (as a theoretical value/virtue) 270, 421n, 422–4, 430, 435–6, 494
constructive empiricism: *see* empiricism: constructive
contingent: laws 119
contrary-to-fact conditionals/suppositions: *see* counterfactual conditionals/counterfactuals
conventionalism 298–9, 363
convergent realism: *see* realism: convergent
Copernican: astronomy 164; revolution 455
Copernicus, Nicolas 164, 250, 370, 417, 423–5, 427, 430, 455
correspondence: principle 188, 221, 225; rule(s) 130, 134–40, 188, 200, 203, 236; theory of truth: *see* truth: as correspondence
corroboration 298, 398–9; degree of 231n
cosmology(-ies) 373, 438, 441
counterfactual conditionals/counterfactuals 42–3, 64–7, 106–14, 117, 124n, 126, 181, 348
Cowan, R. S. 441
creationism 24
Crick, Francis 268, 400
criteria of choice (of theories/hypotheses in science) 427, 429–31, 433, 436; objective 428–9, 434–5; subjective 428–9, 434–5
Cudd, Ann 485n

Curran, Libby 461
Cushing, James T. 281n

D'Alembert, Jean La Rond 415
Dalton, John 216, 414
Danto, A. C. 161n
Darwin, Charles 3, 22–3, 34, 57, 69n, 89n, 239, 286, 302–3, 304n, 305–6; 337, 339n, 357, 373, 402, 425, 470; *see also Origin of Species, The*
Darwinism/Darwinians 3, 22–3, 239, 302–4, 403n, 437n; *see also* evolutionary: theory
Davidson, Donald 183n, 184n, 188n
Davy, Humphry 414, 456
Dawkins, Richard 36n, 470
deduction 25, 126, 132, 389, 392
deductive: argument(s): *see* argument(s): deductive; fallacy 392–3; subsumption 93; system(s) 122–4, 130, 132
definiendum 344–5
definiens 344–5
definition 27, 129–30, 157, 261–2, 343–4, 346, 349, 354, 359n, 440; conventional 350; coordinating 156; explicit 40, 136, 200, *see also* explication; implicit 130, 132–5
DeGré, Gerald L. 439
DeLuc, Jean André 185n
demarcation 377, 380; criteria 378; principle 286, 294n; problem 300, 378
Democritus 255–6
Descartes, René 9, 258, 363, 377, 490–1
descriptive: account of science 198, 200; terms 141, 150–3, 157; *see also* observational: term(s); theoretical: term(s)
determinism 30, 115–16, 122–4, 252, 453, 495
Devitt, Michael 184n
Dewey, J. 46, 55n
Diamond, Cora 466–7, 469, 485n
Dirac, Paul 268
discovery 390, 432; the context of 385–9, 391, 397, 402n, 426–7, 465, 470, 472; the logic of 386, 399, 401n, 402n, 445, 456
disposition 86–7; *see also* property(-ies): dispositional
DNA (Deoxyribonucleic Acid) 86, 100, 400, 470
Donnellan, Keith 167,184n
Douglas, Mary 441
Dray, William H. 55n
Dretske, Fred 43, 75
Driesch, H. 24–6, 31, 35; *see also* entelechy; vitalism
Duhem, Pierre 125, 238, 258, 288, 332, 338n, 360n, 382n, 384n, 401n, 403n, 419
Duhem–Quine thesis, the 288, 356–9, 382n, 403n
Dummett, Michael 275, 278n
Durkheim, Emile 439, 442

Earman, John 44, 115–25, 188n, 404n
ecology: mathematical 471–2
economics 4, 474–5, 478, 492
Eddington, Arthur S. 264, 295, 297
Einstein, Albert 16–17, 36n, 94, 96, 98–9, 101, 131, 252–3, 255–7, 268, 295–9, 300n, 357, 370, 375–6, 396, 401n, 416–17, 420, 435
embryology 23–4, 31, 217
Empedocles 299
empirical: adequacy 58–60, 64, 68, 159, 196, 235, 239, 241, 243–4, 248, 251, 259–60, 268–9, 381–2, 460, 470, 482–3; consequence(s) of theory/hypothesis 364–7, 369, 371–2, 375–7, 381–2, 384n; content 357, 367, 369; data 202, 238; equivalence 251, 362–84; evidence 197, 245, 285, 288, 295, 299, 368; fact(s) 236–8; generalization(s): *see* generalization(s): empirical; hypothesis(-es) 481; law(s): *see* law(s): empirical; meaning/significance 29, 300, 356; method 295; observation(s) 456, 481; phenomena 368; regularity(-ies) 129, 239, 455–6; science(s) 438; statement(s) 365, 374; status 364, 383n; success: *see* success: empirical; support 372, 375
empiricism 79–80, 117–22, 130–1, 141,

143, 146–7, 149, 189n, 193–4, 196, 226, 249, 258–61, 287–8, 340–61, 363, 370, 377, 380–1, 407–8, 448–51, 460, 471, 477–8, 482; constructive 196, 234–47, 281, 381, 383n; logical: *see* logical empiricism
Enç, Berent 187n
entelechy, Driesch's theory of 24–6, 31, 35
entrenchment: *see* predicate(s): entrenched
epistemic: character of scientific reasoning 269; import 466; standard(s) 460; status 368; value(s) 479, 482; warrant 223, 382
epistemology 4–5, 7, 14, 31–2, 36, 193, 196, 211, 214, 226, 229, 246, 362–3, 370, 376–80, 382, 383n, 384n, 409, 460, 494, 498, *see also* knowledge: theory of; feminist: *see* feminist: epistemology; naturalized 459–60, 469, 473, 478–84; social 459–60
esthetics 4, 20, 400
ether 195, 250, 257; mechanical (models of) 205, 250; Newton's 251; optical 257; *see also* aether
ethics 20, 264, 467
Euclid 3, 129–30
evidence 316–19, 391, 394–5, 418, 428, 482; *see also* empirical: evidence
evidential: status 377, *see also* empirical: status; support 372, 376–7, *see also* empirical: support; warrant 379
evolutionary: biology 4, 471; theory 22, 302n, 303, 402, 441, 469–471, 473
experiment(s) 132, 408, 467, 490; Cavendish's 95; crucial 427
experimental: concept(s) 136–8; data: theoretical-ladenness of 146; evidence 286; ideas 137–9; law(s): *see* law(s): experimental; notion(s) 135, 203–4
explanandum 46–7, 49–51, 53–4, 60, 64, 93, 142, 158, 215
explanans 47, 49–51, 53–4, 55n, 60, 142, 158
explanation 26–7, 39–126, 129–31, 141–4, 148, 151–3, 158–60, 161n, 162n, 206, 228, 237–8, 248, 250, 275, 281, 303, 368–9, 378, 382, 384n, 443–4, 448, 470, 472, 477; asymmetry(-ies) of 60–6, 74, 76, 79–80, 82, 85–6, 126; causal 41, 48, 55n, 61, 71n, 80, 92–105, 442–3, 446–50, 495; covering-law model(s) of 45–55, 71–3, 126; deductive-nomological (D-N) model of 40, 45–55, 72–3, 77, 93; elliptic 52–5; functional 92, 100–2; historical 167–8, 175–8, 180, 189n; inductive-statistical (I-S) model of 72, 74; mechanical 103, 400; partial 52–5; phenomenological 226; pragmatics of 56–70, 75–9, 88n, 97, 99, 158, 378; probabilistic(-statistical) model of 40, 45–55, 74, 77, *see also* explanation: inductive-statistical model(s) of; as providing understanding of the world 78, 81, 88n, 102–3; sketches 52–5; sociological 446; statistical relevance (S-R) model of 95; structural 196, 266–70, 272, 280n; teleological 93, 100, 400; theoretical 242, 275; as unification 44, 60, 71–105, 125; *see also* inference to the best explanation
explanatory: connection 320n; content 286; model(s) 255, 484; power 57, 60, 63, 68–9, 126, 193, 228, 235–6, 242–3, 250, 266–7, 269, 274, 278n, 296, 302, 423; relevance 58–9, 67–8, 76–80, 82, 84, 88n, 126; range 388; scope 215; status 379; success: *see* success: explanatory
explication 40, 52, 344–5; *see also* rational reconstruction
extrasensory perception (ESP) 395

falsification 26, 28–30, 286–8, 294n, 298–300, 302n, 304n, 378, 399, 402, 403n, 410, 437n
Faraday, Michael 251, 456
Feigl, Herbert 8n, 55n, 81, 92, 162n, 183n, 385n, 401n
feminism 459, 498, 501
feminist: epistemology 408, 459–88, 500; philosophy of science 459n, 497, 501n
Fermi, Enrico 161n

fertility (of theory(-ies)/hypothesis(-es)) 225, 248, 257, 266, 270, 272–4, 280n, 494; *see also* fruitfulness
Feyerabend, Paul 131, 141–62, 163–5, 170, 172, 174, 182, 183n, 184n, 185n, 186n, 187n, 188n, 189, 262, 321n, 403n, 407, 410, 418
Fichte, Johann Gottlieb 12
Fidell, L. S. 462
Field, Hartry 185n, 188n
Fine, Arthur 188n, 277n, 279n, 281n, 382n
Fizeau, Armand-Hippolyte-Louis 427
Flax, Jane 460
Forman, Paul 442
Foucault, Jean-Bernard-Léon 427
Fox, Mary Frank 461
Frängsmyr, T. 499n
Frank, Philipp 8n
Frankel, Henry 271, 280n
Franklin, Benjamin 185n
Frege, Gottlob 164, 180–1, 185n, 187n, 261, 341, 354
Fresnel, Augustin 57, 69n, 218
Freud, Sigmund 53, 55n, 101, 295–7, 299, 301n, 338
Frey, Ray 466–8
Friedman, Michael 35n, 60, 71n, 81, 88n, 89n, 94–7, 211
fruitfulness (of theories, hypothesis) (as a theoretical virtue/value) 234, 421n, 422, 428, 433, 436, 437n, 460, 470, 482–3; *see also* fertility

Galen 337
Galileo Galilei 3, 47–9, 61, 72, 131, 144, 158, 187n, 194, 248n, 280n, 301n, 337, 424, 430, 490–1
Galton, Francis 441
Garfinkel, Alan 75
Gauss, Carl Friedrich 456
Geach, Peter 66
Gemes, Kenneth 104
gender symbolism 464–73, 484
generalization(s) 315; accidental: *see* law(s): and accidental generalization(s); empirical 129–30, 144, 161n, 232n, 236, 238, 240–1, 379–380, 457; of fact 119; lawlike 118; of law 119; inductive 236–7, 240, 307–8, 363; statistical 236; uninstanced 119; vacuous 122
genetics 51, 72, 267, 441, 470, 472
geography 356
geology 216, 224, 226, 254, 257–8, 261, 266–7, 269, 271–3, 280n, 291, 441
geometry 363; Euclidean 129–30
Gettier, Edmund 379
Gibbins, Peter 277n
Giere, R. 35n
Gilligan, Carol 464
Gingerich, Otto 277n
Glymour, Clark 220, 230n, 277n, 381
God 22, 250, 263, 276n, 358
Gödel, Kurt 8n
Goethe, Johann Wolfgang 334, 337–8
Goldberg, Steven 479
Goldberger, Nancy 464
Goldman, Alan 382n
Goodman, Nelson 73, 117, 287, 307–9, 311, 313–16, 318, 319n, 320n, 363, 379, 398–9, 402n, 403n
Grandy, Richard 174, 185n
Greeks 3, 6, 263, 337
Greeno, J. 60
Greenwood, John 404n
Gregory, J. C. 414
grue paradox, the 321–39, 363, 402, 403n
Gruenberg, Barry 500n
Grünbaum, Adolf 60, 230n
Grünwald, Ernst 452
Gunderson, Keith 184n
Gutting, Gary 196, 230n, 234–47, 280n, 281, 501n

Hacking, Ian 262, 278n, 281n, 501n
Hahn, Hans 8n
Hamilton, William Rowan 415
Hamlyn, D. W. 443
Hannson, B. 63–4, 67, 70n
Hanson, N. Russell 93, 287, 321–39, 399, 401n, 410, 418
Haraway, Donna 469–70, 476, 482, 484
Hardin, Clyde L. 281n
Harding, Sandra 404n, 460, 501n
Hardy, G. H. 401n

Harman, Gilbert 57, 184n, 384n
Harrah, D. 67
Hartley, David 219
Haslanger, Sally 485n
Hatchett, Shirley 463
Hayek, Friedrich 304n
Healey, Richard 277n
Hearne, Vicki 466
Hegel, G. W. F. 6, 12
Heidegger, Martin 499
Heisenberg, Werner 29, 36n, 255–6, 277n
Hempel, Carl G. 35n, 40, 41, 45–55, 60, 69n, 71n, 72–5, 77–81, 88n, 92–7, 126, 142–3, 155, 158, 161n, 179–80, 187n, 189n, 276n, 360n, 374, 379–80
Hermeticism 425
Hertz, Heinrich 415
Herzog, Don 485n
Hess, Harry 271
hidden variable(s) (in quantum theory) 237, 253, 273
Hills, David 485n
Hintikka, Jaakko 398, 401n
history: of chemistry 184n; external 444–5; internal 444–5; of mathematics 495; of philosophy 8–10, 35, 256; of the philosophy of science 81; of science 3–4, 6, 36n, 57, 78, 115–16, 156, 163–4, 172, 179, 182, 185n, 188n, 189, 194–5, 211n, 217, 224–5, 248n, 251, 256–8, 262, 264–5, 268–9, 272–3, 275, 280n, 281, 288, 385–402, 402, 407, 410–11, 414, 416, 418–19, 433, 443, 455, 495–6, 500n
Hoefer, Carl 404n
holism/holist/holistic 365, 382n, 383n, 403
Homer 299, 358
Hooke, Robert 94, 391, 393
Hooker, Clifford 281n
Horwich, Paul 279n, 404n, 501n
Howson, Colin 404n
Hrdy, Sarah 476
Hubbard, Ruth 459
Hull, David 36n
Hullett, James 320n
Hume, David 80, 93, 96, 115–19, 121, 124n, 227, 259, 285, 286, 288, 340, 354, 356, 363, 390, 403n
Humphreys, Paul 88n,
Hutton, James 218, 267
Huxley, Thomas Henry 81, 89n
Huygens, C. 57, 69n
hypothesis(-es) 258, 286–7, 373–4, 389, 482; ad hoc: *see* ad hoc: hypothesis(-es); auxiliary 28, 287–8, 373–4, 391–2, 401n, 481, *see also* assumption(s): auxiliary; background 481, *see also* theory: background; empirical: *see* empirical: hypothesis(-es)
hypothetico-deductive (HD) method 122, 381, 391–3, 395–7, 399

idealism 256
impetus theory of motion 147–52, 154, 161n, 186n, 187n
incommensurability of theories 151, 153–5, 158–160, 170, 189, 416, 418, 420
indeterminism 62, 253–4, 273
induction 286, 308, 289–93, 343, 392, 490; the new riddle of 286–7, 307, 320n, 398, 402, 403n; "pessimistic" 196; the principle of 291–3; the problem of 285, 402, 403n, 431–3
inductive: argument(s): *see* argument(s): inductive; evidence 113–14; generalization(s): *see* generalization(s): inductive; inference 393; justification 238; logic: *see* logic: inductive; method 401n; reasoning 113; subsumption 93; support 51, 220, 242, 392
inductivism 444, 494
inference to the best explanation 69, 242, 256, 259, 263–4
initial condition(s) 148, 153, 160, 391–2, 453; *see also* boundary and/or initial condition(s)
instrumentalism 194–6, 197n, 201–7, 224, 227–8, 256, 264–5, 276n, 279n, 281, 285, 368–9, 371, 382n, 384n

Jacklyn, Carol 474
James, William 264, 335

Janis, Allen 230n
Jeffrey, Richard 74
Jeffreys, Harold 402n
Johnson, M. L. 339n
justification 249, 279n, 377, 379–80, 390, 407, 433; the context of 385–9, 391, 396–7, 400, 402n, 426–7, 465, 470, 472; criteria of 472; the logic of 399

Kant, Immanuel 9, 19, 81, 89n, 259, 263–4, 277n, 278n, 340–1, 434
Karim, Wazir 464
Kekulé, Friedrich August 385
Keller, Evelyn Fox 462, 466, 469–72, 501n
Kelvin, Lord (William Thomson) 205, 441
Kent, Dale 188n
Kepler, Johannes 3, 47–9, 94, 131, 158–9, 195, 301n, 357, 385, 391, 393, 423–5, 430, 444
Kilmister, Clive 403n
kinetic theory (of gases) 133–4, 138, 154–6, 217, 238, 373, 432; *see also* molecular(-kinetic) theory of gases
Kirwan, Richard 174, 184n, 187n
Kitcher, Patricia 188n
Kitcher, Philip 41, 44, 70n, 71–91, 95–7, 103, 126n, 131, 163–88
knowledge: background 76; growth of 267, 445, 462; justification of 4, 25, 27, 39; theory of 4, 14, 246, 362; *see also* epistemology
Kochen, Simon 277n
Koehler, Wolfgang 322, 335
Koertge, Noretta 501n
Kordig, Carl 183n, 188n
Körner, Stephan 70n
Kosso, Peter 36n
Kourany, Janet 501n
Koyré, Alexandre 187n
Krajewski, Wladyslaw 224–5
Kripke, Saul 184n, 187n, 189n
Krüger, Lorenz 230n
Kuhn, Thomas S. 163–5, 170, 172, 174, 182, 183n, 184n, 185n, 186n, 188n, 189, 249, 256, 262, 270, 277n, 280n, 321n, 340n, 401n, 403n, 407–20, 421–37, 441, 455, 457, 493–4, 500, 501n
Kyburg, Henry 74

Lagrange, Joseph Louis de 415
Lakatos, Imre 184n, 303n, 304n, 403n, 437n, 439, 444–6, 495
Lambert, J. Karel 70n
Langer, Susan K. 17
language 169, 200, 207, 217, 263, 271, 313, 338, 345, 347, 350, 352, 354, 356, 359, 360n, 378, 383n, 413, 436, 455, 466, 471; artificial 349–52; extensional 348; formal 397–8; game 338; of immediate experience 199, 355; natural 166–7, 351; of observation(s)/observation(al) 200, 202, 236–8, 249, 364, 378, 380–1; ordinary 349, 352, 434; phenomenalistic 200; philosophy of: *see* philosophy: of language; physicalistic 235–6; of physics 467; of sense data 354–5; theoretical 171–2, 174–5, 179, 182–3, 186n, 187n, 188n, 193, 242, 364, 403
Laplace, Pierre Simon 218
Latour, Bruno 496, 500n, 501n
Laudan, Larry 88n, 189, 195, 196, 211–33, 248n, 257–8, 265, 276n, 277n, 281, 288, 362–84, 403, 407
Laudan, Rachel 271, 280n
Lavoisier, Antoine 57, 69n, 164, 179, 187n, 218, 433, 456
law(s) 79, 96, 113, 126, 293, 374, 412–14; and accidental generalizations 42, 106, 110–15, 117, 122–3; Boyle's/Boyle-Charles's 49, 72, 156, 238, 432, 456; causal 110–14; conservation (of energy, momentum, etc.) 99, 425, 437n, 456; the Dretske–Tooley–Armstrong theory of 43, 126n; empirical 130, 236, 239, 257, 457; experimental 125, 129–31, 132–40, 161n, 197, 202; fundamental 94, 99, 103, 455; Galileo's 47–9, 72, 95, 144, 148; historical 453–4; Hooke's 94, 391, 393; of ideal gases 287; Kepler's 47–9,

94, 159, 391, 393; logical 343, 356–7; mathematical 356; mechanical 453; Mendel's 72; the Mill–Ramsey–Lewis theory of 44, 115n, 121–6; of nature 25–7, 40–3, 94, 106n, 115–24, 126; necessity of 42; Newton's 3, 29, 47–8, 72, 94, 118, 125n, 144, 148, 150–1, 288, 416–17; Ohm's 28–30, 456; as part of explanation 47, 51, 72–3, 77, 82, 88n, 93, 103–4; physical 43, 116, 358; probabilistic 51; regularity theory/analysis/account of 117–25, 126n; scientific 73, 126n, 220, 286–8, 290; Snell's 49; social 455; statistical 51, 237; theoretical 130–1, 141–2, 150–3, 158–9, 228–9, 230n, 231n; of thermodynamics 155; van der Waals' 238
Leacock, Eleanor 463
Lefkowitz, M. R. 462
Leibniz, Gottfried Willhelm 20, 250, 340, 343, 346, 456
Lennox, James 230n
Leplin, Jarrett 277n, 281n, 288, 362–84, 403
Lesage, George Lewis 219
Leverrier, Urbain 168
Levi, Isaac 88n
Levins, R. 36n
Lewis, Clarence Irving 359, 360n
Lewis, David 44, 65, 121–6, 188n
Lewontin, Richard 36n
light: ballistic theory of 300n; classical theory of 195; wave theory of 204–5, 217
likelihood (of theories/hypotheses) 394–5
Lindman, Harold 402n
Lister, Joseph 456
Locke, John 22, 258, 334, 339n, 354, 356
logic 4, 6, 27, 29–31, 40, 129–30, 132, 136, 138, 141–2, 195, 207, 288, 345, 354–5, 358, 360n, 366, 384n, 385, 390, 392, 440, 443–4, 446, 494; first-order 367–8; inductive 215, 389, 397–9, 401n; mathematical 360n
logical empiricism 5, 23–4, 117–18, 143, 164, 340n, 377–8, 380, 407; *see also* logical positivism
logical positivism 5–7, 23–7, 29–31, 35n, 40, 44, 124, 132n, 197n, 249, 259, 378, 380, 386, 401n, 493; *see also* logical empiricism
Longino, Helen 460, 470, 480–4, 500n, 501n
Lorentz, Hendrik Antoon 370, 373
Lovejoy, Arthur O. 452
Lowenheim, Leopold 367–8
Lowenheim–Skolem theorem, the 367–8
Lowinger, Armand 360n
Lugg, Andrew 230n, 232n
Lycan, William 382n, 384n
Lysenko, T. D. 331, 445

McClintock, Barbara 462
Maccoby, Eleanor 474
Mace, A. C. 300n
Mach, Ernst 250, 381, 415
Machamer, Peter 230n
MacKenzie, D. 441
Mackie, J. L. 43, 106–14
McMullin, Ernan 36n, 196, 230n, 231n, 248–80, 281, 408, 489–99
Maffie, James 382n
Maher, Patrick 404n
Maier, Anneliese 161n, 187n
Malthus, Thomas 470
manifest image (of the world) 240, 247n
Mann, Ida 335, 339n
Mannheim, Karl 442, 446, 491, 493
Martin, R. M. 360n
Marx, Karl 295–6, 299–300, 495, 497; theory of history of 295–6, 299–300
Marxism 378, 495, 497
mass 26
materialism 32, 264, 280n
mathematics 12, 25, 27, 29, 52, 138, 166, 255, 280n, 345, 348, 354, 358, 390, 433, 438, 440, 446, 456, 461, 467, 471, 481, 490, 495
Maull, Nancy 230n
Mauskopf, Seymour H. 280n
Maxwell, Grover 183n, 260, 266, 278n, 281n, 401n
Maxwell, James Clerk 28, 219, 251, 257, 373, 375, 384n; Maxwell's

electrodynamics 93; Maxwell's equations 28, 30
Maynard, Patrick 186n
Mayo, Deborah 404n
meaning 5, 15–21, 25–6, 32, 35, 164, 180, 184n, 185n, 199–200, 241, 340–2, 345, 348–9, 353–4, 360n, 377, 380, 417, 437n; change of 261, 263, 416–7; empirical 195; invariance of 152, 158, 161n; observational 143; of observational terms/statements 159, 161n; of theoretical terms/ statements 141–2, 148, 150–4, 156–7, 160, 164, 183n, 188, 193, 196, 199, 206; variance 418, 420; verification theory of 353–5, 360n, 384n
measurement 252, 287; apparatus/ apparata 262
mechanics 12, 156–7, 251, 268, 273, 369–70, 440, 493; analytical 138; celestial 144, 149; classical 149, 188, 250–2, 254, 415, 433, *see also* Newtonian: mechanics; matrix 255, 416; quantum: *see* quantum mechanics/physics/theory; relativistic 188, 416, *see also* relativity theory; statistical 51, 373–4; wave 416
mechanism(s) 95–6, 103–4, 226, 229, 266, 273, 303; *see also* causal: mechanism(s)
medicine 12, 224, 456
Mendel, Gregor 72, 216
Mensh, Elaine 481
Mensh, Harry 481
Mermin, N. David 101
Merton, Robert K. 439, 456, 492–4, 497, 500n
metaphor(s) 257, 270–2, 274–5, 280n
metaphysics 4, 7, 14, 23–7, 29–33, 35, 41, 44, 87–8, 125n, 198, 251–2, 259, 264–5, 276n, 294n, 295, 303, 340, 378, 409, 411, 415–16, 447
methodology 412, 439, 444–5
Meyerson, Émile 360n
Mill, John Stuart 44, 121–6, 382n
Milne, Edward 416
Milton, John 430
modality 348, 360n
model(s) 132, 135–6, 205, 230n, 248, 251, 255, 268–71, 275, 280n, 367, 371, 381–2, 383n, 412, 455–6, 470–3, 476–7, 482; explanatory: *see* explanatory: model(s); structural 267; theoretical 266
molecular(-kinetic) theory (of gases) 93, 202–4
Moravcsik, Julius 70n
Morgenbesser, Sidney 161n
Munitz, M. 70n
Musgrave, Alan 184n, 304n, 403n, 437n
myth(s) 299, 301n, 358, 448

Nagel, Ernest 35n, 129–30, 132–40, 141–2, 144, 148, 150–1, 154–8, 161n, 189n, 194–5, 197–210, 259, 407
natural kind(s) 363, 478
natural ontological attitude 279n, 281n
natural selection, theory of 22, 57, 99, 100, 286, 302, 305–6, 469–70, 472, 476; *see also* Darwin, Darwinism, evolutionary: theory
necessity: logical 43; natural 43, 110, 363; nomological 43–4; physical 43
Nelson, Lynn 478
neoplatonism 425
neopositivism: *see* logical positivism
Neurath, Otto 8n
Newton, Isaac 3, 16, 28–9, 57, 59, 72, 94, 131, 139, 144, 148–53, 158–9, 186n, 187n, 198, 216, 221, 250–2, 254, 257, 268, 276n, 288, 295, 301n, 357, 369–70, 381, 383n, 415–17, 420, 423, 456, 495
Newtonian/Newton's: dynamics 164, 186n, 187n, 416, 420; law(s): *see* law(s): Newton's; mechanics 28–9, 131, 139, 148–9, 152, 198, 221–2, 250, 288, 369, 383n, 415, 423, 495; physics 29, 149–50, 152, 251, 254, 259, 417; revolution 3; theory 57, 59, 139, 148–51, 153, 158–9, 186n, 251, 295, 301n, 370; rule(s) of method 253
Newton-Smith, William 211–12, 214, 222–3, 227, 229, 231n, 266
Nicod's criterion 374, 380
Niiniluoto, Ilkka 211, 230n, 231n

Niven, W. D. 384n
nominalism 256, 259, 274
normal science 411, 415–16
novelty 270; *see also* prediction(s): novel

objectivity: of science/scientific 287–8, 408, 421–37, 438, 443, 452, 466, 469, 483, 494–5, 499, 500
observable: consequence(s) of a theory 222, 224, 232n, 235, 241, 364, 366, 455; data 235; entity(-ies) 120, 130, 134, 188, 193, 196, 198–9, 201–2, 220, 222, 240, 243–6, 259, 261, 273, 278n, 381–2, 383n; fact(s) 236, 240, 455, 490; phenomena 235, 238–9, 241, 260, 266, 274, 364, 380; quantity(-ies) 255; regularities 269
observation(s) 25, 130, 132, 196, 240, 246, 252, 260–1, 285–8, 324, 338, 364, 383n, 403, 408, 464, 467, 481–2, 494; theory-ladenness of: *see* theory-ladenness: of observation(s)
observational: consequence(s) of theory/hypothesis 227, 364, 365, 383n, 391–3; data 145, 201, 286, 288; entity(-ies) 236, 364; evidence 164, 238, 287, 393, 407; fact(s) 240; framework 237, 239–41; statement(s) 200, 202, 220, 238, 365; status 365; term(s) 130, 143, 159, 161n, 193–4, 213, 262, 379; test(s) 407
Occam, William 259
Occam's razor 259
Ohm, Georg Simon 28–30, 456
ontological: claim(s) 258, 267; implication(s) 260; status 260, 265, 273, 281n; structure(s) 273
ontology(-ies) 251–5, 257, 354, 358–9, 363, 464, 484, 498–9
operationalism 147, 179, 187n, 301n
Oppenheim, Paul 45n, 55n, 71n, 92–6, 142, 155, 158, 161n
optics 392, 490; wave 222
Origin of Species, The 3, 22, 69n, 286, 305–6
Ostwald, Wilhelm 414

Pai, Hyung 463
Paludi, Michele Antoinette 462
Papineau, David 88n
paradigm(s) 411–21, 484, 493–4
Parmenides 299
Parsons, Kathryn Pyne 185n
partial interpretation 130, 132n, 188
Partington, James R. 184n, 401n
Pasteur, Louis 456
Pearce, Glenn 186n
Peirce, Charles Sanders 269, 353, 493
Peters, R. S. 443
philosophy: analytic 443; of biology 22–33; history of: *see* history: of philosophy; of language 7, 26, 31, 131, 163n, 195, 249, 409, 468; of mind 468; natural 489, 491; nature and method of 3–7, 8–21; of physics 36n; political 467; as pursuit of meaning 5, 15–21, 26, 35; and science 1–36; as the search for truth 12–15; unavoidability of 6
phlogiston 171–9, 184n, 185n, 187n, 195, 216, 257, 427; hypothesis of 57; theory 163–4, 171–9, 184n, 185n, 195, 217, 224, 419, 423, 427, 456
physics 13, 17, 25, 32, 72, 131, 198, 202, 209, 218, 225–6, 253, 264, 267, 279n, 303, 358, 390, 400, 408, 433, 442, 453–4, 456, 467, 493; classical 35, 209, 256; elementary-particle 253–4, 277n; mathematical 138, 225, 296; modern 116; quantum: *see* quantum mechanics/physics/theory
physiology 440, 482
Pickering, Andrew 501n
Pirie, Antoinette 339n
Pitt, Joseph 126n, 280n
Planck, Max 134–5, 137, 188
plate tectonic: model 253, 258, 271–3; theory 280n
Plato 11, 14, 39, 234n, 263, 495
Podolsky, Boris 101
Polanyi, Michael 466
Ponnekoek, Anton 401n
Popper, Karl R. 26, 142, 158, 161n, 162n, 215, 222, 231n, 286, 294–301, 302–4, 376, 378, 380–1, 398–9, 402–3, 439, 453–4
positivism 117, 219, 227, 244, 249, 255, 276n, 321n, 378, 410, 416, 419,

479–81; *see also* logical positivism, logical empiricism
possible world(s) 65, 67, 111, 115n, 119–21, 124, 241, 273, 340, 343
post-positivism 321n, 340n, 407
postulate(s) 351; fundamental 139; *see also* axiom(s)
Potter, Elizabeth 480
pragmatics: *see* explanation: pragmatics of
pragmatism 227, 263, 275, 278n, 279n, 340, 359, 371
predicate(s) 341, 347, 349, 360n; entrenched 314–15, 363, 398; Goodmanian 83; disjunctive 320n; projectible 73, 87
prediction(s) 58, 143, 160, 193, 206, 213, 215, 226, 267, 269–71, 286–8, 297–9, 301n, 303, 304n, 358, 368, 374, 392, 399, 402, 431, 453–5, 482; novel 68, 242, 270, *see also* novelty
predictive: accuracy 270, 421n, 422, 423; power 131, 193–5, 228, 273, 302, 423; strength 64; success: *see* success: predictive
Price, Derek 497
Priestley, Joseph 163, 171, 173–82, 184n, 185n, 186n, 421, 456
primatology 475–6, 497
probability 58–60, 62, 72, 74, 121, 292, 343, 392–5; calculus of 393, 401n; conditional 60; as degree of belief 50; as degree of confirmation 51, 428; distribution 96; frequency theory of 121, 398–400; inductive 50; logical 50, 397–8; posterior 394–5, 398; prior 62, 394–400; propensity theory of 121; statistical 49–50; subjective interpretation of 398; theory of 288, 402n
progress: scientific/of science 131, 214, 407, 419, 432
projectability: *see* projectible
projectible: predicate(s): *see* predicate(s): projectible; property(-ies): *see* property(-ies): projectible
property(-ies): disjunctive 308–13, 315, 320n; dispositional 86–7, 240, *see also* disposition; projectible 308–11, 313–14, 316–17; temporal 308–10
Prout, William 216
pseudo-problems 19
pseudo-questions 4–5; *see also* pseudo-problems
pseudo-science 24, 27, 141n, 286, 294–5, 300, 301n, 408
psychoanalysis 101, 295–6, 299, 300n
psychology 20, 93, 100, 296, 299, 333, 337, 390, 407, 422, 438–9, 443–5, 449–50, 456, 462, 467–8, 474; individual 295; mob 432, 500
Ptolemaic: astronomy: *see* astronomy: Ptolemaic; theory(-ies) 194
Ptolemy 164, 194, 250, 258, 270, 357, 370, 381, 423–4, 427, 430
Putnam, Hilary 59, 64, 184n, 186n, 189n, 211–12, 214–16, 218–19, 221, 223–5, 227, 229, 230n, 231n, 232n, 242, 251, 262–4, 266, 275, 276n, 277n, 278n, 279n, 280n, 281n
Pythagoras 255

quantum mechanics/physics/theory 29–30, 35, 36n, 51, 62, 96, 101, 134–5, 138, 188, 203–4, 209, 224, 237, 252–6, 261, 269, 273, 275, 277n, 281n, 331, 357, 375–6, 423, 435, 442; the Copenhagen interpretation of 252
Quine, Willard van Orman 261, 288, 340–61, 362–3, 376, 381, 382n, 384n, 403, 478, 480, 499

Radner, Michael 183n
Railton, Peter 87, 88n, 97–8, 103, 485n
Ramanujan, Srinivasa 385, 401n
Ramsey, Frank 44, 121–6
rational reconstruction 52, 386–9, 444–5
rationalism 198
rationality 379; of science/scientific 288, 402, 408, 426, 431, 438, 442–7, 456, 460, 472, 483
Rawley, William 499n
realism 87; and anti-realism 103, 241, 243, 249–65, 270, 273, 275, 281, 382n; convergent (epistemological) 195, 211–33, 277n, 281; entity 281n;

internal 278n, 281n; metaphysical 251, 263–5, 276n, 279n; naive 277n; scientific 193–281, 285, 371, 381
Redhead, Michael 277n
reduction 354–5; theoretical 93, 131, 141–62, 186n, 189n, 407
reductionism 131, 189n, 264, 280n, 340, 354–5
reference 164–70, 172–80, 182, 184n, 185n, 186n, 187n, 188n, 199, 213–21, 223–6, 228–9, 230n, 248n, 249, 254, 257, 261–2, 265, 274, 342, 358, 359n, 360n, 368, 416; inscrutability of 363, 382n; potential 175, 179–83, 187n, 189, 211n; shift 261; theory(-ies) of 163n, 164, 166–70, 175–8, 180, 184n, 189n, 261–2, 359n, 368
referential: change 165; stability 165
refutation(s) 286, 294–301
regularity(-ies): Humean 118, 120; empirical: *see* empirical: regularity(-ies)
Reichenbach, Hans 119, 123, 125n, 363, 385–6, 398, 400n, 401n, 402n; principle of the common cause of 61–2
relativism 363, 408, 418–20, 451–2, 499; conceptual, *see* conceptual: relativism
relativity theory 36n, 131, 164, 204, 255, 295–8, 331, 370, 373, 396, 416–17, 420
religion 378, 492
Rescher, Nicholas 114n, 117, 230n
research program(s): metaphysical 302–4; scientific 303n, 403n, 437n
Reskin, Barbara 461
retroduction 269
retroductive: argument(s): *see* argument(s): retroductive; inference 269; reasoning 258
revolution(s): scientific, *see* scientific: revolution(s); social 299
Richardson, Alan 35n
RNA (Ribonucleic Acid) 100
Rooney, Phyllis 466
Rootselaar, B. van 280n
Rorty, Richard 261–4, 278n, 362
Rose, Hilary 459–60, 464
Rose, Suzanna 461
Rosen, Nathan 101
Rosenberg, Alex 22–33, 35, 36n, 189n, 281n, 404n
Rosenfeld, Rachel 461
Rossiter, Margaret 461
Ruben, David-Hillel 126n
Rudwick, Martin J. S. 442
Rumford, Count (Benjamin Thompson) 218
Ruse, Michael 36n, 403n
Russell, Bertrand 286, 289–93, 341, 354
Rutherford, Ernest 25
Ryle, Gilbert 443, 446

Salmon, Wesley 41, 59–62, 67, 69n, 70n, 71n, 74, 76–7, 88n, 92–105, 126n, 266, 288, 385–401
salva veritate: interchangeability of terms 346–9
Sanford, David H. 320n
Savage, Leonard J. 398, 402n
saving the phenomena 227–8, 380–2, 384n
Schaffer, Simon 496, 501n
Schaffner, Kenneth 88n, 89n, 230n
Scheffler, Israel 93, 165, 188n, 437n
Schelling, Friedrich 12
Schiebinger, Londa 459, 461
Schilpp, Paul 36n, 401n
Schlick, M. 5–6, 8–21, 34–5
Schock, R. 65
Schofield, Robert 184n, 185n
Schopenhauer, Arthur 21
Schrödinger, Erwin 203
Schuman, Howard 463
Schwartz, Judith 320n
Schwartz, Justin 485n
scientific: change 249, 280n, 386, 407–8, 410, 415, 419, 437n, *see also* conceptual: change, theoretical: change; choice 421, 427, 431, *see also* theory(-ies): choice; community 180, 416, 421, 429, 435, 462, 482–4, 493–4, 496; method 415, 433; revolution(s) 164–5, 256, 277n, 386, 401n, 407, 410–20, 437n, 457

scientometrics 497
scope (as a theoretical value/virtue) 270, 421n, 422–3, 425, 430–1, 433, 436
screening-off 62
Scriven, Michael 73, 75, 93, 96–7, 100
Sellars, Wilfrid 69n, 215n, 230n, 231n, 234, 240, 242, 247n
semantical rule(s): *see* semantics
semantics 23, 26, 68, 164, 169, 184n, 214, 242, 349–52, 354, 366, 368, 371, 377–9, 382, 383n, 384n
sense 164, 181, 187n
set theory 354
sexism in scientific theories 479–83
Sextus Empiricus 227
Shapere, Dudley 183n, 188n, 189, 408, 410–20, 437n, 500
Shapin, Steven 496, 500n, 501n
Sher, George 188n
Sherif, Carolyn 463, 468–9, 485n
Sherrington, Charles 334
Shimony, Abner 211, 277n
Simon, Michael 268, 280n
simplicity 203, 234, 359, 399–400; as a theoretical value/virtue 58, 63, 270, 421n, 422–4, 428–9, 436, 470, 472, 482
simultaneity, Einstein's analysis of 16–17
skepticism 245, 286, 363, 370, 452, 492
Skolem, Thoralf Albert 367–8
Smart, J. J. C. 231n
Smith, Dorothy 459–60, 464
Sneed, Joseph D. 363
Snell, W. 49
social constructionism/constructivism 438n, 495, 498–9, 501n; *see also* sociology: of scientific knowledge
social dimensions of science 489–99; *see also* sociology: of science/of scientific knowledge, social constructionism/constructivism
social sciences 93, 100, 400, 408, 453–4, 463, 467, 473, 477–8, 484
sociobiology 471, 474, 482
sociology 60, 102, 295, 407, 438n, 444, 491, 493, 495, 497; reflexive 464; of science 408, 491, 493–8, 500n; of scientific knowledge (SSK) 408, 438–58, 492, 494, 496, 498–500, 501n
Socrates 12, 14, 39, 168
Socratic method 14–16
Sokal, Alan 501n
space-time 384n
Specker, Ernst P. 277n
Spedding, James 499n
Spinoza, B. 9
Stahl, Georg 172, 174, 176–7, 184n, 186n
Stalker, Douglas 320n, 403n, 455
Stalnaker, Robert 63, 65
Stark, Werner 439
statistical: mechanics: *see* mechanics: statistical; theory 157, 495
statistics 398, 441, 474, 477
Stich, Stephen 88n
Stiehm, Judith 474
Storer, N. W. 456
Stump, David 382n
subjunctive conditionals 119; *see also* counterfactual conditionals/ counterfactuals
success: empirical, of science 196, 211, 213–29, 230n, 231n, 232n, 239, 241–2, 261–6, 269, 370–1, 408, 431; explanatory 221–2, 229, 243–4, 257, 262, 266, 473; instrumental 194; predictive 160, 194, 222–3, 225, 231n, 243, 245
Suchtung, W. A. 118, 124n, 125n
Suppe, Frederick 35n, 69n, 189n, 247n
Suppes, Patrick 401n
synonymy 342–9, 353, 359n, 360n
synthetic: knowledge 478; statement(s) 343, 349, 352, 355–7, 359

Tarski, Alfred 379–80
Tarule, Jill 464
Tavris, Carol 464, 474
tectonic plate(s) 226, 253–4, 271–3; *see also* plate tectonic
teleology/teleological 31–3, 395, 400, 445–9, 451, 472, 478
Terman, Lewis 481
testability 59, 286, 298–9, 399
Thagard, Paul 69n

theism 303
theology 492
theorem(s) 130–1
theoretical: change 163–87, 261, 457, *see also* conceptual change; concept(s) 137, 237; construct(s) 249–50, 258–9, 476; entity(-ies) 26, 198, 206, 220, 230n, 235, 241–2, 251–5, 257, 259–62, 265–71, 273, 275, 281n, 384n; framework 240; model(s): *see* model(s): theoretical; notion(s) 134–9; postulate(s) 239; statement(s) 200–1, 358; structure(s) 266; term(s) 130, 133, 165–6, 171, 187n, 188, 189n, 193–4, 198, 200, 211n, 213, 228, 238, 249, 265, 368, *see also* descriptive: term(s); value(s) 270, 421n, 429–36, 437n, 483, 494; virtues: *see* theory(-ies): virtue(s) of
theory(-ies): acceptance of 57; background 28; choice 407–8, 421–37, 494, 500; confirmation of: *see* confirmation; fundamental 131; physical 203; scientific 129–89; structure of 130–1, 133; testing of 285–404, 426; value(s) of: see theoretical: value(s); virtue(s) of 56, 58, 64, 69, 263
theory-ladenness: of language 172, 187n; of observations 146, 321n, 321–39, 403; of scientific facts 435; of scientific statements/terms 178–9, 418, 420
thermodynamics 154–7, 222, 374, 419, 441
Thomson, J. J. 268, 373
Thomson, Judith 320n
Tidball, M. Elizabeth 461
Tiles, Mary 382n, 480
time 4, 16, 268
Tondl, Ladislav 125n
Tooke, Eyton 354
Tooley, Michael 43
topology/topological 384n
Toulmin, Stephen 321n, 410, 456
translation 164–5, 170, 174, 179, 185n, 186n, 187n, 188n, 199–202, 345, 351–5, 436; indeterminacy of 363, 382n
truth 9–13, 17, 27, 131, 194–8, 203–10, 212, 214, 216, 218, 222, 227, 230n, 235, 238–9, 241, 243–4, 246, 249, 264–5, 269, 274–5, 279n, 287, 296, 300, 316, 350–3, 356–7, 360n, 377, 379–80, 407–8, 418–20, 438, 440, 442–53, 479, 494; analytic 340, 351; approximate 212–14, 218, 220–9, 230n, 231n, 232n, 238, 265, 275, *see also* verisimilitude, truthlikeness; conditions 379, 381; as correspondence 248n, 263–5, 278n; logical 342–3, 347, 353; necessary 27, 29; synthetic 340; Tarski's theory of 379–80; as a theoretical virtue 56, 63, 68; value(s) 343, 346, 355
truthlikeness 223, 226, 229, 231n, *see also* truth: approximate, verisimilitude
Tuana, Nancy 484

uncertainty principle, Heisenberg's 29
underdetermination of theory by data/experience/evidence/observation 235, 288, 358, 362–84, 403n, 407, 480
understanding: scientific, of the world 93, 102–4, 274
universal(s) 43–4, 360n, 363
unobservable: consequence(s) 365–6, 383n; entity(-ies) 25, 103, 120, 130, 134–6, 196, 198, 226, 235, 238, 241, 243–6, 258–61, 265, 273, 331, 368, 371; fact(s) 274; mechanisms 235
Urbach, Peter 404n

value judgment(s) in science 421–37, 477, 481–2, 500; *see also* theoretical: value(s)
van der Waals forces 193
van Fraassen, B. 40, 56–70, 73, 75–8, 88n, 126n, 196, 234, 247n, 259–61, 265, 273–4, 277n, 278n, 280n, 281n, 363, 369, 381–2, 383n
Velikovsky, Immanuel 395–6
verification 16, 25, 286, 296–8, 353–4, 410, 432; principle of 6, 25–6, 378; theory of 353; *see also* meaning: verification theory of

verisimilitude 221–2, 227, 231n; *see also* truth: approximate, truthlikeness
Vetter, Betty 461
Vienna Circle 8n, 378
vitalism 31

Ward, Edwards 402n
Warhaft, S. 499n
Waring, Marilyn 474–5, 478
Wartofsky, Marx 183n
Watkins, John 159, 161n
Watson, James 268, 400, 402n
Watt, James 456
Weber, Max 493
Wedgwood, Josiah 185n
Wegener, Alfred 216, 271
Weinert, Friedel 126n
Wheelwright, Philip 280n
Whewell, William 382n, 498–9
White, Morton 360n
Whitehead, Alfred North 416
why-questions, explanatory 63, 67, 75–9, 103
Wignere, Eugene 253, 277n
Wilson, John Rowan 338n
Winokur, Stephen 183n
Wittgenstein, Ludwig 20, 21n, 323–4, 327–8, 331, 333–5, 338, 339n, 495, 499
Woolgar, Steve 496, 498, 500n, 501n
Worrall, John 230n
Wrangham, Richard 476
Wright, Larry 100, 103
Wykstra, Steven 230n

Young, Arthur 185n
Young, Thomas 57, 69n

Zita, Jacquelyn 480
Znaniecki, F. 442
zoology 17